Titles in This Series

Volume

1 **Markov random fields and their applications,** Ross Kindermann and J. Laurie Snell

2 **Proceedings of the conference on integration, topology, and geometry in linear spaces,** William H. Graves, Editor

3 **The closed graph and P-closed graph properties in general topology,** T. R. Hamlett and L. L. Herrington

4 **Problems of elastic stability and vibrations,** Vadim Komkov, Editor

5 **Rational constructions of modules for simple Lie algebras,** George B. Seligman

6 **Umbral calculus and Hopf algebras,** Robert Morris, Editor

7 **Complex contour integral representation of cardinal spline functions,** Walter Schempp

8 **Ordered fields and real algebraic geometry,** D. W. Dubois and T. Recio, Editors

9 **Papers in algebra, analysis and statistics,** R. Lidl, Editor

10 **Operator algebras and K-theory,** Ronald G. Douglas and Claude Schochet, Editors

11 **Plane ellipticity and related problems,** Robert P. Gilbert, Editor

12 **Symposium on algebraic topology in honor of José Adem,** Samuel Gitler, Editor

13 **Algebraists' homage: Papers in ring theory and related topics,** S. A. Amitsur, D. J. Saltman, and G. B. Seligman, Editors

14 **Lectures on Nielsen fixed point theory,** Boju Jiang

15 **Advanced analytic number theory. Part I: Ramification theoretic methods,** Carlos J. Moreno

16 **Complex representations of GL(2, K) for finite fields K,** Ilya Piatetski-Shapiro

17 **Nonlinear partial differential equations,** Joel A. Smoller, Editor

18 **Fixed points and nonexpansive mappings,** Robert C. Sine, Editor

19 **Proceedings of the Northwestern homotopy theory conference,** Haynes R. Miller and Stewart B. Priddy, Editors

20 **Low dimensional topology,** Samuel J. Lomonaco, Jr., Editor

21 **Topological methods in nonlinear functional analysis,** S. P. Singh, S. Thomeier, and B. Watson, Editors

22 **Factorizations of $b^n \pm 1$, $b =$ 2, 3, 5, 6, 7, 10, 11, 12 up to high powers,** John Brillhart, D. H. Lehmer, J. L. Selfridge, Bryant Tuckerman, and S. S. Wagstaff, Jr.

23 **Chapter 9 of Ramanujan's second notebook—Infinite series identities, transformations, and evaluations,** Bruce C. Berndt and Padmini T. Joshi

24 **Central extensions, Galois groups, and ideal class groups of number fields,** A. Fröhlich

25 **Value distribution theory and its applications,** Chung-Chun Yang, Editor

26 **Conference in modern analysis and probability,** Richard Beals, Anatole Beck, Alexandra Bellow, and Arshag Hajian, Editors

27 **Microlocal analysis,** M. Salah Baouendi, Richard Beals, and Linda Preiss Rothschild, Editors

28 **Fluids and plasmas: geometry and dynamics,** Jerrold E. Marsden, Editor

29 **Automated theorem proving,** W. W. Bledsoe and Donald Loveland, Editors

30 **Mathematical applications of category theory,** J. W. Gray, Editor

31 **Axiomatic set theory,** James E. Baumgartner, Donald A. Martin, and Saharon Shelah, Editors

32 **Proceedings of the conference on Banach algebras and several complex variables,** F. Greenleaf and D. Gulick, Editors

33 **Contributions to group theory,** Kenneth I. Appel, John G. Ratcliffe, and Paul E. Schupp, Editors

34 **Combinatorics and algebra,** Curtis Greene, Editor

Titles in This Series

Titles in This Series

Volume

Representation Theory, Group Rings, and Coding Theory

S. D. Berman
1922-1987

CONTEMPORARY MATHEMATICS

Volume 93

Representation Theory, Group Rings, and Coding Theory

Papers in Honor of
S. D. Berman
(1922–1987)

M. Isaacs, A. Lichtman,
D. Passman, S. Sehgal,
N. J. A. Sloane, and
H. Zassenhaus, Editors

AMERICAN MATHEMATICAL SOCIETY
Providence · Rhode Island

1980 *Mathematics Subject Classification* (1985 *Revision*). Primary 16, 20, 94.

Library of Congress Cataloging-in-Publication Data

Representation theory, group rings, and coding theory: papers in honor of S. D. Berman (1922-1987)/M. Isaacs...[et al.], editors.

p. cm. – (Contemporary mathematics, ISSN 0271-4132; v. 93)

Bibliography: p.

ISBN 0-8218-5098-9 (alk. paper)

1. Representations of groups 2. Group rings. 3. Coding theory. 4. Berman, S. D. (Samuil Davidovich), 1922–1987. I. Berman, S. D. (Samuil Davidovich), 1922–1987. Isaacs, M. III. Series: Contemporary mathematics (American Mathematical Society); v. 93.

QA171.R428 1989 — 89-6942
512′.2-dc20 — CIP

Dedicated to the memory of

S. D. Berman

(1922–1987)

Contents

Preface

This volume is dedicated to the memory of the outstanding Soviet mathematician S. D. Berman (1922–1987).

S. D. Berman worked, for the most part, in representation theory, group rings and coding theory. His important mathematical contributions are discussed here in a number of review articles. First B. N. Gartshtein and I. I. Grushko describe Berman's achievements in coding theory, including his pioneering work on abelian codes and his results on the theory of threshold functions, produced during the last ten years of his life. G. Karpilovskii considers Berman's contributions to the representation theory of groups over fields. The article by A. I. Lichtman discusses Berman's work on integral representations of groups, especially his fundamental theorem on groups of finite representation type, proved in the period 1959–1962. Next W. May describes Berman's accomplishments in infinite abelian group rings, in particular the results of his famous Debrecen paper (1967). The article by R. Sandling is devoted to Berman's fundamental results on units in integral group rings. Finally the biographical sketch of S. D. Berman is written by A. I. Lichtman.

This volume also contains twenty-two research articles in the areas of S. D. Berman's major interests. These papers are presented by mathematicians from many countries: United States, Soviet Union, England, Canada, Australia, Hungary, Bulgaria, South Africa—a sign of honor from the international mathematical community to a mathematician who contributed so much.

We would like to thank all the authors, the editors of the American Mathematical Society and all the people who made this collection possible.

M. Isaacs, University of Wisconsin-Madison
A. Lichtman, University of Wisconsin-Parkside
D. Passman, University of Wisconsin-Madison
S. Sehgal, University of Alberta, Edmonton, Canada
N. J. A. Sloane, AT & T Bell Labs
H. Zassenhaus, Ohio State University

S. D. Berman (1922–1987): A Biographical Note

A. I. LICHTMAN

The contributions of S. D. Berman to mathematics are widely recognized. His theorems in the theory of representations of finite groups over a field, integral representations and integral group rings are among the most important and deep ones. His work on the group rings of infinite abelian groups is pioneering and the results are fundamental. He was also a pioneer in the study of abelian codes. The review articles in this volume will discuss all these results in detail, but almost no one among the mathematicians in the West met S. D. Berman personally. I would like to tell here about the life of this remarkable man, to describe him as he was seen in the eyes of his friends and students.

Samuil Davidovich Berman was born on January 3, 1922 in a small town in the Ukraine. In 1939 he became a student at Moscow University, but his studies did not continue for long: the greatest and the most fierce war in Russian history broke out and the nineteen-year-old boy became for five years (1941–1946) a soldier and later a sergeant in the Red Army, serving in a combat unit. He took part in the great battle on the Volga River (Stalingrad) and was wounded there, recovered and returned to the army. Many years later, when he talked about the war he always recalled not only the tragedy and horror of the war but spoke with great warmth and sympathy about the people who fought together with him and became his friends in those difficult years. But few of them came back from the war!

He continued serving in a combat unit until the end of the war, fought in Hungary, the Ukraine, Czechoslovakia, and Bulgaria, took part in the battles for Budapest and Vienna, and was awarded a few medals. It was not until 1946 that he was demobilized and could renew his studies, first in the University of Moscow and then in Lvov University.

His Ph.D. thesis advisor was Ya. B. Lopatinsky, an outstanding Soviet mathematician, whose main interests at that time were in differential equations. Berman, however, was interested in differential equations only for a short time; he published one article and then became involved in the theory of representatons of finite groups and in the properties of integral group rings. His Ph.D. thesis and his first publications **[1]**–**[4]** (1952–1953) include many of the fundamental results and ideas about the group of units in integral group rings: the group of torsion units of the integral group ring ZG is $\pm G$ if and only if the group G is abelian or Hamiltonian of order 2^m; the result that the order of a normalized torsion unit is a divisor of the order of the group; the important fact that for every automorphism ϕ of ZG and a class sum k the element $\phi(k)$ is, up to sign, a class sum; the first studies (after Higman's unpublished thesis, 1940) on the isomorphism problem; the properties and the applications of K-conjugacy classes.

Here, in Lvov, he met Bella Naumovna Gartshtein; she was a graduate student in probability theory and she was also a veteran of the war. She became his wife and later mother of their three children. It was a family full of love and devotion.

In 1952 the couple moved to Uzgorod, a small town near the Carpathian mountains, where they spent fifteen years. The university had no graduate school; the two young candidates of science (the equivalent of the Ph.D. degree in the Soviet Union) taught a lot of courses; in total, they taught more than 1000 hours every year. Samuil Davidovich spent an enormous amount of time and energy working with many of us, the undergraduate students, who were interested in algebra. I recall how in 1956–1960 a few of us (P. Gudivok, B. Bovdi, E. Drobotenko, V. Drobotenko and I) used to wonder how he could find time and energy more than twenty hours of classes per week, service on many committees (he became chairman of the department of algebra) and long discussions with us—sometimes at the department of algebra, sometimes in his apartment, sometimes at the river bank where he often sat, holding in his hands a sheet of paper with formulas—to work on his theorems. He was a man of unlimited generosity and patience, always ready to offer us mathematical problems and ideas and to listen carefully, not interrupting, when we came to him with our theorems, whose proofs almost always contained gaps or mistakes!

His work during the fifties was mostly on representations of finite groups. The theorem on the number of irreducible representations over an arbitrary field and the theorem about the character ring of a finite group over an arbitrary field (the Berman-Witt Theorem) were proven at this time (see Berman **[1]**, **[5]**). These deep and important theorems, which generalize Brauer's results, became classical theorems in representation theory. During these years he also developed methods, based on the theory of induced representations, for effective constructions of representations of different classes of groups: representations of p-groups over an arbitrary field of characteristic zero, abelian-by-supersoluble groups over an algebraically closed field of arbitrary characteristic. Later he continued this work and obtained a number of important results on the Schur index (see **[6]**–**[10]**).

He began working on integral representations of finite groups in 1959. He believed that the theory of integral representatons of groups would have important applications in algebra, especially in the theory of groups. The theory of relation modules (see Gruenberg [**11**]) can now be considered as an example of such an application.

During 1959–1960 most of the discussions and the talks in the algebra seminar were on integral representations. As usual, Samuil Davidovich was interested in many other things—modular representations of p-groups, group algebras of infinite abelian groups, but he became fascinated by the depth and difficulty of the problems related to integral representations of groups. I will describe his work in detail in the review paper devoted to his contribution to the theory of integral representations. Here I will mention only some of his results.

He worked in 1959–1960 with P. Gudivok, his graduate student, and during 1960 they showed that a cyclic group of order p^2 ($p > 2$) has a finite number of indecomposable representations over the ring of p-adic integers, and they gave a description of all of these representations (see [**12**]–[**14**]). (For the cyclic group of order 4 this was done earlier by A. Roiter [**15**]; for cyclic groups of prime order by Diederichsen [**16**] and Reiner [**17**].) Berman and Gudivok also showed that if a p-group is not cyclic or is cyclic of order greater than p^2, then it has an infinite number of indecomposable representations (see [**12**]–[**14**]).

But an important problem was open: what is the class of groups which have only a finite number of indecomposable integral representations? It was in the Fall of 1961 when S. D. Berman proved one of the most remarkable theorems in the theory of integral representations: A finite group G has a finite number of indecomposable integral representations if and only if for each prime p, dividing the order of G, the Sylow p-subgroups of G are cyclic of order p or p^2 (see [**13**], [**18**]).

He was on leave then in Moscow and Leningrad, completing his doctoral dissertation; it included mainly his recent results on integral representations—a breakthrough in this area. The impression created by his dissertation was expressed by the opinion of I. R. Shafarevich, one of the judges; I. R. Shafarevich believed that S. D. Berman moved the theory of integral representations far ahead and he called Berman's result on finiteness of the number of integral indecomposable representations "brilliant." S. D. Berman became a Doctor of Science in 1963. This is the highest degree in science in the Soviet Union; there is no equivalent for it in the United States.

The same breakthrough in the theory of integral representations was made independently, at the same time, by a group of mathematicians in the USA (see Heller and Reiner [**19**], [**20**], Jones [**21**], and Dade [**22**]). Although Berman and Reiner were always very interested in each other's work, these two outstanding mathematicians met only once—in 1966, at the Moscow Mathematical Congress. They spent a lot of time together; it appeared that at this time French was the most convenient language for their conversations. Later Samuil Davidovich improved his English to such an extent that he could enjoy English poetry.

After 1967 S. D. Berman worked at the Kharkov Institute of Radioelectronics. He continued working in different areas in algebra and was at the center of algebraic life in Kharkov. But in addition to this he became involved in another area of mathematics—coding theory. An account of S. D. Berman's results on abelian codes, published in **[23]**, **[24]** (1967) is given in the book, *The Mathematical Theory of Coding* by I. Blake and R. Mullin. They consider his work on abelian codes "very significant" and then discuss his results on the minimum distance of any cyclic code of length p^n, rating them very high. His work in coding theory is described in detail in the review article by B. N. Gartshtein and I. I. Grushko.

S. D. Berman was a pioneer in the study of group rings of infinite abelian groups. His work on this topic began in 1960(**[25]**). In 1967 he published his two famous papers which included a number of fundamental results (see **[26]**, **[27]**). I will mention two of them here.

If K is a field of characteristic p and G_1 and G_2 are countable abelian p-groups such that KG_1 is isomorphic to KG_2, then G_1 is isomorphic to G_2.

If K is an algebraically closed field and G_1 and G_2 are torsion abelian groups of the same cardinality which have no elements of order equal to $\operatorname{char} K$, then KG_1 is isomorphic to KG_2.

S. D. Berman had many students; twenty of them became Ph.D. (A list of these students is included in this volume.) But the list of mathematicians who worked on problems posed by S. D. Berman or were influenced by his ideas is much longer; it includes many mathematicians from different cities of the Soviet Union: Moscow, Leningrad, Kiev, Kharkov, Erevan, Lvov and Uzgorod. He worked also with mathematicians from Hungary and Bulgaria. In many cases his ideas were picked up quickly by mathematicians from the Soviet Union but sometimes it took many years before they reached mathematicians in the West. This happened with his results on integral group rings and with some of his results on the Schur index. One of the main reasons for this is that few mathematicians in the West know the Russian language, and the systematic translation of Russian mathematical journals began only in the middle of the 1960's.

I believe that some exceedingly interesting ideas of S. D. Berman still have not been followed up as much as they deserve. Most important of them are his results on representations of discrete infinite groups. One of his articles on this topic (see Berman and Saraya **[28]**) gives a method for construction of series of irreducible representations (over an algebraically closed field K) for abelian-by-supersoluble groups; these series contain all the finite dimensional irreducible representations. In particular, when G is a finitely generated supersoluble group and $\operatorname{char} K = p > 0$, this result implies, via Roseblade's solution of Hall's problem, that all the irreducible represenations of G over K are obtained by this method.

I have already mentioned that S. D. Berman never traveled to the West and communicated only occasionally with Western mathematicians. He maintained, however, scientific contacts with many Soviet mathematicians and especially with I. R. Shafarevich, who was always very interested in Berman's work in representation theory. S. D. Berman's interest in integral representations was inspired by D. K. Faddeev, who was a pioneer in this field in the Soviet Union.

Reading S. D. Berman's papers gives a strong impression of his remarkable intuition which made it possible for him to begin working in areas completely new to him several times in his life. He also had an excellent memory, not only for mathematical theorems but for poetry as well, mostly Russian classical poetry and above all the great Pushkin. He loved the novels of Tolstoy and Dostoyevsky very much, a counterexample to the infamous opinion that one person cannot love both of them at the same time (see Veresaev **[29]**). One of his favorite Western authors was Thomas Mann. During discussions on literature and history, which he enjoyed very much, he was willing to listen to opinions which were very different from his point of view, but it was felt that he was disgusted by any kind of extremism or dishonesty. Another great love of his, or perhaps his first love, was classical music.

During the last ten years of his life he was deeply interested in philosophy especially in its connections with physics and cybernetics. Even his love of art became inseparable from his fascination with philosophy. In both cases it was probably his infatuation with the search for truth that nurtured such deep interest.

S. D. Berman was an exceptionally generous and kind man, and at the same time he had a strong will and character. He was not broken after he was fired in 1976 from his position as professor in the Kharkov Institute of Radioelectronics, but continued his research as energetically as before. It would be impossible to describe in a few pages all the harassment this outstanding scientist endured in his struggle against lawlessness and despotism—a struggle that lasted over ten years. During this difficult time his principles and high morale remained the same as throughout his whole life, so did his courage, and love and respect for his friends and students.

His main interests in mathematics during the last ten years were in threshold partitions; he worked towards creating a general theory of them. Most of the results which he obtained in this area have not been published. Article **[30]** and the review paper by B. N. Gartshtein and I. I. Grushko list only a small part of these results.

Samuil Davidovich Berman died on February 18, 1987. We will never see his kind eyes and will never hear his voice again. But his theorems and ideas are alive; the mark which he left on the lives of many of us, his friends and students, will not fade, and we will never forget him.

REFERENCES

1. S. D. Berman, *On the theory of representations of finite groups* (Russian), Dokl. Akad. Nauk SSSR **86** (1952), 885–888.

2. S. D. Berman, *On certain properties of integral group rings* (Russian), Dokl. Akad. Nauk SSSR **91** (1953), 7–9.

3. S. D. Berman, *On isomorphism of the centers of group rings of p-groups* (Russian) Dokl. Akad. Nauk SSSR **91** (1953), 185–1987.

4. S. D. Berman, *On a necessary condition for isomorphism of integral group rings* (Ukrainian), Dopovidi Akad. Nauk USSR, (1953), 313–316.

5. S. D. Berman, *The number of irreducible representations of finite group over an arbitrary field* (Russian), Dokl. Akad. Nauk SSSR **106** (1956), 767–769.

6. S. D. Berman, *Group algebras of abelian extensions of finite groups* (Russian), Dokl. Akad. Nauk SSSR **102** (1955), 431–434.

7. S. D. Berman, *The characters of linear representations of finite groups over an arbitrary field* (Russian), Mat. Sbornik **44** (1958), 409–456.

8. S. D. Berman, *The smallest field in which can be realized all complex representations of a p-group of odd order* (Russian), Uspehi Math. Nauk, **16**, N3 (1961), 151–153.

9. S. D. Berman, *Representations of finite groups over an arbitrary field and over the ring of integers* (Russian), Izv. Akad. Nauk SSSR **30** (1966), 69–132.

10. S. D. Berman, *Representations of groups of order* 2^m *over an arbitrary field of characteristic zero* (Ukrainian), Dop. Akad. Nauk USSR **3** (1958), 243–246.

11. K. W. Gruenberg, *Relation modules of finite groups*, AMS Publications, 1976.

12. S. D. Berman and P. M. Gudivok, *On integral representations of finite groups* (Russian), Dokl. Akad. Nauk SSSR **145** (1962), 1199–1201.

13. S. D. Berman and P. M. Gudivok, *On integral representations of finite groups* (Russian), Dokl. Uzgorod Univ. **5** (1962), 79–176.

14. S. D. Berman and P. M. Gudivok, *Indecomposable representations of finite groups over the ring of p-adic integers* (Russian), Izv. Akad. SSSR **28** (1964), 875–910.

15. A. V. Roiter, *On representations of cyclic group of order four by integral matrices*, (Russian), Vestnik Leningrad Univ. **15** (1960), 65–74.

16. F. E. Diederichsen, *Über die Ausreduktion ganzzahliger Gruppen-darstellengen bei arithmetischer Aquivalenz*, Hamb. Abh **17** (1938), 357–412.

17. I. Reiner, *Integral representations of cyclic groups of prime order*, Proc. Amer. Math. Soc. **8** (1957), 142–146.

18. S. D. Berman, *Representations of finite groups over an arbitrary field and over rings of integers* (Russian), Izv. Akad. Nauk SSSR **30** (1966), 69–132.

19. A. Heller and I. Reiner, *Representations of cyclic groups in rings of integers*. I, Ann. of Math. (2) **76** (1962), 72–92.

20. A. Heller an dI. Reiner, *Representations of cyclic groups in rings of integers*. II, Ann. of Math. **77** (1963), 318–328.

21. A. Jones, *Groups with finite number of indecomposable integral representations*, Michigan Math. J. **10** (1963), 257–261.

22. E. C. Dade, *Some indecomposable group representations*, Ann. of Math. (2) **77** (1963), 406–412.

23. S. D. Berman, *On the theory of group codes* (Russian), Cybernetics **1** (1967), 31–39.

24. S. D. Berman, *Semisimple cyclic and abelian codes* (Russian), Cybernetics **3** (1967), 21–30.

25. S. D. Berman, *On isomorphism of the group algebras of countable abelian groups* (Russian), Dokl. Uzgorod Univ. **3** (1960), 56–57.

26. S. D. Berman, *The group algebras of countable abelian p-groups* (Russian), Dokl. Akad. Nauk SSSR **175** (1967), 514–516.

27. S. D. Berman, *The group algebras of countable abelian p-group* (Russian), Publ. Math. Debrecen, 14:1–4 (1967, 365–405.

28. S. D. Berman and V. V. Šaraya, *On irreducible complex representations of finitely generated nilpotent groups* (Russian), Ukr. J. Math. **29**, N4, (1967, 435–442.

29. V. Veresaev, Zhivaia zhizn, O Dostoevskom i L′ve Tolstom (Russian).

30. S. D. Berman and N. G. Sarnavsky, *The combinatorial structure of threshold basis*, Cybernetics **2** (1985), 24–32.

DEPARTMENT OF MATHEMATICS
UNIVERSITY OF WISCONSIN–PARKSIDE
KENOSHA, WISCONSIN 53141

Publications of S. D. Berman

1. Integral representations of finite groups, Dokl. Akad. Nauk SSSR 86 (1952), 885–888.

2. On a necesary condition for isomorphism of integral group rings, Dopovidi Akad. Nauk Ukrain. RSR 5 (1953), 313–316.

3. On certain properties of integral group rings, Dokl. Akad. Nauk SSSR 91 (1953), 7–9.

4. On isomorphism of the centers of group rings of p-groups, Dokl. Akad. Nauk SSSR 91 (1953), 185–187.

5. On reduction of a system of differential equations to the passive form, Proc. of Lvov Univ. 11 (1953), 27–39.

6. On representations of a semidirect product of abelian groups, Dokl. Akad. Nauk SSSR 98 (1954), 177–180.

7. On certain properties of group rings over the field of rational number, Dokl. Uzgorod. Univ. 12 (1955), 88–110.

8. On the equation $x^m = 1$ in integral group ring, Ukrain. J. Math. 7 (1955), 253–261.

9. Group algebras of abelian extensions of finite groups, Dokl. Akad. Nauk SSSR 102 (1955), 431–434.

10. The p-adic ring of characters, Dokl. Akad. Nauk SSSR, 106 (1956), 583–586.

11. The number of irreducible representations of a finite group over an arbitrary field, Dokl. Akad. Nauk SSSR 106 (1956), 767–769.

12. The representations of soluble groups, Proceedings of the 3rd Math Congress of the USSR, vol. 2.

13. The generalized characters of finite groups, Dokl. Akad. Nauk SSSR 2 (1957), 112–115.

14. Groups which admit an arbitrary permutation of quotients of the composition series, Uspehi math. Nauk 12:5, 77 (1957), 181–183 (together with V. V. Lubimov).

15. Groups of which all representations are monomial, Dopovidi Akad. Nauk Ukrain. RSR 6 (1957), 539–542.

16. Certain problems of the theory of representations of finite groups over an arbitrary field, Dokl. Uzgorod Univ. 2 (1958), 21.

17. On the monomial representations of groups, Dokl. Uzgorod Univ. 2 (1958), 21.

18. The representations of the groups of order 2^m over an arbitrary field of characteristic zero, Dokl. Akad. Nauk SSSR 3 (1958), 243–246.

19. The p-blocks for a certain class of finite groups, Dokl. Akad. Nauk SSSR 6 (1958), 606–608 (together with A. A. Bovdi).

20. The characters of the linear representations of finite groups over an arbitrary field, Math. Sb. 44 (1958), 409–456.

21. Modular representations of finite supersoluble groups, Dopovidi Akad. Nauk Ukrain. RSR 5 (1960), 586–589.

22. Normal subgroups of the group of triangular matrices over a finite field, Dokl. Uzgorod. Univ. 3 (1960), 50.

23. On indecomposable modular representations of a certain class of finite groups, Dokl. Uzgorod Univ. 3 (1960), 51–52.

24. On automorphisms of the center of the integral group ring, Dokl. Uzgorod Univ. 3 (1960), 55.

25. On isomorphism of group algebras of countable abelian groups, Dokl. Uzgorod Univ. 3 (1960), 56–57.

26. On induced representations of groups, Dokl. Uzgorod Univ. 3 (1960), 53–54.

27. The smallest field in which can be realized all complex representations of a p-group of odd order, Uspehi Math. Nauk 16 (1961), 151–153.

28. On representations of locally finite groups, Dokl. Uzgorod Univ. 4 (1961), 78–79.

29. On representationof M-groups over an arbitrary field, Dokl. Uzgorod Univ. 4 (1961), 82–83.

30. On isomorphism of group algebras of direct products of primary cyclic groups, Dokl. Uzgorod Univ. 4 (1961), 76–77.

31. On the Schur index, Uspehi Math. Nauk 16 (1961), 95–99.

32. The generalized modular characters of finite groups, Dokl. Uzgorod Univ. 4 (1961), 80–81.

33. On the classes of conjugate elements of cyclic extensions of finite groups, Dokl. Uzgorod Univ. 5 (1962), 76 (with A. P. Shapiro).

34. On integral representations of finite groups, Dokl. Uzgorod Univ. 5 (1962), 74–76 (with P. M. Gudivok).

35. On integral representations of finite groups, Dokl. Akad. Nauk SSSR 145 (1962), 1199–1201 (with P. M. Gudivok).

36. On integral representations of finite groups, Dokl. Akad. Nauk SSSR 152 (1963), 1286–1287.

37. On the theory of integral representations of finite groups, Dokl. Akad. Nauk SSSR 157 (1964), 506–508.

38. Indecomposable representations of finite groups over the ring of p-adic integers, Izv. Akad. Nauk SSSR 28 (1964), 875–910 (together with P. M. Gudivok).

39. Integral representations of cyclic groups containing two irreducible rational components, Volume: N.G. Ĉebotarev, Izdat. Kazan Univ., Kazan (1964), 18–29.

40. On integral monomial representations of finite groups, Uspehi Math. Nauk 20 (1965), 133–134.

41. On integral representation sof finite nilpotent groups, Uspehi Math. Nauk 20 (1965), 186–190 (together with A. I. Lichtman).

42. On group algebras of countable and finite groups, The reports of the International Math. Congress, Section 2, Moscow (1966), 33.

43. On representations of a class of discrete soluble groups over an arbitrary algebraically closed field, Algebra and Logic: Studies in Algebra, Izdat. Kiev Univ., Kiev (1966) 54–61.

44. On integral groups rings of finite and periodic groups, Algebra and Logic: Studies in Algebra, Izdat. Kiev Univ., Kiev (1966), 44–53 (together with A. R.Rossa).

45. Representations of finite groups, Studies in Algebra 1964, Itogi Nauki, Moscow (1966), 83–122.

46. Representations of finite groups over an arbitrary field and over the ring of integers, Izv. Akad. Nauk SSSR 30 (1966), 69–132.

47. Dmitrii Alekseevich Suprunenko, Uspehi Math. Nauk 21 (1966), 170–171 (together with R. T. Volvachev and L. A. Kaluznin).

48. The group algebras of countable abelian p-groups, Dokl. Akad. Nauk SSSR 175 (1967), 514–516.

49. On the theory of group codes, Cybernetics 1 (1967), 31–39.

50. Semisimple cyclic and ableian codes, Cybernetics 1 (1967), 21–30.

51. Group algebras of countable abelian p-groups, Publ. Math. Debrecen, 14:1–4 (1967), 365–405.

52. Yaroslav Borisovich Lopatinsky, Uspehi Math. Nauk 22 (1967), 197–202 (together with A. E. Volpert, T. Ya. Zagorsky, P. S. Kazimirsky, V. G. Kostenko, V. E. Lantze).

53. On irreducible representations of finite direct product of finite groups, Siberian J. Math. 9 (1968), 485–489.

54. The theory of error-correcting codes—a new area of combinatorics and algebra, Uspehi Math. Nauk 23 (1968), 245 (together with R. L. Dobrusin).

55. The Sylow p-subgroup of the group algebra of a countable abelian p-group, Dokl. Akad. Nauk SSSR 10 (1968), 870–872 (together with A. R. Rossa).

56. On the Schur index, Appendix to Russian translation of "Representation Theory of Finite Groups and Associative Algebras" by C. W. Curtis and I. Reiner, Moscow (1969), 610–638.

57. Codes with generalized majority decoding and convolutional codes, Probl. Peredachi Inf. 4 (1970), 6–19 (together with A. B. Yudanina).

58. On group algebras of countable periodic abelian groups, Dokl. Akad. Nauk Ukrain. RSR (1971), 387–390.

59. On simultaneous direct decompositions of a finitely generated abelian group and its subgroup, Dokl. Akad. Nauk SSSR 210 (1973), 1004–1007 (together with Z. P. Zhilinskaya).

60. To the memory of Alexander Gennadievich Kurosh, Trudy Moscov Math. Obš. 29 (1973), 3–4 (together with V. A. Andrunakievich, T. M. Baranovich, V. D. Belousov, L. M. Gluskin, O. N. Volovin, S. T. Zavalo, L. A. Kaluznin, B. I. Plotkin, L. E. Sadovskii, L. A. Skorniakov, D. A. Suprunenko, L. N. Shevrin, A. L. Shmelkin).

61. Irreducible configurations of minimum-weight elements in the Reed-Muller second order codes, Second Int. Symp. on Inf. Theory, Thakkadsor, 1971, Budapest (1973), 115–129 (together with B. Yudanina).

62. On the representations of a group of type p^∞ over a field of characteristic p, Trudy Moscov. Math. Obš. 29 (1973), 87–100 (together with N. I. Vishniakova).

63. On degrees of the indecomposable pairs of finitely generated abelian groups, Trudy XXVIII Conf. Prof.-prep. sostava Uz. Univ., Sect. Math., Uzgorod (1974), 207–226 (together with Z. P. Zhilinskaya).

64. On modules over non-noetherian valuation rings, Dokl. Akad. Nauk Arm. SSR 60 (1975), 114–149 (together with N. I. Vishniakova).

65. On group rings of mixed abelian groups, Dokl. Akad. Nauk 28 (1975), 879–881 (together with T. Zh. Mollov).

66. The Sylow p-subgroup of the group algebra of a countable abelian p-group, Trudy XXIX conf. Prof-prep. sostava Uz. Univ., Sect. Math., Uzgorod (1975), 157–176 (together with A. A. Rossa).

67. A majority algorithm for decoding a certain optimal code, Coding and transmission of discrete messages, Moscow, Nauka (1976), 58–61 (together with I. I. Grushko).

68. On isomorphism of real group algebras of abelian groups, Math. Zametki 21 (1977), 229–238 (together with V. G. Bogdan).

69. On irreducible complex representations of finitely generated nilpotent groups, Ukrain. J. Math. 29 (1977), 435–442 (together with V. V. Saraya).

70. On presentations of the group of type p^∞ over a certain class of fields, Ukrain. J. Math., 29 (1977), 632–637 (together with A. G. Podgorny).

71. On representations of the nilpotent torsion free group with two generators, Funct. Anal. and Its Appl. 4 (1977), 70–71 (together with E. Sh. Kerer).

72. On the supports of the elements of the smallest weight in the principal ideals of the group algebra of the group of the type $(2, 2, \dots, 2)$, The 7th Conference on the coding theory and transmitting of information, Vol. 2 (1978), 13–14 (together with I. I. Crushko).

73. Code parameters of principal ideals of group algebra of the group of type $(2, 2, \ldots, 2)$ over a field of characteristic 2, Probl. Peredach. Inf. 14 (1978), 2–12 (together with I. I. Grushko).

74. On B-functions encountered in modular codes, Prob. Peredachi Inf. 17 (1981), 10–18 (together with I. I. Grushko).

75. On representations of the infinite dihedral group, Publ. Math. Debrecen 28 (1981), 173–187 (together with K. Buzasi).

76. On isomorphism of semisimple group algebras of uncountable abelian p-groups, Dokl. Bulg. Akad. Nauk 35 (1982), 869–871 (together with T. Zh. Mollov).

77. On the multiplicative semigroup of the ideals of a valuation ring, Publ. Math. Debrecen 29 (1982) 171–776 (together with N. I. Vishniakova).

78. On representations of groups containing a cyclic subgroup of finite index, Publ. Math. Debrecen 29 (1982), 163–170 (together with K. Buzasi).

79. On splitting of the basis and isomorphism of groups rings of abelian groups, Pliska Blg. Math. Stud. (1981), 6–22 (together with T. Zh. Mollov).

80. To the theory of digital signal processing, Probl. Peredachi Info. (1983), 43–49 (together with I. I. Grushko).

81. On the specter of a valuation ring, Publ. Math. Debrecen 30 (1983), 203–216 (together with N. I. Vishniakova).

82. On modules over group algebras of groups, containing an infinite cyclic subgroup of finite index, Stud. Sci. Math. Hung. 16 (1981), 455–470 (together with K. Buzasi).

83. Classification of all finite-dimensional real representations of groups, containing an infinite cyclic subgroup of finite index, Publ. Math. Debrecen 31 (1984), 133–144 (together with K. Buzasi).

84. On structure of threshold sets, Dopovidi Kharkov Inst. Radioelectronics (1984) (together with N. G. Sarnavsky).

85. The combinatorial structure of threshold sets, Dopovidi Akad. Nauk Ukrain. RSR 1 (1985), 3–6 (together with N. G. Sarnavsky).

86. Combinatorial structure of threshold basis I, Cybernetics 2 (1985), 24–32 (together with N. G. Sarnavsky).

87. Isomorphism of group algebras of uncountable abelian p-groups, Xth Symp. on group theory in USSR (1986), 19 (together with T. Zh. Mollov).

88. Decoding of maximal error in the calculation of a vector by means of its projections, Proc. of 9th Allunion Symp. on Redundancy of Inf. Systems, Leningrad (1986), 125–127 (together with B. P. Gerasimenko, B. P. Zephirov, S. P. Krotkikch).

89. Detecting and estimation of maximal error in the calculation of a vector by meanss of its projections, Probl. Bionic, 38 (1987), 43–50 (together with B. P. Gerasimentko and B. P. Zephirov).

90. The solution of the optimization problem for two classes of real codes, Dokl. Kharkov Univ., 1986.

91. Parameters of abelian codes in the group algebras KG of $G = (a) \times (b)$, $a^p = b^p = 1$, p prime, over a finite field K with primitive pth root of unity and related MDS-codes, Contemporary Mathematics (1989) (with I. E. Grushko).

92. Locally cyclic modules over a valuation ring, Contemporary Mathematics (1989) (together with N. I. Vishniakova).

Ph.D. Students of S. D. Berman

1. Yu. L. Barannik
2. B. Bovdi (partly)
3. K. Buzasi
4. N. N. Eisenberg
5. P. M. Gudivok
6. E. S. Drobotenko
7. V. S. Drobotenko
8. G. B. Karpilovsky
9. G. K. Kladov
10. V. F. Kniazeva
11. E. Sh. Kerer
12. A. I Lichtman (partly)
13. T. Z. Mollov
14. S. I. Ostrovskaya
15. S. S. Poliak
16. V. S. Rudko
17. A. R. Rossa
18. V. S. Šaraya
19. Z. P. Zhilinskaya
20. N. I. Vishniakova

Contemporary Mathematics
Volume **93** (1989)

The S. D. Berman Work on Coding Theory and on Theory of Threshold Functions

B. N. GARTSHTEIN AND I. I. GRUSHKO

S. D. Berman's works form an essential contribution to discrete mathematics, especially to algebraic coding theory and the theory of threshold functions. The first part of the survey written by I. I. Grushko contains the main results on the topic of algebraic coding. The second half written by B. N. Gartshtein contains a small part of the results in the theory of threshold functions that S. D. Berman had an opportunity to publish just before his sudden death.

The most of Berman's papers on the topic of coding is concerned with the creation of the theory of group codes. In the framework of coding theory one usually concentrates on the calculation of *code parameters.* We will only consider a linear code V, i.e. a linear subspace in the n-dimensional linear space over a field F. In this case the code parameters are defined as follows: n is *block length*: the *number* $k(V)$ of *information symbols* is the dimension of V over F; the *Hamming weight* $\mathrm{wt}(x)$ of an element $x \in V$ is the number of nonzero components in the decomposition $x = \sum_{g\in G} \alpha_g g$, $\alpha_g \in F$; the *code distance* is $d(V) = \min_{0\neq x\in V} \mathrm{wt}(x)$; the *rate* of the code V is the ratio $k(V)/n$.

The first paper contains two parts /1–2/ and two significant results. At first, we will give some introduced in /1/ definitions. Let G be a finite group, F an arbitrary field and FG a group algebra of the group G over F. Any ideal V of FG is called an *FG-code.* If G is abelian, then FG-code is called *abelian.* If the characteristic of F does not divide the order of G, then FG-code is called *semisimple*; otherwise it is called *modular.* If G is a p-group and the characteristic of F is p, then V is called a *p-code.*

All these definitions are the natural generalizations of some classes of codes well-known in those days: cyclic codes, i.e. ideals in the group algebra FG of cyclic group G over the field F, and Reed-Muller codes (RM-codes), i.e. ideals in the group algebra FG of the group G of type $(2, 2, \ldots, 2)$ over the field $F = \{0, 1\}$.

The main result in /1/ is a formula for the code distance $d(R^i)$ of p-codes R^i, where R is the radical of FG.

Let $G = (a_1) \times (a_2) \times \cdots \times (a_s)$, where (a_i) is a cyclic group of order p^{α_i}, $\alpha_1 \geq \alpha_2 \geq \cdots \geq \alpha_s > 0$, and let F be a finite field, $\operatorname{char}(F) = p$. Let R denote the radical of FG (that is the set of all nilpotent elements in FG), and let R^j denote the j-th power of R for $1 \leq j \leq (p^{\alpha_1}-1)+\cdots+(p^{\alpha_s}-1)$. Linear spaces R^j are called *radical p-codes.* The dimension $k(R^j)$ is the volume of the basis $\{(a_1-1)^{i_1}\ldots(a_s-1)^{i_s},\ i_1+\cdots+i_s \geq j,\ (i_1,\ldots,i_s) \neq 0,\ i_l = 0,1,\ldots,p^{\alpha_l}-1\}$. The problem is to calculate the code distance $d(R^j)$.

If $G = (a)$, $a^{p^m} = 1$, then the only nontrivial ideals in FG are principal ideals $R^j = ((a-1)^j)$, $j = 1,\ldots,p^m-1$. In this case we have

THEOREM 1 /1/. *Let* $G = (a)$, $a^m = 1$, $\operatorname{char}(F) = p$, $V = ((a-1)^j)$, $1 \leq j \leq p^m - 1$, $i = \alpha_1 p^{m-1} + \cdots + \alpha_m$, $0 \leq \alpha_l < p$. *If* $\alpha_1 = \cdots = \alpha_m = p-1$, *then* $V = (1+a+\cdots+a^{p^{m-1}})$ *is a one-dimensional ideal with* $d(V) = p^m$. *Let* r *be the first index for which* $\alpha_r \neq p-1$; *then*

$$d(V) = \begin{cases} (\alpha_r+2)p^{r-1}, & \text{if } (\alpha_{r+1},\ldots,\alpha_m) \neq (0,\ldots,0), \\ (\alpha_r+1)p^{r-1}, & \text{if } (\alpha_{r+1},\ldots,\alpha_m) = (0,\ldots,0). \end{cases}$$

If $F = GF(p)$ and $m = 1$, then this class of modular p-codes contains the particular subclass of MDS-codes with optimal parameters $(n = p, k, d = n-k+1)$ and very simple coding and decoding algorithms.

Ten lemmas precede the proof of a general result given by the following theorem.

THEOREM 2 /1/. *Let* $\operatorname{char}(F) = p$, *G be an abelian group of type* $(\underbrace{\alpha_1,\ldots,\alpha_1}_{s_1},\ldots,\underbrace{\alpha_m,\ldots,\alpha_m}_{s_m})$, $\alpha_1 > \alpha_2 > \cdots > \alpha_m$, *that is in the direct decomposition of G there are exactly* s_i *cyclic components of the order* p^{α_i} *and* $(G\colon 1) = p^{s_1\alpha_1+\cdots+s_m\alpha_m}$. *We denote* $t = s_1(p^{\alpha_1}-1)+\cdots+s_m(p^{\alpha_m}-1)$ *and* $m_i = l_i(p-1)p^i$, $i = 0,\ldots,\alpha_1$, *where* $l_{\alpha_1} = m_{\alpha_1} = 0$, $l_{\alpha_1-1} = \cdots = l_{\alpha_2} = s_1$, $l_{\alpha_2-1} = \cdots = l_{\alpha_s} = s_1+s_2,\ldots,l_{\alpha_m-1} = \cdots = l_0 = s_1+s_2+\cdots+s_m$. *(It is easy to check that* $m_0+\cdots+m_{\alpha_1} = t$.*) Then for every integer* $1 \leq j < t$ *there exists an index r such that* $J = m_{\alpha_1}+\cdots+m_{\alpha_1-r}+t(p-1)p^{\alpha_1-r-1}+\delta_{\alpha_1-r-1}p^{\alpha_1-r-1}+\cdots+\delta_0$, *where* $0 \leq \delta_{\alpha_1-r-1} < p-1$, $0 \leq \delta_j \leq p-1$. *The code distance* $d(R^t)$ *is* $(G\colon 1)$ *and* $d(R^j)$ *for* $1 \leq j < t$ *is given by the formula*

$$d(R^j) = p^{l_{\alpha_1}+\cdots+l_{\alpha_1-r}+t}(\delta_{l_\alpha-r-1}+\gamma),$$

where $\gamma = 2$, *if* $(\delta_{\alpha_1-r-2},\ldots,\delta_0) \neq (0,\ldots,0)$ *and* $\gamma = 1$ *otherwise.*

A well known formula for the distance of RM-codes and their generalizations follows from this one as a particular case.

In /2/ the code parameters of ideals of the group algebra FG of an abelian group G over the field F are studied for the case where the characteristic of F does not divide the order of G. In this case the algebra FG is semisimple and decomposes into the direct sum of minimal ideals,

(1) $$FG = I_1 + \cdots + I_s,$$

where $I_j = FGe_j$, $j = 1, \dots, s$, and $\mathfrak{S} = \{e\}_j^s$ is the set of all minimal pairwise orthogonal idempotents. Every ideal $V \subseteq FG$ is a direct sum of minimal ideals from (1). Let us denote by $\mathfrak{U}$ the subset of minimal idempotents from $\mathfrak{S}$, which do not generate minimal ideals that enter in decomposition of V. Then V can be defined as

(2) $$V = \{x \in FG\colon xe_{j_t} = 0, e_{j_t} \in \mathfrak{U}\}.$$

At first, the lower bound for the code distance $d(V)$ of the code V given by (2) is investigated and for the case of cyclic group G the well known BCH-bound is proved as a sequence from the general case.

The most part of /2/ is devoted to the upper bound for the code distance of the semisimple cyclic code V. The main result is formulated as follows.

THEOREM 3. /2/. *Let $p_1, p_2, \dots, p_s$ be fixed primes and $M = \{V\}$ be the family of semisimple cyclic codes over the field F of the characteristic $p \neq p_k$, $k = 1, \dots, s$, with the block length $n = p_1^{\alpha_1} \dots p_s^{\alpha_s}$ (numbers $\alpha_1, \dots, \alpha_s$ depend on V). If the rate $(k(V)/n) \geq b > 0$ for all $V \in M$, then there exist such constant C, that $d(V)$ does not exceed C for every $V \in M$.*

In a particular case of the cyclic code V with block length $n = p^\alpha$ over the field F, char$(F) \neq p$, Theorem 3 puts up that $d(V) \leq (1/b)l$, where number l is defined from the condition: $F(\zeta_1) = F(\zeta_2) = \cdots = F(\zeta_l) \subset F(\zeta_{l+1})$ for $p \neq 2$ and $F(\zeta_2) = \cdots = F(\zeta_l) \subset F(\zeta_{l+1})$ for $p = 2$, where ζ_i is primitive i-th root of unity for $i = 1, 2, \dots$.

In opposite to the cyclic case in /2/ the following result is proved.

THEOREM 4 /2/. *Let $p_1, \dots, p_s$ be fixed primes, $p_i \neq 2$ for $i = 1, \dots, s$. There exists the class $\{V\}$ of abelian codes with block length $n = p_1^{\alpha_1} \dots p_s^{\alpha_s}$ such that* $\lim_{\alpha_1 + \cdots + \alpha_s \Rightarrow \infty} (k(V)/n) = 1$ *and simultaneously $d(V) \Rightarrow \infty$.*

The comparison of Theorems 3 and 4 permits one to conclude that in the semisimple case for rates $k(V)/n \geq b > 0$, fixed odd primes $p_1, \dots, p_s$ and going to the infinity block length $n = p_1^{\alpha_1} \dots p_s^{\alpha_s}$ the code parameters of abelian codes are better than code parameters of cyclic codes.

This earliest of Berman's papers on coding theory turned out to be the best in the following sense. Theorem 2 has given an exact formula for the code distance of radical codes and thus the problem was completely solved. The result given by Theorem 3 has not been improved or specified in any form till today. An advantage of abelian codes in comparison with cyclic codes as shown in /2/ stimulated further research in this area. Although there has not been any essential progress in this field, some results for the group of type (p, p) were obtained by K. Buzasi, A. Pete, P. Lakotos ("On code distance of a class of group codes", Problemy Peredachi Informatsii, vol. 17, No. 3,

1981), and B. Brindza ("On code distance of some (p,p)-codes", Problemy Peredachi Informatsii, vol. 21, No. 92, 1985).

In many of his works S. D. Berman has numerously shown interest in radical codes. His first return to the object in /3/ is connected with a problem of "good" decoding. Codes with superior parameters are of a little interest to practice if no good decoding algorithm is known for them. Good codes need good decoders, and good decoding algorithms have been difficult to find. Majority decoding is a method of decoding that is relatively simple to implement and the history of majority decoding codes (M-codes) can usually be traced back to the RM-codes—the subclass of the radical codes.

A natural way to build codes with inherited properties is the tensor product of codes. This construction was described simultaneously by several researchers, in particular, in /3/:

THEOREM 5 /5/. *The tensor product of t M-codes with the code distances d_i, $i = 1,\dots,t$, is an M-code with the code distance $d = d_1 \cdots d_t$.*

A generalization of tensor product of codes and a construction of convolutional codes connected with this generalization are described in /3/. These constructions are the so-called codes with majority decoding in several steps. We shall not stop on this construction in detail, because very soon these results were absorbed by the constructions of concatenated codes of G. Forney (G. D., Jr. Forney, Concatenated codes, MIT Press Research Monograph, 1966) and generalized concatenated codes of E. Blokh and V. Zyablov (E. L. Blokh, V. V. Zyablov, "Coding and decoding for generalized concatenated codes", The Third Intern. Symp. on Inform. Theory, part II, Moscow-Tallin, 1973).

The next return to radical codes in /4–6/ is connected with a problem of calculation of more refined characteristics of the correcting abilities of linear codes, namely, with the weight distribution of linear codes. Let A_l denote the number of codewords of weight l in a linear (n,k)-code. The $(n+1)$-tuple $(A_0, A_1, \dots, A_n)$ is called the *weight distribution* of the code. Obviously, $A_0 = 1$, and if code distance is d, then $A_1 = \cdots = A_{d-1} = 0$. This problem is still not solved completely not only for radical codes, but even for the third-order RM-codes. Some results in this problem were obtained in /4-6/ and now we shall describe them.

As it was shown in /1/, the binary RM-codes permit the following description. Let G be a direct product of n cyclic groups of second order: $G = (a_1)\cdots(a_n)$, $a_i^2 = 1$, $\pi = \{0,1\}$, $\pi G = \{u = \sum_{g\in G} \alpha_g g \colon \alpha_g \in \pi\}$, let again R denote the radical of πG, $R = \{u \in \pi G \colon \sum \alpha_g = 0\}$. Then the RM-code of the order k, $0 \le k \le n$, is the ideal R^{n-k} in πG ($R^0 = \pi G$), the elements of the group G being taken as the fixed basis of the space πG. It is clear that $\pi G = R^0 \supset R \supset R^2 \supset \cdots \supset R^{n-1} = (u_0)$, where $R^n = (u_0)$ is a one-dimensional subspace in πG generated by the vector

$$u_0 = (1 + a_1)\cdots(1 + a_n). \tag{3}$$

The basis of R^{n-k} is given by the vectors

$$u_{i_1,\dots,i_r} = u_0/(1+a_{i_1})\cdots(1+a_{i_r}), \tag{4}$$

where the set of indices $\{i_1,\dots,i_r\}$ ranges through all the possible subsets of r elements from $\{1,2,\dots,n\}$ for $r=0,\dots,k$.

The elements of the minimum weight 2^{n-k} in R^{n-k} coincide with the sum $A_i = \sum_{g\in C_i} g$, where C_i ranges through all cosets of all possible subgroups of the order 2^{n-k} in G.

By (4) each element in R^{n-k} is represented as the sum of the minimum-weight elements, $u = A_1 + \cdots + A_t$, but in general case this representation is not unique. The set $\{A_1,\dots,A_t\}$ of the minimum-weight elements in R^{n-k} and the sum $u = A_1 + \cdots + A_t$ are called an *irreducible configuration*, if u cannot be written as a sum of less than t minimum-weight elements (where $\{A_i\}$ are allowed to be replaced by other cosets).

It is clear that all the irreducible configurations in R^{n-1} simply are the cosets A_i and the element u_0 being the sum of two non-intersecting cosets of any subgroup of order 2^{n-1} in G.

In /4/ all the irreducible configurations of the minimum-weight elements in R^{n-2} are found. In contrast to the case $k=1$ there exist quite a number of nonequivalent types of irreducible configurations.

The traditional interpretation of RM-codes involves the representation of elements of the codes as the polynomials $f(x_1,\dots,x_n)$ over the field π in Zhegalkin algebra u_n. It is the algebra of polynomials of n variables over the field π with the conditions $x_i^2 = x_i$ for $i=1,\dots,n$. R^{n-k} are the set of polynomials whose degrees do not exceed k.

Since in /4,5,6/ S. D. Berman uses essentially both ways of specifying codes R^{n-k}, we will introduce a one-to-one mapping from the group algebra πG to the Zhegalkin algebra $\mathfrak{U}_n$. We choose a set of all monomials as the basis of $\mathfrak{U}$. Then the correspondence between this basis in $\mathfrak{U}_n$ and the basis (4) in πG is given by

$$\tau_n\colon u_0 \Rightarrow 1,\ u_{i_1,\dots,i_r} \Rightarrow x_{i_1}\cdots x_{i_r}. \tag{5}$$

τ_n preserves the addition. The product of polynomials under this correspondence becomes the operation $u_1 \wedge u_2$, where $u = u_1 \wedge u_2$ is an element in πG whose expression in the basis $\{g\colon g\in G\}$ include those elements that appear simultaneously in the notations for u_1 and u_2.

It is easy to show that a homogeneous affine transformation

$$x_i = \sum_{j=1}^{n} \alpha_{ij} y_j + \beta_i, \alpha_{ij}, \beta_j \in \pi, \tag{6}$$

corresponds to an automorphism of G and the transformation $x_i \Rightarrow x_i + \beta_i$ $(i=1,\dots,n)$ corresponds to a multiplication of all of the elements $u \in \pi G$ by the fixed element $a_1^{\beta_1}\cdots a_n^{\beta_n}$ (this will be referred to as a *translation along the group*). In other words, if $u\in\pi G$ corresponds to the polynomial

$f(x_1, \dots, x_n)$, then the element $a_1^{\beta_1} \cdots a_n^{\beta_n} u$ corresponds to the polynomial $f(x_1 + \beta_1, \dots, x_n + \beta_n)$.

Each automorphism of G and each translation along G transforms a coset A_i into a coset. Therefore, the group of one-to-one mappings of G (the group being generated by automorphisms of G and translations along G) breaks the set of all irreducible configurations into classes of equivalency.

The full description of the configurations $\{A_1, \dots, A_t\}$ from one equivalency class in R^{n-2} is done in terms of a system of integer invariants defined by intersections of A_i. This description indicates a certain system of affine coordinates in which the cosets A_i can be written in the simplest form. The configurations from one equivalency class in an appropriate coordinate system permit identical notations for corresponding cosets.

The element from the R^{n-2} can be written as

$$f(x_1, \dots, x_n) = \sum_{\substack{i,j=1 \\ i \neq j}}^{n} \alpha_{ij} x_i x_j + \sum_{i=1}^{n} \beta_i x_i + \gamma, \quad (\alpha_{ij}, \beta_i, \gamma \in \pi). \tag{7}$$

The minimum-weight elements in R^{n-2} and only those can be presented as a product $f_1 f_2$ of two linear non-homogeneous forms in variables $x_1, \dots, x_n$, $f_1 = \sum_1^n \alpha_i + \delta$, $f_2 = \sum_1^n \beta_i + \gamma$, with non-coinciding homogeneous parts $\sum_1^n \alpha_i x_i$ and $\sum_1^n \beta_i x_i$.

The set of cosets $\{A_1, \dots, A_t\}$, $A_i \in R^{n-2}$, is called the *canonical configuration*, if for some affine coordinate system $A_1 = x_1 x_2, \dots, A_t = x_{2t-1} x_{2t}$. Variables x_{2i-1} and x_{2i} of each pair are called *complementary*.

The following proposition gives the full description of the irreducible configurations of minimum-weight elements in the code R^{n-2}.

THEOREM 6 /4/. *Let $\{A_1, \dots, A_t\}$ be an irreducible configuration in R^{n-2}. If $r = 2, 3, 4$, there can be singular cases. The other types are fundamental and in corresponding affine coordinate systems are given by the following formulas:*

I. $A_1 = x_1 x_2, \dots, A_r = x_{2r-1} x_{2r}$, $2r \le n$ *(canonical configuration).*

II. $A_1 = x_1 x_2, \dots, A_t = x_{2t-1} x_{2t}$, $A_{2t+1} = f_1 f_2$, $t > 1$, *where* $f_1 = 1 + \sum_i x_i$, $f_2 = 1 + \sum_j x_j$, $i, j \in \{1, \dots, 2t\}$, *the number of variables in f_1 with complementions in f_2 is even. In the expression of the forms f_1 or f_2 there are no complementary variables with respect to the canonical configuration* $\{x_1, x_2, \dots, x_{2t-1} x_{2t}\}$.

III. $A_1 = x_1 x_2, \dots, A_t = x_{2t-1} x_{2t}$; $A_{2t+1} = x_{2t+1}(1 + \sum_j x_j)$, $t > 1$, *where* $j \in \{1, \dots, 2t\}$ *and among the variables x_j in the last sum there are no complementary ones.*

The configurations of types I, II and III are mutually nonequivalent. All other types of configurations are reducible.

The paper /5/ is devoted to the calculation of the code parameters of some principal ideals of the group algebra πG. If $u \notin R$, then the principal ideal $(u) = \pi G$ is a trivial code. For every $u \in R$ we can specify a smaller index k,

$1 \le k \le n$, such that $u \in R^{n-k}$ and $u \notin R^{n-r+1}$. In what follows, the notation $u \in R^{n-k}$ means this specific k.

The principal ideal (u) generated by $u \in R^{n-k}$ is a binary linear code of length 2^m, with code distance $d(u)$, $2^{m-k} \le d(u) \le \mathrm{wt}(u)$, where $\mathrm{wt}(u)$ is the Hamming weight of u. (Recall that $d(u)$ is the number of nonzero components in the decomposition $u = \sum_{g \in G} \lambda_g g$, $\lambda_g \in \pi$, or, equivalently, the number of binary vectors $(x_1, \dots, x_n)$ such that $f(x_1, \dots, x_n) = 1$, where f corresponds to u in the Zhegalkin algebra $\mathfrak{U}$.) It is interesting when the upper bound is achieved. An element $u \in R$ with $d(u) = \mathrm{wt}(u)$ is called *nondecreasable.*

In /5/ the necessary and sufficient conditions of nondecreasibility are described for all elements of a second-order RM-code and those elements of a third-order RM-code that are specified by a pair of quadratic forms over π. For RM-codes of arbitrary order it is also shown that an element with weight less than or equal to twice the code distance of the smallest RM-code containing the given element is nondecreasable.

THEOREM 7 /5/. *Each nonzero element $u \in R^{n-k}$, $k \ge 2$, with weight* $\mathrm{wt}(u) \le 2^{n-k+1}$ *is nondecreasable.*

As a corollary, we immediately obtain that an element $0 \ne u \in R^{n-2}$ is nondecreasable if and only if $\mathrm{wt}(u) \le 2^{n-1}$ or $u = u_0$ (3). First-order RM-code $R^{n-1} \subset R^{n-2}$ is $(2^n, n+1, 2^{n-1})$-code and is not a principal ideal. By Theorem 7 for $n = 2s+1$ the principal ideal $(x_1x_2 + \cdots + x_{2s-1}x_{2s} + x_{2s+1})$ has the same parameters.

To describe the next result we need a convenient notation. We fix some generator of G, say $a = a_1$. Then $G = (a) \times H$, $(H\colon 1) = 2^{n-1}$, and every element $u \in \pi G$ can be written in the form $u = \phi_1 + a\phi_2 = (\phi_1 + \phi_2) + (1+a)\phi_2$, where $\phi_1, \phi_2 \in \pi H$. If $u \in R_G^{n-3}$ in group algebra πG, then obviously $\phi_1 \in R_H^{n-3}$ and $\phi_2 \in R_H^{n-3}$, while $\Phi = \phi_1 + \phi_2 \in R_H^{n-2}$. We consider only the case when $\phi_1, \phi_2 \in R_H^{n-2}$ under the natural restriction $\Phi \notin R_H^{n-1}$, since otherwise $u \in R_G^{n-2}$ (the case of Theorem 7).

We would like to emphasize that the notation "uf" above, where $u \in \pi G$ and $f \in \mathfrak{U}$, means the product uv in πG, v being the image of f under the map (5).

Let ϕ be a quadratic form as in (7) and $\bar{\phi} = \sum_{i,j=1}^{n} \alpha_{ij} x_i x_j$, $i \ne j$. We set $\alpha_{ij} = \alpha_{ji}$ for $i \ne j$ and call the symmetric matrix $A_\phi = [\alpha_{ij}]_1^n$ the *matrix of the form* ϕ. If the rank of ϕ is $2r$ and the form $\bar{\phi}$ depends only on $x_1, \cdots, x_{2r}$, then A_ϕ is called the *accompanying matrix* of ϕ.

THEOREM 8 /6/. *Let $u = \phi_1 + \phi_2 \in R_G^{n-3}$, $\phi_1, \phi_2 \in R_G^{n-2}$ and $\phi_1 + \phi_2 \notin R_G^{n-2}$. Assume that the rank of ϕ_2 is $2r$ and the form $\bar{\phi}_2$ depends only on $2r$ variables $x_1, \dots, x_{2r}$. (By means of an affine transformation of variables as in* (6) *the form $\bar{\phi}_2$ always can be written in this way. Neither affine transformation, nor a translation along group change the Hamming weight of an element in πG.)*

An element (7) is nondecreasable if and only if the following conditions are simultaneously satisfied:

1). Accompanying matrices A_{ϕ_1} and A_{ϕ_2} of forms ϕ_1 and ϕ_2 satisfy the condition $(A_{\phi_1} A_{\phi_2}^{-1}) = E$;

2). Either a) the linear parts of ϕ_1 and ϕ_2 contain different variables x_{2r+1} and x_{2r+2} respectively, or (b) the linear part of one of the forms includes the variable x_{2r+1} and the weight of the second form is strictly less than 2^{n-2}.

In /6/ the same mathematical tools—a translation along a group and the affine transform (6)—are used to advance the problem of classification of Bent-functions. Bent-functions arise in many problems of discrete mathematics, in particular, in coding theory and in the problem of constructing elementary Hadamard difference sets, and many papers are devoted to them.

Boolean function $f(x) = f(x_1, \dots, x_n)$ is called the *Bent-function* if coefficients of the Fourier-Hadamard transform

$$\sum_{u \in \pi^n} (-1)^{u \cdot x + f(x)}$$

of $f(x)$ are equal to $\pm 2^{n/2}$.

This definition of Bent-functions in the Zhegalkin algebra $\mathfrak{U}_n$ corresponds to the following definition in the group algebra πG of the group $G = (a_1) \dots (a_n)$, $a_i^2 = 1$, over the field $\pi = \{0, 1\}$: $f(x)$ is a Bent-function if and only if the Hamming weight of the formal derivative $f_v^{(1)}(x) = f(x_1 + v_1, \dots, x_n + v_n) + f(x)$ with respect to any nonzero direction $v = (v_1, \dots, v_n) \in \pi^n$ is 2^{n-1}.

Bent-functions specified by second-degree polynomials in $\mathfrak{U}_n$ have been completely described, but in general case the problem of classification of Bent-functions is not solved even for third-degree polynomials. In /6/ a criterion for a particular class of third-degree Bent-functions is given.

Recall that the degree of an element $u \in \pi G$ is defined as the degree of its image under the map τ_n given by (11). Let us fix one of the generators of the group $G = (a_1) \cdots (a_n)$, say, $a = a_1$, so that $G = (a) \times (H)$. Then every element from πG can be written in the form $u = u_1 + a u_2 = \Phi + (1 + a) u_2$, where $u_1, u_2, \Phi \in \pi H$. If the degree of an element $u \in \pi G$ is 3, then the degrees of the elements $u_1, u_2 \in \pi H$ are not greater than 3, while the degree of $\Phi = u_1 + u_2 \in \pi H$ is not greater than 2. From this immediately an important conclusion follows: if degree $(u \in \pi G) = 3$, then the cubic forms of the polynomials $\tau_{n-1}(u_1)$ and $\tau_{n-1}(u_2)$ in the corresponding to πH Zhegalkin algebra are the same: $\tau_{n-1}(u_1) = q + f_1$ and $\tau_{n-1}(u_2) = q + f_2$. The above-mentioned criterion refers to the case $q = 0$, namely

THEOREM 9 /6/. *Let $G = (a) \times H$, $H = (a_2) \cdots (a_n)$, $a^2 = a_i^2 = 1$. An element $u = u_1 + a u_2 \in \pi G$ of degree 3, where* $\deg(u_1 \in \pi H) = \deg(u_2 \in \pi H) = 2$, *is Bent-function if and only if in the corresponding affine coordinate system $u = f_1(y_1, \dots, y_{n-2}) + y + a f_2(y_1, \dots, y_{n-2})$, where f_1 and f_2 are Bent-functions in the same $n - 2$ variables.*

The next natural case is $q = x_1x_2x_3$. It turns out that in this case the polynomials f_1 and f_2 have to satisfy the following strong necessary conditions.

THEOREM 10 /6/. *If $u = u_1 + au_2$ is a Bent-function in n variables of degree 3 and $q = x_1x_2x_3$, then for $l = 1, 2$ the rank of $f_l = \tau_{n-1}(u_l) + q$ is $n - 2$.*

This necessary condition on the rank of quadratic forms in the Bent-function under consideration can be generalized to the case $q = x_{2t+1}(x_1x_2 + \cdots + x_{2t-1}x_{2t})$.

Most of the papers on the subject contain results of the kind of sufficient conditions, which make it possible to build new Bent-functions from known ones. One such result is also contained in /6/. It is given by the following theorem.

THEOREM 11 /6/. *If f_1 and f_2 are two arbitrary Bent-functions in the n variables, then the function $F(x_1, \ldots, x_n, x_{n+1}, x_{n+2}) = x_{n+2}\{f_1(x_1, \ldots, x_n) + f_2(x_1, \ldots, x_n) + x_{n+1}\} + f_2(x_1, \ldots, x_n) + \lambda x_{n+1}$, $\lambda \in \{0, 1\}$, is the Bent-function in $n + 2$ variables.*

The search for good decoding algorithms at the last time led to the advanced area of digital signal processing. The paper /7/ is done in this field: a new definition of G-spectrum of a discrete signal is given, that permits one to generalize the well known families of orthogonal transforms such as the Discrete Fourier Transform and the Hadamard-Walsh Transform. It is known that these transforms can be formalized as follows. (Though all the definitions and results can be done in a natural way for finite fields too, we consider the signals over the complex field $\mathbf{C}$.) The given block length n of discrete signal lets us choose an abelian group G of the order n and describe a signal $x = (x_1, \ldots, x_n)$, $x_i \in \mathbf{C}$, as a function $f(g)$ on the group G: $f(g_i) = x_i$, $g_i \in G$. Let $\chi_i(g)$, $i = 1, \ldots, n$, be the set of all complex characters of G and $S = [\chi_i(g_j)]_1^n$; then the vector $X = Sx$ is called the *G-spectrum* of signal x. If G is a cyclic group, then G-spectrum is the Discrete Fourier Transform. If G is a direct sum of the second-order groups, then G-spectrum is the Hadamard Transform. If G is a direct sum of groups of order p, then G-spectrum is the Walsh Transform. It follows from the orthogonality relations for characters that the G-spectrum is the set of Fourier coefficients in the decomposition of $f(g)$ in the basis $\{(1/n)\chi_i(g),\ i = 1, \ldots, n\}$:

$$f(g) = (1/n)(X_1\chi_1^*(g) + \cdots + X_n\chi_n^*(g)),$$

where a star denotes the complex conjugation.

Now let G be an arbitrary (not necessarily abelian) finite group of order n and $\{D_r(g) = [d_{ij}^{(r)}],\ r = 1, \ldots, s\}$ be the set of all pairwise non-equivalent indecomposable complex unitary representations of G of the degrees $n_1, \ldots, n_s$. It is well known that the coefficients $\{\sqrt{n_r}d_{ij}^{(r)}(g),\ i, j = 1, \ldots, n,\ r = 1, \ldots, s\}$ of the matrices $D_r(g)$ form the system of orthogonal functions on the group

G, which is complete because of Burnside's theorem $n_1^2+\cdots+n_r^2 = n = (G: 1)$:

(8) $$\sum_{g\in G} \sqrt{n_r} d_{ij}^{(r)}(g) \sqrt{n_t} d_{lk}^{(t)}(g) = n\delta_{il}\delta_{jk}\delta_{rt}.$$

The representations $D_r(g)$ can be naturally extended to the representations $D_r(x)$ of the group algebra $\mathbf{C}G$ of the group G over the complex field $\mathbf{C}$, so that the matrix coefficients $d_{ij}^{(r)}(g)$ are defined for every $x \in \mathbf{C}G$. We introduce some linear indexing $\sqrt{n_r} d_{ij}^{(r)}(g) = d_l(g)$ and some fix order on G and form the matrix $S(g)$, in which i-th row is the values of the function $d_i(g)$ in $g_j \in G$. Then the vector $X = S(g)x$ is defined as the *G-spectrum* of the signal x.

Similarly to the abelian case, we consider a signal x as a function $f(g)$ on G: $x_i = f(g_i)$. Then G-spectrum of x is again the set of the coefficients in the decomposition of $f(g)$ in the orthogonal basis $\{\sqrt{n_r} d_{ij}^{(r)}(g),\ i, j = 1, \ldots, n_r,\ r = 1, \ldots, s\}$. The Fourier coefficients of the vector x associated with $x = \sum_{g\in G} x_i g_i \in \mathbf{C}G$ are given by formulas

$$X_{i,j}^{(r)} = d_{i,j}^{(r)}(x),\ i, j = 1, \ldots, n_r,\ r = 1, \ldots, s.$$

Similarly to the abelian case, it follows from (14) that the powers of the signal x and its G-spectrum X are related by equality $\sum_{i=1}^n |x_i|^2 = n\sum_{i=1}^n |X_i|^2$ and the convolution of signals is the multiplication of elements in the group algebra $\mathbf{C}G$, which now is no more commutative. For abelian case the known fast algorithms of convolution of signals x and y (the fast algorithms of calculation the product $z = x*y$ in $\mathbf{C}G$) use the transition to the G-spectrums X and Y (the calculation of G-spectrum contains about $n \log n$ multiplications in $\mathbf{C}$), the calculation of vector $Z = (X_1Y_1, \ldots, X_nY_n)$ (that needs only n multiplications in $\mathbf{C}$) and the transition from Z to z. In the non-abelian case the G-spectrum of a convolution $z = x * y$ of signals x and y is calculated by the formula

$$Z_{ij}^{(r)} = \sum_{i=1}^{n_r} X_{ij}^{(r)} Y_{ij}^{(r)},\ i, j = 1, \ldots, n_r,\ r = 1, \ldots, s,$$

where $X = \{X_{ij}^{(r)}\}$ and $Y = \{Y_{ij}^{(r)}\}$ are the G-spectrum of x and y. Consequently, contrary to the abelian case the calculation of the G-spectrum Z as a product of G-spectra X and Y requires in general $\sum_{r=1}^s n_r^3 \neq n = \sum_{r=1}^s n_r^2$ multiplications in $\mathbf{C}$. Nevertheless, there are many classes of non-abelian groups, for which n multiplications are sufficient. The family of M-groups (all complex indecomposable representations are monomials one) in an example.

To show the reason of the given generalization of G-spectrum the infinite family $\{G_n\}$ of non-abelian groups is given for which the calculation of G-spectrum is in $(3/4)$ times faster, than FFT for the same length. More precisely, we have

THEOREM 12 /7/. *Let G_n the group of order 2^{2k+1}, $k \geq 1$, with generators $a, b_1, \dots, b_{2k}$ and relations $a^2 = b_i^2 = 1$, $ab_i = b_i a$, $i = 1, \dots, 2k$, $b_{2i-1}b_{2i} = b_{2i}b_{2i-1}a$, $i = 1, \dots, k$, $b_l b_t = b_t b_l$ for the rest indices t and l. The calculation of the G_n-spectrum of a complex signal x of length $n = 2^{2k+1}$ requires at most $(3/4)n \log_2 n + O(n)$ operations.*

The last Berman paper in coding theory was /8/. It is devoted to the problem of recovering a vector in real 3-dimensional space R^3 from the given n-tuple of its projections on the n fixed axes. It is assumed that all components of a given n-tuple can be erroneous and any error-vector $e = (e_1, \dots, e_n)$ satisfies the conditions $|e_i|/|e_j| < \delta$ for all $i \neq j$. The problem is to choose n axes in such a way that for any given n-tuple the most erroneous component e_j can be identically defined with δ as maximal as possible.

For practically important case $n = 5$ in /8/ it is proved that the minimum for all $\delta_{\max} = 1/(5+2\sqrt{5})$, the optimal 5 axes are indicated and the algorithm for calculation of the most erroneous component in the given 5-tuple are described.

This result was generalized by V. I. Panchenko ("The correction of errors in mass of real data", The Proceedings of XII Allunion Seminar on Computer Networks, part 2, Moscow-Odessa, 1987, pp. 125–128). For the real space R^k with arbitrary k he proved that $\delta_{\max} = tg^2(\pi/2n)$ and gave the algorithm for choosing $(k+2)$ axes that permit one to indicate the most erroneous component.

Berman's most preferred occupation in the last ten years was so-called threshold partitions. The first and so far the only results for vector spaces of arbitrary dimension on this area were obtained by S. D. Berman. The problem of classification of the partitions is very important in the theory of pattern recognition, and all known before Berman's papers algorithms for discovering of threshold-property of partitions were empirical ones and referred only to small n.

Let $B = \{(\alpha_1, \alpha_2, \dots, \alpha_n)\}$ be the set of all vertices of the unit cube in n-dimensional Euclidean space R_n and $B = B_1 \cup B_2$, $B_1 \cap B_2 = \varnothing$. If there exist such a hyperplane

$$a_1 x_1 + a_2 x_2 + \cdots + a_n x_n = T,$$

that

$$(15) \qquad \begin{cases} B_1 = \{(\alpha_1, \dots, \alpha_n) \in B, & a_1\alpha_1 + \cdots + a_1\alpha_1 \leq T\}, \\ B_2 = \{(\alpha_1, \dots, \alpha_n) \in B, & a_1\alpha_1 + \cdots + a_n\alpha_n > T\}, \end{cases}$$

then this partition is called *threshold partitioning*. The boolean function f given by (15) also is called *threshold*.

It is known that the hyperoctahedral group G (generated by all permutations on the set $\{x_1, \dots, x_n\}$ and all logical negations $x_i \Rightarrow \bar{x}_i$, $i = 1, \dots, n$) acts on the set of all threshold functions in such a way, that every threshold function is G-equivalent to the function with properties

$$0 \leq a_1 \leq \cdots \leq a_n, \quad T > 0. \tag{16}$$

In what follows these properties are always assumed fulfilling, and, besides, the origin is put in the set B_1.

In addition to the Hamming weight $\mathrm{wt}(u)$ of a binary vector $u = (u_1, \ldots, u_n)$ we define the *weight* $\mathrm{wt}_i(u)$ *of* u *in coordinate* i as a number of unit components in the binary vector $(u_i, u_{i+1}, \ldots, u_n)$, so that $\mathrm{wt}(u) = \mathrm{wt}_1(u)$. Let us set on B the following partial ordering: $u \leq v$, if $\mathrm{wt}_i(u) \leq \mathrm{wt}_i(v)$ for $i = 1, \ldots, n$. This partial order permits one to define monotonicity of subsets in B: $A \subset B$ is called *monotonic to the left* (*to the right*), if the conditions $u \in A$, $v \in B$ and $v \leq u$ ($v \geq u$) imply the condition $v \in A$.

Any subset $A \subset B$ is contained in monotonic to the left (to the right) subset $\overline{A}_l$ ($\overline{A}_r$), that is obtained by joining all $v \leq u$ for all $u \in A$ to A. A subset C of the monotonic to the left subset $A \subset B$ is called the *base* of A, if $\overline{C}_l = A$. Every subset contains the *minimal base* consisting of pairwise non-comparable vectors. The minimal base of a threshold subset is called the *threshold base.* Let $\langle u_1, \ldots, u_s\rangle$ denote $\overline{A}_l$ for $A = \{u_1, \ldots, u_s\}$. The subset $\langle u\rangle = \{x \in B, x \leq u\}$ is called *cyclic* with *generator* u. If a cyclic subset is threshold, then u is called *threshold*.

THEOREM 13 /9/. *Let* $B_1 \subset B$ *be monotonic to the left subset with the base* $u_1, \ldots, u_s$ *and let* $\{v_1, \ldots, v_q\}$ *be the base of the monotonic to the right subset* $B_2 = B \backslash B_1$. *The partition is threshold of the type* (15)–(16) *if and only if for* $r_{ij} = \mathrm{wt}_i(u_j)$, *and* $m_{ij} = \mathrm{wt}_i(v_j)$, $i = 1, \ldots, n$, $j = 1, \ldots, s$, *the system of linear inequalities*

$$\begin{cases} r_{11}x_1 + \cdots + r_{1s}x_s - m_{11}y_1 - \cdots - m_{1q}y_q \geq 0, \\ r_{n1}x_1 + \cdots + r_{ns}x_s - m_{n1}y_1 - \cdots - m_{nq}y_q \geq 0, \\ x_i \geq 0, \ i = 1, \ldots, s, \\ y_i \geq 0, \ i = 1, \ldots, q, \end{cases} \tag{17}$$

has only zero solution in real variables (x_i, y_j), $i = 1, \ldots, s$, $j = 1, \ldots, q$.

Any binary vector u can be given as the integer vector $u = (i_1, \ldots, i_n)$, where $\{i_l\}$ are the locators of unit components. The numbers i_1 and i_n in this notation are called the *minimum* and the *maximum* of vector u accordingly. Let $A = \langle u_1, \ldots, u_s\rangle$ be monotonic to the left subset with minimal base $u_1, \ldots, u_s$ and $\beta_i = \max u_i$. We choose in $\{\beta_i\}$ the maximal subset of different integers $\gamma_1 > \gamma_2 > \cdots > \gamma_q$. Let $\{u_1, \ldots, u_r\}$ be the subset of vectors u with $\max u_i \geq \gamma_j$ for fixed γ_j. The subset $\langle u_1, \ldots, u_r\rangle$ is called the j-th *flag* of A. The following important structural theorem takes place.

THEOREM 14 (unpublished). *Every flag of monotonic to the left threshold subset* $A = \langle u_1, \ldots, u_s\rangle$ *is the threshold subset in the set of all boolean vectors* B.

By Theorem 14 every threshold subset A contains several flags, $A_1 \subset A_2 \subset \cdots \subset A_q = A$. However, A_1 is a particular one: the structure of the base (i.e. the distribution of zeros and units in the basis vectors) depends on A_1 in an essential way. If $A = A_1$, then threshold monotonic to the left subset A is called simply *the flag*.

To go further we need several additional definitions. A consecutive set of l units in a binary vector u is called a conservative *component of weight* l. Vector u is recognized as *vector of the type* $[r_1, r_2, \dots, r_q]$ if it contains q consecutive components of weights $r_1, r_2, \dots, r_q$ accordingly.

The weight of the last consecutive component of u is called *the character* of u: $\mathrm{char}(u) = r_q$. If $A = \{u_1, \dots, u_s\}$, then $\mathrm{char}(A) = \max \mathrm{char}(u_i)$. If a vector $u_i \in A$ satisfies the conditions $\max(u_i) = \max(A)$ and $\mathrm{char}(u_i) = \mathrm{char}(A)$, then it is called *the characteristic vector of the set A*.

If $\min(u) > 1$ for a vector $0 \neq u \in B$ with consecutive components $u_1, \dots, u_q$ then the coordinates 1, $\max(u_1) + 1, \dots, \max(u_q) + 1$ are named *critical points of* u; if $\min(u) = 1$, then critical points are $\max(u_1) + 1, \dots, \max(u_q) + 1$. If, besides, $\max(u_q) = n$, then point $\max(u_q) + 1$ is excluded.

If a weight of characteristic vector of a flag A is maximal in the set of all vectors of A, then it is named *the main vector of the flag.* Characteristic and characteristic vector of any subset $A = \langle u_1, \dots, u_s \rangle$ coincide accordingly with characteristic and characteristic vector of the main flag of A.

If there are no threshold subsets in a threshold set, then it is called *simple.*

A threshold set is called *homogeneous,* if all its elements have the same weight. A threshold set is called *completely homogeneous,* if all vectors of its minimal base have the same type as the main vector.

THEOREM 15 /9/. *The homogeneous threshold set is simple if and only if it is cyclic.*

The statement is not correct if threshold set is not homogeneous.

A vector u is called the vector of *finite type* if for an arbitrary space dimension there exists a constant C such that the minimal base of any completely homogeneous set of the same type as u contains not more than C elements.

THEOREM 16 /9/. *A threshold vector is the vector of finite type if and only if its type is* $[r_1, r_2]$ *for* $r_1 > 1$, *or* $[r_1, r_2, r_3]$ *for* $r_2 > 1$.

S. D. Berman described all completely homogeneous subsets of finite type. They are:

1). Cyclic sets $\langle u \rangle$ with generators in one from two modes $u = ([r_1][^r])$, $r_1 > 1$ or $u = ([_{r_1}] \cdot [_{r_2}][^r])$, $r_2 > 2$.

2). Sets with base $\langle u_1, u_2 \rangle$, where the types of the vectors u_1 and u_2 either are given by Theorem 16, or by equalities $u_1 = ({}_1[\ ]_r \cdot {}_{r+2}[^{r_1}][^{r_2}])$, $u_2 = ({}_2[\ ]_r \cdot {}_{r+3}[^{r_1}][^{r_2}])$.

A set $A = \langle u_1, u_2, \dots, u_s \rangle$ is called *indecomposable* if it contains a vector $u_j \neq u_1$ with the critical point 1.

Let i_t be the minimal critical point of vectors $u_j \neq u_1$ in $A = \langle u_1, \dots, u_s \rangle$; then A is named *non-degenerate*, if for $i_j = 1$ vectors with this critical point contain three consecutive components, and for $i_j > 1$ they contain four consecutive components.

THEOREM 17 (unpublished). *A non-degenerate indecomposable set* $A = \langle u, \dots, u \rangle$ *of equal characteristics is threshold if and only if*

(1). *The critical points form the segment* $[0, i]$;

(2). $\max(u_j)$ *for any pair vectors* u_j *and* $u_k, u_j, u_k \neq u_1$, *is less than or equal to last but one critical points of* u_k, *if the critical points of* u_j *that belong to* $[1, i]$ *are less than critical points of* u_k *that belong to* $[1, i]$.

The full description of homogeneous threshold subsets of equal characteristics with the main vector of type $[i, r]$, $i > 1$, $\min(u_1) = 1$, is given by the following statement.

THEOREM 18 (unpublished). *Let* $\{u_1, \dots, u_s\}$ *be the minimal base of noncyclic homogeneous set* A *of equal characteristic with the main vector* $u_1 = ({}_1[\ ]_i[^r]$. *A is threshold if and only if*:

(1) *Every vector* $u_q \neq u_1$ *has in segment* $[1, i]$ *exactly one critical point* j_q.

(2) *If* $i_q \neq i$ *for all* $u_q \neq u_1$, *then all vectors* u_q *with* $i_q \neq 1$ *consist of three consecutive components and all vectors with* $i_q = 1$ *consist of two consecutive components.*

If A contains vectors with $j_q = 1$, *then*

(3) *There exists a segment* $[t, i]$, $1 \leq t \leq\ll i$, *such that every* $j \in [t, i]$ *is a critical point of some* $u_q \in A$.

(4) *Every vector with* $j_q \in [t, i]$ *consists of three consecutive components for* $j_q = i$ *and of four consecutive components for* $j_q < i$.

(5) *If* $j_q, j_p \in [t, i]$ *and* $j_q < j_p$, *then* $\max(u_q) < j_p$.

(6) *If* $j_q < t$, *then* $\max(u_q) < j_t$ *and* u_q *consists of three consecutive components for* $j_q > 1$ *and of two consecutive components for* $j_q = 1$.

The threshold homogeneous sets of equal characteristics with the main vector of the other types are described too.

The large part of the research is occupied by so named linear filtering of threshold sets. The main idea in this field is to use the main consecutive component of the vectors in the minimal base to reduce the system (17) of linear inequalities to a more simple one.

REFERENCES

1. S. D. Berman, *On the theory of group codes*, Cybernetics **1** (1967), 31–39. (In Russian)

2. ——, *Semisimple cyclic and abelian codes*, Cybernetics **3** (1967), 21–30. (In Russian)

3. S. D. Berman and A. B. Yudanina, *Codes with generalized majority decoding and convolutional codes*, Problemy Peredachi Informatsii, vol. 6, No. 1, 1970, pp. 6–19. (= "Problems of Information Transmission", pp. 1–12, December 1972, Consultants Bureau, New York.)

4. ——, *Irreducible configurations of minimum-weight elements in the Reed-Muller second order codes*, Second Intern. Symp. on Inform. Theory, Tsahkadsor, 1971, Budapest (1973), 115–129.

5. S. D. Berman and I. I. Grushko, *Code parameters of principal ideals of group algebra of group* $(2, 2, \ldots, 2)$ *over the field of characteristic 2*, Problemy Peredachi Informatsii, vol. 14, No. 4, 1978, pp. 3–12. (= "Problems of Information Transmission", pp. 239–246, April 1978, Consultants Bureau, New York.)

6. ——, *B-functions encounted in modular codes*, Problemy Peredachi Informatsii, vol. 17, No. 2, 1981, pp. 10–18. (= "Problems of Information Transmission", pp. 82–89, Consultants Bureau, New York.)

7. ——, *To the theory of digital signal processing*, Problemy Peredachi Informatsii, vol. 19, No. 4, 1983, pp. 43–49. (= "Problems of Information Transmission", pp. 284–288, April 1984, Consultants Bureau, New York.)

8. S. D. Berman, B. P. Gerasimenko, B. P. Zephirov, S. P. Krotkikch, *Decoding of maximal error in the calculation of a vector by means of its projections*, Proceedings of 9th Allunion Symposium on Redundancy of Information Systems, Leningrad, 1986, 125–127.

9. S. D. Berman and N. G. Sarnavsky, *The combinatorial structure of threshold basis*, Cybernetics **2** 1985, 24–32.

Contemporary Mathematics
Volume **93** (1989)

Berman's Contributions to Group Representations Over Fields

G. KARPILOVSKY

In memory of my teacher S. D. Berman

1. Introduction. During 1963–1965 I was extremely fortunate to have been able to attend the lectures of Prof. S. D. Berman on group representations over fields. These lectures were given to graduate students at Uzhgorod State University (U.S.S.R.). Afterwards I commenced my research work under his stimulating guidance. All who knew Berman remember his joie de vivre, his exceptional kindness towards people, his thoughtfulness, his extraordinary conscientiousness, and the thoroughness which is to be found throughout his work. He was a generous mathematician, who always strove to help his students, and encouraged them to sharpen their ideas. At all times he was unselfish with his time and advice. One particular aspect of Prof. Berman that I wish to mention was his ability to pick out what was important in the argument, and this feature, which was apparent in his research, had a very deep effect on his students. Prof. Berman was an excellent and inspiring teacher who made important contributions to group representations over fields, which is the subject of this article. Our review of Berman's work is by no means exhaustive and highlights some of his results which are not too technical.

2. Counting irreducible and indecomposable representations. We begin our review by discussing Berman's contribution to the problem of counting irreducible and indecomposable representations of finite groups.

Let A be a finite-dimensional algebra over a field and let $J(A)$ be the Jacobson radical of A. Then there exists $r = r(A) \geq 1$, $n_i \geq 1$ and a division

0271-4132/89 $1.00 + $.25 per page

ring D_i, $1 \leq i \leq r$, such that

$$A/J(A) \cong \prod_{i=1}^{r} M_{n_i}(D_i).$$

The significance of the natural invariant $r(A)$ of A stems from the fact that it coincides with the number of nonisomorphic irreducible A-modules. The special case where A is the group algebra FG of a finite group G over a field F is of particular importance. This is a consequence of the well-known fact that representations of G over F can be interpreted as FG-modules.

Let F be a field of characteristic $p \geq 0$. An element $g \in G$ is said to be p-regular if $p = 0$ or if $p > 0$ and the order of g is not divisible by p. Let n be the least common multiple of p-regular elements of G and let ε be a primitive nth root of 1 over F. Two p-regular elements $a, b \in G$ are called F-conjugate if $b = g^{-1}a^{\mu}g$ for some $g \in G$ and some μ such that $\varepsilon \mapsto \varepsilon^{\mu}$ determines an automorphism of $F(\varepsilon)$ over F. It is easily seen that F-conjugacy is an equivalence relation, and so the set of p-regular elements of G is a disjoint union of p-regular F-conjugacy classes.

The problem of determining $r(FG)$ exclusively in terms of F and G was solved by Brauer in the case where F is algebraically closed. Brauer proved that if F is algebraically closed, then $r(FG)$ is equal to the number of p-regular conjugacy classes of G. The corresponding problem in the general case was solved by Berman (1952, 1956a) and Witt (1952) independently. This result, known as the Witt-Berman theorem, asserts that $r(FG)$ is equal to the number of p-regular F-conjugacy classes of G. A related result established by Berman (1958a) is the following. Assume that $\operatorname{char} F \nmid |G|$, that $f \in \operatorname{Aut}(G)$ and that Γ is an irreducible F-representation of G. The representation Γ is said to be f-invariant if Γ is equivalent to Γ_f where Γ_f is defined by $\Gamma_f(g) = \Gamma(f(g))$, $g \in G$. It is natural to ask: What is the number of nonequivalent f-invariant irreducible F-representations of G? It was shown by Berman that this number coincides with the number of F-conjugacy classes of G which are sent into themselves by f.

Let F be a field of characteristic p. It is well known that G has only finitely many nonequivalent indecomposable F-representations if and only if the Sylow p-subgroups of G are cyclic. Assume that a Sylow p-subgroup P of G is cyclic. It is natural to enquire about the precise number of nonequivalent indecomposable F-representations of G. This problem was solved by Berman (1960c) in case $P \lhd G$, who showed that the required number is equal to $|P|r$, where r is the number of p-regular F-conjugacy classes of G.

3. Induced representations. Our aim here is to highlight some of Berman's results pertaining to the theory of induced representations. Let N be a normal subgroup of a finite group G and let Λ be an irreducible representation of N over an algebraically closed field F with $\operatorname{char} F \nmid |G|$. Then a necessary and sufficient condition for the irreducibility of the induced representation Λ^G of

G is the following: For all $g \in G - N$, the representations $x \mapsto \Lambda(x)$ and $x \mapsto \Lambda(g^{-1}xg)$ of N are nonequivalent. If F is an arbitrary field, then the given condition is no longer necessary. Nevertheless the mentioned criterion can be extended to arbitrary fields in case G/N is torsion-free. Namely, Berman (1960d) proved the following interesting result:

THEOREM. *Let G be an arbitrary group and let N be a normal subgroup of G such that G/N is torsion-free. Assume that Λ is an arbitrary (possibly infinite-dimensional) irreducible representation of N over an arbitrary field F. Then Λ^G is irreducible if and only if for all $g \in G - N$, the representations $x \mapsto \Lambda(x)$ and $x \mapsto \Lambda(g^{-1}xg)$ of N are nonequivalent.*

Possibly the most often used result on induced representations is the so-called Frobenius Reciprocity Theorem. This theorem was generalized by Berman (1958a), who established the following property.

THEOREM. *Let F be an arbitrary field of characteristic 0, let H be a subgroup of a finite group G and let Γ and γ be irreducible F-representations of G and H, respectively. Let D and D' be the division rings corresponding to the minimal ideals of FG and FH associated with Γ and γ, respectively. If the restriction of Γ to H contains γ n times, then γ^G contains Γ $n(\dim_F D')/(\dim_F D)$ times.*

Let G be a finite group. By an elementary subgroup of G we understand one which is a direct product of a cyclic group and a p-group for some prime p. A classical result of Brauer asserts that every complex character of G is a Z-linear combination of characters induced from linear characters of elementary subgroups of G. Using this theorem, Brauer was able to give new and simpler proofs of some of his earlier basic results on group representations. By making use of Roquette's simplification of Brauer's proof, Berman (1956b) and Witt (1952) independently obtained the generalized result known as the Witt-Berman Induction Theorem. In order to state the Witt-Berman theorem, we shall introduce some notation and definitions.

Let K be a subfield of $\mathbf{C}$. Given a positive integer m, let ε_m denote a primitive mth root of 1 over K and let I_m be the multiplicative group consisting of those integers μ, taken modulo m, for which $\varepsilon_m \mapsto \varepsilon_m^{\mu}$ defines an automorphism of $K(\varepsilon_m)$ over K. A subgroup H of a finite group G is said to be K-elementary at p, which is a prime, if

(i) H is a semidirect product $\langle a \rangle B$ of a p-group B and a cyclic group $\langle a \rangle$ whose order m is not divisible by p, and

(ii) For each $b \in B$ there exists a $\mu \in I_m$ such that $bab^{-1} = a^{\mu}$.

Thus the $\mathbf{C}$-elementary subgroups of G are precisely the elementary subgroups of G defined before. We can now state the following result.

THEOREM (The Witt-Berman Induction Theorem). *Every K-character of G is a Z-linear combination $\sum n_i \chi_i^G$, $n_i \in Z$, where the χ_i are K-characters of K-elementary subgroups of G.*

Let G be a finite group. Then G is said to be an M-group if all irreducible $\mathbf{C}$-representations of G are induced from one-dimensional representations. The following two facts are standard.

(i) All M-groups are solvable (Taketa).

(ii) If G is a solvable group that has a normal subgroup N such that G/N is supersolvable and all Sylow subgroups of N are abelian, then G is an M-group (Huppert).

Berman (1961e) discovered the following class of M-groups containing the one described in (ii). Let G be a solvable group and let $1 \subset G_1 \subset \cdots \subset G_s = G$ be an invariant series with abelian factors. Then G is an M-group, provided the following property of extension of linear characters holds:

If χ is a linear character of a subgroup H of G_k, $k \geq 1$, with $H = G_1$ if $k = 1$, and

$$N(\chi) = \{g \in G_{k+1} \mid g^{-1}G_k g = G_k,\ \chi(g^{-1}hg) = \chi(h) \text{ for all } h \in H\}$$

then χ can be extended to a character of $N(\chi)$.

Let F be an algebraically closed field of arbitrary characteristic, and let G be a (not necessarily finite) group. It is natural to investigate conditions under which all finite-dimensional irreducible F-representations of G are monomial, i.e. induced from one-dimensional representations of subgroups of G. This problem was approached by Berman (1960b) under the condition that G is finite. Subsequently in a joint paper with Šaraya (see Berman (1966c)) the following was discovered.

Let G be a group with a finite invariant series

$$1 \subset G_1 \subset G_2 \subset \cdots \subset G_s = G$$

where G_1 is abelian and the factors $G_2/G_1, \dots, G_s/G_{s-1}$ are cyclic. Assume, further, that F is an algebraically closed field of arbitrary characteristic. Then the following properties hold:

(i) There exists a family $\{\Gamma_i\}$ of irreducible representations Γ_i where for each $i, \Gamma_i = \chi_i^G$ for some linear character χ_i of a subgroup N_i of G.

(ii) The set of linear characters $\{\chi_i\}$ is invariant with respect to the group of inner automorphisms of G.

(iii) The family $\{\Gamma_i\}$ contains all finite-dimensional irreducible F-representations of G.

(iv) The representations Γ_i and Γ_j are equivalent if and only if the characters χ_i and χ_j are mapped into each other by an inner automorphism of G, i.e. if there exists some $g \in G$ such that $g^{-1}N_i g = N_j$ and $\chi_j(g^{-1}ag) = \chi_i(a)$, for all $a \in N_i$.

(v) If $\operatorname{char} F = 0$, then for any $1 \neq g \in G$, there exists Γ_i such that $\Gamma_i(g) \neq I$, where I is the identity operator.

In fact, as was pointed out by the authors, the above properties hold under more general circumstances, namely for any group G which has a transfinite invariant series

$$1 \subset G_1 \subset \cdots \subset G_\alpha \subset \cdots \subset G$$

where G_1 is abelian and each $G_{\alpha+1}/G_\alpha$ is cyclic.

4. Effective constructions of representations. The group representation theory is an instance of a theory whose constructive version is at least as complicated as its classical counterpart. For various applications the constructive results are of special importance. Thus we are confronted with the following problem: given a finite group G and a field F, find an effective method for the construction of irreducible F-representations of G. Our aim here is to survey some of Berman's results pertaining to the above problem.

Let G be a finite solvable group with composition series

$$1 = G_0 \subset G_1 \subset \cdots \subset G_{s-1} \subset G_s = G.$$

In what follows we describe a method for constructing irreducible $\mathbf{C}$-representations of G which is exhibited in Berman (1955).

Since G_1 is abelian, all minimal idempotents of $\mathbf{C}G_1$ are well known. Assume that we have already constructed all minimal central idempotents $e_1, e_2, \dots, e_n$ of $\mathbf{C}G_t$, $2 \le t \le s$, and their corresponding minimal idempotents $e'_1, e'_2, \dots, e'_n$. Then a similar system of idempotents of $\mathbf{C}G_{t+1}$ can be calculated as follows. Let e_q, $1 \le q \le n$, be an arbitrary minimal central idempotent of $\mathbf{C}G_t$. We distinguish two cases:

(i) $g^{-1}e_q g = e_q$ for all $g \in G_{t+1}$;

(ii) $g^{-1}e_q g \neq e_q$ for some $g \in G_{t+1}$.

Consider the latter case. Among the idempotents $g^{-1}e_q g (g \in G_{t+1})$ there are exactly $p = (G_{t+1} \colon G_t)$ distinct:

$$e_q^{(1)} = e_q, e_q^{(2)}, \dots, e_q^{(p)}.$$

Let us put

$$\tilde{e} = e_q^{(1)} + e_q^{(2)} + \cdots + e_q^{(p)}.$$

Then the following holds:

(iii) $\tilde{e}$ is a minimal central idempotent of $\mathbf{C}G_{t+1}$ and e'_q is a corresponding minimal idempotent of $\mathbf{C}G_{t+1}$.

Now consider the case (i). Let $C_1, C_2, \dots, C_m$ be all conjugacy classes of G_{t+1} and, for each $i \in \{1, 2, \dots, m\}$, put $C_i^+ = \sum_{x \in C_i} x$. Then there exists a conjugacy class C_i, $1 \le i \le m$, such that the following two conditions hold:

(iv) $C_i^+ \notin \mathbf{C}G_t$.

(v) $C_i^+ e_q \neq 0$.

It follows that

$$(C_i^+)^p e_q = \lambda e_q \quad \text{for some } 0 \neq \lambda \in \mathbf{C}.$$

Consider the following element

$$b = (C_i^+ e_q)/\sqrt[p]{\lambda}$$

and put

$$u_i = (1/p)(e_q + \varepsilon^i b e_q + \varepsilon^{2i} b^2 e_q + \cdots + \varepsilon^{i(p-1)} b^{p-1} e_q) \quad (1 \le i \le p)$$

where ε is a primitive complex pth root of 1. Then

$$e_q = u_1 + u_2 + \cdots + u_p$$

and each u_i is a minimal central idempotent of $\mathbf{C}G_{t+1}$. Furthermore, the corresponding minimal idempotents of G_{t+1} are given by

$$v_i = e_q' u_i \quad (1 \leq i \leq p).$$

The next result due to Berman (1966b) gives a very convenient effective method to classifying all absolutely irreducible representations of a certain class of groups G. This method reduces to a succession of extensions of linear characters such that at each inductive step the problem reduces to a standard procedure of extending a character of a subgroup of a finite abelian group to a character of the group itself.

THEOREM (Berman (1966b)). *Let G be a finite solvable group having a normal subgroup N such that G/N is supersolvable and all Sylow subgroups of N are abelian. Let F be a field containing a primitive nth root of* 1 *where $n = |G|$ and let*

$$1 = G_0 \subset G_1 \subset \cdots \subset G_k = N \subset \cdots \subset G_s = G$$

be a chief series of G passing through N. Then there exists a system of linear characters $\chi_1, \ldots, \chi_r$ of subgroups $N_1, \ldots, N_r$ of G such that

(i) *The elements*

$$e_i = |N_i|^{-1} \sum_{a \in N_i} \chi_i(a) a \quad (1 \leq i \leq r)$$

form a complete system of minimal orthogonal idempotents of FG.

(ii) *Every inner automorphism of G and every transformation $g \mapsto g^\mu ((\mu, n) = 1)$ induce a bijection of the set $\{e_1, \ldots, e_r\}$.*

(iii) *Minimal central idempotents of FG are the sums of orbits under the action of the group of inner automorphisms of G on the set $\{e_1, \ldots, e_r\}$.*

The result above enables us to state (in group-theoretic terms) necessary and sufficient conditions for the isomorphism of the group algebras FG and FG_1, where G and G_1 satisfy the hypothesis of the theorem and F is an algebraically closed field whose characteristic does not divide the orders of G and G_1. To do this, it is enough to observe that the isomorphism class of FG is determined by the degrees of irreducible F-representations of G (counting their multiplicities) and the latter are the indices of the subgroups N_i. For the precise formulation of the corresponding result we refer to the original paper of Berman (1966b).

For the rest of this section G denotes a group satisfying the hypothesis of the proceeding theorem and K is an arbitrary field whose characteristic does not divide the order of G. Put $F = K(\varepsilon)$ where ε is a primitive $n = |G|$th root of 1 over K. Our next aim is to discuss another result of Berman (1966b) which reduces the task of classifying all irreducible K-representations of G to the same task for a group which is an abelian extension of a self-centralizing cyclic normal subgroup. We employ the results of the previous theorem and write

$$e_i = |N_i|^{-1} \sum_{a \in N_i} \chi_i(a)a \quad (1 \leq i \leq r)$$

for a complete system of minimal orthogonal idempotents of FG. Let Φ be the group of transformations $g \mapsto g^\mu$, $g \in G$, where μ ranges those integers, taken modulo n, for which $\varepsilon \mapsto \varepsilon^\mu$ is an automorphism of F over K. Since the group Φ acts on the set $M = \{e_1, \dots, e_r\}$, we may write M as a disjoint union of orbits:

$$M = M_1 \cup \cdots \cup M_t, \ \ M_i = \{e_{i1}, \dots, e_{im_i}\} \quad (1 \leq i \leq t).$$

Let $\chi_{i1}, \dots, \chi_{im_i}$ be the linear characters of the subgroups $N_{i1}, \dots, N_{im_i}$ which correspond to the idempotents $e_{i1}, \dots, e_{im_i}$ of M_i, $1 \leq i \leq t$.

Clearly we may assume that after a suitable renumbering of the N_j we have

$$N_{i1} = \cdots = N_{im_i} = N_i \quad (1 \leq i \leq t)$$

where the characters $\chi_{i1}, \dots, \chi_{im_i}$ have the same kernel, say H_i, and the factor group $N_i \mid H_i$ is cyclic, say generated by $a_i H_i$. The characters χ_{ij} $(1 \leq j \leq m_i)$ are completely determined by their values on $a_i \in G$, and these characters have the same K-normalizer

$$\tilde{N}_i = N_K(\chi_{i1}) = \cdots = N_K(\chi_{im_i}) \quad (1 \leq i \leq t).$$

Here by the K-normalizer $N_K(\chi)$ of a linear character χ of a subgroup H of G we mean the set of all $g \in G$ for which $g^{-1}Hg = H$ and $\chi(g^{-1}hg) = \chi(h^\mu)$ for all $h \in H$, where $\mu = \mu(g)$ is such that $\varepsilon \mapsto \varepsilon^\mu$ is an automorphism of $K(\varepsilon)$ over K. Note that $\tilde{N}_i$ consists precisely of those elements $g \in G$ for which

$$g^{-1}H_i g = H_i, g^{-1}N_i g = N_i, g^{-1}(a_i H_i)g = a_i^\mu H_i$$

where $g \mapsto g^\mu$ is a mapping from Φ. The centralizer of the cyclic normal subgroup N_i/H_i of $\tilde{N}_i/H_i$ coincides with N_i/H_i. Therefore the corresponding factor group $\tilde{N}_i/N_i$ is abelian and is isomorphic to a subgroup of the automorphism group of N_i/H_i. Put

$$u_i = e_{i1} + \cdots + e_{im_i} \quad (1 \leq i \leq t).$$

THEOREM (Berman (1966b)). *With the notation above, the following properties hold*:

(i) *Each* u_i *is an idempotent of* KG *and*

$$1 = u_1 + \cdots + u_t, u_i u_j = 0 \quad \text{for } i \neq j.$$

(ii) *Each* u_i *is a minimal central idempotent of* $K\tilde{N}_i$.

(iii) *If* e *is a minimal idempotent of* $K\tilde{N}_i$ *belonging to* $K\tilde{N}_i u_i$, *then the endomorphism rings of* KGe *and* $K\tilde{N}_i e$ *are isomorphic. In particular,* e *is a minimal idempotent of* KG.

5. The Schur index. Here we briefly discuss some of Berman's results concerning the Schur index of absolutely irreducible representations (or characters). Most of the mentioned properties below are extracted from Berman (1966b).

It is well known that the Schur index of any absolutely irreducible character of a finite p-group, p odd, with respect to an arbitrary field of characteristic 0 is equal to 1. This fact was generalized by Berman who proved the following result.

THEOREM (Berman (1966b)). *Let G be a finite solvable group with a normal subgroup N such that G/N is supersolvable and all Sylow subgroups of N are abelian. Let K be a field of characteristic* 0 *and let Γ be an absolutely irreducible K-representation of G of degree m. Assume that at least one of the following two conditions hold*:

(i) *$|G|/m$ is a power of an odd prime.*

(ii) $(|G|/m, m) = 1$.

Then the Schur index of Γ with respect to any subfield F of K is equal to 1.

In what follows, we denote by Q_p the rational p-adic field. Given an absolutely irreducible character χ of G, we write $m_{Q_p}(\chi)$ for the Schur index of χ with respect to Q_p. One aspect of behaviour of $m_{Q_p}(\chi)$ is described by the following result.

THEOREM (Berman (1966b)). *Let G be a finite group, let p be a prime and let χ be an absolutely irreducible character of G. Then $m_{Q_p}(\chi) \equiv 0 \pmod p$ if $p \neq 2$ and $m_{Q_2}(\chi) \equiv 0 \pmod 4$ if $p = 2$.*

The next result generalizes a theorem due to Serre which states that if a Sylow p-subgroup P of G is normal and G/P is a cyclic p'-group, then $m_{Q_l}(\chi) = 1$, where $l \neq p$ and χ is any absolutely irreducible character of G.

THEOREM (Berman (1966b)). *Let N be a normal p'-subgroup of a finite group G such that G/N is nilpotent and $(|N|, |G\colon N|) = 1$. Let χ be any absolutely irreducible character of G. Then*

(i) *$m_{Q_p}(\chi) = 1$ for $p \neq 2$ and $m_{Q_2}(\chi) = 1$ or 2.*

(ii) *$m_{Q_2}(\chi) = 1$ if the Sylow 2-subgroup of G/N is abelian.*

We next provide an explicit formula for $m_{Q_p}(\chi)$ in a very important special case where G is an abelian extension of a cyclic group. In what follows, given a character χ of a subgroup of G, we write $Q_p(\chi)$ for the field obtained from Q_p by adjoining the values of χ.

In view of the discussion in the preceding section, it is important to investigate the case where G contains a self-centralizing cyclic normal subgroup $N = \langle a \rangle$ of order qp^α (p prime, $(q,p) = 1$) with $|G\colon N| = n \not\equiv 0 \pmod p$ and such that the Q_p-normalizer of N in G is equal to G. The latter means that for any $g \in G$, $g^{-1}ag = a^\mu$ where $\mu = \mu(g)$ is such that $\varepsilon \mapsto \varepsilon^\mu$ is an automorphism of $Q_p(\varepsilon)$ over Q_p (ε is a primitive $|G|$th root of 1). Let χ be an absolutely irreducible character of G which is induced by the character ψ of N given by

$$\psi(a) = \varepsilon_1\varepsilon_2$$

where ε_1 and ε_2 are primitive p^αth and qth roots of 1, respectively.

THEOREM (Berman (1966b)). *With the notation above, write* $a = a_1a_2$ *with* $a_1a_2 = a_2a_1$ *and* $a_1^{p^\alpha} = a_2^q = 1$, *and let* H *be a complement to* $\langle a_1\rangle$ *in* G. *Denote by* λ *any linear character of* $C_H(a_2)$ *whose restriction to* $\langle a_2\rangle$ *is equal to* μ, *where* μ *is the restriction of* ψ *to* $\langle a_2\rangle$. *Finally, let* S *be the* Q_p*-normalizer of* λ *in* H. *Then*

$$m_{Q_p}(\chi) = (Q_p(\lambda)\colon Q_p(\mu))|H\colon S|.$$

The result above can be applied to prove the following.

THEOREM (Berman (1966b)). *Let* m *be a natural number and let* p *be a prime. Then the following conditions are equivalent:*

(i) *There exists a finite group* G *and an absolutely irreducible character* χ *of* G *such that* $m = m_{Q_p}(\chi)$.

(ii) $p - 1 \equiv 0 \pmod{m}$ *if* $p \neq 2$ *and* $m \in \{1, 2\}$ *if* $p = 2$.

We close by quoting the following interesting result which can be obtained as an easy consequence of the preceding theorem.

THEOREM (Berman (1961a)). *For any natural number* m, *there exists a finite group* G *and an absolutely irreducible character* χ *of* G *such that* $m = m_Q(\chi)$.

BIBLIOGRAPHY OF S. D. BERMAN ON GROUP REPRESENTATIONS OVER FIELDS

[1952] On the theory of representations of finite groups, Dokl. Akad. Nauk SSSR 86, 885–888.

[1954] On representations of semi-direct product of abelian groups, Dokl. Akad. Nauk SSSR 98, 177–180.

[1955] Group algebras of abelian extensions of finite groups, Dokl. Akad. Nauk SSSR 102, 431–434.

[1956a] The number of irreducible representations of a finite group over an arbitrary field, Dokl. Akad. Nauk SSSR, 767–769.

[1956b] p-adic ring of characters, Dokl. Akad. Nauk SSSR 106, 583–586.

[1957a] Generalized characters of finite groups, Dopovidi Akad. Nauk Ukrain. RSR, 112–115.

[1957b] Groups of which all representations are monomial, Dopovidi Akad. Nauk Ukrain. RSR, 539–542.

[1958a] Characters of linear representations of finite groups over an arbitrary field, Mat. Sb. **44(86)**, 409–456.

[1958b] Representations of groups of order 2^m over an arbitrary field of zero characteristic, Dopovidi Akad. Nauk Ukrain. RSR, 243–246.

[1958c] (With Bovdi, A. A.) P-blocks for some classes of groups, Dopovidi Akad. Nauk. Ukrain. RSR 6, 606–609.

[1959a] On characters of finite nilpotent groups, Uspehi Mat. Nauk. **14**, No. 5, 217–218.

[1959b] On indecomposable modular representations of finite groups, Uspehi Mat. Nauk **14**, No. 5, 218.

[1960a] Modular representations of finite supersolvable groups, Dopovidi Akad. Nauk Ukrain. RSR, 586–589.

[1960b] Modular representations of finite groups whose chief factors has cyclic factors, Dopovidi Akad. Nauk RSR, No. 5, 586–589.

[1960c] On indecomposable modular representations of a class of finite groups, Dokl. Uzhgorod. Univ., No. 3, 51–52.

[1960d] On induced representations of groups, Dokl. Uzhgorod Univ., No. 3, 53–54.

[1960e] Normal subgroups of the group of triangular matrices over a finite field, Dokl. Uzhgorod Univ. No. 3, 50.

[1961a] On Schur's index, Uspehi Mat. Nauk 16, No. 2(98), 95–99.

[1961b] The smallest field in which can be realized all complex representations of a p-group of odd order, Uspehi Mat. Nauk 16, No. 3(99), 151–153.

[1961c] On representations of locally finite groups, Dokl. Uzhgorod Univ. 4, 78–79.

[1961d] Generalized modular characters of finite groups, Dokl. Uzhgorod Univ. 80–81.

[1961e] On representations of M-groups over an arbitrary field, Dokl. Uzhgorod Univ. 4, 82–83.

[1966a] Representations of finite groups, Algebra 1964, Akad. Nauk SSSR Inst. Naucn. Informacii, Moscow, 83–122.

[1966b] Representations of finite groups over an arbitrary field and over rings of integers, Izv. Akad. Nauk SSSR Ser. Mat. 30, 69–132.

[1966c] (with Šaraja, V. V.) Representation of a class of discrete solvable groups over an arbitrary algebraically closed field, Algebra and Math. Logic: Studies in Algebra, Izdat. Kiev Univ., Kiev, 54–61.

[1968] (with Šaraja, V. V.) The irreducible representations of an infinite direct product of finite groups, Sibirsk. Mat. Ž. 9, 485–489.

[1969] On the Schur index, Appendix to Russian translation of "Representation Theory of Finite Groups and Associative Algebras" by C. W. Curtis and I. Reiner, Moscow, pp. 610–638.

[1973] (with Višnjakova, N. I.) The representations of a group of type p^∞ over a field of characteristic p, Trudy Moscov. Mat. Obšv. 29, 87–100.

[1977a] (with Kerer, E. Š.) Representations of a torsion-free nilpotent group of class two with two generators, (Russian), Funkcional Anal. i Priložen. 11, No. 4, 70–71.

[1977b] (with Šaraja, V. V.) Irreducible complex representations of finitely generated nilpotent groups, Ukrain. Mat. Ž. 29, No. 4, 435–442.

[1977c] (with Pidgornii, O. G.) The representations of a group of type p^∞ over a certain class of fields, Ukrain. Mat. Ž. 29, No. 5, 632–637.

[1981a] (with Buzaši, K.) Representations of the infinite dihedral group, Publ. Math. Debrecen 28, No. 1-2, 173–187.

[1981b] (with Buzaši, K.) On modules over group algebras of groups containing an infinite cyclic subgroup of finite index, Studio Sci. Math. Hungarica 16, 455–470.

[1982] (with Buzaši, K.) On representations of a group containing infinite cyclic subgroup of finite index, Publ. Math. (Debrecen) 29, 163–170.

[1984] (with Buzaši, K.) A description of all finite-dimensional real representations of groups containing an infinite cyclic subgroup of finite index, Publ. Math. (Debrecen) 31, 133–144.

Department of Mathematics
California State University
Chico, CA 95929

Contemporary Mathematics
Volume 93 (1989)

S. D. Berman's Contribution to the Theory of Integral Representations of Finite Groups

A. I. LICHTMAN

1. Introduction. S. D. Berman's work on integral representations of finite groups began in 1959. During 1959–1962 he proved a number of interesting and important theorems which became classical in the theory of integral representations. These results are published in articles **[1]**–**[6]**. The most famous of them is the theorem that a finite group G has a finite number of indecomposable integral representations if and only if for each prime p, dividing the order of G, the Sylow p-subgroups of G are cyclic of order p or p^2. He proved this theorem in the Fall of 1961 in his Doctoral Dissertation (see **[2]**, **[3]**, **[6]**). In 1959–1960 he and P. M. Gudivok obtained the classification of the indecomposable representations of cyclic groups of order p^2. We discuss these results in Sections 2 and 3 of this article. (These results were obtained independently and simultaneously by a group of mathematicians in the USA: Heller and Reiner **[10]**–**[11]**, Jones **[12]**, Dade **[13]**; their methods are essentially different from Berman's methods.)

In Section 4 of this article we discuss S. D. Berman's theorems on the integral representations of a group, which are afforded by the ideals of its group ring ZG (see **[3]**, **[7]**). We review the results on the functor Ext for modules over integral group rings of nilpotent groups in Section 5 (see **[8]**).

2. Representations of Cyclic Groups. Let G be a finite group, R be a principal ideal domain of characteristic zero, T be the field of fractions of R. Every R-representation Γ of G is R-equivalent to a representation of the form

$$g \to \Gamma(g) = \begin{pmatrix} \Gamma_1(g) & & * \\ & \ddots & \\ 0 & & \Gamma_s(g) \end{pmatrix}, \tag{2.1}$$

where Γ_j $(j = 1, 2, \ldots, s)$ are R-representations of G such that all the irreducible T-components of Γ_i are equivalent and the representations Γ_i and Γ_j have distinct irreducible components when $i \neq j$.

Supported In part by NSF Grant DMS 8802634.

Now let $R = I_p$ be the ring of p-adic integers, Ω_p be the field of p-adic numbers and G be a cyclic group of order p^h. Let ξ_i $(i = 1, 2, \dots, h)$ be a primitive root from 1 of degree p^i. It is known that every irreducible representation of the group G is given in this case by

$$a \to \tilde{\xi}_i,$$

where $a \to \tilde{\xi}_i$ is the operator of multiplication by ξ_i in the ring $I_p[\xi_i]$; the matrix of this operator in the basis $1, \xi_i, \xi_i^2, \dots, \xi_i^{\phi(p^i)-1}$ is also denoted by $\tilde{\xi}_i$. Furthermore, every representation Γ_j $(i = 1, 2, \dots, s)$ in (2.1) is now completely reducible by Steinitz's Theorem and hence representation (2.1) can be written, after a renumeration of the roots ξ_i, in a form

$$\Gamma(a) = \begin{pmatrix} \tilde{\xi}_1^{(n_1)} & & & \\ & \tilde{\xi}_2^{(n_2)} & & * \\ & & \ddots & \\ 0 & & & \tilde{\xi}_\kappa^{(n_\kappa)} \end{pmatrix} \quad (k \le h) \tag{2.2}$$

where n_i $(i = 1, 2, \dots, k)$ is the multiplicity of the representation $\tilde{\xi}_i$.

The case when in (2.2) $k = 2$ is of special importance in the study of representations of cyclic groups of order p^2. Let thus

$$\Gamma(a) = \begin{pmatrix} \tilde{\xi}^{(n_1)} & * \\ 0 & \tilde{\xi}^{(n_2)} \end{pmatrix} \tag{2.3}$$

where $\xi^{p^{m_1}} = 1$, $\varepsilon^{p^{m_2}} = 1$, $0 \le m_i \le h$, $m_i \ne m_2$. It is proven (see **[5]**, Lemma 2.3) that the representation (2.3) is equivalent to

$$\Gamma(\Delta) = \begin{pmatrix} \tilde{\xi}^{(n_1)} & \langle \Delta \rangle \\ 0 & \tilde{\varepsilon}^{(n_2)} \end{pmatrix} \tag{2.4}$$

where Δ is a block matrix over I_p; each block δ_{ij} of Δ has $\phi(p^{m_1}) - 1$ rows and $\phi(p^{m_2}) - 1$ columns; every column of δ_{ij}, except the last one, is zero and the last column defines the coordinates of an element of the ring $I_p[\xi]$ in the basis

$$1, \xi, \xi^2, \dots, \xi^{\phi(p^{m_1})-1}.$$

It is convenient to denote this element of $I_p[\xi]$ by δ_{ij} and to assume thus that Δ is a matrix over $I_p[\xi]$.

The further study of the representations (2.4) is based on the following Theorem (see Berman and Gudivok **[5]**, Theorem 2.1).

THEOREM 1. *Let $G = (a)$ be a cyclic group of order p^h, ξ and ε be primitive roots of 1 of orders p^{m_1} and p^{m_2} correspondingly $(m_1 \ne m_2)$. Then the representations*

$$\Gamma = \begin{pmatrix} \tilde{\xi}^{(n_1)} & \langle \|\mu_{ij}\| \rangle \\ 0 & \tilde{\varepsilon}^{(n_2)} \end{pmatrix} \text{ and } \Gamma_1 = \begin{pmatrix} \tilde{\xi}^{(n_2)} & \langle \|\mu_{ij}\| \rangle \\ 0 & \tilde{\varepsilon}^{(n_2)} \end{pmatrix}$$

of G over I_p are equivalent if and only if the matrices $\|\mu_{ij}\|$ and $\|\lambda_{ij}\|$ are equivalent modulo the ideal $(1-\xi)^t$, where $t = \{\phi(p^{m_1}), \phi(p^{m_2})\}$.

This important theorem implies easily the following Corollary:

Every representation (2.3) *is a sum of indecomposable representations of the form*

$$\tilde{\xi}, \tilde{\varepsilon}, \begin{pmatrix} \tilde{\xi} & \langle\{i\}\rangle \\ 0 & \tilde{\varepsilon} \end{pmatrix} \tag{2.5}$$

where $\{i\} = (1-\xi)^i$ $(i = 0, 1, \ldots, t-1;\ t = \min\{\phi(p^{m_1}), \phi(p^{m_2})\})$. *All the indecomposable representations of* G *over* I_p, *whose irreducible components belong to the set* $\{\tilde{\xi}, \tilde{\varepsilon}\}$, *are given by* (2.5).

Since the cyclic group of order p has only two irreducible representations over I_p, namely 1 and $\tilde{\varepsilon}$, one can obtain immediately from (2.5) all the $p+1$ indecomposable I_p-representations of this group. These representations were described earlier by Diederichsen in **[14]** and Reiner in **[15]**.

Now once again let $G = (a)$ be a cyclic group of order p^h $(h \geq 2)$ and

$$\{1, \tilde{\varepsilon}, \tilde{\xi}\}(\xi^{p^{m_1}} = 1; \varepsilon^{p^{m_2}} = 1;\ (0 < m_2 < m_1 < h)) \tag{2.6}$$

be a set which contains three irreducible I_p-representations of G. It is worth remarking that when G is a cyclic group of order p^2 then (2.6) contains all the irreducible I_p-representations of G.

Define the following representations of G.

$$\begin{aligned}
&\Gamma_1 : a \to \tilde{\xi};\ \Gamma_2 : \alpha \to \tilde{\varepsilon};\ \Gamma_3 : a \to 1; \\
&\Gamma_4^{(i)} : a \to \begin{pmatrix} \tilde{\xi} & \langle\{i\}\rangle \\ 0 & \tilde{\varepsilon} \end{pmatrix}; \\
&\Gamma_5 : a \to \begin{pmatrix} \tilde{\xi} & \langle 1\rangle \\ 0 & 1 \end{pmatrix};\ \Gamma_6 : a \to \begin{pmatrix} \tilde{\varepsilon} & \langle 1\rangle \\ 0 & 1 \end{pmatrix}; \\
&\Gamma_7^{(i)} : a \to \begin{pmatrix} \tilde{\xi} & \langle\{i\}\rangle & 0 \\ & \tilde{\varepsilon} & \langle 1\rangle \\ 0 & & 1 \end{pmatrix}; \\
&\Gamma_8^{(i)} : a \to \begin{pmatrix} \tilde{\xi} & \langle\{i\}\rangle & \langle 1\rangle \\ & \tilde{\varepsilon} & 0 \\ 0 & & 1 \end{pmatrix}; \\
&\Gamma_9 : a \to \begin{pmatrix} \tilde{\xi} & 0 & \langle 1\rangle \\ & \tilde{\varepsilon} & \langle 1\rangle \\ 0 & & 1 \end{pmatrix}; \\
&(i = 0, 1, \ldots, \phi(p^{m_2} - 1).
\end{aligned} \tag{2.7}$$

Furthermore, if $p^{m_2} \neq 2$ the following additional representation is defined:

$$\Gamma_{10}^{(i)} : a \to \begin{pmatrix} \tilde{\xi} & \langle\{i\}\rangle & 0 & \langle 1\rangle \\ & \tilde{\varepsilon} & \langle 1\rangle & 0 \\ & & 1 & 0 \\ 0 & & & 1 \end{pmatrix} \quad (i = 1, 2, \ldots, \phi(p^{m_2}) - 1). \tag{2.8}$$

The following theorem is one of the main results of the article **[5]** by Berman and Gudivok (see **[5]**, Theorem 4.1).

THEOREM 2. *If* $p^{m_2} \neq 2$ *then the representations* (2.7) *and* (2.8) *are the only indecomposable* I_p*-representations of the group* $G = (a)$ $(a^{p^h} = 1, h \geq m_1)$, *whose irreducible components are taken from the set* (2.6). *If* $p^{m_2} = 2$ *then all the indecomposable representations of the group* G *with the irreducible components from the set* (2.6) *are given by* (2.7).

The power of this theorem can be illustrated by the fact that the classification of all the indecomposable I_p-representations of cyclic groups of order p^2 is obtained immediately as an easy corollary of this theorem. It follows immediately from Theorem 2 via the fact that all the irreducible representations of such groups are given by (2.6).

COROLLARY. *The cyclic group of order* p^2 *has* $4p + 1$ *indecomposable* I_p*-representations. When* $p \neq 2$ *these representations are given by* (2.7) *and* (2.8); *when* $p = 2$ *they are given by* (2.7).

REMARK 1. The classification of all the indecomposable integral representations of the cyclic group of order 4 was obtained earlier by A. V. Roiter (see **[16]**).

REMARK 2. It is worth remarking that if $p^2 \neq 4$ then the degree of the representation (2.8) is $p^2 + 1$; it means, in particular, that it is not afforded by an ideal of the group ring ZG.

Now let $G = (a)$ a cyclic group of order p^m $(m \geq 3)$, ξ_s be a primitive root of 1 of degree p^{m_s} $(s = 1, 2, 3;\ m_1 > m_2 > m_3)$, n be an arbitrary natural number. Slightly changing the notations of Berman and Gudivok we denote

$$\Gamma_n(\theta) = \begin{pmatrix} \tilde{\xi}_1^{(n_1)} & 0 & \langle E \rangle & \langle \theta \rangle \\ & \tilde{\xi}_2^{(n)} & \langle E \rangle & \langle E \rangle \\ & & \tilde{\xi}_3^{(n)} & 0 \\ 0 & & & E_n \end{pmatrix} \tag{2.9}$$

where θ is an invertible matrix over $I_p[\xi_1]$. The following result is a very special corollary of Theorem 3.2 in **[5]**.

THEOREM 3. *Two representations* $\Gamma_n(\theta)$ *and* $\Gamma_n(\overline{\theta})$ *are equivalent over* I_p *if and only if the matrices* θ *and* $\overline{\theta}$ *are similar modulo the ideal* $(1 - \xi_1)$. *A representation* $\Gamma_n(\theta)$ *is indecomposable over* I_p *if and only if the matrix* θ *is indecomposable modulo the ideal* $(1 - \xi_1)$.

Since every cyclic G group of order p^m $(m \geq 3)$ contains at least 4 different irreducible I_p-representations one can pick for every natural n a matrix $\theta \in \mathrm{GL}_n(Z[\xi_1])$, indecomposable modulo the ideal $(1 - \xi_1)$, and to obtain a Z-representation $\Gamma_n(\theta)$ of G which is indecomposable over I_p. Thus the following result is established.

COROLLARY. *A cyclic group of order* p^m $(m > 3)$ *has* Z*-representations of unbounded degree which are indecomposable over* I_p.

The existence of indecomposable I_p-representations of unbounded degree for an arbitrary abelian group of type (p, p) was first proven by Borevich and

Faddeev in **[17]**. The methods of Berman's and Gudivok's article **[5]** made it possible to construct explicit series of such representations. These series are obtained from the following theorem (see **[5]**, Theorem 3.3).

THEOREM 4. *Let G be an abelian group of type (p,p):*

$$a^p = 1, \ b^p = 1, \ ab = ba$$

and $\Gamma(\Lambda_n)$ be the following I_p-representations of G:

$$\Gamma_a(\Lambda_n) = \begin{pmatrix} \tilde{\varepsilon}^{(n)} & 0 & 0 & \langle\Lambda_n\rangle \\ & E_{sn} & E_s & 0 \\ & & \tilde{\varepsilon}^{(n)} & 0 \\ \mathbf{0} & & & E_n \end{pmatrix}$$
$$(2.10)$$
$$\Gamma_b(\Lambda_n) = \begin{pmatrix} E_{sn} & 0 & E_{sn} & 0 \\ & \tilde{\varepsilon}^{(n)} & 0 & \langle E_n\rangle \\ & & \tilde{\varepsilon}^{(n)} & 0 \\ \mathbf{0} & & & E_n \end{pmatrix}$$

where ε is a primitive root of degree p from 1, $s = p-1$, and $\Lambda_n = \|\lambda_{ij}\|$ $(\lambda_{ij} \in I_p[\varepsilon])$ is an arbitrary $n \times n$ matrix over the ring $I_p[\varepsilon]$.

Two representations $\Gamma(\Lambda_n)$ and $\Gamma(\Lambda'_n)$ are I_p-equivalent if and only if the matrices Λ_n and Λ'_n are similar modulo the ideal $(1-\varepsilon)$.

By letting Λ_n in (2.10) be a Jordan block of degree n with units on the main diagonal the following result is obtained.

COROLLARY. *An abelian group of type (p,p) has Z representations of unbounded degree, which are indecomposable over I_p.*

The classification of the indecomposable I_p-representations made it possible to answer two following questions posed by I. Reiner. It was proven in **[17]** and **[18]** that the Krull-Schmidt theorem remains true for I_p-representations of finite groups. In relation to this fact I. Reiner posed in **[19]** a question whether the same result is true for the representations of G over the ring of rational p-integers R_p. The negative answer is given by Berman and Gudivok in **[5]** for the representations of cyclic groups with an order not divisible by a square.

The second question of I. Reiner is whether non-equivalent Z-representations of G can become equivalent over some ring of algebraic integers R (see **[19]**); this question is related to the classical fact that if ρ_1 and ρ_2 are two representations of a finite group G over a field K and they become equivalent over a field $K_1 \supseteq K$, then these representations are equivalent over K. The negative answer is given by Berman and Gudivok for the representations of a cyclic group G of order p^2, where $p = 4m+1$ and m is a quadratic nonresidue modulo p; they pick two representations of G which are not equivalent over Z but become equivalent over $Z[\varepsilon]$, where $\varepsilon^p = 1$.

3. The representations over the ring of rational p-integers. We will discuss now the results of S. D. Berman on representations of cyclic group G of order mp^2 over the ring of rational p-integers R_p. These representations are constructed from the following irreducible representations of G

$$\Gamma^{(1)} = \widetilde{\eta\xi} : \Gamma^{(2)} = \widetilde{\eta\varepsilon};\ \Gamma^{(3)} = \tilde{\eta}, \tag{3.1}$$

where η is an arbitrary root of 1 of degree t, $t|m$, and ξ and ε are primitive roots of 1 of degree p^2 and p correspondingly; it is important that every indecomposable R_p-representation Γ of G has the following decomposition over the field of rationals:

$$\Gamma = \alpha\Gamma^{(1)} + \beta\Gamma^{(2)} + \gamma\Gamma^{(3)}.$$

The vector (α, β, γ) is called the type of the representation Γ. S. D. Berman obtained in **[4]** an interesting result which gives a complete classification of all the possible types of indecomposable R_p-representations of a cyclic group of order mp^2, $(m, p) = 1$.

THEOREM 5. (See Berman **[4]**, Theorem 1.) *Let G be a cyclic group of order mp^2, $(m,p) = 1$, δ be the order of p modulo m; then the types of all the indecomposable R_p-representations of G are given by:*

i) *The vectors*

$$(1,0,0);\ (0,1,0);\ (0,0,1);\ (1,1,0);\ (0,1,1);\ (1,0,1);\ (1,1,1);\ (1,1,2),\ \text{if } \delta = \phi(m); \tag{3.2}$$

ii) *The vectors* (3.2) *and vectors*

$$(2,1,1),\ (1,2,1),\ (2,1,2),\ (1,2,2),\ (2,2,2),\ (2,3,3),\ (3,2,3),\ (3,2,2),\ (2,3,2),\ \text{if } \delta = \phi(m)/2; \tag{3.3}$$

iii) *The vectors* (3.2), (3.3) *and*

$$(3,1,2),\ (1,3,2),\ (4,2,2),\ (2,4,2),\ (2,4,4),\ (4,2,4),\ \text{if } \frac{\phi(m)}{\delta} \geq 3. \tag{3.4}$$

When $p = 2$ the corresponding types are

$$\text{iii) } (1,0,0),\ (0,1,0),\ (0,0,1),\ (1,1,0),\ (1,0,1),\ (0,1,1),\ (1,1,1),\ \text{if } \delta = \phi(m); \tag{3.5}$$

$$\text{iv) } (2,1,1),\ (1,2,1),\ (1,1,2),\ (1,2,2),\ (2,1,2),\ (2,2,2),\ \text{if } \delta \geq \frac{\phi(m)}{2}. \tag{3.6}$$

Let for an arbitrary R_p-representation F of G, $n(\Gamma)$ denotes the number of all the irreducible rational components and $m(\Gamma)$ be the maximal number of components, equivalent to each other. Theorem 5 immediately yields the following corollary.

An R_p representation Γ of G is decomposable if at least one of the following conditions hold:

$$n(\Gamma) > 10;\ m(\Gamma) > 4\ (\text{if } p \neq 2);$$
$$n(\Gamma) > 6;\ m(\Gamma) > 2\ (\text{if } p = 2).$$

4. Groups of finite representation type. A group G has a finite representation type over a ring R if it has a finite number of indecomposable R-representa-
tions. When $R = Z$ or $R = I_p$ the Jordan-Zassenhaus theorem implies that this is equivalent to the existence of a bound for the degrees of the indecomposable R-representations.

First let $R = I_p$. Since the Krull-Schmidt theorem holds for I_p-representations of G one can apply Higman's method **[20]** and to obtain all the indecomposable representations of G as components of the representations, induced from its Sylow p-subgroup. This implies that the group G and its Sylow p-subgroup H simultaneously have a finite or infinite representation type. Corollaries of Theorems 3 and 4 now imply that the only p-groups which have a finite representation type over I_p are the cyclic groups of order p or p^2. This yields the following theorem.

THEOREM 6. *A group G has a finite representation type over I_p if and only if its Sylow p-subgroup is cyclic and its order is not divided by p^3.*

The case when $R = Z$ is much more complicated. The complete solution of the problem is given by the following remarkable theorem of Berman.

THEOREM 7. *A group G has a finite representation type over Z if and only if for each prime p dividing the order of G the Sylow p-subgroups of G are cyclic of order p or p^2.*

This theorem was announced by S. D. Berman in 1962 in **[2]** and **[3]**; its proof was given in his Doctoral Dissertation (1961) and then in his fundamental article **[6]**.

We will now describe the main steps of Berman's proof; some of them are of considerable independent interest.

First, corollaries of Theorems 3 and 4 imply, as in the proof of Theorem 6, that if G satisfies at least one of two following conditions

1) *The order of G is divisible by a cube of a prime number;*

2) *A certain Sylow p-subgroup of G is noncyclic.*

Then G has indecomposable representations of unbounded degree.

This proves the necessity of the conditions of Theorem 7.

In order to prove the sufficiency of the conditions the following results are obtained (see Berman **[3]**, **[6]**).

The irreducible I_p-representations of a finite group G, whose Sylow p-subgroup H is a cyclic normal subgroup, are classified and it follows in particular that if M_1 and M_2 are two irreducible $G-I_p$-modules such that $\Omega_p \otimes M_i$ $(i = 1, 2)$ are irreducible $G - \Omega_p$-components of the same $G - Q$-module M then

$$\operatorname{Ext}^1(M_1, M_2) = 0. \tag{4.1}$$

These results together with some of Berman's results on induced representations and on Schur's index are used to obtain the following theorem (see **[6]**, Theorem 5.1).

THEOREM 8. *Let G be a group such that all its Sylow p-subgroups are cyclic, M and N be two irreducible $G-Z$-module, which are equivalent over the field of rationals Q. Then*

$$\operatorname{Ext}^1(M, N) = 0. \tag{4.2}$$

The condition (4.2) is called in **[6]** condition I.

The next important step in the proof of Theorem 7 is based on the study of $G - I_p$-modules which satisfy the following condition (condition II).

Let

$$M = M_1 + M_2 + \cdots + M_s$$

where $M_1, \dots, M_s$ are irreducible $G - I_p$-modules, which are equivalent over Ω_p. Then every $G - I_p$-automorphism of the module M can be represented as a product

$$\theta = \theta_1 \theta_{i_1 j_1} \theta_{i_2 j_2} \dots \theta_{i_t j_t} \tag{4.3}$$

where θ_1 is a diagonal automorphism of M: $\theta_1(M_i) = M_i$ $(i = 1, 2, \dots, s)$, and $\theta_{i_r j_r}$ are elementary triangular automorphisms of this module: $\theta_{i_r j_r}(x) = x$ if $x \in M_j$ and $j \neq j_r$, and $\theta_{i_r j_r}(x) = x + y$, where $y \in M_{i_r}$, if $x \in M_{j_r}$ (the possibility $(i_r, j_r) = (i_m, j_m)$ for $r \neq m$ is not excluded.)

THEOREM 9. (See **[6]**, Theorem 5.2.) *Let G be a group such that for every prime p dividing the order of G the Sylow p-subgroups of G are cyclic of order p or p^2. Then G satisfies condition* II *for each p, dividing $(G: 1)$.*

The proof of Theorem 7 follows then from Theorems 8 and 9 together with some technical results.

Heller and Reiner proved in **[10]** and **[11]** that a cyclic group H of order p or p^2 has a finite number of indecomposable Z-representations; if H is noncyclic, or cyclic of order greater than p^2, then it has indecomposable Z-representations of unbounded degree. These results together with the results of Jones' article **[12]** yield an alternative proof of Theorem 7.

4. The representations which are afforded by the ideals of integral group ring *ZG*. The results, described in Section 3, imply that G has indecomposable representations of unbounded degree if it has a Sylow p-subgroup which is not a cyclic group of order p or p^2. Furthermore, if G has a Sylow p-subgroup H, which is cyclic of order p^2 $(p \neq 2)$ then H has an indecomposable representation Γ of degree $p^2 + 1$. Once again, Higman's method **[20]** implies that Γ can be used to obtain an indecomposable representation of G which is not afforded by an ideal of ZG. This is the idea of the proof of the necessary conditions in the following theorem of S. D. Berman (see **[3]**, Theorem 3).

THEOREM 10. *Let G be a group. Then all the indecomposable Z-representations of G are afforded by the ideals of ZG if and only if for every prime p dividing $(G: 1)$ the Sylow p-subgroups of G are cyclic and the order of an arbitrary cyclic subgroup of G is either 4 or is a product of distinct odd prime numbers.*

The proof of sufficiency is based on a few results, which are of independent interest. The main step in the proof of Theorem 10 is the following result (see **[3]**, Theorem 4).

THEOREM 11. *Let G be a group such that for each p dividing the order of G the Sylow p-subgroup is cyclic and Γ be a Z-representation of G such that all its Q-irreducible components are equivalent. Then Γ is a sum of irreducible Z-representations*

$$\Gamma = \Gamma_1 + \Gamma_2 + \cdots + \Gamma_m. \tag{4.4}$$

It is well known that the integral group ring ZG is indecomposable in a direct sum of left ideal. S. D. Berman obtained (see **[3]**, Theorem 5 and **[7]**) the following result, which is a far reaching generalization of this fact.

THEOREM 12. *Let G be an arbitrary finite group, H be a subgroup of G, χ be linear character of H, Γ be a representation of the group G over the ring $Z[\chi]$ which is induced by this linear character. Then the representation Γ is indecomposable over the ring $\overline{Z}$ of algebraic integers. In particular, every transitive representation of G by permutations, (considered as a Z-representation), is indecomposable.*

5. The functor Ext^1 over nilpotent group rings. Let G be a group, U be a left ideal in the rational group ring QG or in the p-adic group ring $\Omega_p G$. The intersection $U \cap ZG$ (or, correspondingly, $U \cap I_p G$) is called the integral part of U and is denoted by $\tilde{U}$. The following theorem is proven in the article **[8]** by S. D. Berman and A. I. Lichtman.

THEOREM 13. *Let G be a p-group, U be a left ideal in QG (or in $\Omega_p G$). Then its integral part $\tilde{U}$ is an indecomposable left ideal of ZG (or $I_p G$, correspondingly).*

The proof makes an essential use of the fact that if K is a field of characteristic p and G is a p-group then two arbitrary ideals in the group ring KG have a non-zero intersection.

This theorem implies, in particular, that if G is a p-group and M and N are two left ideals of ZG such that QM and QN are distinct left ideals in QG, then there exists $G - Z$-modules M_1 and N_1, rationally equivalent to M and N correspondingly, and such that

$$\mathrm{Ext}^1(M_1, N_1) \neq 0. \tag{5.1}$$

Now let the group G be nilpotent. Thus,

$$G = G_1 \times G_2 \times \cdots \times G_s,$$

where G_i has order $p_i^{d_i}$ $(i = 1, 2, \ldots, s)$. Let e be a minimal central idempotent of the group ring QG. It is well known that

$$e = e^{(1)}e^{(2)} \cdots e^{(s)} \tag{5.2}$$

where $e^{(i)}$ is a minimal central idempotent in the group ring QG_i $(i = 1, 2, \ldots, s)$. Let M and N be two rationally nonisomorphic irreducible $G-Z$-modules, and u and v be the minimal central idempotent in QG which act identically on M and N correspondingly and

$$u = u^{(1)}u^{(2)} \cdots u^{(s)}, \quad v = v^{(1)}v^{(2)} \cdots v^{(s)}. \tag{5.3}$$

THEOREM 14. *The condition* (5.1) *holds for every two* $G - Z$*-modules* M_1, N_1, *rationally equivalent to* M *and* N *correspondingly, if and only if the idempotents* (5.3) *differ in only one factor.*

Some more results on integral representations containing two irreducible components are obtained in Berman's article **[9]**.

REFERENCES

1. S. D. Berman and P. M. Gudivok, *On integral representations of finite groups*, (Russian), Dokl. Akad. Nauk SSSR **145** (1962), 1199–1201.

2. ——, *On integral representations of finite groups*, (Russian), Dokl. Uzgorod Univ., **5** (1962), 79–76.

3. S. D. Berman, *On integral representations of finite groups*, Dokl. Akad. Nauk SSSR **152** (1963), 1286–1287.

4. ——, *On the theory of integral representations of finite groups*, Dokl. Akad. Nauk SSSR **157** (1964), 506–508.

5. S. D. Berman and P. M. Gudivok, *Indecomposable representations of finite groups over the ring of p-adic integers*, (Russian), Izv. Akad. Nauk SSSR **28** (1964), 875–910.

6. S. D. Berman, *Representations of finite groups over an arbitrary field and over rings of integers*, (Russian), Izv. Akad. Nauk SSSR **30** (1966), 69–132.

7. ——, *On integral monomial representations of finite groups*, Uspehi Mat. Nauk **20** (1965), 133–134.

8. S. D. Berman and A. I. Lichtman, *On integral representations of finite nilpotent groups*, Uspehi Mat. Nauk **20** (1965), 186–188.

9. S. D. Berman, *Integral representations of cyclic groups containing two irreducible rational components*, In Memoriam: N. G. Čebotarev, Izdat. Kazan Univ., Kazan, 1964, 18–29.

10. A. Heller and I. Reiner, *Representations of cyclic groups in rings of integers I*, Ann. of Math. (2) **76** (1962), 72–92.

11. ——, *Representations of cyclic groups in rings of integers II*, Ann. of Math. **77** (1963), 318–328.

12. A. Jones, *Groups with finite number of indecomposable integral representations*, Michigan Math. J. **10** (1963), 257–261.

13. E. C. Dade, *Some indecomposable group representations*, Ann. of Math. (2) **77** (1963), 406–412.

14. F. E. Diederichsen, *Über die Ausreduktion ganzzahliger Gruppen-darstellengen bei arithmetischer Aquivalenz*, Hamb. Abh **17** (1938), 357–412.

15. I. Reiner, *Integral representations of cyclic groups of prime order*, Proc. Amer. Math. Soc. **8** (1957), 142–146.

16. A. V. Roiter, *On representations of cyclic group of order four by integral matrices*, Vestnik Leningrad. Univ., (Russian), **15** (1960), 65–74.

17. Z. I. Borevich and D. K. Faddeev, *The theory of homology of groups II*, Vestnik Leningrad. Univ. **7** (1959), 72–87.

18. I. U. Maranda, *On p-adic integral representations of finite groups*, Canad. J. Math. (1953), 344–355.

19. I. Reiner, *The Krull-Schmidt theorem for integral group representations*, Bull. Amer. Math. Soc. (1961).

20. D. G. Higman, *Indecomposable representations at characteristic p*, Duke Math. J. **21** (1954), 377–381.

Contemporary Mathematics
Volume 93, 1989

The Work of S. D. Berman on Infinite Abelian Group Rings

WARREN MAY

It is fair to say that the study of group algebras of infinite abelian groups began twenty years ago with the Debrecen paper of S. D. Berman. Berman's work in this area was largely devoted to aspects of the isomorphism problem in which the coefficient ring was usually a field, or could just as well be taken to be one. Earlier, Perlis and Walker [17] and Deskins [13] had solved the isomorphism problem for finite abelian p-groups over fields. In the Debrecen paper [2], Berman gave a complete solution for when the group algebras of two countably infinite abelian p-groups were isomorphic over a given field. In addition, a number of related results were also proved.

In reviewing Berman's contributions, one is struck by the thoroughness with which a problem is examined. Ironically, this comprehensiveness may have hindered propagation of his solutions. The Debrecen paper is frequently cited when referring to the isomorphism problem, and deservedly so. Nevertheless, the results have usually been summarized, and complete proofs have not been given due to length and intricacy. There is the prospect, however, that new proofs will soon appear based upon a different point of view [11,12].

In this overview of Berman's accomplishments, we think it best to present clearly the main points of his work rather than to aim for completeness. Thus we shall resort to abbreviating the theorems on countable p-groups by omitting more complicated special cases, or by restricting consideration to those p-groups with the most structural interest. The omitted results from the Debrecen paper will then be restored in an appendix which forms the last section of this review. We remark that the appendix corrects a few minor errors of statement which have persisted.

Beyond the countable p-groups case solved by Berman, there has been only limited progress. There are some joint papers by Berman and several coauthors addressing various topics. We shall discuss these in two brief sections devoted to uncountable p-groups and to mixed groups.

1980 *Mathematics Subject Classification* (1985 *Revision*). Primary 16A27; Secondary 20C07.

1. Countable p-Groups.

In this section we shall review the main parts of [**2**]. We let K always denote a field and T will denote a countably infinite abelian p-group. Isomorphisms of group algebras will always be algebra isomorphisms. First we distinguish between the semisimple case in which the characteristic of K is different from p, and the indecomposable case in which the characteristic of K equals p. In the indecomposable case, there are no nontrivial idempotents, thus KT preserves many of the properties of T. For example, Berman established that the Ulm invariants of T are inherent in the group algebra, thus one obtains a strong isomorphism result.

THEOREM 1. ([**1**,**2**]). *Let K be a field of characteristic p and T a countably infinite abelian p-group. If $\tilde{T}$ is a group such that $KT \cong K\tilde{T}$, then $T \cong \tilde{T}$.*

Now let K be of characteristic different from p. In the semisimple case, by contrast, one expects that much of the structure of T will be lost in KT, particularly if many roots of unity are present. To clarify this situation, Berman introduced two classes of fields. We say that K is of the first kind with respect to p if the extension field gotten by adjoining all p^nth roots of unity for all n is of infinite degree over K. Otherwise we say that K is of the second kind with respect to p. Since a field of the second kind is so close to containing all the p^nth roots of unity, one expects the antithesis of Theorem 1. For simplicity of statement we shall exclude the case that $p = 2$ and K does not contain a primitive 4th root of unity.

THEOREM 2. ([**1**,**2**]). *Let K be a field of the second kind with respect to p and let T be a countably infinite abelian p-group. If $p = 2$, assume that K contains a primitive 4th root of unity. If $\tilde{T}$ is another countably infinite abelian p-group, then $KT \cong K\tilde{T}$.*

In the excluded case, there is a countable family of isomorphism types for KT. We refer the reader to the appendix.

For fields of the first kind, less of the structure of T is lost in the group algebra, but still quite a bit. Berman divided the countable p-groups into nine classes to state his classification over fields of the first kind. Passman [**16**] reduced this to seven by allowing the trivial group in a few places. Groups in different classes have nonisomorphic algebras, and conditions are given, if necessary, to decide whether groups in the same class have isomorphic algebras. Let $T^1 = \cap_n T^n$ denote the first Ulm subgroup of T. Note that if T is a p-group, then $T^1 = \cap_n T^{p^n}$. We wish to restrict attention to those T such that T/T^1 is unbounded. These are the interesting p-groups structurally since if T is not of this type, then T is isomorphic to the direct product of a divisible and a bounded p-group.

THEOREM 3. ([**1**,**2**]). *Let K be a field of the first kind with respect to p, and let T be a countable abelian p-group such that T/T^1 is unbounded. Then*

$KT \cong K\tilde{T}$ if and only if $\tilde{T}$ is a countable abelian p-group such that $\tilde{T}/\tilde{T}^1$ is unbounded, and such that $|T^1| = |\tilde{T}^1|$ if either T^1 or $\tilde{T}^1$ is infinite or trivial.

An example of a field of the first kind is the field of rational numbers. Perlis and Walker [17] showed that the group algebra of a finite abelian group over the rational field determines the group. By contrast, we see in the infinite case that very little is determined. Partly, this is because T^1 is contained in the maximal divisible subgroup of the normalized unit group $V(KT)$. Thus elements of T of p-height at least ω become of p-height ∞ in $V(KT)$ and much of the structure of T^1 is lost.

Since there are relatively few isomorphism classes of group algebras in the semisimple case, one might ask for descriptions of these algebras other than as group algebras. Passman poses this question in his book [16] in terms of certain building blocks $A(K)$. By using ideas of Arens and Kaplansky on topological representation of algebras, Chatzidakis and Pappas [11] have given representations of the semisimple algebras KT as rings of continuous functions. Moreover, they have retrieved the theorems of Berman for fields of the first and second kinds by applying these representations.

The Debrecen paper also considered the semisimple case for countable torsion abelian groups over fields with sufficiently many roots of unity, for example, algebraically closed fields. The result is what one would expect for fields that are of the second kind with respect to every prime different from the characteristic.

THEOREM 4. ([1,2]) *Let K be an algebraically closed field and let T and $\tilde{T}$ be countably infinite torsion abelian groups having no element of order equal to the characteristic of K. Then $KT \cong K\tilde{T}$.*

There was also a classification of the real group algebras $\mathbf{R}T$ for T as above. This was later superceded by a theorem of Berman and Bogdan which we shall discuss in Section 3.

Finally, Berman initiated study of the normalized unit group $V(KT)$ for T a p-group. He described $V(KT)$ when K has characteristic p and K and T are no more than countable. In this case, the normalized unit group is a p-group and is countable, hence its structure is determined by Ulm-Kaplansky invariants. A representation in these terms was given later by Mollov and Nachev, and we shall use their formulation. If A is an abelian group and σ is an ordinal or ∞, let $f_A(\sigma)$ denote the σ-th Ulm-Kaplansky p-invariant of A.

THEOREM 5. ([1,2,15]). *Let K be a countable field of characteristic p and let T be a countable abelian p-group of p-length λ. Let $V = V(KT)$ be the normalized unit group of KT, and let $V_\sigma = V(K^{p^\sigma}T^{p^\sigma})$ for σ any ordinal.*

(1) *Let $\sigma < \lambda$. If both K^{p^σ} and T^{p^σ} are finite, then $f_V(\sigma) = f_{V_\sigma}(0)$; otherwise $f_V(\sigma) = \aleph_0$.*

(2) *If $\sigma \geq \lambda$, then $f_V(\sigma) = 0$.*

(3) $f_V(\infty) = 0$ *or* $\aleph_0$, *depending on whether* T *has trivial maximal divisible subgroup or not, respectively.*

The structure of the p-component $V_p(KT)$ of $V(KT)$ in the semisimple case was announced in the following year by Berman and Rossa. We shall exclude certain special cases in stating the theorem. First we must define an index $f = f(K)$ for K a field of the first kind with respect to p. Let ε_n denote a primitive p^nth root of unity over K. Then $f(K)$ is defined to be the smallest integer f such that $f \geq 1$ if p is odd, $f \geq 2$ if $p = 2$, and $K(\varepsilon_f) \neq K(\varepsilon_{f+1})$.

THEOREM 6. ([**9**]). *Let* K *be a field of characteristic different from* p, *and let* T *be a countably infinite abelian* p-*group. If* $p = 2$, *assume that* K *contains a primitive* 4*th root of unity.*

(1) *If* K *is of the second kind with respect to* p, *then* $V_p(KT) \cong D = \oplus_\omega \mathbf{Z}(p^\infty)$.

(2) *If* K *is of the first kind with respect to* p *and if* T/T^1 *is unbounded, then* $V_p(KT) \cong A \times D$, *where* $A = \oplus_{i \geq f}(\oplus_\omega \mathbf{Z}(p^i))$.

The representation in the semisimple case by Chatzidakis and Pappas of KT as a ring of continuous functions allows the structure of $V(KT)$ to be given in [**12**] in terms of the multiplicative groups of various cyclotomic extensions of K. For the p-component, the description reduces to the one above.

2. Uncountable p-Groups.

Let T now denote an uncountable abelian p-group. Little is known about KT since little is known about uncountable abelian p-groups except for a few restricted families. In the indecomposable case, Berman had shown that the Ulm invariants of T were determined by KT; however, these invariants do not determine uncountable T unless T is simply presented. Even when T is simply presented, it was unknown whether an isomorphism $KT \cong K\tilde{T}$ implies that $\tilde{T}$ was also simply presented. With countable p-groups there is no problem since $\tilde{T}$ is also countable, hence simply presented.

In a paper in 1969, Berman and Mollov conjectured that if T is an abelian p-group of arbitrary cardinality, and if K has characteristic p, then KT determines T. They showed the following special case.

THEOREM 7. ([**4**]). *Let* K *be a field of characteristic* p *and* T *a direct sum of cyclic* p-*groups. If* $\tilde{T}$ *is a group such that* $KT \cong K\tilde{T}$, *then* $T \cong \tilde{T}$.

Over a decade later, Berman and Mollov returned to uncountable groups, this time examining the semisimple case. From the evidence on countable p-groups, one might hope to resolve the problem for fields of the second kind more easily than for fields of the first kind. In fact, they did this for mixed groups whose torsion subgroups are p-torsion. We shall discuss this result in the next section.

For fields of the first kind with respect to p, there are difficulties. For example, T/T^1 may no longer be a direct sum of cyclic p-groups. In the case that T/T^1 is such a direct sum, Berman and Mollov gave a complete set of invariants for KT. To do this, each direct sum of cyclic p-groups $\oplus_{i\geq 1}(\oplus_{\alpha_i}\mathbf{Z}(p^i))$ is assigned an "accompanying vector" in a certain fashion which we shall not describe. This is a finite sequence of cardinal numbers that tells not only the cardinality of the sum, but also which α_i, if any, are large compared to the ones that follow. Over fields of the first kind, the accompanying vector is a complete set of invariants for the structure of the group algebra if the sum of cyclic p-groups is infinite. If the sum is finite, one applies [**17**]. Berman and Mollov then showed

THEOREM 8. ([**7,8**]). *Let K be a field of the first kind with respect to p and let T be an infinite abelian p-group such that T/T^1 is a direct sum of cyclic p-groups. If $\tilde{T}$ is another p-group, then $KT \cong K\tilde{T}$ if and only if the following hold.*

(1) *$\tilde{T}/\tilde{T}^1$ is a direct sum of cyclic p-groups such that $K(T/T^1) \cong K(\tilde{T}/\tilde{T}^1)$.*

(2) *$|T^1| = |\tilde{T}^1|$ if either T^1 or $\tilde{T}^1$ is infinite or trivial.*

(3) *If T/T^1 is finite, then $T^1 \cong \mathbf{Z}(p^\infty)$ if and only if $\tilde{T}^1 \cong \mathbf{Z}(p^\infty)$.*

If T/T^1 is infinite, then the condition $K(T/T^1) \cong K(\tilde{T}/\tilde{T}^1)$ in (1) is equivalent to the accompanying vectors of T/T^1 and $\tilde{T}/\tilde{T}^1$ being the same.

3. Mixed Groups.

With the exception of an omitted case, Theorem 6 shows that $V_p(KT)$ is a divisible group for K a field of the second kind and T a countable p-group. It follows that the same is true for uncountable T. Hence, if G is a mixed abelian group whose only torsion is p-torsion, then there exists a subgroup of $V(KG)$ which concsists of coset representatives for G modulo $V_p(KT)$. Consequently, KG is isomorphic to $K((G/T) \times T)$, and the structure of KG depends largely on the structure of KT. Berman and Mollow proved

THEOREM 9. ([**7,8**]). *Let K be a field of the second kind with respect to p, and let G and $\tilde{G}$ be abelian groups with torsion subgroups T and $\tilde{T}$, respectively, that are p-groups. Then $KG \cong K\tilde{G}$ if and only if the following hold.*

(1) *$G/T \cong \tilde{G}/\tilde{T}$.*

(2) *$|T| = |\tilde{T}|$.*

(3) *Suppose $p = 2$ and K does not contain a primitive 4th root of unity. Then $(T : T^2) = (\tilde{T} : \tilde{T}^2)$, and if either T^2 or $\tilde{T}^2$ is infinite or trivial, then $|T^2| = |\tilde{T}^2|$.*

Actually, this result was preceded by a paper of Berman and Bogdan for the specific field of real numbers $\mathbf{R}$, and without the restriction to p-tortion.

THEOREM 10. ([**3**]). *Let G and $\tilde{G}$ be abelian groups with respective torsion subgroups T and $\tilde{T}$. Then $\mathbf{R}G \cong \mathbf{R}\tilde{G}$ if and only if $G/T \cong \tilde{G}/\tilde{T}$, $|T| = |\tilde{T}|$, $(T : T^2) = (\tilde{T} : \tilde{T}^2)$, and if either T^2 or $\tilde{T}^2$ is infinite or trivial, then $|T^2| = |\tilde{T}^2|$.*

Finally, we summarize the work of Berman and Mollov on the splitting problem, namely, when is $KG \cong K((G/T) \times T)$? As we have observed, this occurs over fields of the second kind, therefore the interest in the question is over fields of the first kind, or in the indecomposable case. In the indecomposable case, we have

THEOREM 11. ([**6**]). *Let K be a field of characteristic p and let G be an abelian group whose torsion subgroup T is a p-group. Assume that G/T is countable. Then KG splits if and only if G splits (i.e., T is a direct factor of G).*

In the semisimple case, Berman and Mollov considered finite extensions of the rational numbers and finite fields.

THEOREM 12. ([**6**]). *Let K be either a finite extension field of the rational numbers or a finite field, and let G be an abelian group with torsion subgroup T. If K is finite, assume that T has no element of order equal to the characteristic of K, and only finitely many nonzero primary components. Assume further that either T or G/T is countable. Then KG splits if and only if G/T^1 splits (over its torsion subgroup T/T^1).*

4. Appendix.

We state Berman's complete theorem for fields of the first kind, and the omitted case for fields of the second kind (see Theorems 3 and 2). Let ε_n denote a primitive p^nth root of unity over K. If T is a countably infinite abelian p-group, then $T \cong R \times D$, where R is reduced and D is divisible. Then T is exactly one of the following types.

(1) R is an unbounded direct sum of cyclic p-groups, $D = 1$.
(2) R is infinite bounded, $D = 1$.
(3) R is finite, $\operatorname{rank} D = 1$.
(4) R is finite, $\operatorname{rank} D \geq 2$.
(5) R is infinite bounded, $D \neq 1$.
(6) $T^1 \neq 1$ is finite.
(7) T^1 is infinite, T/T^1 is unbounded.

THEOREM 13. ([**1,2**]). *Let K be a field of the first kind with respect to p, and let $T \cong R \times D$ and $\tilde{T} \cong \tilde{R} \times D$ be countably infinite abelian p-groups.*

(a) *If T and $\tilde{T}$ are of different types, then $KT \ncong K\tilde{T}$.*

(b) *If T and $\tilde{T}$ are of the same type* 1, 6 *or* 7, *then $KT \cong K\tilde{T}$.*

(c) *If T and $\tilde{T}$ are of type* 2, *let p^n be the exponent of T, and let m be minimal such that T^{p^m} is finite. Similarly define $\tilde{n}$ and $\tilde{m}$. Then $KT \cong K\tilde{T}$ if and only if $K(\varepsilon_n) = K(\varepsilon_{\tilde{n}})$ and $K(\varepsilon_m) = K(\varepsilon_{\tilde{m}})$.*

(d) *If T and $\tilde{T}$ are of the same type* 3, 4 *or* 5, *then $KT \cong K\tilde{T}$ if and only if $KR \cong K\tilde{R}$ (this being decided by* [**17**] *if R is finite and by* (c) *if R is infinite).*

In order to state the omitted case for fields of the second kind, we define special groups $G_1 = \oplus_\omega \mathbf{Z}(2)$, $G_2 = \mathbf{Z}(4) \oplus G_1$, $G_3 = \oplus_\omega \mathbf{Z}(4)$, $G_4 = \mathbf{Z}(p^\infty)$, and $G_{5_s} = \mathbf{Z}(p^\infty) \oplus (\oplus_s \mathbf{Z}(2))$ $(s \geq 1)$.

THEOREM 14. ([**1**,**2**]). *Let K be a field of the second kind with respect to $p = 2$, and assume that K does not contain a primitive 4th root of unity. Let T be a countably finite abelian 2-group. Then $KT \cong KG_i$ for exactly one $i = 1, 2, 3, 4$ or 5^s. More precisely:*

(1) *$KT \cong KG_1$ if and only if $T \cong G_1$.*

(2) *$KT \cong KG_2$ if and only if $T \cong G_1 \times A$, where A is a finite abelian 2-group of exponent at least 4.*

(3) *$KT \cong KG_3$ if and only if either $T^1 \neq 1$ and T/T^1 is infinite, or else $T^1 = 1$ and T^2 is infinite.*

(4) *$KT \cong KG_4$ if and only if T is divisible.*

(5) *$KT \cong KG_{5^s}$ if and only if $T \cong R \times D$, where T is a direct product of s nontrivial cyclic 2-groups and D is divisible.*

Finally, we mention that in [**1**, Theorem 6], the statement for the case $\mathbf{R}G \cong \mathbf{R}G_2$ should read that $R \not\cong G_1$ if $N \times P = 1$. Also, there should either be a case $\mathbf{R}T \cong \mathbf{R}G_4$ if T is divisible, or else $s = 0$ should be allowed for G_{5^s}.

REFERENCES

1. S. D. Berman, Group algebras of countable abelian p-groups, Soviet Math. Dokl. 8 (1967), 871–873.

2. S. D. Berman, Group algebras of countable abelian p-groups, Publ. Math. Debrecen 14 (1967), 365–405.

3. S. D. Berman and V. G. Bogdan, The isomorphism of real group algebras of abelian groups, Mat. Zametki 21 (1977), 229–238.

4. S. D. Berman and T. Zh. Mollov, The group rings of abelian p-groups of arbitrary power, Mat. Zametki 6 (1969), 381–392.

5. S. D. Berman and T. Zh. Mollov, Group rings of mixed abelian groups, C. R. Acad. Bulgare Sci. 28 (1975), 879–881.

6. S. D. Berman and T. Zh. Mollov, Decomposability of the bases and isomorphisms of group rings of abelian groups, PLISKA Stud. Math. Bulgar. 2 (1981), 6–22.

7. S. D. Berman and T. Zh. Mollov, Isomorphism of semisimple group algebras of uncountable abelian p-groups, C. R. Acad. Bulgare Sci. 35 (1982), 869–871.

8. S. D. Berman and T. Zh. Mollov, Isomorphism of semisimple group algebras of abelian groups, PLISKA Stud. Math. Bulgar. 8 (1986), 107–121.

9. S. D. Berman and G. R. Rossa, The Sylow p-subgroup of a group algebra over a countable abelian p-group, Dopovidi Akad. Nauk Ukrain. RSR Ser. A (1968), 870–872.

10. S. D Berman and G. R. Rossa, Group algebras of countable periodic abelian groups, Dopovidi Akad. Nauk Ukrain. RSR Ser. A (1971), 387–390.

11. Z. Chatzidakis and P. Pappas, Group rings of countably infinite abelian p-groups, preprint.

12. Z. Chatzkdakis and P. Pappas, Units in abelian group rings, preprint.

13. W. E. Deskins, Finite abelian groups with isomorphic group algebras, Duke Math. J. 23 (1956), 35–40.

14. G. Karpilovsky, "Commutative Group Algebras," Marcel Dekker, New York, 1983.

15. T. Zh. Mollov and N. A. Nachev, Ulm-Kaplansky invariants of the group of normalized units of modular group rings of primary abelian groups, Serdika 6 (1980), 258–263.

16. D. S. Passman, "The Algebraic Structure of Group Rings," Wiley-Interscience, New York, 1977.
17. S. Perlis and G. L. Walker, Abelian group algebras of finite order, Trans. Amer. Math. Soc. 68 (1950), 420–426.

DEPARTMENT OF MATHEMATICS
UNIVERSITY OF ARIZONA
TUCSON, ARIZONA 85721

Contemporary Mathematics
Volume **93** (1989)

Berman's Work on Units in Group Rings

ROBERT SANDLING

Many of the fundamental results about units of group rings of finite groups and about the isomorphism problem for such group rings were obtained by S. D. Berman. Among them: the fact that the coefficient of the identity in a nontrivial unit of finite order is zero; the characterisation of groups whose integral group rings have only trivial torsion units; the class sum correspondence. These can all be found in his earliest papers. A later concern of his and of his student A. R. Rossa has turned out to be particularly insightful: the question of whether group bases of the p-adic group ring of a p-group are conjugate. The deepest work in the subject to date [Roggenkamp-Scott 87; Weiss 88; Scott 87] has answered this question affirmatively, thereby settling the group ring problem for finite p-groups and for certain other soluble groups.

With some items, such as that about the coefficient of the identity, Berman had been anticipated by Higman in the unpublished and little known thesis [Higman 40b]. With others, such as the class sum correspondence, recognition of Berman's contribution was limited. I have already examined in [Sandling 81, 85], in a sketchy manner, the relation of Berman's work to that of Higman and to the current state of knowledge as regards the isomorphism problem. In this paper Berman's work on units and on the group ring problem is analysed in detail. With most points, the attempt is made to describe the setting in which they came to light both in Berman's papers and in independent sources. It is not without significance that various facts discovered by Berman were discovered by others as well, in some cases more than once. To measure how well Berman captured a given fact with respect to the best version of it known to date, the latter is also mentioned at times although

1980 *Mathematics Subject Classification* (1985 *Revision*). Primary 20C05; Secondary 01A60, 16-02, 16A25, 16A26, 20-02.

Key words and phrases. Group rings and their history, isomorphism problem, unit group, finite group, p-group, S. D. Berman.

"This paper is in final form and no version of it will be submitted for publication elsewhere."

this is not done systematically to avoid excessive repetition of material in my earlier studies.

The analysis of Berman's contributions in these areas proceeds here largely through his own words. Excerpts and statements of results are quoted in full (the translations are my own) despite the problems of notation which this poses (Berman was not consistent in his usage nor was it that which is common nowadays; at least G is a finite group in almost all of this paper!). My intention has been to give the reader a flavour of Berman's style and perspective. This seemed particularly desirable in that various of his original papers on these themes are mainly known through their reviews (these are also commented on occasionally). Some papers are in journals not widely distributed, some are in Ukrainian. Few readers are likely to have had the opportunity to have heard his lectures as his travelling was limited. In the Soviet Union his approach to these subjects is better known; it can be appreciated in various surveys [Zalesskiĭ-Mihalev 73; Bovdi 74; Karpilovsky 80; Mollov 84] where the contributions of other Soviet authors such as Saksonov which deserve wider attention, are also described.

Berman's initial attacks on these topics come in four of his earliest papers [53a,b,c, 55a]; these are reviewed here in detail. His approach is usually representation-theoretic, embedding an integral group ring in a group algebra which is then decomposed. Although some of the ideas derive from his previous paper [52], there is little indication of how he came to study these topics. In each of the papers [52, 53a,b,c, 55a, 58b] he thanks Ya. B. Lopatinskiĭ for observations (during this period the latter had been chairman of the Department of Differential Equations in L'vov); Berman later joined in a jubilee article for Lopatinskiĭ's 60th birthday [Berman et al 67]. He also thanks I. R. Shafarevich in [52, 53b, 58b].

Material from other papers from various stages of his career is also presented but only one, [Berman-Rossa 66] which he wrote with his student A. R. Rossa, is treated comprehensively. Two inaccessible brief notes, his [60] and that [Polyak 60] of his student S. S. Polyak, appear in their entirety in translation as appendices. As reviews can serve to indicate points of view, these are also included when relevant. However, although Berman did a good deal of reviewing for the Referativnyĭ Zhurnal. Matematika on representation theory, he reviewed few of the articles which concern units or the group ring problem. He drew up the survey [66b] in the Itogi Nauki Algebra-Topologiya-Geometriya series, and he is thanked for "useful consultation" in [Zalesskiĭ-Mihalev 73].

Section 1. Units of finite order. The groups of units of the integral group ring of a finite group has applications in many areas both in algebra and elsewhere. Despite its importance, few facts are known about it. Some of the most basic ones about torsion units entered the literature through Berman's work. He first broaches the topic in [53a], in which the ring **Z** of rational integers is used as ring of coefficients. The more detailed paper [55a], written about a year later, has much the same emphasis as [53a] but is broader in scope; for example, many of its theorems allow as coefficients the ring of algebraic integers of an algebraic number field. This distinction was not drawn in its review in Math. Reviews which simply refers back to the review of [53a] and which was classified in the section Rings, Fields, Algebras (so that [55a] was omitted from [Gorenstein 74] (but included in [Baumslag 74])). In any event it became an influential paper as will be noted in the next section.

The first key fact about units in these two papers is the fact that the coefficient of the identity of a nontrivial unit of finite order must be 0. In Berman's words:

"LEMMA 2. *If* $u = \alpha_0 e_0 + \cdots + \alpha_{n-1} e_{n-1}$ *is a root of unity in* $R(G, C)$ *and* $\alpha_0 \neq 0$, *then* $u = \pm e_0$". [53a, p. 7]

The notation requires some comment. In his early papers Berman uses the same notation that Higman used for a group ring: $R(G, C)$, with C denoting the ring $\mathbf{Z}$ of rational integers (in the 1960's, he used, e.g., GK, and, lastly, as in [Berman-Rossa 71], KG). The elements of the group G are, in [53a], $\{e_0, e_1, \ldots, e_{n-1}\}$ with e_0 the identity element. The phrase "root of unity" instead of, for example, unit of finite order, is deliberate; in [55a, p. 254], it is defined as "an element $u \in R(G, K)$ for which there is a natural number m such that $u^m = 1$" (here K is an algebraic number field); furthermore, "the order of a root of unity is the smallest natural number m for which $u^m = \varepsilon \cdot 1$ where ε is a root of 1 in K". The definitions in [53a] are the same, *mutatis mutandis*, given that only $K = \mathbf{Q}$ is considered.

In his papers, Berman rarely takes the point of view that the group of all units $U(RG) \approx U(R) \times V(RG)$, the latter being the group of normalised units, those of augmentation 1. In fact the augmentation homomorphism is never formalised in his early papers although it is used occasionally and the fact that it is multiplicative is stated as Lemma 3 of [55a].

The proof of Lemma 2 [53a] is only sketched. The corresponding lemma in [55a] is also Lemma 2. It is stated there for the ring of algebraic integers of an algebraic number field. The proof is given in full. It is the proof that is presently familiar, using the trace of the regular representation and an inequality involving the absolute value of a sum of roots of unity in $\mathbf{C}$. The fact itself, with this proof and in this context, appeared in [Higman 40b]; it also surfaced in [Takahashi 65].

In [55a] the well-known extension of Lemma 2 (also in [Higman 40b]) to coefficients of central elements is drawn as:

"COROLLARY 1. *Let* $u = \alpha_0 \cdot 1 + \alpha_1 a_1 + \cdots + \alpha_{n-1} a_{n-1}$ *be a root of* 1 *in* $R(G, K)$, a_i *an element of the centre of* G *and* $\alpha_i \neq 0$. *Then* $u = \varepsilon a_i$ *where* ε *is some root of* 1 *of* K (*as before* K *is assumed to be the ring of integers of a number field*)." [55a, p. 254]

The corresponding result for the group ring of a p-group over the ring of p-adic integers is obtained in [Berman-Rossa 66], its Lemma 3 being the analogue of Lemma 2 [53a, 55a]. This corollary does not appear as such in [53a]. Instead there is a theorem about normal elements, the theorem for which Lemma 2 is ancillary (interestingly, the fact of Lemma 2 is also a lemma on its first appearance in [Higman 40b] but is, in a later chapter, elevated to the status of theorem, which is how it is viewed at present).

The theorem on normal elements is presented as one of the main contributions in [53a]. Its statement is as follows:

"THEOREM 1. *Up to sign, every normal element of finite order in* $R(G, C)$ *coincides with an element of the group.*" [53a, p. 8]

With $(\ )^*$ denoting the involution on the group ring induced by sending a group element to its inverse, "(an) element x of the group ring is called normal if it satisfies the condition $xx^* = x^*$" in [53a]. In [55a] where the range of coefficient rings has been broadened, the definition of the involution, and consequently of normal element, is also broadened: "if $x = \alpha_0 \cdot 1 + \alpha_1 a_1 + \cdots + \alpha_{n-1} a_{n-1}$, then $x^* = \bar{\alpha}_0 \cdot 1 + \bar{\alpha}_1 a_1^{-1} + \cdots + \bar{\alpha}_{n-1} a_{n-1}^{-1}$" [55a, p. 256]. Theorem 1 [53a] is not broadened, however, remaining with $\mathbf{Z}$ as the ring of coefficients and reduced in status to Lemma 5 [55a]. Neither the involution nor the concept of normal element appears in [Higman 40a,b].

One application of the theorem on normal elements is the following:

"COROLLARY 1. *If* G *is an abelian group or a Hamiltonian group of order* 2^m, *then* $R(G, C)$ *contains only trivial roots of unity.*" [53a, p. 8]

With the different range of coefficient rings, this appears in [55a] as two items: as Corollary 2 for G abelian and for the ring of integers of a number field as coefficients; as Lemma 6 for G a Hamiltonian 2-group and for $\mathbf{Z}$ as coefficients. In [Higman 40b], using considerations involving maximal orders, Higman had found that it is only for the ring $C = \mathbf{Z}$ that Corollary 1 [53a] holds for a Hamiltonian 2-group.

The above results, together with detailed manipulations involving primitive central idempotents in $R(G, K)$, G Hamiltonian, $K = \mathbf{Q}$, culminate in one of the best known of Berman's theorems, that which characterises groups in whose integral group rings all torsion units are trivial:

"THEOREM 4. *Let* G *be a finite group. The following statements are equivalent*:

I. *All roots of unity in* $R(G, C)$ *are trivial.*

II. *All elements of* $R(G, C)$ *are normal.*

III. G *is either an abelian group or Hamiltonian of order* 2^m." [53a, p. 9]

In [53a], the proof is only sketched. As Theorem 2 [55a], all the details of its proof are given in full, taking up most of pp. 256–261.

What remains for the proof is the implication "I implies III". The detailed version in [55a] begins by showing that if all torsion units are trivial, then

$R(G, R')$ has no nilpotent elements (here $R' = \mathbf{Q}$); for this he decomposes $R(G, R')$ as a direct sum of full matrix rings. He can then apply the following result which had appeared as Theorem 3 in [53a] in slightly more detail:

"LEMMA 8. *If the group ring $R(G, R')$ does not contain nilpotent elements, then G is either an abelian group or is a Hamiltonian group of order not divisible by* 3". [55a, p. 257]

He next restricts attention to a cyclic subgroup H of G of order $4p$, p a prime number, $p \geq 5$. Writing H as $\langle a, c\rangle$, a of order p, c of order 4, he uses the fact that $p > 3$ to obtain a unit $y = e_1 + \cdots + e_5 + (1 + ac + a^2c^2)e_6$ of $R(H, R')$; here the e_i are the primitive central idempotents of $R(H, R')$ with $e_6 = 1/2p(1 - c^2)((p-1) - a - \cdots - a^{p-1})$. Then $u = y^s \in R(H, C)$ which s is "the number of residue classes modulo $4p$ in the field $R'(\varepsilon)$ which are relatively prime to $4p$" and ε is "a primitive root of unity of degree $4p$" [55a, p. 258] (this is obtained from [Higman 40a] and refers to the residue classes in the ring $\mathbf{Z}[\varepsilon]$ of algebraic integers of the field $\mathbf{Q}(\varepsilon)$). All this, as well as a description of y in the group ring $(\mathbf{Q}\langle a\rangle)\langle c\rangle$, is accomplished in two pages as Lemma 9 [55a]. In the context of the proof of Theorem 2 [55a], the final point of the proof is the demonstration that $u^{-1}cu$ is a nontrivial unit, a lengthy case-by-case analysis.

There is one further result for which Berman is known. It is that which concerns the order of a torsion unit. He viewed it as "a generalisation of the theorem on the order of elements of a finite group" [53a, p. 8] while later he refers to this theorem of Lagrange as "an immediate corollary" [55a, p. 256] of his result. In [53a] it is stated, as Theorem 2, only for the coefficient ring $\mathbf{Z}$. In [55a] it became

"THEOREM 1. *The order of a root of* 1 *in $R(G, K)$, where K is the ring of integers of a number field P, is a divisor of the order of the group G.*" [55a, p. 255]

The proof is straightforward; it is based on Lemma 2 [53a, 55a] and on the following:

"LEMMA 4. *Let $e = \alpha_0 \cdot 1 + \alpha_1 a_1 + \cdots + \alpha_{n-1} a_{n-1}$ be an idempotent element of the group ring $R(G, K)$ of the group G over the field of complex numbers,* I *the left ideal which is generated by e in $R(G, K)$, and p the number of eigenvalues equal to* 1 *of the matrix corresponding to e in the regular representation of $R(G, K)$.*

Then the degree of the algebra I *over K is equal to p and $\alpha_0 = p/n$.*" [55a, p. 255]

Without the statement involving I, this had appeared as Lemma 1 [53a], where a further conclusion had been drawn, that $R(G, C)$ is indecomposable.

With the extra restriction that G be soluble, this theorem is proved in [Higman 40b]. Berman returned to the theme later on; with I_p denoting the ring of p-adic integers, he and Rossa proved:

"THEOREM 1. *Let GI_p be the group ring of the finite p-group G over the ring I_p. For every p-element x in the ring GI_p, the number $o(x)$ divides the order of the group.*" [Berman-Rossa 66, p. 46]

The best possible result, that the order of a normalised torsion unit divides the exponent of G, appears as Corollary 4.2 of [Cohn-Livingstone 65]. It is a corollary to Theorem 4.1 but this theorem is not correct as Berman and Bovdi noted in their review [Berman-Bovdi 67]. Saksonov also noted this mistake (in [Saksonov 66]) and gives (in [Saksonov 67a, 71]) a proof for an even more general coefficient ring.

Section 2. The group ring problem. Higman stated the isomorphism problem for integral group rings of a finite group in a forthright manner and in more than one version [Higman 40b]. Berman is more hesitant in addressing it. In his early study of group rings, the main objectives seem the elucidation of the nature of units and the exploitation of his concept of K-division (defined below). Consequences for the group ring problem seem of lesser importance.

Isomorphism issues enter explicitly into three of his earliest papers, each with a distinct context. In [53a] he draws the following corollary to his theorem characterising normal units as trivial:

"COROLLARY 2. *If the integral group rings of G_1 and G_2 are isomorphic, then the centres of these groups are isomorphic.*" [53a, p. 8]

Although, in [55a], the theorem is generalised to rings of algebraic integers, Corollary 2 is one of few items in [53a] without parallel in [55a], a paper which does not mention isomorphism questions. For integral group rings, the first published references to isomorphism questions are this corollary in [53a] and two statements in [Cohn 52], one of which being:

> "By Theorem 5 an abelian group is determined to within isomorphism by its Magnus-ring, and hence by Theorem 3 it is determined (to within isomorphism) by its group-ring. This is a special case of a theorem due to G. Higman (5)." [Cohn 52, p. 304]

(Reference (5) is [Higman, 40b], the unpublished thesis in which the isomorphism problem is stated explicitly; the published paper [Higman 40a] does not mention the issue. It might also be noted that Phillip Hall served as supervisor both to Higman and to Cohn and that Cohn attributes the other of his statements concerning isomorphism questions to Higman in [Cohn 51, 52].)

In [53b] the isomorphism problem is framed in the terms of the question Thrall had originally asked:

> "In what follows we will assume that G is a finite group, whose order is not divisible by the characteristic of K.
>
> The following question is unsolved: given a group G and a field K, what are all the groups H for which $R(G,K) \approx R(H,K)$? Perlis and Walker (5) showed how to find, for fixed abelian groups G and H, all fields K such that $R(G,K) \approx R(H,K)$.
>
> In this note a necessary and sufficient condition for the isomorphism of the centres of group rings of p-groups ($p \neq 2$) is established. For this, the concept of K-division in a group, introduced in (2), is used. The corresponding theorem turns out to be correct for the group rings of abelian p-groups for an arbitrary prime p." [53b, p. 185]

(The references here are to: (5) =[Perlis-Walker 50], (2) = [52].)

The methods of [53b] are those of field theory and of character theory. It contains two results described as theorems:

"THEOREM 1. *The centres of the group rings of the p-groups G and H* ($p \neq 2$) *over the field K are isomorphic if and only if one can establish between the K-divisions of G and H a one-to-one correspondence, in which corresponding K-divisions of G and H contain the same number of classes.*" [53b, p. 187]

This theorem was extended to a result for finite groups in general in [58b], a paper which contains all the material of [52, 53b] in a generalised, definitive and detailed form as its §1, pp. 411–433. The later version is as follows:

"THEOREM 7.1. *Let G and H be finite groups; m the least common multiple of the orders of the elements of G and H; K′ an arbitrary field whose characteristic does not divide m;* ε *a primitive root of* 1 *of degree m. If the Galois group of the field* $K'(\varepsilon)$ *over K′ is cyclic, then the centres of the group algebras* $R(G,K')$ *and* $R(H,K')$ *are isomorphic if and only if there is a one-to-one correspondence between the K′-divisions of the groups G and H for which corresponding K′-divisions of these groups contain one and the same number of classes of conjugate elements.*" [58b, p. 430]

A corollary gives conditions under which the hypothesis will hold: G and H of order p^α or $2p^\alpha$, p odd; K' of characteristic $p > 0$; $K' = \mathbf{R}$. In [53b], counterexamples were mentioned:

> "Theorem 1 is not true for groups of order 2^m, as one can show by the example of the rational group rings of degree 16." [53b, p. 187]

In [58b, pp. 430–431] the details justifying this remark are spelled out; the groups are the dihedral and semidihedral groups of order 16. Necessary conditions for the isomorphism of the centres of group algebras are given in

Theorem 9.1 of [58b]: using the notation of Theorem 7.1 above and introducing an ascending chain of fields $K' = P_0 \subset P_1 \subset \cdots \subset P_r = K'(\varepsilon)$ with P_{i+1} a cyclic extension of P_i, he states the conclusion as:

> "If the centres of the group algebras $R(G, K')$ and $R(H, K')$ are isomorphic, then there is a one-to-one correspondence between the P_i-divisions of the groups G and H, for which corresponding P_i-divisions of these groups contain one and the same number of P_{i+1}-divisions ($i = 0, \dots, r-1$)." [58b, p. 432]

Necessary and sufficient conditions for such an isomorphism in the case of M-groups appear in [57].

The second theorem of [53b], reduced to a corollary (to Theorem 8.1) in [58b], is as follows:

"THEOREM 2. *The group rings $R(G, K)$ and $R(H, K)$, where G and H are primary abelian groups, are isomorphic if and only if corresponding K-divisions of the groups contain the same number of elements.*" [53b, p. 187]

The concept of K-division, central to [53b], played no role in [53a]. In Berman's papers it made its appearance in [52] whose main result is that which is now called the Berman-Witt theorem:

"THEOREM. *The number of irreducible representations of a group G over a field K is equal to the number of K-divisions of the group and coincides with the number of K-divisions of characters.*" [52, p. 888]

(In [52], K is a field whose characteristic does not divide the order of G nor the degrees of its absolutely irreducible representations.)

Berman uses the term K-отдел (K-otdel) for what elsewhere is termed a K-conjugacy class of a group (e.g., [Curtis-Reiner 62, p. 306]). He uses the same term for the equivalence classes of characters, and it is generalised in [58b]. He was still using it in, e.g., [69, p. 630]. The term, meaning "section", is one unusual in mathematics; it is not listed in [Lohwater 61], for example. I have translated it "K-division" to convey Berman's usage. To define it, let n be the exponent of G, ε a primitive nth root of unity and Γ the Galois group of $K(\varepsilon)$ over K; elements $a, b \in G$ are K-conjugate if there are $c \in G$ and $\mu \in \Gamma$ such that $b = c^{-1}a^{\mu}c$; a K-division of G is an equivalence class of the equivalence relation of K-conjugacy.

The correspondence of K-divisions in Theorem 1 [53b] prefigures a similar result in [53c]. The latter carries with it, however, in its proof, one of the basic facts in the theory of the isomorphism problem for integral group rings: the class sum correspondence. This asserts that the class sums are canonical elements of an integral group ring in the following sense: let $k = x_1 + \cdots + x_q$ be the class sum in $\mathbf{Z}G$ of the conjugacy class $C = \{x_1, \dots, x_q\}$ of G and let H be a group basis for $\mathbf{Z}G$; then there is a conjugacy class $C' = \{x'_1, \dots, x'_q\}$ of H whose class sum $k' = \pm k$.

Berman gives the class sum correspondence little emphasis. Instead he stresses two aspects which have received less attention subsequently: he views the correspondence as one of cardinalities, just as in his earlier papers (the class sum correspondence is merely a step in the proof); he formulates the result for K-divisions (although the special case of ordinary conjugacy classes is an essential part of the proof). The class sum correspondence itself he views as

> "(generalising) the fact that in the integral group ring of a finite abelian group the elements of the group basis are determined exactly up to sign [3]" [53c, p. 316]
> (Here [3] = [Higman 40a].)

His article being in Ukrainian, he had prepared an abstract in Russian which underlines these points:

> "In this article the concept of a K-division of a group, introduced in [1], is used to prove the theorem that the group ring of a finite group over the ring of rational integers uniquely determines the orders of the K-divisions and the orders of the classes contained in corresponding K-divisions, for an arbitrary number field K." [53c, p. 316] (Here [1] = [52].)

As the ideas of his proof are by now familiar, a brief summary will suffice to convey his style of presentation. Its two main elements, the use of the orthogonality relations for characters and a combinatorial argument, appear in almost all of the many proofs which have been given over the years. (Character theory has been avoided in [Passman 77, p. 655] and also in [Sandling 88] which uses the methods of Hattori based on the fact that the involution $(\ )^*$ is positive on a real group algebra. In what was perhaps his last paper, the late Akira Hattori returned to this theme putting it into the context of a more general result about strongly separable algebras [Hattori 85].)

Much of the proof is given before the theorem itself is stated. The setting is

> "Let G and H be finite groups of order h, C the ring of rational integers, K an arbitrary number field.
> Assume that $R(G,C) \approx R(H,C)$. Then it is clear that $R(G,K) \approx R(H,K)$ and that, in $R(G,C)$, there is a basis H' whose elements form a group with respect to multiplication which is isomorphic to H." [53c, p. 313]

With the notation t_i and t'_i for the sums of the K-divisions of G and of H' respectively, he expresses

$$t'_j = \sum_{1\le m\le r} \alpha_{jm} t_m \ (j = 1,\dots,r) \text{ and } t_j = \sum_{1\le m\le r} \beta_{jm} t'_m \ (j = 1,\dots,r),$$

the α_{jm} and β_{jm} being rational integers.

Manipulation of character identities from the orthogonality relations is then used to deduce the equations

$$l'_i = \sum_{1\leq m\leq r} |\alpha_{im}|^2 l_m \ (i = 1,\dots,r) \text{ and } l_i = \sum_{1\leq m\leq r} |\beta_{im}|^2 l'_m \ (i = 1,\dots,r)$$

where l_i and l'_i are the cardinalities of the K-divisions. At this point he states his theorem, the only one of the paper:

> "*Theorem. If G and H are finite groups and* $R(G,C) \approx R(H,C)$, *then G and H have the same orders of K-divisions and the same orders of the classes which belong to corresponding K-divisions, for an arbitrary number field K.*" [53c, p. 315]

What follows, described as the proof, is combinatorial, involving dimension arguments in $R(G,K)$. It begins with an induction to establish the fact that the number of K-divisions of given cardinality is the same in G and in H; for this he shows that, to each i, there is an m with $l_m = l'_i$ and concludes that $\alpha_{im} = \pm 1$ and $\alpha_{ij} = 0$, $j \neq m$. Renumbering he expresses this as $t_i = \varepsilon t'_i$ $(i = 1,\dots,r)$, $\varepsilon = \pm 1$. As K is arbitrary, he concludes that $k_i = \varepsilon k'_i$ where k_i, k'_i are the class sums; combined with the previous identities, he obtains the statement of the theorem.

The impact of the class sum correspondence on the isomorphism problem was not felt for over a decade; it is to be found principally in the work of Passman and of Saksonov. The independent formulation of the correspondence in [Passman 65] is direct:

> "THEOREM C. *The class sums in* $R[G]$ *can be obtained canonically from the group ring up to factors of roots of unity in R. Moreover a consistent set of such class sums can be chosen.*" [Passman 65, p. 566]

(Here R is a ring of algebraic integers in an algebraic number field.) Theorem C is described there as a "result of G. Glauberman".

In the interim, S. S. Polyak, a student of Berman in Uzhgorod, had generalised Berman's theorem to such a setting. His result is announced in [Polyak 60], a translation of which is provided here as Appendix A; no details seem to have been published.

Saksonov also announces the extension to the algebraic integer case [Saksonov 66] and describes this extension as having been "independently effected" [Saksonov 71, p. 197]; in the latter paper he proves what is probably the best that can be hoped for, the extension of the class sum correspondence to the type of ring most appropriate for the isomorphism problem, an integral domain of characteristic 0 in which no prime of the group order is invertible.

With Berman's point of view and the way in which he formulated his result, it is not surprising that its significance for the isomorphism problem was not picked up more quickly. That his article was written in the Ukrainian literature is another factor. The reviews of [53c] follow Berman's emphases (as does Berman's review [62] of Polyak's note). The review in Zentralblatt is just a translation of Berman's abstract; that in Math. Reviews gives the necessary definitions and states its theorem but without reference to ordinary classes. In the event Glauberman was moved to have included in [Smith 79]: "(Prof. Glauberman points out that this result, usually attributed to him, appeared earlier in work of S. S. Polyak" [Polyak 60] "and Berman.)" The priority of Berman and Polyak in the matter had been championed in the review [Gudivok 66].

At one point [60], Berman asserted that the class sum correspondence could be obtained merely from an isomorphism between the centres of two integral group rings; this would represent an amalgam of the ideas of [53b] and [53c]. The note [60], consisting of only a few paragraphs, is translated here as Appendix B. No further details were published. A proof of its lemma that such an isomorphism commutes with the involutions induced by ( )* can be found in [Bovdi 74, p. 107]. For abelian groups, the theorem is no more than the fact that torsion units are trivial. For nilpotent groups of class 2, Bovdi reports that Berman (and Saksonov) had a proof [Bovdi 73]. For groups of class 3, Bovdi himself gives a proof [Bovdi 74, p. 107]. In [Saksonov 71], results on the class sum correspondence are formulated in terms of isomorphisms between centres to as great an extent as possible.

Much of the impact of the class sum correspondence on the isomorphism problem is through the intermediacy of the normal subgroup correspondence, that an isomorphism of integral group rings induces an isomorphism of normal subgroup lattices. This is emphasised in [Passman 65]. The normal subgroup correspondence can be approached, however, in quite different ways (reviewed in [Sehgal 78, III.4]). On its first appearance in the Soviet literature in [Saksonov 66, 71] and in [Zhmud′-Kurennoĭ 67], it is so presented. The latter paper was published in a journal of Khar′kov State University at around the time that Berman had taken up residence there. It makes numerous references to Berman. Its statement of the normal subgroup correspondence (with coefficient ring the ring of integers of a number field) is accompanied by the footnote:

> "This result is not new: it is a simple consequence of the coincidence, proven by S. D. Berman, of the multiplication tables of the conjugacy classes of G and H." [Zhmud′-Kurennoĭ 67, p. 26]

Another approach to the normal subgroup correspondence can be found in [Cohn-Livingstone 65]; its review [Berman-Bovdi 67] relates it to Passman's work but not to that of Berman.

The discoveries about group rings which Berman describes in these early papers [53a,b,c, 55a] have become central facts for the study of units and of the isomorphism problem. Their influence can be seen in the fact that [55a] and [Cohn-Livingstone 65] were identified, by a partially "automatic procedure", as the core documents for the CompuMath Research Front 80-0633 "Isomorphism-problem for integral group-rings". This identification is based on the number of citations of these papers in papers in the Institute for Scientific Information's CompuMath database which appeared in 1976–1980 and which are judged to deal with the topic. As the database includes relatively little Soviet material, this finding presumably understates Berman's influence.

In the period immediately following their publication, however, there was little mention of them. An exception is that of [Hewitt-Zuckerman 55] who cite [53a,b] (their paper was reviewed by Berman [56]). It is a paper which provides a connection with the literature on the isomorphism problem of the group algebras of analysis; indeed, in it, the context of the problem is given as that of [Kawada 48] with only a footnote reference to [Perlis-Walker 50] and [53b]. Its concern is with the group-algebras of finite semigroups; one incidental result is that, for a group G, there is a semigroup S which is not a group and yet $\mathbf{C}G \approx \mathbf{C}S$.

It is convenient here to record some notes on the early history of the isomorphism problem which have come to light since [Sandling 85]. Thrall proposed his version of the problem in his opening remarks to the Michigan Algebra Conference in 1947 but gave it no special emphasis. The first talk was given by Hewitt who stood in for Eilenberg who had been injured (Eilenberg might have provided a link with Higman's work in the school of topology centred on J. H. C. Whitehead in England). It seems that neither Philip Hall nor Richard Brauer left any items in their unpublished papers concerning the isomorphism problem for group rings.

Many other topics concerning the isomorphism problem for group rings were touched on by Berman over his career. In [55b] he gave the example of the two nonabelian groups of order p^3, p an odd prime, as one of nonisomorphic groups having isomorphic rational group algebras. In this he had been anticipated in [Higman 40b]. Distinguishing between the integral group rings of these groups was an early concern (this case is covered by results for groups of class 2 in [Higman 40b; Passman 65; Saksonov 66, 71]; in [Col 62], it is asserted without proof—a possible reason found in [Col 61] is incorrect). Berman noted a mistake in the attempt to do this in [Cohn-Livingstone 63]. In his "reviewer's note" he wrote:

> "Nevertheless, applying other arguments, based on the results of the reviewer (RZh Mat, 1954, 2872), it can be proven that the theorem formulated by the authors is correct." [64, p. 26]

It is not clear what he had in mind as it follows from the isomorphism of their rational group algebras and Theorem 2.1 of [Col 62] that the appropriate correspondence of K-divisions holds for all non-modular fields K. The distinction can be made on the basis of the exponents of the groups, information available in the versions of the class sum correspondence which were formulated in [Passman 65] and in [Saksonov 66, 71] who show that the power map on classes is determined; this aspect does not appear in Berman's papers but his remarks here may indicate that he was aware of it. Berman later returned to this example: that their group rings over the ring of p-adic integers are not isomorphic is shown in [Berman-Rossa 66].

Necessary and sufficient conditions for the isomorphism of group algebras of soluble groups over fields K of characteristic not dividing the group order are given in a string of highly representation-theoretic papers [57, 58a, 66a]. In [57] results are announced for M-groups with K algebraically closed or of positive characteristic; the conditions are complicated, involving correspondence between indices of subgroups from which the monomial representations are induced. A footnote added "in proof" [58b, p. 430] reports further progress and, in particular, mentions results for such M-groups as "p-groups and groups of square free order" with no extra restrictions on K. In [66b], however, where detailed proofs are given, the results are only stated for K algebraically closed and for M-groups which satisfy Huppert's condition (that G have a normal subgroup $\hat{G}$ for which $G/\hat{G}$ is supersoluble and $\hat{G}$ has all its Sylow subgroups abelian), a condition that figures strongly in the proofs. In [58a], a paper which, as with most of his Doklady and Dopovīdī notes, gives definitions and statements of results but no proofs, similar results are stated as Theorem 2 for groups of order 2^m and for $K \subseteq \mathbf{R}$ or K of characteristic 0. The conditions again involve indices but for subgroups from series too complicated to describe here; the conditions also involve an interplay between these subgroups and the Galois groups of various associated field extensions.

In [Berman-Rossa 66] there is the recognition that the integral group ring determines upper central factors [Col 62]:

> "*Remark.* A theorem analogous to Theorem 3 can be proved; the integral group ring GZ of a periodic group G determines the factors of the upper central series of the group uniquely." [Berman-Rossa 66, p. 47]

Theorem 3 of the paper states this fact for (not necessarily finite) p-groups with the ring of p-adic integers as coefficients.

One topic that he might have pursued is an implication of his results about normal elements [53a, 55a] for the following weak form of the isomorphism problem: an involution and augmentation preserving isomorphism $\mathbf{Z}G \xrightarrow{\varphi} \mathbf{Z}H$ induces an isomorphism from G to H. This follows readily from Theorem 1 of [53a] as, an element $x \in G$ being normal in $\mathbf{Z}G$, its image $\varphi(x)$ is normal in $\mathbf{Z}H$ so that $\varphi(x) \in H$. Although Berman does not seem to have made this observation, it was pointed out in [Takahashi 72]. The result itself first appeared in [Banaschewski 67]. Further approaches to the problem which make use of the involution were considered in [Hattori 71; Obayashi 74].

Isomorphism questions about the group rings of abelian groups were one of Berman's enduring interests. The later work, particularly that done with such collaborators as Bogdan, Mollov and Rossa, concentrated on the infinite case (a summary may be found in [Mollov 84]). Nevertheless, items concerning the finite case appear in many of his papers. That a finite abelian group is determined by its integral group ring is an implication of both [53a] and [53c] but it is a conclusion he never draws as such, the only specific mention of the abelian case being to aspects already in [Higman 40a]. With a field as ring of coefficients, he mainly treats the case of p-groups. The reduction from the nonprimary to the p-group case is referred to [Perlis-Walker 50] in his early papers. That their proof of the reduction is incorrect seems to have been overlooked by many authors. Berman first makes mention of its "obvious mistake" in [Berman-Rossa 71]. The reduction is stated without proof as Theorem 1 of this paper; it is remarked that

> "a correct proof can be given based on special lemmas concerning characters of finite abelian groups." [Berman-Rossa 71, p. 387]

Around the same time (note the submission date of this paper), other proofs were given for this same purpose in [Bautista 67; Cohen 68].

The criterion of [53b] for the isomorphism of the group algebras of two abelian p-groups over a field of characteristic $\neq p$ has been quoted earlier. This result also figures in [58b, 67b]. The modular case where K is a field of characteristic p is the subject of the first section of [67b]. This seems the only place where Berman entertained the modular isomorphism problem for finite p-groups (his review [67c] concerns this topic). The first section of [67b] contains two theorems which are relevant to the case of finite abelian groups:

"THEOREM 1.1. *The group algebras GK and G_1K of two countable abelian p-groups G and G_1 over a field K of characteristic p are isomorphic if and only if the groups G and G_1 are isomorphic.*" [67b, p. 365]

This is mentioned in [67a] as well. It is a result in [Deskins 56] but Berman distinguishes his proof from that of Deskins. Many proofs of the result have been given (a list of references appears in [Sandling 84]); most of them use the same ideas as that of Berman.

The second theorem marks a conjunction of his interests in the isomorphism problem and in units. After noting that the multiplicative group $M(G)$ of GK is "$= K^* \times S(G)$ where K^* is the multiplicative group of the field K and $S(G)$ is the Sylow p-subgroup of the group $M(G)$: $S(G) = \{\sum_{g\in G}\alpha_g g, \sum_{g\in G}\alpha_g = 1\}$", he states and proves:

"THEOREM 1.2. *Let $K = GF(q)$ be a finite field. Finite abelian p-groups G and G_1 are isomorphic if and only if the groups $S(G)$ and $S(G_1)$ are isomorphic.*" [67b, p. 369]

(The Sylow p-subgroup in the nonmodular case is described in [Berman-Rossa 68, 75].)

The final topic concerning the group ring problem to which Berman turned his attention has proved to be among the most fruitful. This is the issue of conjugacy, particularly for group rings of p-groups over the ring of p-adic integers. He viewed the issue in relation to the Sylow theorems. It is first mentioned in [66b]:

> "In [6] it was proven that for any finite group G, distinct from abelian groups or the group of quaternions, there are nontrivial units of finite order, which are conjugate to trivial units. A. R. Rossa proved [64] that in GZ there can be units of finite order not conjugate to trivial units. Thus in the group of units of GZ Sylow's Second Theorem does not generally hold." [66b, p. 113] (Here [6] = [55a] and [64] =[Rossa 64].)

Higman had known of such an example [Higman 40b] (see [Sandling 81] for further discussion). In [Berman-Rossa 66] there are full details of such an example for the dihedral group G of order 8. With the definitions "G: $a^4 = 1$; $b^2 = 1$; $b^{-1}ab = a^3$", "$a' = a - 2a^3 + b + ab - a^2b - a^3b$; $b' = a - a^3 + ab - a^2b - a^3b$", that b' is an element of order 2 in the integral group ring GZ which is not conjugate to a trivial unit, and that a' and b' generate a group basis not conjugate to G, are the examples used to justify:

"THEOREM 4. *There is a finite p-group G satisfying the following conditions:*

1. *In the ring GZ there is a unit of finite order not conjugate to trivial units.*
2. *In the ring GZ there are two group bases* $g_1, \dots, g_s$ *and* $g'_1, \dots, g'_s$ *not conjugate in the group of units of the ring GZ.*" [Berman-Rossa 66, p. 48]

In [Whitcomb 68] the same group is used for this same purpose; the generators that Whitcomb gives for his group basis not conjugate to G in $\mathbf{Z}G$ are $a + (1 - a^2)(a + b + ba)$ and $b + (1 - a^2)(a + ba)$, which are just $-\varphi(a')$ and $-\varphi(b')$ where φ is the automorphism of G sending a to a^{-1} and b to a^2b. In [Obayashi 70] there is a description of all group bases for the integral group ring of the dihedral group of order 8; one in particular is given as an example; it has generators $a + 2(1 - a^2)(2a + b + 2ab)$ and $b + 2(1 - a^2)(a + ab)$ and is cited as not being conjugate to G. In [Weller 72], it is shown that there are precisely two conjugacy classes of normalised group bases over $\mathbf{Z}$ in this case.

Earlier in [Berman-Rossa 66] the following theorem was deduced as an immediate consequence of their Lemma 3, the p-adic analog of Lemma 2 of [53a].

"THEOREM 2. *Let* GI_p *be the group ring of a finite p-group over the ring* I_p *of integral p-adic numbers. Then the Sylow p-subgroup H of the group E of units of the ring* GI_p *coincides with the group G for* $p \neq 2$*, and consists of all elements of the form* $\pm g$, $(g \in G)$, *for* $p = 2$." [Berman-Rossa 66, p. 46] [They assume $G \subseteq H$.]

After the proof of Theorem 4, the two theorems are combined for the following observations:

> "*Remark.* It follows from Theorems 2 and 4 that, generally speaking, the second Sylow theorem does not hold in the group of units of the integral group ring GZ of a finite p-group G. For the ring GI_p (G a finite p-group) the question of the second Sylow theorem remains open.
>
> Let G be the group of order 8 considered in Theorem 4. Then in the ring GI_2 the Sylow 2-subgroups $\{\pm a^i b^j\}$ and $\{\pm(a')^i(b')^j\}$ are conjugate. $(1-b+ab)^{-1}a'(1-b+ab) = -a$; $(1-b+ab)^{-1}b'(1-b+ab) = -b$. We note that in the ring GI_p (G a p-group) the elements $\sum_i a_i g_i (a_i \in I_p; g_i \in G)$ for which $\sum_i a_i = 1$ form a compact topological p-group E with finite Sylow p-subgroups. From the validity of the second Sylow theorem for the group of units of the ring GI_p it would follow, in particular, that the group ring GI_p determines the p-group G up to isomorphism." [Berman-Rossa 66, pp. 51–52]

In [Whitcomb 68], the unit $1 + a^3 + a^3 b$ is used to conjugate his group basis to G in $\mathbf{Q}G$ and in I_2G. In [Hattori 71], the unit $1/4 - a/2(1+b)$ is given to conjugate G to Obayashi's group basis in $\mathbf{Q}G$.

In the p-group case, Whitcomb, with John Thompson, had explored the possibility of conjugacy of group bases over $\mathbf{Q}$ and over I_p and was bold enough to pose the issue as a conjecture instead of as an "open question":

"CONJECTURE C. *Let H be a group basis for $Z(G)$, then $H^u = G$ for some unit $u \in I_p(G)$." [Whitcomb 68, p. 20]*

Whether conjecture or open question, it is this that has now been settled affirmatively for normalised group bases in [Roggenkamp-Scott 87]. It served the purpose envisaged by Berman and Rossa of establishing one of the strongest results on the isomorphism problem known to date, that a finite p-group is determined by its integral group ring.

Acknowledgments. In addition to those thanked in [Sandling 81, 85], I am grateful to Jonathan Alperin, Gordon James, Alex Lichtman, Todor Mollov, Cecil Nesbitt, Sam Perlis, Jim Roseblade and Robert Thrall for further information and bibliography.

Appendix A. [Polyak 60] A group G uniquely defines a group ring $R(G, M)$ over a ring M, but a single group ring may be produced from nonisomorphic groups.

In article [2], a necessary condition for the isomorphism of group rings of finite groups G and H over the ring of rational integers was given. In this report the statement of the problem is generalised and we follow up the relation between a group and its group ring over the ring P_c of integers of a number field $P = R(\theta)$ being a finite algebraic extension of the field of rational numbers. The theorem proved sharpens the result of the above mentioned article and is formulated in terms of K-divisions, as defined in [1].

Let G be a finite group, n the least common multiple of the orders of elements of G, K a field. The partitioning of G into K-divisions is effected by defining the following equivalence relation: $a \sim b$ if $a = c^{-1}b^{\mu}c$, where $\varepsilon \to \varepsilon^{\mu}$ is an automorphism in the Galois group of the field $K(\varepsilon)$ over K, ε being a primitive nth root of unity.

THEOREM. *If $R(G, P_c) \cong R(H, P_c)$, then in the groups G and H the orders of the K-divisions and the order of the classes contained in corresponding K-divisions coincide, for any number field K.*

In the course of the proof the relation

$$(*) \qquad t_i' = \xi_i t_i$$

is obtained where t_i (t_i') is the sum of the elements of the K-division T_i (T_i') of the group $G(H_1)$, H_1 is a basis of $R(G, P_c)$, the elements of which form a group with respect to multiplication, isomorphic to H, ξ_i is an mth root of unity where m is the degree of P over R. (*) generalizes the analogous relation proven in [2].

(Here [1] = [52], [2] =[53c].)

Appendix B. [Berman 60] The centre Z of the group ring Γ of the finite group G over the ring of rational integers C consists of all linear combinations $\sum_{i=1}^{s} \alpha_i \kappa_i$ where $\alpha_i \in C$, and $\kappa_1, \dots, \kappa_s$ correspond to the sums of the elements of the conjugacy classes $C_1, \dots, C_s$ of the group G in Γ.

The following theorem is valid:

Every automorphism ψ of the centre Z of the ring Γ effects a generalized permutation of the conjugacy classes of the group:

$$(1) \qquad \psi(\kappa_i) = \pm\kappa_{j_i} \ (i = 1, \dots, s),$$

and for a group G of odd order

$$(2) \qquad \psi(\kappa_i) = \kappa_{j_i} \ (i = 1, \dots, s).$$

The proof of this theorem is based on the following lemma:

Any automorphism of the centre Z commutes with the transformation of involution: $\alpha_1\kappa_1 + \dots + \alpha_s\kappa_s \to \alpha_1\kappa_1' + \dots + \alpha_s\kappa_s'$, $(\alpha_i \in C)$, where κ_i' is the sum *of the elements of the class C_i', containing the elements inverse to those of the class C_i.*

Bibliography. References are treated in several different ways. Items by Berman alone are referred to in the text by year of publication (with *a*, *b*, *c* if necessary), his joint articles by names of authors and date. They are listed below (with the date codes) in three sections: the first contains his principal mathematical articles which are referenced here, the second his papers with Rossa, the third his reviews and so on. When available, note is made of the location of the reviews of each paper in Mathematical Reviews, Zentralblatt für Mathematik and Referativnyĭ Zhurnal. Matematika, and of the date of submission. Reviews which are taken from the authors' abstract are marked by *.

The last part of the bibliography contains **some** of the items by other authors which are cited in the text (with similar referencing conventions). Not all items cited appear, however. To have included them all would have made the bibliography disproportionately lengthy. As most of the items cited already appear in the bibliography of my survey [Sandling 85], it seemed adequate to use this as their source; it has been reprinted in abbreviated form in such an ISI publication as CompuMath Source Index and is available on their database.

Mathematical Papers By S. D. Berman

[52] *On the theory of representations of finite groups* (Russian) Dokl. Akad. Nauk SSSR **86** (1952), 885–888. MR 14, 619 Zbl. 47, 28 29.IV.1952

[53a] *On certain properties of integral group rings.* (Russian) Dokl. Akad. Nauk SSSR **91** (1953), 7–9. MR 15, 99 Zbl. 50, 255 RZh1954,1574 29.IV.1952

[53b] *On isomorphism of the centers of group rings of p-groups.* (Russian) Dokl. Akad. Nauk SSSR **91** (1953), 185–187. MR 15,99 Zbl. 50, 256 RZh1954,1082 29.XI.1952

[53c] *On a necessary condition for isomorphism of integral group rings.* (Ukrainian) Dopovīdī Akad. Nauk Ukraïn. RSR 1953, 313–316. MR 15, 599 Zbl. 53, 211* RZh1954,2872 29.IV.1952

[55a] *On the equation $x^m = 1$ in an integral group ring.* (Russian) Ukrain. Mat. Zh. **7** (1955), 253–261. MR 17, 1048 Zbl. 66, 19 RZh1956,4378 12.V.1953

[55b] *On certain properties of group rings over the field of rational numbers.* (Russian) Uzhgorod. Gos. Univ. Nauchn. Zap. Him. Fiz. Mat. **12** (1955), 88–110. MR 20 #3920 RZh1957,8451

[57] *Groups of which all representations are monomial.* (Ukrainian) Dopovīdī Akad. Nauk Ukraïn. RSR 1957, 539–542. MR 20 #5241* Zbl. 77, 251* RZh1958,6492 15.II.1957

[58a] *Representations of groups of order 2^m over an arbitrary field of zero characteristic.* (Ukrainian) Dopovīdī Akad. Nauk Ukraïn. RSR 1958, 243–246. MR 20 #6472 Zbl. 80, 250* RZh1959,1259 23.V.1957

[58b] *Characters of linear representations of finite groups over an arbitrary field.* (Russian) Mat. Sb. (N.S.) **44**(**86**) (1958), 409–465. MR 20 #3221 Zbl. 80, 21 RZh1959,3529 17.V.1956

[60] *On automorphisms of the centre of an integral group ring.* (Russian) Dokl. Soobshch. Uzhgorod. Gos. Univ. Ser. Fiz.-Mat. Nauk 1960, no. 3, 55. RZh1961,8A199

[66a] *Representations of finite groups over an arbitrary field and over rings of integers.* (Russian) Izv. Akad. Nauk SSSR Ser. Mat. **30** (1966), 69–132. Amer. Math. Soc. Transl. (2) 64, 147–215. MR 33 #5747 Zbl. 203,31 RZh1966,7A206* 19.III.1965

[67a] *Group algebras of countable abelian p-groups.* (Russian) Dokl. Akad. Nauk SSSR **175** (1967), 514–516. Soviet Math. Dokl. **8** (1967), 871–873. MR 35 #5504 Zbl. 178,27 RZh1968, 1A198* 19.IX.1966

[67b] *Group algebras of countable abelian p-groups.* (Russian) Publ. Math. Debrecen **14** (1967), 365–405. MR 37 #1488 Zbl. 281.20006 RZh1968,12A168 20.XII. 1966

[69] *On the Schur index.* (Russian) (Supplement to the Russian translation of [Curtis-Reiner 62]), 610–638 Izdat. "Nauka", Moscow, 1969. MR 40 #1490 (RZh1969,6A210)

Mathematical Papers By S. D. Berman and A. R. Rossa:

[BR66] *Integral group-rings of finite and periodic groups.* (Russian) Algebra and Math. Logic: Studies in Algebra, 44–53. Izdat. Kiev. Univ., Kiev, 1966. MR 35 #265 Zbl. 248.20005* RZh1967, 10A183

[BR68] *The Sylow p-subgroup of a group algebra over a countable abelian p-group.* (Ukrainian) Dopovīdī Akad. Nauk. Ukraï n. RSR Ser. A 1968, 870–872. MR 40 #1479 Zbl. 185,70* RZh1969, 3A155* 22.II.1968

[BR71] *Group algebras of countable periodic abelian groups.* (Ukrainian) Dopovīdī Akad. Nauk Ukraïn. RSR Ser. A 1971, 387–390, 476. MR 45 #400 Zbl. 275.20010* RZh1971, 12A337 18.XII.1968

[BR75] *The Sylow p-subgroups of the group algebras of countable abelian p-groups.* (Russian) Material from the XXIX Scientific Conference of Professors and Instructors of the Staff of Uzhgorod University, 158–176. Dept. of Mathematical Sciences, Uzhgorod, 1975. RZh1976,7A296*

Miscellaneous Items By S. D. Berman

[56] Review 7170 (of [Hewitt-Zuckerman 55]). RZh.Mat.1956, vol. 10, 23–24.

[62] Review 3A208 (of [Polyak 60]). RZh.Mat.1962, vol. 3, 30.

[64] Review 7A235 of ([Cohn-Livingstone 63]). RZh.Mat.1964, vol. 7, 26.

[66b] *Representations of finite groups.* (Russian) Algebra 1964, 83–122, Akad. Nauk SSSR Inst. Nauchn. Informacii, Moscow, 1966 = Itogi Nauki-Seriya Matematika. Alg. Top. Geom. 3 MR 35 #2973 Zbl. 209, 58 RZh1967,5A237

[BB67] (with Bovdi, A.) Review 2A207 (of [Cohn-Livingstone 65]). RZh.Mat.1967, vol. 2, 33.

[B et al **67]** (with Vol'pert, A. I.; Zagorskiĭ, T. Ya.; Kazimirskiĭ, P. S.; Kostenko, V. G.; Lyantse, V. È.) *Yaroslav Borisovich Lopatinskiĭ (on his 60th birthday).* Uspekhi Mat. Nauk **22** (1967), 5, 197–202. Russian Math. Surveys **22** (1967), 5, 181–186. Zbl. 162,9. RZh1968,3A38

[67c] Review 10A148 (of [Sehgal 67]). RZh.Mat.1967, vol. 10, 26.

Papers By Other Authors

Cohn, J. A.; Livingstone, D. *On groups of the order p^3.* Canad. J. Math. **15** (1963), 622–624. MR 27 #3700

——, *On the structure of group algebras.* I. Canad. J. Math. **17** (1965), 583–593. MR 31 #3514

Gudivok, P. Review 6A166 (of [Passman 65]). RZh.Mat. 1966, vol. 6, 23–24.

Hattori, Akira. *Strong separability and group algebras.* Sci. Papers College Arts Sci. Univ. Tokyo **35** (1985), 1–10. MR 87g:16007

Hewitt, E.; Zuckerman, H. S. *Finite dimensional convolution algebras.* Acta Math. **93** (1955), 67–119. MR 17, 1048

Higman, G. *The units of group-rings.* Proc. London Math. Soc. (2) **46** (1940), 231–248. MR 2, 5

——, *Units in group rings.* D. Phil. thesis, Oxford Univ., 1940.

Lohwater, A. J. *Russian-English Dictionary of the Mathematical Sciences.* Amer. Math. Soc., Providence, 1961. MR 23 #A3044

Mollov, T. Z. *The invariants and the groups of units of group algebras.* (Russian) (Author's summary of his dissertation) Plovdiv Univ., Plovdiv, Bulgaria, 1984.

Passman, D. S. *Isomorphic groups and group rings.* Pacific J. Math. **15** (1965), 561–583. MR 33 #1381

Polyak, S. S. *A necessary condition for the isomorphism of group rings over a ring.* (Russian) Dokl. Soobshch. Uzhgorod. Gos. Univ. Ser. Fiz.-Mat. Nauk 1960, no. 3, 62.

Roggenkamp, K. W.; Scott, L. L. *Isomorphisms of p-adic group rings.* Ann. of Math. **126** (1987), 593–647. Zbl. 633.20003

Rossa, A. R. *On integral group rings of finite groups.* (Russian) Tezicy Dokl. Soobshch. Uzhgorod. Gos. Univ. Ser. Fiz.-Mat. Nauk 1964, 4–5.

Sandling, R, *Graham Higman's thesis "Units in group rings".* Integral representations and applications (Oberwolfach, 1980), 93–116, Lecture Notes in Math., 882, Springer, Berlin, 1981. MR 83g:20009

——, *The isomorphism problem for group rings: a survey.* Orders and their applications (Oberwolfach, 1984), 256–288, Lecture Notes in Math., 1142, Springer, Berlin, 1985. MR 87b:20007

——, *A proof of the class sum correspondence using the real group algebra.* Ring theory (Granada, 1986), 237–244, Lecture Notes in Math., Springer, Berlin, 1988.

Scott, L. L. *Recent progress on the isomorphism problem*, 259–273, The Arcata conference on representations of finite groups, Part 1. Proc. Sympos. Pure Math. 47, Amer. Math. Soc., Providence, 1987.

Sehgal, Sudarshan K. *On the isomorphism of group algebras.* Math. Z. **95** (1967), 71–75. MR 34 #5950

Takahashi, S. *Some properties of the group ring over rational integers of a finite group.* Notices Amer. Math. Soc. **12** (1965), 463.

Weiss, A. *Rigidity of p-adic p-torsion.* Ann. of Math. **127** (1987), 317–332.

MATHEMATICS DEPARTMENT,
THE UNIVERSITY,
MANCHESTER, M13 9PL
ENGLAND

Contemporary Mathematics
Volume **93** (1989)

On p-subgroups of Finite Symmetric and Alternating Groups

JA. G. BERKOVIC

In memory of my dear friend Samuil Davidovic Berman (1922–1987)

1. Let $M = \{1, 2, \ldots, n\}$ be a finite set, S_n be the symmetric group on M, A_n be the alternating group on M. For a prime p let $E_n^{(p)} = E_n$ be a Sylow p-subgroup of S_n, $N_n^{(p)} = N_n$ be its normalizer in S_n. If $n < p$ then we assume $E_n = 1$, $N_n = S_n$. If $M_1 \subseteq M$ we assume that $S(M_1) \leq S(M)$. For $\pi \in S_n$ we set $T(\pi) = \{x \in M | \pi(x) \neq x\}$, a support of π. For $H \subseteq S_n$ we set $T(H) = \bigcup_{\pi \in H} T(\pi)$. For a group G and a positive integer t let G^t be a direct product of t copies of G, $G^0 = 1$. Let C_m denote a cyclic group of order m.

In this section we describe the structure of $N_n^{(p)} = N_n$. Our presentation will be elementary and will not require bulky calculations. Specific case of Theorem 2 has been considered in **[1–3]**.

LEMMA 1. *Let π be a permutation of type $(b_1, b_2, \ldots)$ in $S(M)$, $|M| = n$. Let M_i be the set of all points which are permuted by i-cycles from the canonic decomposition of π, $i > 1$, $M_1 = M - T(\pi)$. Let Z be the centralizer of π in S_n. For $\sigma \in Z$ we shall write $\sigma_i = \sigma | M_i$, a restriction of σ on M_i. Then $Z = Z_1 \times \cdots \times Z_n$, where $Z_i = \{\sigma_i | \sigma \in Z\}$. We have $T(Z_i) = M_i$, $Z_i \cong C_i \operatorname{wr} S_{b_i}$, $M = \bigcup_{i=1}^{n} M_i$ be a partition of M. Base of wreath product Z_i is generated by the cycles from the canonic decomposition of $\pi_i = \pi | M_i$, $Z_1 = S(M_1)$.*

We omitted an easy proof of this result.

THEOREM 2. *Let $n = a_0 + a_1 p + \cdots + a_t p^t$ be a p-adic decomposition of a natural number n, $a_t \neq 0$. Then*

(a) $N_n \neq N_{a_0} \times N_{a_1 p} \times \cdots \times N_{a_t p} t$, $|T(N_{a_i p} i)| = a_i p^i$, $T(N_{a_i p} i) \cap T(N_{a_j p} j) = \varnothing$ *for* $i \neq j$, $i, j \in \overline{1, t}$.

(b) *If* $1 < a < p$ *then* $N_{ap^m} \cong N_{p^m} \operatorname{wr} S_a$ *(this makes sense only when p is odd).*

(c) $N_{p^m} / E_{p^m} \cong C_{p-1}^m$.

(d) $[N_{p^m}, N_{p^m}] = E_p m$ *for* $p > 2$. *If* $p > 2$ *then* $E_n \leq [N_n, N_n]$.

(e) *Let* $m > 1$ *and* $W = N_{p^m} - 1 \operatorname{wr} N_p$. *Let* Q *be a Sylow* p*-subgroup of the base of wreath product* W *and* D *be diagonal of this base. Then* $N_{p^m} = QDN_p$.

PROOF. (1) At first we assume that $n = p^m$. Notice that E_{p^m} is then transitive. Therefore according to Schur-Zassenhaus theorem there exists a p-complement K in N_{p^m}, and K acts on E_{p^m} faithfully since K is a subgroup of a point stabilizer in transitive group N_{p^m}.

Let $m = 1$. Then $E_p = \langle\pi\rangle$ where $\pi = (1, 2, \ldots, p)$ be a p-cycle. If g is a primitive root modulo p then we define a permutation σ as follows: $\sigma(i) = \bar{i}$ where $\bar{i}$ is the minimal positive residue of $1 + (i - 1)g$ modulo p. Then $\sigma\pi\sigma^{-1} = \pi^g$ and $\langle\pi, \sigma\rangle = \langle\pi, \sigma | \pi^p = \sigma^{p-1} = 1, \sigma\pi\sigma^{-1} = \pi^g\rangle$ is isomorphic to the holomorph of cyclic group $C_p = \langle\pi\rangle$. Since $C_{N_p}(E_p) = E_p$ we have $N_p = \langle\pi, \sigma\rangle$, $N_p/E_p \cong C_{p-1}$ and $[N_p, N_p] = E_p$. For $n = p$ the proof is complete.

Let $m > 1$. In this case we shall use induction on m. A permutation π, generating the centre of E_{p^m} takes form $\pi = (1, 2, \ldots, p) \ldots (p^m - p + 1, p^m - p + 2, \ldots, p^m)$. Let Z be the centralizer of π im S_pm. Then $Z = Z^{(2)}[Z^{(1)}$ (a semidirect product with the core $Z^{(1)}) \cong C_p \operatorname{wr} S_{p^m} - 1$ (here $Z^{(1)}$ is the base of wreath product Z; see Lemma 1).

Let T be the normalizer of $\langle\pi\rangle$ in S_{p^m}. Obviously Z is normal in T and T/Z is isomorphic to a subgroup of $\operatorname{Aut}(\langle\pi\rangle) \cong C_{p-1}$. Like before for $m = 1$ one can easily build a permutation σ of order $p - 1$ such that $\sigma\pi\sigma^{-1} = \pi^g$ where g is a primitive root modulo p. It is clear that $\sigma \in T$, $\langle\sigma\rangle \cap Z = 1$. Hence $T = \langle\sigma\rangle$ $[Z$. Since $\sigma\tau\sigma^{-1} = \tau^g$ for all $\tau \in Z^{(1)}$ then $Z^{(1)}$ is normal in $T = \langle Z, \sigma\rangle$. Since $\langle\pi\rangle$ is characteristic in S_pm then $N_n \leq T$. Since $Z/Z^{(1)} \cong S_{p^m} - 1$, $E_{p^m}/Z^{(1)} \cong E_{p^m} - 1$ then the normalizer of $E_{p^m}/Z^{(1)}$ in $Z/Z^{(1)}$ is isomorphic to $N_{p^m} - 1$, and the structure of latter is known by induction. We have $T/Z^{(1)} \cong Z/Z^{(1)} \times C_{p-1}$. (Since $Z/Z^{(1)} \cong S_{p^m} - 1$ is complete by Hölder's theorem.) So the normalizer of $E_{p^m}/Z^{(1)}$ in $T/Z^{(1)} \cong S_{p^m} - 1 \times C_{p-1}$ is isomorphic to the $N_{p^m} - 1 \times C_{p-1}$ on the one hand and to the $N_{p^m}/Z^{(1)}$ on the other hand. Thus by induction we obtain $N_{p^m}/E_{p^m} \cong C_{p-1} \times N_{p^m} - 1/E_{p^m} - 1 \cong C_{p-1} \times C_{p-1}^{m-1} \cong C_{p-1}^m$, and assertion (c) is proved.

We shall show that for $p > 2$

$$[N_{p^m}, N_{p^m}] = E_{p^m}.$$

Since N_{p^m}/E_{p^m} is abelian by (c) it is sufficient to show that

$$E_{p^m} \leq [N_{p^m}, N_{p^m}].$$

The proof is by induction on m. Let it have already been proved that

$$E_{p^m} - 1 = [N_{p^m} - 1, N_{p^m} - 1].$$

The proved above implies

$$N_{p^m}/Z^{(1)} \cong N_{p^m} - 1 \times C_{p-1},$$

and by induction

$$[N_{p^m}/Z^{(1)}, N_{p^m}/Z^{(1)}] = E_{p^m}/Z^{(1)}.$$

The above-constructed permutation σ satisfies the equality $\sigma\tau\sigma^{-1} = \tau^g$ for all $\tau \in Z^{(1)}$ (here g is a primitive root modulo p). Hence $[\langle Z^{(1)}, \sigma\rangle, \langle Z^{(1)}, \sigma\rangle] = Z^{(1)}$ and $Z^{(1)} \leq [N_{p^m}, N_{p^m}]$. Therefore $[N_{p^m}, N_{p^m}] \geq E_{p^m}$, and the first assertion of (d) is proved.

(2) Let now $n = ap^m$, $1 < a < p$, $p > 2$.

We shall partition the set $M = \{1, 2, \dots, ap^m\}$ into sets of equal cardinality:

$$R_i = \{(i-1)p^m + 1, \dots, ip^m\}, \quad i \in \overline{1, a}.$$

On each of these R_i we construct a symmetric group $S(R_i)$ and take a Sylow p-subgroup $E_{p^m}^{(i)}$ in it. Then as it is well-known

$$E_{ap^m} = E_{p^m}^{(1)} \times \cdots \times E_{p^m}^{(a)}.$$

The permutation

$$\pi_i = ((i-1)p^m + 1, \dots (i-1)p^m + p) \dots (ip^m - p + 1, \dots, ip^m)$$

generates the centre of $E_{p^m}^{(i)}$. Let $\pi = \pi_1 \cdot \pi_2 \cdot \dots \cdot \pi_a$. Obviously π belongs to the centre of $E_n = E_{ap^m}$. Let $\sigma \in N_n = N_{ap^m}$. Then $\sigma\langle\pi\rangle\sigma^{-1} = \langle\pi\rangle$. To prove this equality it is sufficient to show that $\sigma\langle\pi_i\rangle\sigma^{-1} = \langle\pi_j\rangle$ for $i \in \overline{1, a}$ and suitable $j = j(i) \in \overline{1, a}$. We set $L = \langle\pi_1\rangle \cup \langle\pi_2\rangle \cup \dots \cup \langle\pi_a\rangle$. Then any non-identity permutation from L is a product of p^{m-1} pairwise independent p-cycles, and every permutation from $C - L$ where $C = \langle\pi_1\rangle \times \cdots \times \langle\pi_a\rangle$ has no such form. Since $C = Z(E_n)$, the centre of E_n, then $\sigma C \sigma^{-1} = C$, and equality $\sigma\langle\pi_i\rangle\sigma^{-1} = \langle\pi_j\rangle$ is proved. From $\sigma\langle\pi_i\rangle\sigma^{-1} = \langle\pi_j\rangle$ we deduce that $\sigma(R_i) = R_j$ for the same i, j. Thus, σ can be considered as a permutation of a set system $\{R_i\}_1^a$. Since S_{ap^m} contains the permutations which give all possible permutations of this set system then $N_{ap^m} = N_n = N_{S(R_1)}(E_{p^m}^{(1)}) \operatorname{wr} S_a \cong N_{p^m} \operatorname{wr} S_a$, $T(N_n) = M$. We proved the assertion (b).

(3) Let now $n = a_0 + a_1 p + \cdots + a_t p^t$ be a p-adic decomposition of n, and let at least two coefficients in this decomposition differ from 0. The set M can be partitioned into $t + 1$ subsets $Q_0, Q_1, \dots, Q_t$ with $|Q_i| = a_i p^i$, $i \in \overline{0, t}$. Let

$$H = S(Q_0) \times S(Q_1) \times \cdots \times S(Q_t).$$

Then

$$E_n = E_{a_1 p} \times \cdots \times E_{a_t p^t} \cong E_p^{a_1} \times \cdots \times E_{p^t}^{a_t}.$$

Suppose that $E_{a_i p^i} \in \mathrm{Syl}_p(S(Q_i))$, $i \in \overline{0, t}$. Let F_i be the centre of $E_{a_i p^i}$, $i \in \overline{1, t}$, and $F = F_1 \times \cdots \times F_t$. Then F is the centre of E_n. If $\sigma \in N_n$ then $\sigma F \sigma^{-1} = F$. Since the permutations of F_1 are of definite type among the elements of F, then $\sigma F_i \sigma^{-1} = F_i$ for all i. This means that $\sigma(Q_i) = Q_i$ for all $i > 0$ and now it is clear that $\sigma(Q_0) = Q_0$. Thus $\sigma \in S(Q_0) \times \cdots \times S(Q_t) = H$, i.e.

$$N_n = N_H(E_n) = N_{a_0} \times N_{a_1 p} \times \cdots \times N_{a_t p^t}.$$

The statement about $T(N_{a_i p^i})$ is obvious now, and the proof of assertion (a) is completed.

(4) Let us show that $[N_n, N_n] \geq E_n$ for $p > 2$. The obtained decomposition of N_n in (a) and first assertion of (d) imply that it is sufficient to consider the case when $n = ap^m$, $1 < a < p$. We have

$$E_{ap^m}(\cong E^a_{p^m}) \leq N^a_{p^m} < N_{ap^m}$$

and

$$E_{ap^m} = [N^a_{p^m}, N^a_{p^m}] \leq [N_{ap^m}, N_{ap^m}]$$

by the first assertion of (d).

(5) Now we pass to (e). Let $m > 1$, $W = N_{p^m} - 1 \operatorname{wr} N_p$. Let Q be a Sylow p-subgroup of the base K of wreath product W and D is a diagonal of this base. Since Q is characteristic in K then Q is normal in W. Since $D \leq C_W(N_p)$ then QDN_p is the subgroup of W. We have $QDN_p/Q \cong C^{m-1}_{p-1} \times N_p$ hence QDN_p has normal Sylow p-subgroup. The equality $N_{p^m} = QDN_p$ follows from the equality of orders, and (e) is proved.

COROLLARY 3. *$E^{(p)}_n = N^{(p)}_n$ if and only if $p = 2$.*

2. In this section $p = 2$. Let

$$\overline{E}_n = E_n \cap A_n \in \mathrm{Syl}_2(A_n), \overline{N}_n = N_{A_n}(\overline{E}_n).$$

In this section we shall calculate $\overline{N}_n$.

At first we shall study a structure of $\overline{E}_n$ in a more detailed way. Let us begin with $n = 2^m$, $m \geq 1$. $\overline{E}_2 = 1$, $\overline{E}_4 = C^2_2$. Now let $m \geq 3$. $\overline{E}_{2^m}$ consists of all even permutations of the group $E_{2^m} = E_{2^m} - 1 \operatorname{wr} C_2$, $C_2 = \langle\sigma\rangle$,

$$\sigma = (1, 1 + 2^{m-1})(2, 2 + 2^{m-1}) \ldots (2^{m-1}, 2^m) \in E_{2^m}.$$

The base of wreath product E_{2^m} is equal to $E^{(1)}_{2^{m-1}} \times E^{(2)}_{2^{m-1}}$, i.e. cartesian square of $E_{2^m} - 1$, with $(E^{(1)}_{2^{m-1}})\sigma = E^{(2)}_{2^{m-1}}$. Let $E^{(i)}_{2^{m-1}} \cap A_n = \overline{E}^{(i)}_{2^{m-1}}$, $i = 1, 2$, and ρ_0 be a transposition in $E^{(1)}_{2^{m-1}}$. Then $\rho = \rho_0\rho_0^\sigma \in \overline{E}_{2^m}$ and

$$\overline{E}_{2^m} = \langle\sigma, \rho\rangle[(\overline{E}^{(1)}_{2^{m-1}} \times \overline{E}^{(2)}_{2^{m-1}}),$$

a semi-direct product. Furthermore for $i = 1, 2$,

$$\langle\overline{E}^{(i)}_{2^{m-1}}, \sigma\rangle \cong \overline{E}^{(i)}_{2^{m-1}} \operatorname{wr} C_2,$$
$$\langle\overline{E}^{(i)}_{2^{m-1}}, \rho\rangle \cong E_{2^m} - 1.$$

These two isomorphisms show that the group

$$\langle\overline{E}^{(1)}_{2^{m-1}}, \overline{E}^{(2)}_{2^{m-1}}, \rho\rangle$$

is isomorphic to direct product of two copies of $E_{2^{m-1}}$ with the amalgamated quotient-group of order 2, and σ transfers the factors of this product. The description of the group $\overline{E}_{2^m}$ shows that its centre is equal to the centre of the group E_{2^m}, in particular $\overline{E}_{2^m}$ contains a characteristic subgroup of order 2 which is generated by permutation that is a product of 2^{m-1} pairwise independent transpositions.

Let now n be not a power of 2,

$$n = a_0 + 2a_1 + \cdots + 2^t a_t$$

is 2-adic decomposition of n, $a_t = 1$. If $a_1 = \cdots = a_{t-1} = 0$ then $\overline{E}_n = \overline{E}_{2^t}$, thus we shall assume that at least one of $a_1, \ldots, a_{t-1}$ is equal to 1.

Let coefficients $a_{i_1}, \ldots, a_{i_k}$ be equal to 1 and the other a_j's equal to 0. By assumption $k > 1$. Let ρ_m be a transposition E_{2^m}, $m \in \{i_1, \ldots, i_k\}$. Then the permutations $\rho_{i_1}, \ldots, \rho_{i_k}$ are commute pairwise. Here

$$\overline{E}_n = \langle \rho_{i_1}, \rho_{i_2}, \ldots, \rho_{i_1}\rho_{i_k}\rangle[(\overline{E}_{2^{i}1} \times \cdots \times \overline{E}_{2^{i}k})$$

is a semi-direct product (it is obvious that this product lies in $\overline{E}_n$). It is easy to see that $\langle \overline{E}_{2^s}, \rho_{i_1}\rho_{i_s}\rangle \cong E_{2^s}$, $\langle \overline{E}_{2^{i}1}, \rho_{i_1}\rho_{i_s}\rangle \cong E_{2^{i}1}$, $s \in \{i_2, \ldots, i_k\}$.

The key fact for calculation of $\overline{N}_n$ is

LEMMA 4. *If $n > 5$ then $\overline{E}_n$ contain a characteristic subgroup $R = \langle \pi \rangle$ of order 2, π is equal to a product of 2^{t-1} pairwise independent transpositions, where $2^t \leq n < 2^{t+1}$.*

PROOF. If $n = 2^t$ then, as we have seen, the centre of E_n is the desired subgroup.

Now let n be not a power of 2. If $n = 6 = 2 + 4$ then by the proved above $\overline{E}_n$ is the dihedral group of order 8, so lemma holds true in this case as well. Obviously lemma is also true for $n = 7$ since $\overline{E}_7 = \overline{E}_6$. Assume $n > 7$ and $n \neq 2^t, 2^{t+1}$. Let $n = a_0 + 2a_1 + \cdots + 2^t a_t$ is a 2-adic decomposition of n, $a_t = 1$. Let R be the centre of group $L = \langle \overline{E}_{2^t}, \rho_{i_1}\rho_{i_t}\rangle$ (see previous considerations). Let $|\overline{E}_n| = 2^a$ and $|L| = 2^b$. Then $a < 2b - 1$, and $|\overline{E}_n| < |I, E_{2^t}|$. So $L \cap (\overline{E}_{2^t})^\varphi > 1$ for all $\varphi \in \mathrm{Aut}(\overline{E}_n)$. Since $L \cap (\overline{E}_{2^t})^\varphi$ is normal in L and R is the only normal subgroup of order 2 in L, we have $R \leq L \cap (\overline{E}_{2^t})^\varphi$, $R \leq L \cap L^\varphi$ for all $\varphi \in \mathrm{Aut}(\overline{E}_n)$. Hence $R^\varphi = R$ for all φ and R is a characteristic in $\overline{E}_n$. Last assertion follows from considerations in the previous lemma.

For the subgroup R constructed above let $R = \langle \pi \rangle$. We know that π is a product of 2^{t-1} pairwise independent transpositions. Let us calculate $\overline{Z} = C_{A_n}(\pi)$. We set $G = S_n$, $\overline{G} = A_n$, $Z = C_G(\pi)$. We have $Z = (C_2 \operatorname{wr} S_{2^{t-1}}) \times S_{n-2^t} = Z^{(2)}Z^{(1)}$ (see Lemma 1). Denote the factor in parentheses by A and the complementary direct factor by B, then $Z = A \times B$. Let $\overline{Z}^{(1)} = Z^{(1)} \cap \overline{G}$, then $|Z^{(1)} : \overline{Z}^{(1)}| = 2$. If $B > 1$ then let B_0 be a subgroup of index 2 in B. If $B = 1$ then assume $B_0 = 1$. Obviously $B_0 < \overline{Z}$. If ρ_1 is a transposition in $Z^{(1)}$, ρ_2 is a transposition in B for $B > 1$ (if $B = 1$ we set $\rho_2 = 1$ but do not consider this case since it is easy), then $\rho = \rho_1\rho_2 \in \overline{Z}$. Let D be an active factor of the wreath product A. Then $D < \overline{Z}$ (because any permutation from D can be represented as product of even number of transpositions). We have

$$\overline{Z} = \langle Z^{(1)} \times B_0, \rho \rangle,$$

$\overline{Z}^{(1)}$ is normal in $\overline{Z}$ and ρ centralizes $\overline{Z}^{(2)} = Z^{(2)} \cap A_n$ modulo $\overline{Z}^{(1)}$. Besides, $\langle B_0, \rho \rangle \cong S_{n-2}t$, $D \cong S_{2^{t-1}}$. Since $D \cong Z^{(1)}D/Z^{(1)}$ is complete by Hölder's theorem and normal in $\overline{Z}/\overline{Z}^{(1)}$ this leads to

LEMMA 5. $\overline{Z}/\overline{Z}^{(1)} \cong S_{n-2^t} \times S_{2^{t-1}}$.

Let us prove the main result of this section.

THEOREM 6. *If $n > 5$ then $\overline{E}_n = \overline{N}_n$.*

PROOF. Let $\langle \pi \rangle$ be a characteristic subgroup of order 2 in $\overline{E}_n$ built in Lemma 4. By virtue of Lemma 5 $\overline{Z}/\overline{Z}^{(1)} \cong S_{2^{t-1}} \times S_{n-2^t}$. Let $\overline{Z} = C_{\overline{G}}(\pi)$. Since $\langle \pi \rangle$ is characteristic in E_n we have $\overline{N}_n \leq \overline{Z}$. By Corollary 3 Sylow 2-subgroup is equal to its normalizer in $\overline{Z}/\overline{Z}^{(1)}$. Hence owing $\overline{Z}^{(1)} < \overline{N}_n$ we obtain $\overline{N}_n = \overline{E}_n$.

3. Let G be a finite group, p be a prime. Introduce the following denotions:

$$p^{a_p(G)} = \max\{|A|| \, A \text{ is an abelian } p\text{-subgroup in } G\},$$
$$p^{e_p(G)} = \max\{|H|, | \, H < G, \ \exp(H) = p\}.$$

Let $d(H)$ be the minimal number of generators of p-group H. We set $p^{d_p(G)} = \max\{d(H) | \, H \text{ is a } p\text{-subgroup in } G\}$.

LEMMA 7. *Let A be a maximal abelian subgroup or a maximal subgroup of exponent p in $G = B \times C$. Then*

$$A = (A \cap B) \times (A \cap C).$$

PROOF. Both assertions are proved similarly. We prove the second assertion. Let A_1, A_2 be projections of A on B and C respectively. Then $A \leq A_1 \times A_2$, $\exp(A_1 \times A_2) = p$ so $A = A_1 \times A_2$, $A_1 = A \cap B$, $A_2 = A \cap C$.

Recall an important property of regular wreath product $W = G \operatorname{wr} C_p$ with the base $K = G_1 \times \cdots \times G_p$, $G_i \cong G$. If $1 \neq x_i \in G_i$ then $C_W(x_i) \leq K$ since any element from $W - K$ induces a p-cycle on $\{G_1, \dots, G_p\}$.

LEMMA 8. *Let $W = G \operatorname{wr} C_p$ be a finite p-group, $K = G_1 \times \cdots \times G_p$ is a base of W with $G_i \cong G$. Then*

(a) *If A is an abelian subgroup of W with $|A| > p^{1+a_p(G)}$ then $A \leq K$. In particular $a_p(W) = pa_p(G)$.*

(b) *If M is a subgroup of exponent p in W and $|M| = p^{pe_p(G)}$ then $M \leq K$, if $e_p(G) > 1$.*

PROOF. (a) Let A be an abelian subgroup in W and $|A| > p^{1+a_p(G)}$.

Suppose that $A \not\leq K$. If $A \cap (G_1 \dots G_{p-1}) = 1$ then $K \cap A$ is isomorphic to an abelian subgroup of G_p and $a_p(G_p) = a_p(G) \geq -1 + \log_p(|A|)$, which is a contradiction. Let $1 \neq y \in A \cap (G_1 \times \cdots \times G_{p-1})$, $C_p = \langle c \rangle$, $A = \langle A \cap K, x \rangle$, $x = kc$, $k \in K$.

We assume that $y \in G_1 \times \cdots \times G_t$ with a minimal t. Then, if $y = g_1 \dots g_t$ with $g_i \in G_i$, then $y^{kc} = y^x = y = g_1^{kc} \dots g_t^{kc}$, $g_1, g_i^k \in G_i$. Since c induces p-cycle on $\{G_1, \dots, G_p\}$ so

$$G_1^c \times \cdots \times G_t^c G_1^c \times G_2^c \times \cdots \times G_1^c \neq G_1 \times G_2 \times \cdots \times G_t$$

and consequently $y = y^{kc} \notin G_1^c \dots G_t^c$ which is a contradiction leading to $A \le K$.

(b) Lemma 7 implies $e_p(K) = pe_p(G)$, hence $e_p(W) \le 1 + pe_p(G)$. Let M be a subgroup of exponent p with order no less than $p^{pe_p(G)}$ and $e_p(G) > 1$. Suppose that $M \not\le K$. If $M \cap G_1 = 1$ then $M \cap K$ can be isomorphically imbedded in $G_2 \times \cdots \times G_p \cong K/G_1$. Therefore $pe_p(G) - 1 \le (p-1)e_p(G)$ which is impossible for $e_p(G) > 1$. This implies the existence of non-identical y in $M \cap G_1$. Consider $x \in M - K$, $yx \in M$. Then

$$(yx)^p = yy^{x^{-1}} \dots y^{x^{-(p-1)}}.$$

This product is not equal to 1 as a product of non-identical elements each of which lie in different G_i. The contradiction shows that $M \le K$ and $e_p(W) = e_p(G)p$.

When $e_p(G) = 1$, W has a subgroup of order p^p and of exponent p which does not belong to K.

COROLLARY 9. $a_p(S_n) = [n/p] = e_p(S_n)$.

A p-group G is called a tame group if it contains only one subgroup H with $d(H) = d_p(G)$ and H is elementary abelian. For a positive integer n and a tame p-group G the group G^n is tame also.

THEOREM 10. *Let $W = G \operatorname{wr} C_p$ with a tame p-group G, $p > 2$. Then $d_p(W) = pd_p(G)$. If $H \le W$ and $d(H) = pd_p(G)$ then $H \le K = G_1 \times \cdots \times G_p$, a base of W, $(G_i \cong G)$. In particular W is a tame.*

PROOF. It is easy to check that

$$d_p(A \times B) = d_p(A) + d_p(B).$$

So $d_p(K) = pd_p(G)$ and $d_p(W) \le 1 + pd_p(G)$. Then if M is a subgroup in W with $d(M) = 1 + pd_p(G)$ then $d(M \cap K) = pd_p(G)$ and $M \cap K$ is elementary abelian by above. If $x \in M - (M \cap K)$ then $M = \langle M \cap K, x\rangle$ is non-abelian by Lemma 8 so $|M| = p^{1+pd_p(G)}$, $d(M) \le pd_p(G)$, which is a contradiction.

Let H be a subgroup in W with $d(H) = pd_p(G)$. If $H \le K$ we are done. Let $H \not\le K$. If $d(H \cap K) = pd_p(G)$ then $H \cap K$ is elementary abelian normal subgroup by induction. In this case $H/C \cong L \operatorname{wr} C_p$ where L is elementary of order $p^{d_p(G)}$, C is cyclic, $d(H) \le 1 + d_p(G) < pd_p(G)$, which is a contradiction. Thus $d(H \cap K) = pd_p(G) - 1$.

If $d(G_1 \cap H) < d_p(G)$ then as $d(H \cap K/H \cap G_1) \ge (p-1)d_p(G)$ and K/G_1 is a tame, $H \cap K/H \cap G_1$ is elementary. In this case $d(H \cap G_i) = d_p(G)$ for $i > 1$. Let $x \in H - (H \cap K)$. Then $H = \langle H \cap K, x\rangle$. Since $(H \cap G_1)^x \le G_1^x$ and $H \cap G_1^x$ is elementary of order $p^{d_p(G)}$ then $|H \cap G_1| = p^{d_p(G)-1}$ and H is elementary of order $p^{pd_p(G)}$. This contradicts Lemma 8(a).

Hence $d(H \cap G_i) = d_p(G)$ for all i. By supposition all $H \cap G_i$ are elementary of order $p^{d_p(G)}$ and for $x \in H - K$ we have $H = \langle H \cap K, x\rangle$, $\langle H \cap K, x\rangle / x^p \cong (H \cap G_1) \operatorname{wr} C_p$ and $d(H) \le 1 + d_p(G) < pd_p(G)$, which is a contradiction. We obtain $H \le K$.

Theorem 10 is fail for $p = 2$.

COROLLARY 11. $d_p(S_n) = [n/p]$, E_n *is tame for* $p > 2$.

COROLLARY 12. *Let a finite p-group G contain a subgroup H such that* $C_G(H) \leq H$. *If* $|H| = p^n$ *then every normal abelian subgroup of G is generated by* p^{n-1} *elements.*

PROOF. Let A be a normal abelian subgroup of G, $A^0 = \langle x \in A | x^p = 1\rangle$. Since $d(A^0) = d(A)$ and A^0 is normal in G then without loss of generality we may assume $A^0 = A$. Let A^1 be a complement to $A \cap H$ in A. We have $D = AH = A^1H$. Since $C_G(H) \leq H$ then $C_{A^1}(H) = 1$ and the representation of D given by permutations of left cosets of A^1 is faithful. This means that D can be isomorphically embedded in $S_{|D:\,A^1|} = S_{|H|} = S_{p^n}$. Now our statement follows from Corollary 11.

THEOREM 13. $d_2(S_{2^m}) = 2^{m-1}$. *Specifically* $d_2(S_n) = [n/2]$.

PROOF. For $m = 1, 2$ it is obvious. Let us consider the construction of the maximal abelian subgroup $B_m \cong C_4^{2^{m-2}}$ in E_{2^m}. When $m = 3$ the base $K_3 = E_2^{(1)} \times E_2^{(2)}$ contains $B_3 = C_4^{(1)} \times C_4^{(2)}$, an abelian normal subgroup of E_8 with $C_4^{(i)} < E_4^{(i)}$, $i = 1, 2$. Then

$$E_8/B_8 \cong (E_4^{(1)}/C_4^{(1)}) \operatorname{wr} C_2 \cong E_4.$$

Suppose $E_{2^{m-1}/B_{m-1}} \cong E_{2^{m-2}}$ to be proved and let us prove $E_{2^m}/B_m \cong E_{2^{m-1}}$. We have $K_m = E_{2^{m-1}}^{(1)} \times E_{2^{m-1}}^{(2)}$ and $B_m = B_{m-1}^{(1)} \times B_{m-1}^{(2)}$ is normal in E_{2^m}, $B_{m-1}^{(i)} < E_{2^{m-1}}^{(i)}$, $i = 1, 2$, and by induction

$$E_{2^{m-1}}^{(i)}/B_{m-1}^{(i)} \cong E_{m-2}.$$

Then $E_{2^m}/B_m \cong E_{2^{m-2}} \operatorname{wr} C_2 \cong E_{2^{m-1}}$. Obviously

$$d_1(E_{2^m}) \leq d_2(B_m) + d_2(E_{2^m}/B_m) = 2^{m-1}.$$

Since $d_2(E_{2^m}) \geq 2^{2^{m-1}}$ the proof is complete.

Notice the difference in the proof when $p > 2$ and $p = 2$ (Theorems 10 and 13). A more difficult task is a description of structure of 2-subgroup H in S_n with $d(H) = [n/2]$. We assume that such a subgroup is isomorphic to the quotient-group of the direct product of dihedral groups of order 8.

LEMMA 14. *Let G be a finite p-group,* $W = G \operatorname{wr} C_p$. *Then* $\phi(W)$ *contains the diagonal of the base K of wreath product W.*

PROOF. $K = G_1 \times \cdots \times G_p$, $G_i \cong G$. If $\langle b \rangle = C_p$ then

$$D = \{xx^{b^{-1}} \ldots x^{b-(p-1)} | x \in G_1\} \text{ and}$$

$$xx^{b^{-1}} \ldots x^{b^{-(p-1)}} = (xb)^p b^{-p} \in \phi(W).$$

LEMMA 15. *Let* $n = 2^m$ *and* $E_{2^m} \in \mathrm{Syl}_2(S_{2^m})$. *Then there exist subgroups in* E_{2^m}, $\phi_1, \phi_2, \ldots, \phi_{m-1}$ *such that*

$$\phi(E_{2^m}) = \phi_{m-1}[\phi_{m-2}[\ldots[\phi_1$$

with $\phi_i \cong E_{2^i}$, $i \in \overline{1, m-1}$.

PROOF. $E_{2^m} = C_2[(E^{(1)}_{2^{m-1}} \times E^{(2)}_{2^{m-1}})$ where $E^{(i)}_{2^{m-1}} \cong E_{2^{m-1}}$. Diagonal D of the group of parentheses belongs to $\phi(E_{2^m})$ by Lemma 14. Besides that $\phi(E_{2^m}) > \phi(E^{(1)}_{2^{m-1}}) = \phi_{m-2}[\phi_{m-3}[\ldots[\phi_1$ with $\phi_i \cong E_{2^i}$. We have $D \cap (\phi(E^{(1)}_{2^{m-1}}) = 1$ and $D[\phi(E^{(1)}_{2^{m-1}}) = \phi(E_{2^m})$ (compare the orders). Since $D \cong E_{2^{m-1}}$ we obtain the desired result by induction.

THEOREM 16. $d_p(E_{p^m}) > d_p(\phi(E_{p^m}))$.

PROOF. (1) Let $n = 2^m$. By induction $d_2(\phi(E_{2^{m-1}})) < d_2(E_{2^{m-1}})$. By Lemma 15 we have $d_2(\phi(E_{2^m})) \le d_2(E_{2^{m-1}}) + d_2(\phi(E_{2^{m-1}}) < 2^{m-2} + 2^{m-2} = 2^{m-1} = d_2(E_{2^m})$.

(2) Let $n = p^m$ with $p > 2$. Let H be a subgroup in E_{p^m} with $d(H) = p^{m-1}$. Then H is the only elementary subgroup of this order and normal in E_{p^m} and $H < K$, a base of wreath product E_{p^m}. Since $E_{p^m}/H \cong E_{p^{m-1}}$ and $d(E_{p^m}) = m > m - 1 = E_{p^{m-1}}$ then $H \not\le \phi(E_{p^m})$. The proof is completed.

THEOREM 17. *Let* $m > k + 1$, $p > k$, *m and k are natural numbers. Then all the normal subgroups of exponent* p^k *of* $E_{p^m} = E_{p^{m-1}} \mathrm{wr}\, C_p$ *lie in the base K of this wreath product.*

PROOF. Suppose it is not so and let A be normal subgroup of exponent p^k in E_{p^m}, $A \not\le K$.

(1) Assume $m = k + 2$. Then $d(E_{p^m}) = m = k + 2 \le p + 1$, $E_{p^m} = KA$ and $E_{p^m}/K \cap A = K/K \cap A \times A/K \cap A \cong K/K \cap A \times C_p d(E_{p^m}/K \cap A) = 1 + d(K/K \cap A) \Rightarrow d(K/K \cap A) \le m - 1 = k + 1 \le p$. We know that $\exp(K) = p^{m-1} = p^{k+1} > p^k = \exp(A) \ge \exp(K \cap A)$. Let $K = G_1 \times \cdots \times G_p$ with $G_i \cong E_{p^{m-1}}$, A_i is a projection of A on $G_{i, i \in \overline{1,p}}$. Then $\exp(A_i) \le \exp(A) = p^k < \exp(G_i) \Rightarrow |G_i \colon A_i| \ge p$. Furthermore, A_i is normal in G_i, $A \le A_1 \times \cdots \times A_p$. Since K/K' is elementary abelian then, if $|G_i \colon A_i| > p$ for some i then $d(G_i/A_i) \ge 2$ and $D(K/A_1 \times \cdots \times A_p) \ge p + 1 \Rightarrow d(K/K \cap A) \ge p + 1 \Rightarrow d(E_{p^m}/K \cap A) \ge p + 2 \ge k + 3 > m = d(E_{p^m})$, which is a contradiction. Thus $|G_i \colon A_i| = p$ for all i. If $|G_i \colon A \cap G_i| > p$ for some i then $d(K/K \cap A) \ge p + 1$ and we obtain a contradiction again. So $G_i \cap A = A_i$ for all i and $K \cap A = A_1 \times \cdots \times A_p$. We obtain $d(E_{p^m}/K \cap A) = p + 1 \ge k + 2 = m$, $d(E_{p^m}/K \cap A) = m = d(E_{p^m})$. Since $E_{p^m}/K \cap A$ is abelian of order $p^{p+1} = p^m$ we have $K \cap A = \phi(E_{p^m})$. Since $\exp(\phi(E_{p^m})) = p^{m-1} = p^{k+1} > \exp(A) \ge \exp(K \cap A)$ we obtain a contradiction. Thus the theorem is proved for $m = k + 2$.

(2) Let now $m > k + 2$.

Let $c_1, \ldots, c_{p^{m-1}}$ be pairwise independent p-cycles in E_{p^m}, L is the subgroup generated by these cycles. Then L is elementary abelian normal subgroup in

E_{p^m}, $L < K$, $E_{p^m}/L \cong E_{p^{m-1}}$, K/L is a base of wreath product $E_{p^m}/L \cong E_{p^{m-1}} = E_{p^{m-2}} \operatorname{wr} C_p$. Let $\exp(A) = p^k$, $\exp(A/L \cap A) = p^s$. Then $s \leq k$ and $m - 1 \geq k + 2 \geq s + 2$. Since $LA/L \cong A/L \cap A$ we have by induction $LA/L < K/L \Rightarrow A < K$.

This paper is a selection from my unpublished manuscript, prepared in 1976. Theorems 2 and 6 were proved by Suschansky and Dmitruk (after acquaintance with my manuscript, their proof different from above).

Further results on subgroup structure of S_n and A_n see in short note [**4**].

References

1. R. W. Carter, P. Fong, *The Sylow 2-subgroups of the finite classical groups*, J. Algebra **1** (1964), 139–151.

2. H. Cardenas, E. LLuis, *The normalizer of the Sylow p-group of the symmetric group* S_{p^2}, An. Inst. Mat. Nac. Autonoma Mexico **2** (1962), 1–7.

3. ——, *The normalizer of the Sylow p-group of the symmetric group* S_{p^n}, Bol. Soc. Mat. Mexicana **2** (1964), 1–6.

4. Ja. G. Berkovic, *Subgroups of symmetric and alternating groups*, Izvestija Severo-Kavkazskogo Zentra Vyscei Skoli, Estestvennije Nauki, **1**(33), 1981, 6–9.

USSR, 344006
Rostov-on-Don,
pr. Engels 111, apt. 18

Contemporary Mathematics
Volume **93** (1989)

Parameters of Abelian Codes in the Group Algebra KG of $G = (a) \times (b)$, $a^p = b^p = 1$, p Prime, Over a Finite Field K with Primitive pth Root of Unity and Related MDS-codes[1)]

S. D. BERMAN AND I. I. GRUSHKO

According to **[1]**, abelian code is an ideal in the group algebra of abelian group over a finite field. If the characteristic of K does not divide the order of G, the algebra KG is semisimple and KG is a direct sum of minimal ideals

$$KG = I_1 + \cdots + I_s, \tag{1}$$

where $I_j = KGe_j$, $j = 1, \dots, s$, $M = \{e_j\}_{j=1}^{s}$—the set of minimal pairwise orthogonal idempotents. Every ideal $I \subseteq KG$ is a direct sum

$$I = I_{i_1} + \cdots + I_{i_t}, t \leq s, \tag{2}$$

and $KG = I + J$, where $J = I_{j_1} + \cdots + I_{j_{s-t}}$ is the direct sum of minimal ideals from (1) that does not enter in (2). If we denote $\mathfrak{A} = \{e_{j_1}, \dots, e_{j_{s-t}}\}$, the ideal (2) can be given by

$$I = \{u \in KG\colon ue_{j_r} = \theta \ \forall\ e_{j_r} \in \mathfrak{A}\}. \tag{3}$$

This definition, of course, is equivalent to usual in coding theory definition of a code by its parity-check matrix.

Let us recall some notations and definitions of coding theory. The dimension N of KG and the dimension K of $I \subseteq KG$ are named the *code length* and the *number of information symbols* respectively. The (Hamming) *weight* $\mathrm{wt}(u)$ of an element $u = \sum_{g \in G} \lambda_g g \in KG$ is the number of nonzero components $\lambda_g \in K$. The *code distance* $d(I_{\mathfrak{A}})$ is a minimal nonzero weight of its elements: $d(I_{\mathfrak{A}}) = \min_{0 \neq u \in I_{\mathfrak{A}}} \mathrm{wt}(u)$. A code with length N, distance d and K information symbols is named *(N, K, d)-code.* It is clear that *parameters* N, K and d of a code are related by the bound $d \leq N - K + 1$; the best code has $d = N - K + 1$ and is named *MDS*-code.

1) The paper was reported on Third International Workshop on Information Theory "Convolutional codes; multi-user communication" (USSR, Sochi, May 1987).

Our prime interest was to find the connection between the code parameters of code $I_{\mathfrak{A}}$ and the choice of the set $\mathfrak{A}$ in the case $G = (a) \times (b)$, $a^p = b^p = 1$, p prime, $1 \neq \varepsilon \in K$, $\varepsilon^p = 1$. But it turns out that the problem essentially depends on the case for the group algebra of cyclic group of order p over K.

The cyclic code of length N (the ideal in the group algebra of cyclic group of order N) is studied from the very beginning of the coding theory and is an indispensable part of any course or textbook on the topic of error-control codes. But most of the results in this direction are related to the case of "consecutive roots": there exist integer i and primitive pth root $\beta \in K$ of unity, such that all idempotents in $\mathfrak{A}$ are given by the set of characters $\chi(\mathfrak{A}) = \{\beta^i, \beta^{i+1}, \dots, \beta^{i+l-1}\}$. The main result here is the BCH-bound for code distance: $d(I_{\mathfrak{A}}) \geq l + 1$. Some recent results on code distance of cyclic codes (mainly for fields of the characteristic 2) were obtained by J. H. van Lint and R. M. Wilson [2]. We are short of the case of cyclic codes of prime length for arbitrary choice of $\mathfrak{A}$ and the characteristic.

Let $G = (a)$, $a^p = 1$, p prime, $1 \neq \varepsilon \in K$, $\varepsilon^p = 1$; then

$$M = \{e_i = (1/p)(1 + \varepsilon^{-i}a + \cdots + (\varepsilon^{-i}a)^{p-1}),\ i = 1, \dots, p\} \tag{4}$$

and

$$ue_i = \left(\sum_{l=0}^{p-1} \lambda_l \varepsilon^{il}\right) e_i \tag{5}$$

holds for $u = \sum_{l=0}^{p-1} \lambda_l a^l \in KG$.

It is easy to see that $d(I_{\{e_i\}}) = 2$ for any e_i, because there is no element $u = \lambda_g g$, $\lambda_g \neq \theta$, that could be orthogonal to any idempotent (4), and, on the other hand, the equation $1 + \lambda_g \chi_i(g^{-1}) = \theta$ always has nonzero solution, and so $u = 1 + \lambda_g g \in I_{\{e_i\}}$.

It is clear also, that if $|\mathfrak{A}| = 2$, then $d(I_{\mathfrak{A}}) = 3$. Really, let $\mathfrak{A} = \{e_i, e_j\}$, $i \neq j$. The transformation

$$\{\varepsilon^{-i}a = b, \varepsilon^{-(i-j)} = \beta\} \tag{6}$$

preserves the Hamming weight of element in KG and brings us into the known situation of "consecutive roots" $\chi(\mathfrak{A}) = \{1, \beta\}$; now BCH-bound gives $d(I_{\mathfrak{A}}) \geq 3$.

LEMMA 1. *If the characteristic of K is a primitive root modulo p, then $d(I_{\mathfrak{A}}) = |\mathfrak{A}| + 1$ for any $\mathfrak{A} \subset M$.*

PROOF. Let $\chi(\mathfrak{A}) = \{\varepsilon^{-i_1}, \varepsilon^{-i_2}, \dots, \varepsilon^{-i_t}\}$, $i_1 < i_2 < \cdots < i_t$, $t \leq p - 1$ and assume that code $I_{\mathfrak{A}}$ contains the word $u = \lambda_1 a^j + \cdots + \lambda_r a^{j_r}$ with weight $\mathrm{wt}(u) = r \leq t$: $\lambda_i \neq 0$ for all i.

First of all, notice that we have to consider only the case $r = t$, because $I_{\mathfrak{A}} \subset I_{\mathfrak{A}_1}$ for $\mathfrak{A}_1 = \mathfrak{A} \backslash \{e_i\}$ for any $e_i \in \mathfrak{A}$, so the induction can be used for $r < t$ (the cases $t = 1, 2$ are done above).

According to (5), the orthogonality of u to the idempotents from $\mathfrak{A}$ is equivalent to the linear system:

$$(7)\qquad \begin{cases} \lambda_1\varepsilon^{i_1j_1}+\cdots+\lambda_t\varepsilon^{i_1j_r}=0,\\ \qquad\cdots\\ \lambda_1\varepsilon^{i_tj_1}+\cdots+\lambda_t\varepsilon^{i_tj_t}=0.\end{cases}$$

Let us denote $\varepsilon^{j_s}=z_s$, $s=1,\dots,t$. By assumption the system (7) has a nonzero solution $(\lambda_1,\dots,\lambda_t)$, so the equation

$$(8)\qquad \begin{vmatrix} z_1^{i_1} & \dots & z_t^{i_1}\\ & \dots & \\ z_1^{i_t} & \dots & z_t^{i_t}\end{vmatrix}=0$$

holds and there exists a set of nonzero coefficients $\{c_j\}_{j=1}^t$ such that $\sum_{j=1}^t c_jz_s^{i_j}$ for $s=1,\dots,t$.

Denoting $f(x)=\sum_{j=1}^t c_jx^{i_j}$, we get $f(z_s)=0$ for $s=1,\dots,t$ and the assumption of the characteristic of K implies $(1+x+x^2+\cdots+x^{p-1})|f(x)$. As $i_1<i_2<\cdots<i_t\le p-1$, there are two possibilities: (1) $f(x)\equiv 0$; (2) $f(x)=\text{const}(1+x+\cdots+x^{p-1})$. The first one contradicts to linear dependence of rows of (8) and our assumption of existing in $I_{\mathfrak{A}}$ the word with weight t. The second one converts (8) into Vandermonde-determinant and the contradiction arises again.

REMARKS. (1) Lemma 1 gives the new family of cyclic MDS-codes. The best investigated and widely spread family of MDS-codes is the family of Reed-Solomon codes and their shortening; q-ary RS-codes are constructed by the set of consecutive roots in $GF(q)$. Another example of MDS-codes is radical p-codes in KG for $G=(a)$, $a^p=1$, over the field K of the characteristic p **[3]**.

(2) It seems that the condition of the characteristic of K in Lemma 1 is very strong. There are many cases in which it does not need. But the following example shows that in general case it is necessary.

Let $|\mathfrak{A}|=3$, $\mathfrak{A}=\{1,\varepsilon,\varepsilon^i\}$, $2\le i\le p-1$. If $i=2$, without any restrictions on the characteristic of K the BCH-bound tells that $d(I_{\mathfrak{A}})\ge 4$. Let $i>2$ and $I_{\mathfrak{A}}$ contain the codeword $v=\lambda_1a^{l_1}+\lambda_2a^{l_2}+\lambda_3a^{l_3}$, $\lambda_i\ne\theta$, $i=1,2,3$, $l_1<l_2<l_3\le p-1$. Then $I_{\mathfrak{A}}$ contains the codeword $u=\lambda_1^{-1}a^{-l_1}v=1+\lambda a^s+\mu a^t$, $\lambda\ne 0$, $\mu\ne 0$, $1\le s,t\le p-1$, $\mathrm{wt}(u)=3$, and according to (5) the conditions of orthogonality of u to the idempotents from $\mathfrak{A}$ give the linear system

$$\begin{cases}\lambda(1-\varepsilon^s)+\mu(1-\varepsilon^t)=0,\\ \lambda(1-\varepsilon^{is})+\mu(1-\varepsilon^{it})=0.\end{cases}$$

As $1-\varepsilon^s\ne 0$ and $1-\varepsilon^t\ne 0$ for $s,t\le p-1$, the equation

$$\begin{vmatrix} 1 & 1\\ 1+\varepsilon^s+\cdots+\varepsilon^{(i-1)s} & 1+\varepsilon^t+\cdots+\varepsilon^{(i-1)t}\end{vmatrix}=0$$

takes place. The transformation $\{\varepsilon^s=\beta, t=sl\}$ converts the left part into $\varphi(\beta)=(\beta+\beta^2+\cdots+\beta^{i-1})-(\beta^l+\beta^{2l}+\cdots+\beta^{l(i-1)})$, where β is a primitive

pth root of unity in K and $l \neq 1$. The equality $\varphi(\beta) = 0$ means that either $\varphi(\beta) \equiv 0 \pmod{\operatorname{char} K}$, or $\varphi(\beta) = 0 \pmod{\mathrm{m}_\beta(\beta)}$, where $\mathrm{m}_\beta(x)$ is minimal polynomial of the element β. The first congruence is equivalent to coincidence in $\mathbf{Z}_p$ the sets $\{1, 2, \dots, i-1\}$ and $\{l, 2l, \dots, (i-1)l\}$ for $2 \leq i < p$. The last is possible only for $l = 1$ and we receive the contradiction to the assumption $\operatorname{wt}(u) = 3$ $(s \neq t)$.

The second congruence can take place in many cases. For $i = 3$, for example, it becomes

$$\varphi(x) = x + x^2 - (x^l + x^{2l}) = x(1 - x^{l-1})(1 + x + x^l) \equiv 0 \pmod{\mathrm{m}_\beta(x)}$$

and if $\operatorname{char}(K) = 2$ the congruence is satisfied in infinite number of cases, the first ones of which are $(p, l) = (3,2), (7,3), (61,6), (73,9), (127,7)$.

Now let $G = (a) \times (b)$, $a^p = b^p = 1$, $1 \neq \varepsilon \in K$, $\varepsilon^p = 1$; then

$$\text{(9)} \qquad M = \begin{cases} e_k = (1/p^2) \sum_{g \in G} \chi_k(g) g = (1/p^2) \left(\sum_{s=0}^{p-1} \varepsilon^{-is} a^s \right) \left(\sum_{t=0}^{p-1} \varepsilon^{-jt} b^t \right) \\ (k = 1, \dots, p^2;\ i, j = 1, \dots, p) \end{cases}$$

and code $I_{\mathfrak{A}}$ is given by the set $\mathfrak{A} \subset M$, or, equivalently, by the set $\chi(\mathfrak{A})$ of corresponding characters. Recall that

$$\text{(10)} \qquad u e_k = \left(\sum_{g \in G} \lambda_g \chi_k(g^{-1}) \right) e_k$$

for $u = \sum_g \lambda_g g \in GK$.

It is easy to see that $G = \bigcup_{i=0} H_i$, where

$$\text{(11)} \qquad H_0 = (a), H_{i+1} = (a^i b),\ i = 0, \dots, p-1,\ H_i \cap H_j = \{1\},\ i \neq j.$$

LEMMA 2. *There is one and only one subgroup* (11) *such that any pair* (e_1, e_2) *of different idempotents* (9) *induces on it the same character.*

PROOF. Let $\chi(\mathfrak{A}) = \{(i_1, j_1), (i_2, j_2)\}$ and $(i_1, j_1) \neq (i_2, j_2)$. If $i_1 = i_2$, then the subgroup is H_0, and only this one because $j_1 \neq j_2$. If $i_1 \neq i_2$, the number i of the subgroup is defined by $i = x + 1$, where x is (unique in Z_p) solution of congruence $x i_1 + j_1 \equiv x i_2 + j_2 \pmod{p}$.

Now let us return to the problem "What is the connection between $\mathfrak{A}$ and $d(I_{\mathfrak{A}})$?". By the same reason, as in case of cyclic group, it is clear that again $d(I_{\{e_k\}}) = 2$ for any $e_k \in M$. Let $\mathfrak{A} = \{e_1, e_2\}$. According to Lemma 2 there is the subgroup H_i such that $\chi_1(h) = \chi_2(h) = \varphi(h)$ for all $h_1, h_2 \in H_i$, $h_1 \neq h_2$; then element $u = h_1 - \varphi(h_1^{-1} h_2) h_2$ is a word of weight 2 in code $I_{\mathfrak{A}}$.

Thus for $G = (a) \times (b)$, $a^p = b^p = 1$, an ideal $I_{\mathfrak{A}} \subset KG$ can not be MDS-code; the best possible value for code distance is $d(I_{\mathfrak{A}}) - |\mathfrak{A}|$.

Arising on a pair of idempotents from $\mathfrak{A}$ subgroup is called *associated with* $\mathfrak{A}$.

THEOREM 1. *For* $|\mathfrak{A}| = 3$ *there is only one of two possibilities:* (1) *all associated with* $\mathfrak{A}$ *subgroups are different and in this case* $d(I_{\mathfrak{A}}) = 3$; (2) *all associated with* $\mathfrak{A}$ *subgroups coincide and in this case* $d(I_{\mathfrak{A}}) = 2$.

PROOF. Let $\mathfrak{A} = \{e_1, e_2, e_3\}$ and H_{12}, H_{13}, and H_{23} are associated with $\mathfrak{A}$ subgroups (11).

(1) The order of any non-unity element in G is p and all associated with $\mathfrak{A}$ subgroups are different. So, using, if necessary, a transformation of the type $\{\varepsilon^i a \to a, \varepsilon^j b \to b\}$ and replacement of prime pth root of unity, we can take the set $\mathfrak{A}$ in the form

$$\tag{12} \begin{aligned} \mathfrak{A} = \{e_1 &= (1/p^2)(1 + a + \cdots + a^{p-1})(1 + b + \cdots + b^{p-1}), \\ e_2 &= (1/p^2)(1 + a + \cdots + a^{p-1})(1 + \varepsilon^{-1} b + \cdots + (\varepsilon^{-1} b)^{p-1}), \\ e_3 &= (1/p^2)(1 + \varepsilon^{-1} a + \cdots + (\varepsilon^{-1} a)^{p-1})(1 + b + \cdots + b^{p-1})\}. \end{aligned}$$

Since $I_{\mathfrak{A}} \subset I_{\{e_1,e_2\}}$ it is clear that $d(I_{\mathfrak{A}}) \geq 2$, and we have to show that in $I_{\mathfrak{A}}$ there are no elements of weight 2. Let us assume to the contrary that $u = 1 + \lambda a^k b^l \in I_{\mathfrak{A}}$ for $(k, l) \neq (\theta, 0)$. According to (10) the orthogonality conditions take the form of the linear system $\{1 + \lambda = 0, 1 + \lambda\varepsilon^l = 0, 1 + \lambda\varepsilon^k = 0\}$, that under the restrictions $(k, l) \neq (0, 0)$, $0 \leq k \leq l \leq p - 1$ has only zero solution.

In fact $d(I_{\mathfrak{A}}) = 3$ for $p \geq 3$. Really, let us recall that the *support* of an element $u = \sum_{g \in G} \lambda_g g \in KG$ is $\{g \in G\colon \lambda_g \neq 0\}$, and consider the elements from $I_{\mathfrak{A}}$ with support in $H_0 \subset G$. Drop out zero components of words for $g \notin H_0$. Such a restriction of code $I_{\mathfrak{A}}$ produces the cyclic code $I_{\mathfrak{A}'} \subset KH_0$ with $\mathfrak{A}' = \{e_1' = e_2' = (1/p^2)(1 + a + \cdots + a^{p-1}), e_3' = (1/p^2)(1 + \varepsilon^{-1} a + \cdots + (\varepsilon^{-1} a)^{p-1})$, thus $d(I_{\mathfrak{A}}) = 3$ by Lemma 1. The inverse image of any word $u \in I_{\mathfrak{A}'}$ of weight 3 is a codeword of weight 3 in $I_{\mathfrak{A}}$.

The above construction of the restriction of code $I_{\mathfrak{A}} \subset KG$ onto the group algebra KH_0 will be referred to in the following as *cyclic restriction* onto KH_0.

(2) Primarily assume the coincidence of some two of three associated with $\mathfrak{A}$ subgroups, for example, $H_{12} = H_{13}$, and then show the coincidence in all of them. Actually, according to the definition of the associated subgroup and Lemma 2 the assumption $H_{12} = H_{13}$ is followed by the chain of equalities $\chi_3(H_{13}) = \chi_1(H_{13}) = \chi_1(H_{12}) = \chi_2(H_{13})$, and, thus, $H_{12} = H_{13} = H_{23}$. Hence, transforming, if necessary, the generators of direct summands and choosing a primitive pth root of unity in a proper way, we can suppose that

$$\begin{aligned} \mathfrak{A} = \{e_1 &= (1/p^2)(1 + a + \cdots + a^{p-1})(1 + b + \cdots + b^{p-1}), \\ e_2 &= (1/p^2)(1 + a + \cdots + a^{p-1})(1 + \varepsilon^{-i} b + \cdots + (\varepsilon^{-i} b)^{p-1}), \\ e_3 &= (1/p^2)(1 + a + \cdots + a^{p-1})(1 + \varepsilon^{-j} b + \cdots + (\varepsilon^{-j} b)^{p-1})\}. \end{aligned}$$

Now it is clear that $d(I_{\mathfrak{A}}) = 2$ (for example, $u = 1 - a^t \in I_{\mathfrak{A}}$) and the theorem is proved.

LEMMA 3. *Difference of all* $\binom{|\mathfrak{A}|}{2}$ *associated with* $\mathfrak{A}$ *subgroups is a necessary condition for existence* $(p^2, p^2 - |\mathfrak{A}|, |\mathfrak{A}|)$*-code in the inspected class.*

PROOF. The case $|\mathfrak{A}| = 3$ is contained in Theorem 1. Let $|\mathfrak{A}| \geq 4$ and denote by $\mathfrak{A}_{\mathfrak{H}}$ the set of characters that $\mathfrak{A}$ induces on $\mathfrak{H}$. Select e_1, e_2, e_3, e_4 from $\mathfrak{A}$ arbitrary and denote their associated subgroups by H_{ij}, $1 \leq i < j \leq 4$. According to Theorem 1, subgroups H_{12}, H_{13} and H_{23} are different. Assume now, for example, that $H_{14} = H_{12}$. Then $|\mathfrak{A}_{\mathfrak{H}_{12}}| \leq |\mathfrak{a}| - 2$. Hence, if we take the cyclic restriction of $I_{\mathfrak{A}}$ onto KH_{12}, will have, according to Lemma 1, a codeword of the weight $\leq |\mathfrak{A}| - 1$. Arbitrariness in the choice of idempotents from $\mathfrak{A}$ completes the proof.

Lemma 3 shows the lower bound for p in looking for good codes in the class. For example, there is no $(9, 5, 4)$-code in it, because for $p = 3$ there are only $p + 1 = 4 < \binom{4}{2} = 6$ subgroups (11). In the same reason $p = 11$ is minimal prime, when the existence of $(121, 116, 5)$-code in the class is possible.

LEMMA 4. *If the characteristic of K is a primitive root modulo p, then a support of any "weight-3" word in $(p^2, p^2 - 3, 3)$-code $I_{\mathfrak{A}}$ contains the elements only from one associated with $\mathfrak{A}$ subgroup.*

PROOF. As in the proof of Theorem 1 we can suppose that the set $\mathfrak{A}$ is given by (12). Also, it is easy to see that it is sufficient to analyse a weight-3 element of the type $u = 1 + \lambda(a^i b)^j + \mu(a^k b)^l$, $\lambda \neq 0$, $\mu \neq 0$, $(i, j) \neq (k, l)$, $j \neq 0$, $l \neq 0$. Then the orthogonality conditions lead us to the linear system

$$\begin{cases} \lambda(\varepsilon^j - 1) + \mu(\varepsilon^l - 1) = 0, \\ \lambda(\varepsilon^{ij} - 1) + \mu(\varepsilon^{kl} - 1) = 0, \end{cases}$$

where $j \neq 0$, $l \neq 0$. The determinant is zero if and only if

$$(1 + \varepsilon^j + \varepsilon^{2j} + \cdots + \varepsilon^{(i-1)j}) - (1 + \varepsilon^l + \varepsilon^{2l} + \cdots + \varepsilon^{(k-1)l}) = 0. \tag{13}$$

By assumption the characteristic of K is a primitive root modulo p, so the ideal $(1 - \varepsilon)$ is maximal in $Z[\varepsilon]$ and $Z[\varepsilon]/(1 - \varepsilon) \cong Z_p$. This implies that modulo $(1 - \varepsilon)$ the equation (13) is reduced to congruence $k \equiv i \pmod{p}$, and, because $k, i \in Z_p$, to identity $k = i$.

Thus for a weight-3 element $u \in I_{\mathfrak{A}}$ the support contains the elements only from one subgroup, say, $H_t = (a^t b)$. If the H_t is not associated with given $\mathfrak{A}$, then all idempotents $e_j \in \mathfrak{A}$ induce on it three different characters, and so, by Lemma 1, the code distance of the cyclic restriction $I_{\mathfrak{A}}$ onto KH_t is 4 and we receive a contradiction to $\mathrm{wt}(u) = 3$. Thus, H_t is associated with subgroup and the lemma is proved.

REMARK (3). The assumption on the characteristic of K is essential. For example, if $K = GF(11)$ and $p = 5$, then for $\varepsilon = 4$ and $\mathfrak{A}$ given by (12) the element $u = 1 + \lambda a^3 b + \mu(a^2 b)^3$ is a codeword of weight 3 in $I_{\mathfrak{A}}$.

THEOREM 2. *Let $p \geq 5$, the characteristic of K is primitive modulo p and all associated with $\mathfrak{A} = \{e_1, e_2, e_3, e_4\}$ subgroups are different; then $d(I_{\mathfrak{A}}) = 4$.*

PROOF. $I_{\mathfrak{A}} \subset I_{\{e_i, e_j, e_k\}}$ for any triple $\{i, j, k\} \subset \{1, 2, 3, 4\}$; so $d(I_{\mathfrak{A}}) \geq d(I_{\{e_i, e_j, e_k\}}) = 3$ by Theorem 1. Thus we have only to show that $I_{\mathfrak{A}}$ does not contain words of weight-3. Let us, in contradiction, assume that $u \in I_{\mathfrak{A}}$ and $\mathrm{wt}(u) = 3$. Then, for example, $ue_i = 0$, $i = 1, 2, 3$, and $u \in I_{\{e_1,e_2,e_3\}}$. By Lemma 4 the support of u contains the elements only from one associated with $\{e_1, e_2, e_3\}$ subgroups, for definiteness, from H_{12}. If $e_1(e_2)$ and e_3 induce on H_{12} the characters $\chi_1(= \chi_2)$ and χ_3 correspondingly, then because all associated with $\mathfrak{A}$ subgroups are different, e_4 induces on H_{12} the character $\chi_4 \neq \chi_1, \chi_3$. Thus, by Lemma 1, the code distance of the cyclic restriction of $I_{\mathfrak{A}}$ onto KH_{12} is four and $u \notin I_{\{e_1,e_2,e_3\}}$. The proof is ended.

The first code in the class with $d = 4$ is $(25, 21, 4)$-code over $GR(16)$; for comparison, the shortened BCH-code over $GF(16)$ is $(25, 20, 4)$-code.

References

1. Берман С. Д. Полупростые и абелевы коды. II. “Кибернетика”, No. 3, 1967, 21–30.
2. J. H. van Lint and R. M. Wilson, *On the minimum distance of cyclic codes*, IEEE Trans. on Inform. Theory, IT-32, Jan. 1986, 23–40.
3. Берман Ц. Д. К теории групповых кодов. “Кибернетика”, No. I, 1967, 31–39.

Contemporary Mathematics
Volume **93** (1989)

Locally Cyclic Modules Over a Valuation Ring

S. D. BERMAN AND N. I. VISHNYAKOVA

In this article we give a complete characterization of all locally cyclic modules—and, in particular, all faithful and all divisible locally cyclic modules—over a valuation ring of countable type. We also characterize all locally cyclic modules over an arbitrary almost maximal valuation ring and all indecomposable injective modules over such a ring.

We shall use the following concepts and definitions.

DEFINITION. By a valuation ring we mean a commutative ring R with unit having more than one element, all of whose ideals are linearly ordered, i.e., for any two ideals $\mathscr{I}$, $\mathscr{T}$ of R, one is always contained in the other.

A valuation ring is obviously a local ring, and every finitely generated ideal of a valuation ring R is principal, i.e., R is a so-called "arithmetic ring".

Let V be the maximal ideal of a valuation ring R. Every ideal $\mathscr{I} \subseteq V$ of R which is not of the form aV $(a \in R)$ can be represented as the intersection of all principal ideals which properly contain it. If a valuation ring has the property that such ideals $\mathscr{I}$ can be represented as an intersection of countably many principal ideals, then it is called a valuation ring of countable type.

DEFINITION. A valuation ring R is said to be maximal (almost maximal) if, for any family of ideals $\{\mathscr{I}_\alpha\}$ (any family of ideals $\{\mathscr{I}_\alpha\}$ for which $\cap \mathscr{I}_\alpha \neq 0$), a system of congruences $x \equiv x_\alpha \pmod{\mathscr{I}_\alpha}$, where $x_\alpha \in R$, always has a solution in R, provided that any two of the congruences are compatible (see **[1]**).

In **[4]** it is shown that a valuation ring which is almost maximal but not maximal is always an integral domain.

We shall always use the letter R to denote a valuation ring, and the letter V to denote its maximal ideal. In what follows, unless the contrary is specified, we shall assume that R is not a noetherian ring. This means that the ideal V either is non-principal, or else is a principal ideal for which $\bigcap_{i=1}^{\infty} V^i \neq 0$.

1980 *Mathematics Subject Classification* (1985 *Revision*). Primary 13C05, 13C13; Secondary 13H99.

In R every ideal is a union of a (transfinite) non-decreasing sequence of principal ideals.

We shall use the notation $\subset$, $\subseteq$ for set inclusions.

DEFINITION. An element $r \in R$, $r \neq 0$, is called regular if r is not a divisor of zero. The irregular elements of R form a prime ideal A of R. If $A = 0$, then R is an integral domain. The ideal A contains the prime radical of R, which is the ideal B consisting of all nilpotent elements of the ring. It is easy to see that B is a prime ideal, since the intersection of any family of prime ideals of a valuation ring is a prime ideal.

DEFINITION. By a ring of the first kind we mean a valuation ring R all of whose non-invertible elements are divisors of zero (in such a ring we have $A = V$). By a ring of the second kind we mean a ring of the first kind with non-principal ideal V containing the minimal ideal.

DEFINITION. By the stabilizer of an ideal $\mathcal{J} \neq 0$ of R (denoted $S(\mathcal{J})$), we mean the set of all $t \in R$ for which $t\mathcal{J} = \mathcal{J}$. The stabilizer $S(\mathcal{J})$ is obviously a subsemigroup of the multiplicative semigroup of R which does not contain zero.

We shall let $\operatorname{Ann}\mathcal{J}$ denote the annihilator of a subset $\mathcal{J} \subseteq R$.

If $\mathcal{J}$ is an ideal and $\mathcal{J} \subset (a)$, then we let $\mathcal{J}/a$ denote the ideal $\mathcal{T} = \{b \in R, ba \in \mathcal{J}\}$. It is easy to prove that, if $a\mathcal{J} \neq 0$, then

$$\operatorname{Ann} a\mathcal{J} = \frac{\operatorname{Ann}\mathcal{J}}{a}. \tag{1}$$

DEFINITION. Let $\mathcal{J}$ and $\mathcal{T}$ be nonzero ideals of a valuation ring R. We say that these ideals belong to the same class (the same regular class) if there exist elements $a, b \in R$ (regular elements $a, b \in R$) such that $a\mathcal{J} = b\mathcal{T} \neq 0$ ($a\mathcal{J} = b\mathcal{T}$).

The set of all nonzero ideals of R is partitioned into disjoint subsets of ideals, each consisting of ideals in the same class. In fact, suppose that $a\mathcal{J} = b\mathcal{T}$ and $d\mathcal{T} = cT$, where $\mathcal{J}$, $\mathcal{T}$ and T are ideals of R which $b\mathcal{T} \neq 0$ and $d\mathcal{T} \neq 0$. To be definite, we write $d = bs$. Then $cT = as\mathcal{T}$.

A nonzero ideal always forms a regular class. If R is a valuation ring which is not an integral domain, then it follows from (1) that the nonzero annihilators of the nonzero elements of the ring all lie in the same ideal class. We adopt the convention that the zero ideal also belongs to this class. If R is an integral domain, then it is natural to introduce the ideal class consisting only of the zero ideal.

In the sequel we shall make use of the following lemmas, which describe properties of valuation rings and their ideals.

LEMMA 1. *Let R be a valuation ring. If $a, b, c \in R$ and if $ab = ac \neq 0$, then $b = \theta c$, where θ is a unit of R.*

COROLLARY. *All of the nonzero ideals $\mathcal{J}$ in a given ideal class have the same stabilizer $S(\mathcal{J})$.*

LEMMA 2. *Let $\mathcal{J}$ be a non-principal ideal of R, and let T be an ideal such that $L = \mathcal{J} \cdot T \neq 0$. Then L is a non-principal ideal. The product $\mathcal{J} V$ coincides with $\mathcal{J}$.*

LEMMA 3. *A proper ideal $\mathcal{J} \neq 0$ of a valuation ring R is prime if and only if $S(\mathcal{J}) = R \backslash \mathcal{J}$.*

LEMMA 4. *A subsemigroup S of the multiplicative semigroup of the ring R not containing 0 is the stabilizer of a nonzero ideal $\mathcal{J} \subseteq V$ if and only if S is the complement in R of a nonzero prime ideal P of R.*

All of the above lemmas are proved in [**2**].

REMARK. According to Lemma 3, the stabilizer of a prime ideal $P \neq 0$, $P \subseteq V$, is $R \backslash P$.

From Lemma 4 it follows that a nonzero ideal $\mathcal{J} \subseteq V$ corresponds to a uniquely determined prime ideal $P, \mathcal{J} \subseteq P \subseteq V$, such that $S(\mathcal{J}) = S(P) = R \backslash P$. We shall call P the prime ideal corresponding to $\mathcal{J}$, and shall denote it $P(\mathcal{J})$.

DEFINITION. Suppose that $\mathcal{J}$ is a nonzero ideal of R which is contained in the prime ideal $P \subset R$. We shall say that $\mathcal{J}$ is P pure if $S(P) \subseteq S(\mathcal{J})$. If A is the ideal of R consisting of all irregular elements, then an A-pure ideal $\mathcal{J} \subseteq A$ will also be called regularly pure.

It is easy to see that the intersection of any family of P-pure ideals either is zero or else is a P-pure ideal. Hence, given any nonzero ideal $\mathcal{J} \subseteq P$, there exists a uniquely determined minimal P-pure ideal $\mathcal{J}'$ which contains it. $\mathcal{J}'$ is the intersection of all P-pure ideals $\mathcal{T} \supseteq \mathcal{J}$. It is not hard to show that $\mathcal{J}'$ is the union of all of the ideals $\mathcal{J}/a$, where a runs through the elements of $R \backslash P = S(P)$ (see Lemma 3).

DEFINITION. The ideal $\mathcal{J}'$ defined above will be called the P-pure closure of $\mathcal{J}$. An ideal $\mathcal{J} \subseteq P$ which is the P-pure closure of a principal ideal $(a) \subset \mathcal{J}$ $(a \neq 0)$ will be called a P-principal ideal, and will be denoted $(a)_P$. The A-principal ideals will also be called regularly principal.

Let P be a prime ideal of R. As usual, we let R_P denote the ring of fractions of R relative to the multiplicative system $R \backslash P$. R_P is the ring consisting of all fractions of the form a/r, where $r \notin P$. The map $\varphi \colon a \mapsto a/1$ is a ring homomorphism from R to R_P, which is not, in general, injective: it is well known that the kernel is the union of the annihilators of all the elements $a \notin P$. If $P = A$ is the set of all irregular elements of R, then φ is an injective map, and R may be regarded as a subring of R_P.

The following two lemmas are proved in [**2**].

LEMMA 5. *Suppose that R is not an integral domain, and P is a prime ideal of R different from A (A consists of all of the irregular elements of the ring). Then the ideal* Ann P *is a P-principal ideal, and it is the minimal P-pure ideal (i.e.,* Ann P *is contained in every nonzero P-pure ideal). The ideal* Ann P *is the kernel of the natural homomorphism $R \to R_P$.*

REMARK. Let P be a prime ideal of R. Let $\overline{\mathcal{J}}$ denote the holomorphic image in R_P of the ideal $\mathcal{J} \subseteq R$. From Lemma 5 it follows that the P-pure (P-principal) ideals of R can be characterized as the nonzero ideals whose images in R_P are ideals (principal ideals) of that ring. (For the P-principal ideal $Q = \operatorname{Ann} P$, the ideal $\overline{Q}$ is equal to zero.) Here, again using Lemma 5, we see that the P-pure ideals $\mathcal{J}$ of R are the preimages of the ideals $\overline{\mathcal{J}}$.

LEMMA 6. *For any ideal* $\mathcal{J} \subseteq V$, $\mathcal{J} \neq 0$,

$$L = \bigcup_{\lambda \notin \mathcal{J}} \mathcal{J}/\lambda = R \backslash S(\mathcal{J}) = P(\mathcal{J}).$$

We now prove several lemmas about the annihilator of an element or ideal in a valuation ring.

LEMMA 7. *The annihilator of any nonzero ideal* $\mathcal{J}$ *of* R *is either regularly pure or else equal to zero.*

PROOF. Since the intersection of any family of regularly pure ideals is either regularly pure or else equal to zero, it suffices to prove the lemma for the annihilator of a nonzero element of the ring. Let $a \in R$, $a \neq 0$. If a is a regular element, then $\operatorname{Ann} a = 0$. Suppose that a is an irregular element. Let $\mathcal{T} = \operatorname{Ann} a$, and let $\mathcal{T}'$ be the regularly pure closure of $\mathcal{T}$. If $c \in \mathcal{T}'$, then $b = ct$, where $b \in \mathcal{T}$ and t is a regular element. We have $(tc)a = t(ca) = 0$. Since t is regular, we conclude that $ca = 0$ and $c \in \mathcal{T}$. Hence $\mathcal{T} = \mathcal{T}'$, i.e., $\operatorname{Ann} a$ is a regularly pure ideal. The lemma is proved.

The next results about annihilators will first be proved for valuation rings which do not contain nontrivial regular elements, where we do not assume that R is necessarily non-noetherian. If such a ring has a non-principal ideal $\mathcal{J} \subseteq V$, then we call it a ring of the first kind.

We shall use the following formula, proved in **[3]**, which holds for any commutative ring R and any ideal $\mathcal{T}$ of R:

$$\operatorname{Ann} \operatorname{Ann} \operatorname{Ann} \mathcal{T} = \operatorname{Ann} \mathcal{T}. \tag{2}$$

LEMMA 8. *If* R *does not contain nontrivial regular elements, then for any element* $a \in R$

$$\operatorname{Ann} \operatorname{Ann} a = (a). \tag{3}$$

The proof of (3) is given in **[4]**.

LEMMA 9. *Let* R *be a valuation ring with trivial regular elements. If* R *is a ring of the second kind or a ring with principal maximal ideal* V, *then for any ideal* $\mathcal{J} \subseteq R$

$$\operatorname{Ann} \operatorname{Ann} \mathcal{J} = \mathcal{J}. \tag{4}$$

If R *is a ring of the first kind which is not of the second kind, then*

$$\operatorname{Ann} \operatorname{Ann} \mathcal{J} = \begin{cases} \mathcal{J}, & \text{if } \mathcal{J} \neq aV \ (a \neq 0) \\ (a), & \text{if } \mathcal{J} = aV. \end{cases}$$

PROOF. If $\mathscr{I}$ is a principal ideal, then the result follows from Lemma 8. Now suppose that $\mathscr{I}$ can be represented as an intersection of principal ideals $\mathscr{I} = \bigcap_i (\lambda_i)$. We set $L = \bigcup_i \operatorname{Ann} \lambda_i$. Then $\mathscr{I} = \operatorname{Ann} L$. In fact, if $\gamma \in \operatorname{Ann} L$, then $\gamma \operatorname{Ann} \lambda_i = 0$ for all i. Hence, by (3), we have $\gamma \in (\lambda_i)$ $(i = 1, 2, \dots)$, and $\gamma \in \bigcap(\lambda_i) = \mathscr{I}$. On the other hand, $\mathscr{I} L = 0$, since, if $\gamma \in \mathscr{I}$, then $\gamma = \lambda_i \mu_i$ $(i = 1, 2, \dots)$, and we have $\gamma \operatorname{Ann} \lambda_i = \mu_i \lambda_i \operatorname{Ann} \lambda_i = 0$ $(i = 1, 2, \dots)$. By (2), we now obtain: $\operatorname{Ann}\operatorname{Ann} \mathscr{I} = \operatorname{Ann}\operatorname{Ann}\operatorname{Ann} L = \mathscr{I}$. Since aV is a principal ideal when $V = (c)$, to finish the proof it remains only to consider ideals $\mathscr{I} = aV \neq 0$ with V non-principal. If $\operatorname{Ann} V = 0$, then the equality $\gamma a V = 0$ implies that $\gamma a = 0$, and hence $\operatorname{Ann}\operatorname{Ann} aV = (a)$. Let $\mathscr{I} = aV$ $(a \neq 0)$ be an ideal in a ring R of the second kind with minimal ideal (b). Then $\operatorname{Ann} b = V$, $\operatorname{Ann} V = (b)$, and, if $b = as$, then (1) gives us $\operatorname{Ann} a = sV$, and, by (3), $\operatorname{Ann} sV = (a)$. We then have

$$\operatorname{Ann}\operatorname{Ann} aV = \operatorname{Ann} s = aV \ (b = as, a \neq 0).$$

The lemma is proved.

LEMMA 10. *Let $a, b \in R$ and $b = at$. The ideals* $\operatorname{Ann} a$ *and* $\operatorname{Ann} b$ *coincide if and only if t is a regular element.*

PROOF. Let $\operatorname{Ann} b = \operatorname{Ann} a$. Then

$$(b) \cap \operatorname{Ann} t = 0. \tag{5}$$

In fact, suppose that $(bc)t = 0$ and $bc \neq 0$. Since $\operatorname{Ann}(bc) = \operatorname{Ann} b/c = \operatorname{Ann} a/c = \operatorname{Ann}(ac)$, it follows that $act = 0$; but, on the other hand, $act = bc \neq 0$, which gives a contradiction. If b is regular, then t is also a regular element. But if b is irregular, then it follows from (5) that t is again regular. Conversely, if t is regular, then the elements a and b obviously have the same annihilator. The lemma is proved.

LEMMA 11. *Let R be a valuation ring which is not an integral domain; let A be the ideal of R consisting of all of the irregular elements, let $\mathscr{I} \subseteq A$ $(\mathscr{I} \neq 0)$ be an ideal of R, and let $\mathscr{I}'$ be the regularly pure closure of $\mathscr{I}$. If* $\operatorname{Ann} A \neq 0$, *then*

$$\operatorname{Ann}\operatorname{Ann} \mathscr{I} = \mathscr{I}'.$$

If $\operatorname{Ann} A = 0$ *and $\mathscr{I} \subseteq A$ for $A \neq V$, then*

$$\operatorname{Ann}\operatorname{Ann} \mathscr{I} = \begin{cases} \mathscr{I}', & \text{if } \mathscr{I}' \neq aA,\ a \neq 0 \\ (a)_A, & \text{if } \mathscr{I}' = aA, \end{cases}$$

where $(a)_A$ is the regularly pure closure of the ideal (a).

PROOF. Let $\operatorname{Ann} \mathscr{I} = \mathscr{T}$. Then $\operatorname{Ann} \mathscr{I}' = \mathscr{T}$, by Lemma 10. Let T be the full ring of fractions of R. Then the ideals $\mathscr{I}'$ and $\mathscr{T}$ are also ideals of T (see Lemma 7). Here $\mathscr{T}$ is clearly the annihilator of $\mathscr{I}$ in T as well, and A is the maximal ideal of T.

If A is a principal ideal of T (i.e., A is an A-principal ideal) or if $\operatorname{Ann} A \neq 0$ and $A \neq (a)_A$ (i.e., T is a ring of the second kind), then, by Lemma 9, $\operatorname{Ann} \mathscr{T} = \mathscr{I}'$ in the ring T, and hence also in R.

Suppose that T is a ring of the first kind but not of the second kind. If $A = V$, then the lemma is obviously true. Let $A \neq V$ and $\mathscr{I} \subset A$. Since A is not a principal ideal, we have $\mathscr{I} \subset (a) \subset A$ for some principal ideal (a), and hence $\operatorname{Ann} \mathscr{I} \neq 0$. Every maximal ideal in a nonzero principal ideal (a) of T (where $a \in A$) has the form aA. Thus, if $\mathscr{I}' = aA$, then $\operatorname{Ann} \mathscr{T} = (a)_A$; and if $\mathscr{I}' \neq aA$, $a \neq 0$, then $\operatorname{Ann} \mathscr{T} = \mathscr{I}'$. The lemma is proved.

COROLLARY. *If $\mathscr{I}$ is a nonzero ideal, $\mathscr{I} \subset A$, then* $\operatorname{Ann} \mathscr{I} \neq 0$ *and* $S(\mathscr{I}) \subseteq S(\operatorname{Ann} \mathscr{I})$.

PROOF. In Lemma 11 we saw that $\operatorname{Ann} \mathscr{I} \neq 0$. If $a\mathscr{I} = \mathscr{I}$, then $\operatorname{Ann} a\mathscr{I} = \operatorname{Ann} \mathscr{I}$, and, by (1), we have $\operatorname{Ann} \mathscr{I}/a = \operatorname{Ann} \mathscr{I}$, i.e., $a \in S(\operatorname{Ann} \mathscr{I})$.

DEFINITION. We say that a nonzero R-module M is locally cyclic if any finitely generated submodule is cyclic.

Every nonzero ideal $\mathscr{I}$ in a valuation ring R and every quotient module $\mathscr{I}/\mathscr{T}$ (where $\mathscr{I}$ and $\mathscr{T}$ are ideals of $R, \mathscr{I} \supset \mathscr{T}$) are locally cyclic. It is easy to see that an R-module M over a valuation ring is locally cyclic if and only if M can be written as a union of cyclic submodules: $M = \bigcup_\alpha (x_\alpha)$, where the index α runs through some totally ordered set and $(x_\alpha) \supset (x_\beta)$ for $\alpha > \beta$.

DEFINITION. A locally cyclic module will be called regular if it contains a free cyclic submodule. As usual, a module M will be called exact if its annihilator is zero: $\operatorname{Ann} M = 0$.

DEFINITION. We say that a nonzero R-module M is divisible if for any element $u \in M$ the relation $\operatorname{Ann} \lambda \subseteq \operatorname{Ann} u$ (where $\lambda \in R$) implies that the equation $\lambda x = u$ can be solved in M.

In this article all R-modules are assumed to be unitary.

DEFINITION. Let R be a valuation ring, and let M be an R-module. If $x \in M$, $x \neq 0$, then we set

$$H(x) = \{\lambda \in R,\ \lambda y = x,\ y \in M\}.$$

By the height of an element x of a module M we mean the ideal $W_M(x) = W(x) = \bigcap_{\lambda \in H(x)} (\lambda)$.

Let M be a locally cyclic module over a valuation ring R. We shall use the following standard notation: $\mathscr{I} = \mathscr{I}(u)$ is the annihilator of the nonzero element $u \in M$; $H = H(u) = \{\lambda \in R, \lambda x = u, x \in M\}$; $W = W(u) = \bigcap_{\lambda \in H(u)} (\lambda)$ is the height of an element $u \in M$.

It is easy to see that if an element is in $H(u)$, then so are all of its divisors. This immediately implies the following lemma.

LEMMA 12. *If M is a non-cyclic module, then $H(u) = R \backslash W(u)$. If M is a cyclic module, then $W(u) = (\mu)$ is a principal ideal of R, and the set $H(u)$ consists of all of the divisors of μ.*

REMARK. We shall later need the following fact. Let M be a locally cyclic module, $u \in M$, $u \neq 0$. If $\lambda x = \lambda y = u$ (where $x, y \in M$), then $y = \theta x$, where θ is an invertible element of R.

In fact, suppose that we have the strict inclusion $(x) \subset (y)$. Then $x = \gamma y$, where $\gamma \in V$. We have $\lambda \gamma y = \lambda y = u$, so that $\lambda(1-\gamma)y = 0$, which leads to a contradiction, since $1 - \gamma$ is invertible in R.

We now describe some invariants of a locally cyclic module. The annihilators of all nonzero elements of M belong to the same ideal class. In fact, let $a, b \in M$, $a \neq 0$, $b \neq 0$, and $a = \mu b$. If $\operatorname{Ann} b \neq 0$, then $\operatorname{Ann} b = \mu \operatorname{Ann} a$; while if $\operatorname{Ann} b = 0$, then $\operatorname{Ann} a = \operatorname{Ann} \mu$. Hence, $\operatorname{Ann} a$ and $\operatorname{Ann} b$ are in the same ideal class. Thus, our first natural invariant of a locally cyclic module is the ideal class of the annihilators of its nonzero elements.

A second obvious invariant of a locally cyclic module (or any R-module) is its annihilator $\operatorname{Ann} M$.

Finally, the heights of all of the elements in a locally cyclic module M also belong to the same ideal class. This follows from the relation

$$\mu \bigcap_i (\lambda_i) = \bigcap_i (\mu \lambda_i). \tag{6}$$

To prove this equality, we note that clearly

$$\mu \bigcap_i (\lambda_i) \subseteq \bigcap_i (\mu \lambda_i). \tag{7}$$

If the right side of (7) is zero, then we immediately obtain (6). Now suppose that $\bigcap_i (\mu\lambda_i) \neq 0$ and $y \in \bigcap_i (\mu\lambda_i)$, $y \neq 0$. Then

$$y = \delta_i \mu \lambda_i. \tag{8}$$

Since μ divides y, we have the formula

$$y = \mu s, \tag{9}$$

where s is a fixed element of R. If we compare (8) and (9) and use the remark after Lemma 12, we find that $s = \theta_i \delta_i \lambda_i$, where the θ_i are units in R. Hence, $s \in \bigcap_i (\lambda_i)$. Thus, $\mu \bigcap_i (\lambda_i) \supseteq \bigcap_i (\mu\lambda_i)$, and (6) is proved.

Now let a, b be two nonzero elements of a locally cyclic module M, and let $a = \mu b$. We have $W(b) = \bigcap(\lambda_x)$, where $\lambda_x x = b$ for $x \in M$. Then $a = \mu \lambda_x x$ and $W(a) = \bigcap(\mu\lambda_x) = \mu \cap (\lambda_x) = \mu W(b)$ according to (6). Hence, $W(a)$ and $W(b)$ are in the same ideal class.

Thus, if $a, b \in M$, $a \neq 0$, $b \neq 0$ and $a = \mu b$, it follows that

$$\operatorname{Ann} b = \mu \operatorname{Ann} a. \tag{10}$$

$$W(a) = \mu W(b). \tag{11}$$

Next we prove some lemmas.

LEMMA 13. *Let M and N be locally cyclic modules over a valuation ring R of countable type which contain nonzero cyclic submodules (u) and (v), respectively, such that* $\operatorname{Ann} u = \operatorname{Ann} v$ *and* $W_M(u) = W_N(v) = W$. *If the modules M and N are either both cyclic or both non-cyclic, then they are isomorphic.*

PROOF. First suppose that the modules M and N are not cyclic. We write W as the intersection of a strictly decreasing countable chain of principal ideals:

$$W = \bigcap_{i=1}^{\infty}(\lambda_i), \quad (\lambda_1) \supset (\lambda_2) \supset \dots.$$

Let $\lambda_i x_i = u$ and $\lambda_i y_i = v$, where $x_i \in M$, $y_i \in N$. Then, by Lemma 12, $M = \bigcup(x_i)$ and $N = \bigcup(y_i)$. For each $i = 1, 2, \dots$ the cyclic modules (x_i) and (y_i) are isomorphic, since $\operatorname{Ann} x_i = \operatorname{Ann} y_i = \lambda_i \operatorname{Ann} u$. Moreover, any isomorphism ψ_i between the modules (x_i) and (y_i) extends to an isomorphism between (x_{i+1}) and (y_{i+1}). In fact, let $\lambda_{i+1} = \lambda_i \gamma$. Then $\lambda_i(\gamma x_{i+1}) = u$ and $\lambda_i(\gamma y_{i+1}) = v$. Consequently, by the remark following Lemma 12, we have $\gamma x_{i+1} = \theta_1 x_i$ and $\gamma y_{i+1} = \theta_2 y_i$, where θ_1 and θ_2 are units of R. Let $\psi_i(x_i) = \theta y_i$ (with θ a unit of R). Then an isomorphism $\psi_{i+1}\colon (x_{i+1}) \to (y_{i+1})$ extending ψ_i is given by the formula: $\psi_{i+1}(x_{i+1}) = \theta_1\theta\theta_2^{-1} y_{i+1}$. Thus, we can construct an isomorphism $\psi\colon M \to N$ by induction in the obvious way.

Now suppose that M and N are cyclic modules: $M = (x)$, $N = (y)$. Let $\lambda x = u$ and $\mu y = v$. Obviously $W(x) = (\lambda)$ and $W(y) = (\mu)$, from which we have $(\lambda) = (\mu)$. Consequently, the annihilators of x and y coincide, and the cyclic modules M and N are isomorphic. The lemma is proved.

REMARK. Lemma 13 remains true without the assumption that R is a ring of countable type, provided that u is a regular element (i.e., $\operatorname{Ann} u = 0$).

PROOF. It clearly suffices to consider the case of non-cyclic modules M and N. Since u is a regular element, it follows that for every $\lambda \notin W$ there exist exactly one element $x_\lambda \in M$ and one element $y_\lambda \in N$ such that $\lambda x_\lambda = u$ and $\lambda y_\lambda = v$. In fact, if $\lambda x = \lambda x' = u$, then, by the remark following Lemma 12, we have $x' = \theta x$ with θ a unit of R. Then $\theta u = u$, and so $\theta = 1$. It is now easy to give an isomorphism ψ between the modules M and N. For any $\lambda \in H = R \backslash W$ we set $\psi_\lambda(x_\lambda) = y_\lambda$. Then, because of the uniqueness of x_λ and y_λ, for $\lambda \equiv 0 \pmod{\mu}$ the restriction of ψ_λ to (x_μ) gives ψ_μ. Since $M = \bigcup(x_\lambda)$ and $N = \bigcup(y_\lambda)$ (here $\lambda \notin W$), it hence follows that there exists an isomorphism $\psi\colon M \to N$, defined as follows: $\psi(x_\lambda) = y_\lambda$. This proves the claim.

LEMMA 14. *Let $\mathcal{I}$ and W be two ideals of a ring R of countable type, where $\mathcal{I} \neq 0$ or R, $W \neq R$, and W is not an ideal of the form aV $(a \neq 0)$. Suppose that* $\operatorname{Ann}\lambda \subset \mathcal{I}$ *for every $\lambda \in H = R\backslash W$. Then there exists a non-cyclic locally cyclic module M such that $\mathcal{I} = \mathcal{I}(u)$, $W = W(u)$ for some $u \in M$, $u \neq 0$.*

PROOF. The ideal W can be written in the form $W = \bigcap_{i=1}^{\infty}(\lambda_i)$, where $(\lambda_1) \supset (\lambda_2) \supset \dots$. Let $\lambda_1 = \mu_1, \lambda_2 = \lambda_1\mu_2, \dots, \lambda_n = \lambda_{n-1}\mu_n, \dots$. Then $\lambda_n =$

$\mu_1 \dots \mu_n$. Since $\lambda_n \neq 0$ for $n = 1, 2, \dots$, it follows that $\operatorname{Ann} \mu_{i+1} \dots \mu_s \subset (\mu_i)$, and hence

$$\operatorname{Ann} \mu_i \mu_{i+1} \dots \mu_s = \frac{\operatorname{Ann} \mu_{i+1} \dots \mu_s}{\mu_i}. \tag{12}$$

Let S be a free R-module of countable rank with R-basis which we shall denote $u, u_1, u_2, \dots$. We introduce the submodule Q of S which is generated by all of the elements $\alpha u, \alpha \in \mathscr{I}$, and by the elements

$$a_1 = u - \mu_1 u_1;\ a_2 = u_1 - \mu_2 u_2;\ \dots a_n = u_{n-1} - \mu_n u_n; \dots.$$

Then the quotient S/Q will be the desired module. In fact, we show that

$$\operatorname{Ann}(u + Q) = \mathscr{I}. \tag{13}$$

Let $\gamma u \in Q$. Then $\gamma u = \delta u + \delta_1 a_1 + \dots + \delta_s a_s = \delta u + \delta_1(u - \mu_1 u_1) + \dots + \delta_s(u_{s-1} - \mu_s u_s)$, where $\delta \in \mathscr{I}$. We have

$$\gamma u = (\delta + \delta_1)u + (\delta_2 - \mu_1 \delta_1)u_2 + \dots + (\delta_s - \mu_{s-1}\delta_{s-1})u_{s-1} - \delta_s \mu_s u_s. \tag{14}$$

Since $u, u_1, \dots$ form a free basis of S, it follows from (14) that

$$\delta + \delta_1 = \gamma;\ \delta_2 - \mu_1 \delta_1 = 0;\ \dots \delta_s - \mu_{s-1}\delta_{s-1} = 0;\ \delta_s \mu_s = 0. \tag{15}$$

Using (12), we successively obtain from (15):

$$\delta_s \subseteq \operatorname{Ann} \mu_s;\ \delta_{s-1} \subseteq \frac{\operatorname{Ann} \mu_s}{\mu_{s-1}} = \operatorname{Ann} \mu_{s-1}\mu_s;\ \dots,$$

$$\delta_1 \subseteq \frac{\operatorname{Ann} \mu_2 \dots \mu_s}{\mu_1} = \operatorname{Ann} \mu_1 \dots \mu_s = \operatorname{Ann} \lambda_s.$$

Since, by assumption, we have $\operatorname{Ann} \lambda_s \subseteq \mathscr{I}$ and $\delta \in \mathscr{I}$, we finally conclude that $\gamma = \delta + \delta_1 \in \mathscr{I}$, i.e., $\operatorname{Ann}(u + Q) \subseteq \mathscr{I}$. The reverse inclusion is obvious, and thus (13) is proved. It follows from (13) that, in particular, $u + Q \neq Q$.

We set $\overline{u} = u + Q$; $\overline{u}_i = u_i + Q$ $(i = 1, 2, \dots)$. It is clear that the module M is the union of the strictly increasing sequence of cyclic submodules $(\overline{u}) \subset (\overline{u}_1) \subset \dots$. Because $\lambda_i \overline{u}_i = \overline{u}$, we find that the height of the elements $\overline{u}$ in M is equal to $\bigcap(\lambda_i) = W$, and, by what was proved above, $\mathscr{I}(\overline{u}) = \mathscr{I}$. This completes the proof of the lemma.

LEMMA 15. *A non-cyclic locally cyclic module M is divisible if and only if, for any nonzero element $u \in M$ and any nonzero $\mu \in W(u)$, the ideal $\mathscr{I} = \mathscr{I}(u)$ is properly contained in* $\operatorname{Ann} \mu$.

PROOF. Suppose that M is a divisible module. If $\mu \in W$, $\mu \neq 0$, and $\operatorname{Ann} \mu \subseteq \mathscr{I}$, then $\mu x = u$ for some $x \in M$, i.e., $\mu \in H(u) = R \backslash W(u)$, which gives a contradiction. Conversely, if the conditions of the lemma hold, then we have $\operatorname{Ann} \mu \subseteq \mathscr{I}$ only for elements $\mu \in H(u)$; consequently, M is a divisible module.

LEMMA 16. *Suppose that R is a ring of countable type, M is a locally cyclic but not cyclic module, $u \in M$, $u \neq 0$, $\mathscr{I} = \mathscr{I}(u)$, $\mathscr{I} \subseteq V$, $W = W(u)$. Suppose that the ideal W contains an element μ such that* $\operatorname{Ann}\mu \subseteq \mathscr{I}$. *Then the module M is isomorphic to a quotient ideal $\mathscr{I}/\mathscr{T}$ of R.*

PROOF. Let $W = \bigcap_{i=1}^{\infty}(\lambda_i)$; $(\lambda_1) \supset (\lambda_2) \supset \dots$; $\mu = \gamma_i\lambda_i$. We set $\mathscr{N} = R/\mu\mathscr{I}$. Then the annihilator of the element $x = \mu + \mu\mathscr{I} \in \mathscr{N}$ is $\mathscr{I}$. In fact, if $\alpha\mu = \mu t$ with $t \in \mathscr{I}$, then $\alpha \in \operatorname{Ann}\mu \subseteq \mathscr{I}$ if $\mu t = 0$, and if $\mu t \neq 0$, then, by Lemma 1, $\alpha \equiv t \pmod{\operatorname{Ann}\mu}$, and, since $\operatorname{Ann}\mu \subseteq \mathscr{I}$, we have $\alpha \in \mathscr{I}$. We let $\mathscr{N}'$ denote the submodule of $\mathscr{N}$ which is the union of the submodules (x_i), where $x_i = \gamma_i + \mu\mathscr{I}$. Since $\lambda_i x_i = x$, the height of the element x in $\mathscr{N}'$ is equal to $\bigcap(\lambda_i) = W$. From Lemma 13 we conclude that $\mathscr{N}' \cong M$. The lemma is proved.

THEOREM 1. *Every locally cyclic module over a valuation ring R of countable type either is divisible or else is isomorphic to a quotient ideal of R.*

PROOF. If $M = (u)$ is a cyclic module, then M is obviously isomorphic to the quotient $R/\mathscr{I}$, where $\mathscr{I} = \operatorname{Ann} u$. Now suppose that M is not cyclic. Let $u \in M$, $u \neq 0$, $\mathscr{I} = \mathscr{I}(u)$, $W = W(u)$. If W contains an element $\mu \neq 0$ such that $\operatorname{Ann}\mu \subseteq \mathscr{I}$, then, by Lemma 16, the module M is isomorphic to a quotient ideal $L/\mathscr{T}$. But if $\operatorname{Ann}\mu \supset \mathscr{I}$ for all nonzero elements $\mu \in W$ (if such elements exist), then M is a divisible module by Lemma 15. The theorem is proved.

We next prove some further lemmas about locally cyclic modules. We first make the following obvious observation.

If a locally cyclic module M is a proper submodule of a locally cyclic module $\mathscr{N}$, then M is not divisible. In fact, let $x \in M$ and $M \subset (y) \subset \mathscr{N}$. Then $\lambda y = x$ and $\operatorname{Ann}\lambda \subseteq \operatorname{Ann} x$; but, by the remark following Lemma 12, the module M does not contain an element z such that $\lambda z = x$.

LEMMA 17. *A locally cyclic module M over an arbitrary valuation ring R is divisible if and only if there exists a nonzero element $u \in M$ such that $\lambda y = u$ with $y \in M$ for any $\lambda \in \mathscr{I} = \operatorname{Ann} u$.*

PROOF. Necessity follows immediately from the definition of a divisible module. Now suppose that the conditions in the lemma hold for some nonzero $u \in M$, and let x be any nonzero element of M. We consider two cases:

1) $x = \gamma u$, $\gamma \in R$. Then $\operatorname{Ann} x = \mathscr{I}/\gamma$ (if $\mathscr{I} = 0$, the ideal $\mathscr{I}/\gamma$ is $\operatorname{Ann}\gamma$). If $\operatorname{Ann}\mu \subseteq \mathscr{I}/\gamma$, then $\gamma\operatorname{Ann}\mu \subseteq \mathscr{I}$. For $\gamma = \mu\tau$ we have $x = \mu(\tau u)$. Suppose that $\mu = \gamma\tau$. Then $\gamma\operatorname{Ann}\mu = \operatorname{Ann}\tau$ and $\operatorname{Ann}\tau \subseteq \mathscr{I}$. Consequently, $\tau y = u$ ($y \in M$), and hence $\mu y = x$.

2) $\gamma x = u$, $\gamma \in R$. Then $\operatorname{Ann} x = \gamma\mathscr{I}$. Let $\operatorname{Ann}\mu \subseteq \gamma\mathscr{I}$. If $\mathscr{I} \neq 0$, then $\operatorname{Ann}\mu/\gamma \subseteq \mathscr{I}$ and $\operatorname{Ann}(\mu\gamma) \subseteq \mathscr{I}$. But if $\mathscr{I} = 0$, then μ and γ are regular elements, and again $\operatorname{Ann}(\mu\gamma) \subseteq \mathscr{I}$. Thus, $\mu\gamma y = u = \gamma x$ ($y \in M$), and, by

the remark following Lemma 12, we have $\mu y = \theta x$, where θ is a unit of R. The lemma is proved.

THEOREM 2. *Let R be any valuation ring. The quotient $\mathcal{S}/\mathcal{T}$, where $\mathcal{S}$ and $\mathcal{T}$ are ideals of R with $\mathcal{S} \supset \mathcal{T}$, is divisible if and only if $\mathcal{S} = R$, R is a ring of the first kind, and either $\mathcal{T} = 0$ or else $\mathcal{T} = \operatorname{Ann} V$, $\operatorname{Ann} V \neq 0$, $V \neq (a)$.*

PROOF. If $\mathcal{S} \subset R$, then $\mathcal{S}/\mathcal{T}$ is a proper submodule of the module $R/\mathcal{T}$, and so $\mathcal{S}/\mathcal{T}$ is not a divisible module. We now consider the cyclic module $R/\mathcal{T}$. If R contains nontrivial regular elements λ then $R/\mathcal{T}$ is not a divisible module, because $\operatorname{Ann} \lambda \subseteq \mathcal{T} = \operatorname{Ann}(1 + \mathcal{T})$; however, $1 + \mathcal{T} \neq \lambda(y + \mathcal{T})$ ($(1 - \lambda y)$ is a unit of R). Now suppose that R is a ring of the first kind, and $\mathcal{T} \neq 0$, $\mathcal{S} \neq \operatorname{Ann} V$ ($\operatorname{Ann} V$ is the minimal ideal of R). Then $\operatorname{Ann} \mathcal{T} \subset V$, and hence $\operatorname{Ann} \mathcal{T} \subseteq (b) \subset (c)$ for certain principal ideals $(b), (c) \subset V$. We have $\operatorname{Ann} c \subset \operatorname{Ann} b \subset \operatorname{Ann} \operatorname{Ann} \mathcal{T}$; hence, by Lemma 9, $\operatorname{Ann} c \subset \mathcal{T}$. The equality $c(\lambda + \mathcal{T}) = 1 + \mathcal{T}$ is impossible in $R/\mathcal{T}$. Thus, this module is not divisible. It remains to show that, if R is a ring of the first kind and $\mathcal{T} = 0$ or $\mathcal{T} = \operatorname{Ann} V$, $V \neq (a)$, then $R/\mathcal{T}$ is a divisible module. In fact, if $\mathcal{T} = 0$ or $\mathcal{T} = \operatorname{Ann} V$, then $\operatorname{Ann} \lambda \supset \mathcal{T}$ for any $\lambda \in V$, as follows immediately from Lemma 9. Hence, if $\operatorname{Ann} \lambda \subseteq \mathcal{T}$ it follows that λ is an invertible element of R. This proves that $R/\mathcal{T}$ is a divisible module.

THEOREM 3. *Let M be a regular locally cyclic module over a valuation ring R. Then M is isomorphic either to a regular ideal of R or else to the full ring of fractions T of R (regarded as an R-module). M is a divisible module if and only if $M \cong T$.*

PROOF. If M is a cyclic module, then M is obviously free. Suppose that M is not cyclic. Let u be a regular element of M, and let W be the height of u in M. If W contains a regular element μ, then $\operatorname{Ann} \mu = 0 = \mathcal{S}$, and, by Lemma 16 and the remark following Lemma 13, the module M is isomorphic to a quotient ideal $L/\mathcal{T}$. Since all of the elements in the module $L/\mathcal{T}$ are annihilated by multiplication by $\mathcal{T}$, and since M is a regular module, it follows that $\mathcal{T} = 0$, and so M is isomorphic to the regular ideal L of R. Now suppose that W does not contain regular elements. Since every element $\lambda \notin W$ is regular, it follows that W is the ideal of A consisting of all of the irregular elements of R. But the height of the element 1 in the full ring of fractions T is also A. Hence, by the remark following Lemma 13, we have an R-isomorphism $M \cong T$.

We now show that the full ring of fractions T of the ring R is a divisible module. In fact, the annihilator of 1 in T is the zero ideal, and in T the element 1 is divisible by all regular elements. By Lemma 17, T is a divisible module. The theorem is proved.

LEMMA 18. *Let M and $\mathcal{N}$ be locally cyclic modules. If the annihilators of the elements of M and $\mathcal{N}$ lie in the same ideal class, then M and N contain nonzero elements with the same annihilator.*

PROOF. Let $u \in M$, $u_1 \in \mathcal{N}$, $u \neq 0$, $u_1 \neq 0$, $\operatorname{Ann} u = \mathcal{I}$, $\operatorname{Ann} u_1 = \mathcal{T}$. If, for example, $\mathcal{I} = \alpha\mathcal{T}$, $\alpha \neq 0$, then $\operatorname{Ann}(\alpha u) = \mathcal{T}$.

LEMMA 19. *Let M be a non-cyclic locally cyclic module whose nonzero elements have annihilators which are not the annihilators of elements of the ring R. If M is a divisible module, then for every nonzero $u \in M$ we have $W = \operatorname{Ann} \mathcal{I}$. If this equality holds for a single nonzero u, then M is divisible.*

PROOF. Suppose that M is a divisible module satisfying the conditions in the lemma, $u \neq 0$ is any element of M, and $\operatorname{Ann} u = \mathcal{I}$, $W = W(u)$. Then, by Lemma 15, we have $\operatorname{Ann} \mu \supset \mathcal{I}$ for any $\mu \in W$, and hence $\mu\mathcal{I} = 0$ for all $\mu \in W$, or $W\mathcal{I} = 0$, i.e., $W \subseteq \operatorname{Ann} \mathcal{I}$. On the other hand, let $\alpha \in \operatorname{Ann} \mathcal{I}$. If we suppose that $\alpha \in H(u) = R \backslash W$, then, since M is a divisible module, we have $\alpha x = u$, $x \in M$, and hence $\operatorname{Ann} \alpha \subseteq \mathcal{I}$. Then $\mathcal{I} = \operatorname{Ann} \alpha$, which is impossible, by assumption. Thus $\alpha \in W$ and $\operatorname{Ann} \mathcal{I} \subseteq W$. Therefore, $W = \operatorname{Ann} \mathcal{I}$.

We next prove sufficiency. Suppose that $W = \operatorname{Ann} \mathcal{I}$ for some nonzero $u \in M$. We take any element $\mu \in R$, $\operatorname{Ann} \mu \subseteq \mathcal{I}$. If $\mu \in W$, then $\mu\mathcal{I} = 0$, and hence $\mathcal{I} \subset \operatorname{Ann} \mu$, from which $\mathcal{I} = \operatorname{Ann} \mu$, and this is not allowed. Thus, $\mu \in H(u) = R \backslash W$. Applying Lemma 17, we conclude that M is a divisible module.

THEOREM 4. *Let R be a valuation ring of countable type, and let M be a divisible locally cyclic R-module. The ideal class of the annihilators of the nonzero elements of M determine M up to isomorphism. Every ideal class of R is the class of the annihilators of the nonzero elements of some divisible locally cyclic R-module.*

PROOF. We first show that the ideal class $\mathcal{A}$ of the annihilators of the nonzero elements of M determine M up to isomorphism. We consider the following cases.

1) The class $\mathcal{A}$ is the class of the annihilators of the nonzero elements of the ring R. Then the divisible modules M corresponding to this class are isomorphic to the full ring of fractions T of the ring R. In fact, if $\operatorname{Ann} u = \operatorname{Ann} \alpha$ ($u \in M$, $\alpha \in R$, $u \neq 0$, $\alpha \neq 0$), then, since the module M is divisible, we have $\alpha x = u$, $x \in M$. Then $\operatorname{Ann} x = 0$, i.e., M is a regular module. By Theorem 3, $M \cong T$.

2) $\mathcal{A} \neq 0$; the class $\mathcal{A}$ is not the class of the annihilators of the nonzero elements of the ring or the class of the principal ideals, or else $\mathcal{A}$ is the class of the principal ideals but $\operatorname{Ann} V = 0$. The divisible modules corresponding to such a class cannot be regular or cyclic. In fact, if M is cyclic, then, by Theorem 2, R is a ring of the first kind and $M \cong R$ or $M \cong R/\operatorname{Ann} V$, if R is a ring of the second kind. In the first case $\mathcal{A}$ is the class of the

annihilators of the nonzero elements of R, and in the second case it is the class of the nonzero principal ideals. Let M_1 and M_2 be divisible modules corresponding to the class $\mathscr{A}$. By Lemma 18, there exist elements $u_1 \in M_1$ and $u_2 \in M_2$ such that $\operatorname{Ann} u_1 = \operatorname{Ann} u_2 = \mathscr{I} \neq 0$. By Lemma 19, we have $W(u_1) = W(u_2) = \operatorname{Ann} \mathscr{I}$. Now, applying Lemma 13, we find that $M_1 \cong M_2$.

3) R is a ring of the second kind, and $\mathscr{A}$ is the class of all nonzero principal ideals of R. We show that in this case one has $M \cong R/\operatorname{Ann} V$. In fact, M is not a regular module, since, by assumption, the class $\mathscr{A}$ does not contain the zero ideal. Suppose that M is not a cyclic module. Let $\operatorname{Ann} V = (\alpha)$, $\alpha \neq 0$. We take any nonzero element $u \in M$ and let $\operatorname{Ann} u = (\gamma)$, $\gamma \neq 0$. If $\alpha = \gamma\beta$, then, by Lemma 19, we have $W(u) = \operatorname{Ann} \gamma = \beta V$. Since $\beta \neq 0$, we have obtained a contradiction, since the height $W(u)$ cannot be the ideal βV, $\beta \neq 0$. Hence, M is a cyclic module, and, by Theorem 2, $M \cong R/\operatorname{Ann} V$.

Thus, the divisible module M is determined up to isomorphism by the ideal class $\mathscr{A}$ of the annihilators of its nonzero elements.

We next show that any ideal class $\mathscr{A}$ is the class of the annihilators of the nonzero elements of some divisible locally cyclic module M.

If $\mathscr{A}$ is the class of the annihilators of the nonzero elements of the ring R, then it corresponds to the full ring of fractions T of the ring. If R is a ring of the second kind, and if $\mathscr{A}$ is the class of all the nonzero principal ideals of this ring, then we take M to be the module $R/\operatorname{Ann} V$. Otherwise, $\mathscr{A}$ consists only of nonzero ideals $\mathscr{I}$ such that $\operatorname{Ann} \mathscr{I} \neq \alpha V$, $\alpha \neq 0$. In fact, suppose that $\mathscr{I} \neq 0$ and $\operatorname{Ann} \mathscr{I} = \alpha V$, $\alpha \neq 0$. If $\mathscr{I}$ is not a principal ideal, then, by Lemma 2, we have $\mathscr{I} \alpha V = \mathscr{I} \alpha = 0$, which contradicts the assumption that $\operatorname{Ann} \mathscr{I} = \alpha V$. If, on the other hand, $\mathscr{I} = (\beta)$, then $\beta\alpha V = 0$ and $\beta\alpha \neq 0$; hence, R is a ring of the second kind, and $\mathscr{A}$ is the class of the nonzero principal ideals, which we assumed not to be the case.

We now show that the class $\mathscr{A}$ contains an ideal $\mathscr{I}$ such that $\operatorname{Ann} \lambda \subset \mathscr{I}$ for all $\lambda \in W = \operatorname{Ann} \mathscr{I}$. From this the theorem will follow, since, by Lemma 14, there will exist a locally cyclic module M such that $\operatorname{Ann} u = \mathscr{I}$, $W(u) = W$, $u \in M$, $u \neq 0$; and, by Lemma 19, M will be a divisible module. We consider the following cases:

1) The class $\mathscr{A}$ contains an ideal $\mathscr{I}$ which is not regularly pure. Then $P(\mathscr{I}) = P \supset A$. By Lemma 6, we have $\mathscr{I}/\lambda \supset A$ for some $\lambda \in R$, where $\mathscr{I}/\lambda \in \mathscr{A}$. Thus, the class $\mathscr{A}$ contains an ideal $\mathscr{T} \supseteq A$. Then for all $\lambda \in R$ we have $\operatorname{Ann} \lambda \subset \mathscr{T}$ (by assumption, we cannot have the equality $\operatorname{Ann} \lambda = \mathscr{T}$).

2) The class $\mathscr{A}$ consists of regularly pure ideals. If $A \in \mathscr{A}$, then $\operatorname{Ann} \lambda \subset A$ for all $\lambda \notin \operatorname{Ann} A$. Suppose that $\mathscr{I} \subset A$ for all $\mathscr{I} \in \mathscr{A}$. We fix an ideal $\mathscr{I} \in \mathscr{A}$ and let $\lambda \notin W = \operatorname{Ann} \mathscr{I}$. We have $\operatorname{Ann} \lambda \subseteq \operatorname{Ann} W = \operatorname{Ann} \operatorname{Ann} \mathscr{I}$. But $\mathscr{I} \neq aA$, since otherwise we would have $A \in \mathscr{A}$. Hence, by Lemma 11, $\operatorname{Ann} \operatorname{Ann} \mathscr{I} = \mathscr{I}$ and $\operatorname{Ann} \lambda \subseteq \mathscr{I}$; consequently, $\operatorname{Ann} \lambda \subset \mathscr{I}$.

The theorem is proved.

REMARK. If R is an integral domain, then its field of fractions T is obviously the divisible module corresponding to the zero ideal class. The divisible module corresponding to an ideal class $\mathscr{A}$ which contains a nonzero ideal $\mathscr{I}$, is isomorphic to the quotient module $T/\mathscr{I}$. In fact, $\operatorname{Ann}(1+\mathscr{I}) = \mathscr{I}$, and the height of the element $(1+\mathscr{I})$ is 0. Hence, $T/\mathscr{I}$ is a divisible module.

Suppose that R is not an integral domain, and P is a prime ideal such that $P \subseteq A$. Then the ring of fractions R_P is a divisible module. In fact, if $P = A$, then R_A is the full ring of fractions of R, and R_P is a divisible module by Theorem 3. Suppose that $P \subset A$. Then, by Lemma 5, the annihilator of the element 1 in R_P (regarded as an R-module) is $\operatorname{Ann} P$. On the other hand, the equality $1 = \lambda x$, $\lambda \in R$, holds in R_P if and only if $\lambda \notin P$. Thus, $W(1) = P$. But $\operatorname{Ann}\operatorname{Ann} P = P$ (Lemma 11), and so R_P is a divisible R-module, by Lemma 19.

We introduce one more invariant of a locally cyclic module M. Let $\mathscr{I} = \mathscr{I}(u)$ be a nonzero annihilator of a nonzero element $u \in M$, and let $P = P(\mathscr{I}) = R \backslash S(\mathscr{I})$. According to the corollary of Lemma 1, the stabilizer $S(\mathscr{I})$ is an invariant of the ideal class, and so the ideal P is the same for all elements $u \neq 0$ with nonzero annihilator. We shall let $P(M)$ denote this ideal P.

LEMMA 20. *Suppose that M is a non-cyclic, irregular locally cyclic divisible module over a valuation ring R which is not an integral domain. Then* $\operatorname{Ann} M = \operatorname{Ann} P(M)$, *where $P(M)$ is the prime ideal corresponding to the module M.*

PROOF. We let $\mathscr{A}$ denote the ideal class of the annihilators of nonzero elements of M. If $\mathscr{A}$ does not contain a regularly pure ideal, then, as shown in the proof of Theorem 4, there exists an element $u \in M$, $u \neq 0$, such that $\operatorname{Ann} u = \mathscr{I} \supset A$. Then $P = P(M) = P(\mathscr{I})$ is a regular ideal, and consequently $\operatorname{Ann} P = 0$. On the other hand, by Lemma 19, we have $W = W(u) = \operatorname{Ann}\mathscr{I} = 0$. We further have $\operatorname{Ann} M = \bigcap_{\lambda \in R\backslash W} \mathscr{I}\lambda = \bigcap(\lambda) = 0$. Consequently, $\operatorname{Ann} M = \operatorname{Ann} P(M)$. Now suppose that the class $\mathscr{A}$ consists of regularly pure ideals $\mathscr{I} \subseteq A$. If $\mathscr{I} = aA$ for some ideal $\mathscr{I} \in \mathscr{A}$, then $P(M) = A$, $\operatorname{Ann}(au) = A$, and, by Lemma 19, $W(au) = \operatorname{Ann} A$. If $\operatorname{Ann} A \neq 0$, then, by Lemma 5, $\operatorname{Ann} A$ is a regularly principal ideal: $\operatorname{Ann} A = (b)_A$. Because of Lemma 11, we have $A = \operatorname{Ann}\operatorname{Ann} A = \operatorname{Ann} b$. From Theorem 3 we now find that M is a regular module, and hence $\operatorname{Ann} A = 0$, $W(au) = 0$, $\operatorname{Ann} M = 0 = \operatorname{Ann} A = \operatorname{Ann} P(M)$. Finally suppose that $\mathscr{A}$ consists of nonzero regularly pure ideal $\mathscr{I} \neq aA$. We fix an ideal $\mathscr{I} = \operatorname{Ann} u$. Then $W = W(u) = \operatorname{Ann}\mathscr{I}$ (Lemma 19). We set $L = \bigcup_{\lambda \notin W} \operatorname{Ann}(\mathscr{I}\lambda)$. From (1) and Lemma 6 we have

$$L = \bigcup_{\lambda \notin W} \frac{\operatorname{Ann}\mathscr{I}}{\lambda} = \bigcup_{\lambda \notin W} \frac{W}{\lambda} = P(W).$$

Since $\mathscr{I}$ is a regularly pure ideal and $\mathscr{I} \neq aA$, it follows from Lemma 11 that $\operatorname{Ann} W = \mathscr{I}$; hence, by the corollary of Lemma 11, $P(W) = P(\mathscr{I}) = P(M)$. By Lemma 11, we have $\operatorname{Ann}\operatorname{Ann}(\lambda\mathscr{I}) = \lambda\mathscr{I}$ $(\lambda \notin W)$, and consequently $\operatorname{Ann} P(M) = \operatorname{Ann} L = \bigcap_{\lambda \notin W} \lambda\mathscr{I} = \operatorname{Ann} M$. The lemma is proved.

We can now give a complete description of all exact locally cyclic modules over a ring R of countable type. According to Theorem 1, any locally cyclic module M is either isomorphic to a factor $\mathcal{I}/\mathcal{T}$ (with $\mathcal{I}$ and $\mathcal{T}$ ideals of $R, \mathcal{I} \supset \mathcal{T}$), or else is divisible. We now determine when a factor $\mathcal{I}/\mathcal{T}$ is faithful. It is clear that for $\mathcal{T} \neq 0$ the module $\mathcal{I}/\mathcal{T}$ is not faithful, and so it remains to investigate the ideals $\mathcal{I}$ of R. Every regular ideal $\mathcal{I}$ is faithful; and if $\mathcal{I} \subset A$, then $\operatorname{Ann}\mathcal{I} \neq 0$ (see the corollary of Lemma 11). Hence, the faithful quotients $\mathcal{I}/\mathcal{T}$ are: all of the regular ideals, and the ideal A if $\operatorname{Ann} A = 0$. In addition, a regular divisible locally cyclic module, i.e., the full ring of fractions T of R, is faithful (see Theorem 3). It remains to determine which of the irregular non-cyclic divisible locally cyclic modules are faithful. The next theorem answers this question.

THEOREM 5. *Suppose that R is a valuation ring of countable type and M is an irregular non-cyclic divisible locally cyclic module. If R is an integral domain, then M is faithful. If R is not an integral domain, then M is an exact module if and only if $P(M) \supseteq A$ and* $\operatorname{Ann} A = 0$ *in the case $P(M) = A$.*

PROOF. Suppose that R is an integral domain. Then every divisible module M is faithful, since the height of any nonzero element is 0. In fact, it follows immediately from (10) that $\operatorname{Ann} M = \bigcap_{\lambda \in H(u)} \lambda\mathcal{I}$, $u \in M$, $u \neq 0$, $\mathcal{I} = \operatorname{Ann} u$.

Now suppose that R is not an integral domain. Then $\operatorname{Ann} M = \operatorname{Ann} P(M)$ by Lemma 20. By the corollary of Lemma 11, we hence obtain the final assertion of the theorem.

According to Theorem 3, every regular locally cyclic module is isomorphic either to an ideal of R or to the full ring of fractions of R. We now determine which of the ideals of R are isomorphic to one another.

LEMMA 21. *Suppose that R is a valuation ring, and $\mathcal{I} \neq 0$ is an ideal of R. Then every R-homomorphism $\varphi: \mathcal{I} \to R$ is given by a formula of the form*

$$\varphi(\lambda) = \theta_\lambda \delta \lambda, \tag{16}$$

where $\lambda \in \mathcal{I}$, θ_λ is a unit of R, and δ is a fixed element of the full ring of fractions T of the ring R.

PROOF. We show that $\varphi(\lambda) = \delta_\lambda \lambda$, where $\delta_\lambda \in T$. We first suppose that the ideal $\mathcal{I}$ contains a regular element λ. Then all of the elements $\mu \in \mathcal{I}$, $(\mu) \supset (\lambda)$, are also regular. Since all regular elements are invertible in the ring T, it follows that for all $\mu \in \mathcal{I}$, $(\mu) \supset (\lambda)$, we have the formula $\varphi(\mu) = \mu\delta_\mu$ with $\delta_\mu \in T$. Then this formula obviously holds for all $\lambda \in \mathcal{I}$.

We next suppose that $\mathcal{I}$ is an irregular ideal. Let $\lambda \in \mathcal{I}$, $\lambda \neq 0$, and $\mathcal{T} = \operatorname{Ann}\lambda$. Then $\mathcal{T}\varphi(\lambda) = 0$. Applying Lemma 10, we find that either $\varphi(\lambda) = \delta_\lambda\lambda$ with $\delta_\lambda \in R$, or else $\lambda = \gamma\varphi(\lambda)$ with γ a regular element of the ring. In either case $\varphi(\lambda) = \delta_\lambda\lambda$ with $\delta_\lambda \in T$. We fix an element $\lambda \in \mathcal{I}$, $\lambda \neq 0$. Let $\varphi(\lambda) \neq 0$. If $\mu \in \mathcal{I}$, $(\mu) \supseteq (\lambda)$, then $\lambda = \mu\tau$ $(\tau \in R)$ and

$$\varphi(\lambda) = \varphi(\tau\mu) = \tau\varphi(\mu) = \tau\mu\delta_\mu = \lambda\delta_\mu = \lambda\delta_\lambda. \tag{17}$$

According to Lemma 10, it follows from (17) that all of the elements δ_μ are invertible in T if δ_λ is invertible in T, and all of the elements δ_μ are noninvertible elements of R if $\delta_\lambda \in R$ and δ_λ is not invertible in R. Suppose that δ_λ is not invertible in R. Then, by Lemma 1, we conclude from (17) that $\delta_\mu = \delta_\lambda \theta_\mu$ for all $\mu \in \mathscr{I}$, $(\mu) \supseteq (\lambda)$, where θ_μ is a unit of R. Let δ_λ be invertible in T. Then one of δ_λ, $1/\delta_\lambda$ is a regular element of R, and one of δ_μ, $1/\delta_\mu$ is a regular element of R. If we again apply (17) and Lemma 1, we see that one of the elements $\delta_\mu \delta_\lambda^{-1}$, $\delta_\lambda \delta_\mu^{-1}$ is invertible in R. In either case $\delta_\mu = \theta_\mu \delta_\lambda$, where θ_μ is a unit of R. Thus, we have (16) for all $\lambda \in \mathscr{I}$. The lemma is proved.

COROLLARY. *If $\mathscr{I}$ is a regular ideal of R, and if the homomorphism $\varphi\colon \mathscr{I} \to R$ maps $\mathscr{I}$ onto a regular ideal of R, then $\varphi(\lambda) = \delta\lambda$ $(\lambda \in \mathscr{I})$, where δ is a fixed invertible element in T.*

PROOF. Let λ and μ be regular elements of the ideal $\mathscr{I}$ with regular images $\varphi(\lambda)$ and $\varphi(\mu)$. Then, by (17), we have $\lambda\delta_\lambda = \lambda\delta_\mu$, and hence $\delta_\lambda = \delta_\mu$. Thus, the formula (17) takes the form $\varphi(\lambda) = \delta\lambda$, where δ is a fixed invertible element of T.

THEOREM 6. *Ideals $\mathscr{I}$ and $\mathscr{T}$ of R are isomorphic (as R-modules) if and only if they belong to the same regular class.*

PROOF. Let $\mathscr{I}$ and $\mathscr{T}$ be nonzero ideals of R. If they belong to the same regular class, then we can write, for example, $\mathscr{I} = r\mathscr{T}$, where r is a regular element. Then an R-isomorphism between $\mathscr{I}$ and $\mathscr{T}$ is given by the formula $a \to ra$ $(a \in \mathscr{I})$. Now suppose that $\varphi\colon \mathscr{I} \to \mathscr{T}$ is an R-isomorphism. By Lemma 21, we have $\varphi(a) = \delta a \theta_a$, where δ is a fixed element of the full ring of fractions T, θ_a is a unit of R, and $a \in \mathscr{I}$. Since φ is an isomorphism, it follows that $\operatorname{Ann} a = \operatorname{Ann} \varphi(a)$ for all $a \in \mathscr{I}$. Consequently, δ is an invertible element of T, since otherwise we would have a contradiction to Lemma 10 for the nonzero element $a \in \mathscr{I}$. Since φ is an isomorphism, it follows that for $(b) \supseteq (a)$, $b \in \mathscr{I}$, we have $\delta\theta_b a = \delta\theta_a a$; and since δ is an invertible element of T, it follows that $\theta_b a = \theta_a a$. Hence, the map $\psi\colon a \to \theta_a a$ is a homomorphism from $\mathscr{I}$ to itself. Because θ_a is invertible, this homomorphism is injective. In addition, ψ is obviously a surjective map. Thus, $\mathscr{T} = \{\delta\theta_a a, a \in \mathscr{I}\} = \delta\{\theta_a a\} = \delta\mathscr{I}$. The theorem is proved.

COROLLARY. *Let R be a ring of the first kind. Then distinct ideals of R are not isomorphic.*

In fact, every regular class of R consists of a single ideal.

In [**4**] it is proved that every indecomposable injective module over an almost maximal ring R is locally cyclic.

LEMMA 22. *Every injective module over a valuation ring R is divisible.*

PROOF. Suppose that M is an injective R-module, $x \in M$, $x \neq 0$, $\operatorname{Ann} x = \mathscr{I}$, and $\lambda \in R$ is an element such that $\operatorname{Ann} \lambda \subseteq \mathscr{I}$. Then one can construct a cyclic extension $\mathscr{N} = \{M, u\}$ of the module M using the relation $\lambda u = x$. The quotient module $\mathscr{N}/M$ is isomorphic to the module $R/(\lambda)$. Since M is an injective module, we have the direct sum $\mathscr{N} = M \dotplus (y)$, where $(y) \cong \mathscr{N}/M$. Hence, $y = z + \theta u$, $z \in M$, where θ is a unit of R. Since $\operatorname{Ann} y = (\lambda)$, it follows that $\lambda z = -\theta\lambda u = -\theta x$ and $x = \lambda(-\theta^{-1} z)$. Thus, M is a divisible module.

THEOREM 7. *A locally cyclic module M over an almost maximal valuation ring is injective if and only if it is divisible.*

PROOF. Necessity follows from Lemma 22. Suppose that M is a divisible locally cyclic module over R, and $\mathscr{I}$ is an arbitrary nonzero ideal of R. We consider an arbitrary nonzero R-homomorphism $\varphi\colon \mathscr{I} \to M$. Let $\mu \in \mathscr{I}$ and $\varphi(u) \neq 0$. Then for all $\lambda \in \mathscr{I}$, $(\lambda) \subseteq (\mu)$, we have $\varphi(\lambda) = y_2 \neq 0$. Since $\gamma y_\lambda = 0$ for all $\gamma \in \operatorname{Ann}\lambda$, it follows from the fact that M is a divisible module that we have $y_2 = \lambda x_2$ for all $\lambda \in \mathscr{I}$, $(\lambda) \supseteq (\mu)$. Let $\lambda_1 = \gamma\lambda_2$. Then $y_{\lambda_1} = \gamma y_{\lambda_2}$, or $\lambda_1 x_{\lambda_1} = \gamma\lambda_2 x_{\lambda_2} \neq 0$. If, now, one of the modules (x_{λ_1}) or (x_{λ_2}) is a proper submodule of the other, say $x_{\lambda_1} = \tau x_{\lambda_2}$, where $\tau \in V$, then we have $\lambda_1\tau x_{\lambda_2} = \lambda_1 x_{\lambda_2}$, and hence $\lambda_1\theta x_{\lambda_2} = 0$ (with θ a unit of R), which gives a contradiction. Hence, $(x_{\lambda_1}) = (x_{\lambda_2})$ and $x_{\lambda_2} = \theta x_{\lambda_1}$, where θ is a unit of R. Thus, for all $\lambda \in \mathscr{I}$, $(\lambda) \supseteq (\mu)$, we have $x_\lambda = \theta_\lambda x_\mu$ and $\varphi(\lambda) = \lambda\theta_\lambda x_\mu$. Here, if $\lambda_2 \equiv 0 \pmod{\lambda_1}$, then

$$(18) \qquad \lambda_2\theta_{\lambda_2} x_\mu = \lambda_2\theta_{\lambda_1} x_\mu.$$

We consider the system of congruences

$$(19) \qquad \theta \equiv \theta_\lambda \pmod{\operatorname{Ann}\lambda x_\mu}.$$

By (18), the congruences in (19) are pairwise compatible. We show that there exists a unit θ which satisfies this system. This follows from the definition of maximality if R is a maximal ring. Suppose that R is an almost maximal ring but not a maximal ring. Then R is an integral domain. If M is not a regular module, then $\operatorname{Ann} x_\mu = \mathscr{T} \neq 0$, and, since $\bigcap_{(\lambda)\supseteq(\mu)}(\lambda) \neq 0$, it follows that $\bigcap_{(\lambda)\subseteq\mathscr{I},(\lambda)\supseteq(\mu)} \mathscr{T}\lambda \supseteq \mathscr{T}(\bigcap(\lambda)) \neq 0$. Consequently, from the definition of an almost maximal ring it again follows that there exists a unit $\theta \in R$ which satisfies the system (19). Then for all $\lambda \in \mathscr{I}$ we have $\varphi(\lambda) = \lambda\theta x_\mu$, i.e., M satisfies the Baer criterion **[5]**. The theorem is proved.

LEMMA 23. *Let R be an almost maximal valuation ring. For each ideal class $\mathscr{A}$ of R there exists a divisible locally cyclic module M such that the annihilators of its nonzero elements are in the class $\mathscr{A}$.*

PROOF. If R is an integral domain, then the quotient module $T/\mathscr{I}$ can be taken for M, where T is the field of fractions of R. Suppose that R is a maximal ring which is not an integral domain. If the ideal class $\mathscr{A}$ contains the zero ideal, then we set $M = T$, where T is the full ring of fractions of R. If R is a ring of the second kind and $\mathscr{A}$ consists of the nonzero principal ideals, then we set $M = R/\operatorname{Ann} V$ (see the proof of Theorem 4). In the remaining cases the class $\mathscr{A}$ consists of nonzero ideals $\mathscr{I}$ such that for each $\mathscr{I} \in \mathscr{A}$ we have $\operatorname{Ann} \mathscr{I} \neq aV$, $a \neq 0$. We fix an arbitrary ideal $\mathscr{I} \in \mathscr{A}$ and set $W = \operatorname{Ann} \mathscr{I}$. Then $W = \bigcap_i (\lambda_i)$, where $(\lambda_i) \supset W$, the index i runs through a set $\mathscr{M}$ of ordinals, and $\lambda_j = \lambda_i \tau_{ij}$, $\tau_{ij} \in V$ for $i < j$. For any $i < j < k$ we have $\lambda_k = \lambda_i \tau_{ik}$ and $\lambda_k = \lambda_i \tau_{ij} \tau_{jk}$; hence,

$$\tau_{ik} \equiv \tau_{ij} \tau_{jk} \pmod{\operatorname{Ann} \lambda_i}, \tag{20}$$

from which we have

$$\tau_{ik} \equiv \tau_{ij} \tau_{jk} \theta_{ijk}, \tag{21}$$

where θ_{ijk} is a unit of R. We consider the system of congruences

$$\theta \equiv \theta_{ijk} \pmod{\operatorname{Ann} \lambda_i} \tag{22}$$

for all possible triples $i < j < k$. It is easy to see that these congruences are pairwise compatible. Since R is a maximal ring, it follows that there exists a unit $\theta \in R$ which satisfies this system. We set $\tau'_{ij} = \tau_{ij} \theta$. Then from (20), (21), (22) it follows that for any $i < j < k$

$$\tau'_{ik} = \tau'_{ij} \tau'_{jk}. \tag{23}$$

We fix i, and we introduce symbols a_j in one-to-one correspondence with the indices $j \geq i$, $j \in \mathscr{M}$, and also a symbol u. By (23), the desired module M can be given by the generators $u, a_j, j \in \mathscr{M}$, and the defining relations $\alpha u = 0$ if $\alpha \in \mathscr{I}$; $\lambda_i a_i = u$; $\tau'_{jk} a_k = a_j$ for $i \leq j < k$. The equality $\operatorname{Ann} u = \mathscr{I}$ is proved in exactly the same way as the corresponding assertion in Theorem 4. In addition, we obviously have $W(u) = W$. Since $W = \operatorname{Ann} \mathscr{I}$, it follows by Lemma 19 that M is a divisible module. The lemma is proved.

Lemma 24. *Let M and $\mathscr{N}$ be non-cyclic locally cyclic modules over an almost maximal valuation ring R. Suppose that M and $\mathscr{N}$ contain elements u and u', respectively, such that $\mathscr{I}(u) = \mathscr{I}(u') = \mathscr{I}$, $W_M(u) = W_{\mathscr{N}}(u') = W$. Then the modules M and $\mathscr{N}$ are isomorphic.*

Proof. The height W can be represented in the form $W = \bigcap_i (\lambda_i)$, $(\lambda_i) \supset W$, where the index i runs through some set of ordinals and $\lambda_j \equiv 0 \pmod{\lambda_i}$ for $j > i$. Suppose that for all indices $i < j$ we have already constructed isomorphisms $\varphi_i \colon (u_i) \to (u'_i)$ ($u_i \in M, u'_i \in \mathscr{N}$), where $\lambda_i u_i = u$, $\lambda_i u'_i = u'$, $\varphi(u_i) = u'_i$, and for $s < k$ the isomorphism φ_k is an extension of the isomorphism φ_s. Let u_j and $\tilde{u}_j$ be elements of M and $\mathscr{N}$, respectively, such that $\lambda_j u_j = u$, $\lambda_j \tilde{u}_j = u'$. If $\operatorname{Ann} u_j = 0$, then $\operatorname{Ann} \tilde{u}_j = 0$. In addition, $W(u_j) = W(\tilde{u}_j)$ (see (10) and (11)). If we set $u = u_j$, $u' = \tilde{u}_j$, and use the remark following Lemma 13, we obtain $M \cong \mathscr{N}$.

Let $\operatorname{Ann} u_j \neq 0$, $\gamma_{ji} u_j = u_i$ ($i < j$), and $\mathscr{I}_i = \operatorname{Ann} u_i$ ($i \leq j$). Then $\gamma_{ji} \tilde{u}_j = \theta_{ji} u'_i$, where θ_{ji} is a unit of R. We show that

$$\theta_{js} \equiv \theta_{jt} \pmod{\mathscr{I}_t}, \text{ if } s, t < j;\ s > t. \tag{24}$$

In fact, let $u_t = \delta_{st} u_s$ $(t < s < j)$. Then by the induction assumption we have $u'_t = \delta_{st} u'_s$. Since $\gamma_{js} u_j = u_s$, it follows that $\delta_{st} \gamma_{js} u_j = \delta_{st} u_s = u_t$. On the other hand, $\gamma_{jt} u_j = u_t$. Consequently, $\delta_{st} \gamma_{js} \equiv \gamma_{jt} \pmod{\operatorname{Ann} u_j}$, and hence $\delta_{st} \gamma_{js} \equiv \gamma_{jt} \pmod{\operatorname{Ann} u'_j}$. We now have $\gamma_{js} u'_j = \theta_{js} u'_s$, $\delta_{st} \gamma_{js} u'_j = \theta_{js} u'_t$ or $\gamma_{jt} u'_j = \theta_{js} u'_t$. At the same time we have $\gamma_{jt} u'_j = \theta_{jt} u'_t$, which proves (24). We consider the system of congruences

$$(25) \qquad \theta \equiv \theta_{jt} \pmod{\mathscr{I}_t} \; (t < j).$$

The congruences (24) show that the congruences in (25) are pairwise compatible. Since $\bigcap_{t<j} \mathscr{I}_t \supseteq \mathscr{I}_j \neq 0$, it follows that there exists a unit $\theta \in R$ which satisfies the system (25). Setting $u'_j = \theta^{-1} \tilde{u}_j$, we find that the map $\varphi(u_j) = u'_j$ is an isomorphism of the cyclic submodules u_j and u'_j which extends all of the isomorphisms φ_i, $i < j$. Consequently, we can extend the isomorphisms to construct an isomorphism of M and $\mathscr{N}$. The lemma is proved.

THEOREM 8. *Let R be an almost maximal valuation ring. The indecomposable injective R-modules are precisely the divisible locally cyclic modules. A divisible locally cyclic module M is determined up to isomorphism by the ideal class $\mathscr{A}$ containing the annihilators of its nonzero elements, and any ideal class $\mathscr{A}$ corresponds to some divisible locally cyclic module M.*

PROOF. In **[4]** it is shown that every indecomposable injective R-module is locally cyclic. According to Theorem 7, a locally cyclic module is injective if and only if it is divisible. This proves the first part of the theorem. The proof of the second part uses Lemmas 23 and 24, and is completely analogous to the proof of Theorem 4.

REFERENCES

1. I. Kaplansky, *Modules over Dedekind rings and valuation rings*, Trans. Amer. Math. Soc., **72** (1952), 327–340.

2. S. D. Berman and N. I. Vishnyakova, *On the multiplicative semigroup of ideals of a valuation ring*, Publ. Math. Debrecen **29** (1982), 171–176.

3. J. Lambek, *Lectures on Rings and Modules*, 3rd ed., New York: Chelsea, 1986.

4. D. Gill, *Almost maximal valuation rings*, J. London Math. Soc., **4**, no. 1, 1971, 140–149.

5. C. Faith, *Lectures on injective modules and quotient rings*, Lecture Notes in Math., vol. 49, Springer-Verlag, Berlin and New York.

Contemporary Mathematics
Volume 93 (1989)

On The Modules Over A Real Crossed Group Algebra

K. BUZÁSI

ABSTRACT. Let G be an arbitrary group containing infinite cyclic subgroups of finite index and K an arbitrary field such that the group algebra KG is semisimple. S. D. Berman and K. Buzási showed (see **[1]**) that the investigation of the finitely generated KG-modules can be reduced to the study of modules over so called algebras of type E over K. This algebras are crossed group algebras of the infinite cyclic group or the infinite dihedral group.

S. D. Berman and K. Buzási describe in [**2**] all the algebras of type E over the real field R. They determined the finite R-dimensional finitely generated modules over all the algebras A_i of type E over R, and the finitely generated torsion-free modules over algebras A_i which contain zero divisors. It was proved that the algebra

(1) $$A = (R, a, b); \lambda a = a\lambda; \lambda b = b\lambda; b^{-1}ab = a^{-1}; b^2 = 1 \quad (\lambda \in R)$$

of type E over R contains no zero divisors. K. Buzási proved (see **[3]**) that algebra A is not a ring of principal left ideals. It follows from this fact that the finitely generated torsion-free A-modules cannot be described by classical results and by the methods applied in the paper [**2**]. K. Buzási characterized all the left ideals of the algebra A and studied the isomorphism between these left ideals (see **[4]**).

The aim of the present is to describe the structure of finitely generated torsion-free A-modules (up to the A-isomorphism), its direct decomposition and to determine the Grothendieck group $K_0(A)$ of the algebra A.

§1. Preliminary Results

Let A denote the algebra defined by (1) throughout this paper. It is a crossed group algebra of the infinite dihedral group over the real field R. It is clear that the real group algebra $R(a)$ of the infinite cyclic group $\overline{(a)}$ is a subalgebra in A. In **[1]** we defined an R-automorphism of $R(a)$: $\overline{f(a)} = f(a^{-1})$ for the elements $f(a) \in R(a)$. The element $f(a)$ is called symmetric if $\overline{f(a)} = \mu a^k f(a)$ for some $\mu \in R$; $k \in Z$.

It was proved in [**3**] that

LEMMA 1. *Every left ideal I of algebra A can be expected by $I = I_1 d$, where $d \in R(a)$ and I_1 is a left ideal of A generated by elements p, $1 + qb$, where p, $q \in R(a)$; element p is symmetric and generates the ideal $I_1 \cap R(a)$.*

LEMMA 2. *Every element of the left ideal $I \subseteq A$ generated by elements p, $l + qb(p, q \in R(a);\ (p) = R(a) \cap I)$ can be written as*

$$ubp + v(1 + qb),$$

where $u, v \in R(a)$.

LEMMA 3. *Let I_1 and I_2 be two left ideals in A. Let L be the quotient field of $R(a)$. Then the crossed product $T = L(b)$ of the field L with an automorphism of order 2 generated by the element b is a skew field. The left ideals I_1 and I_2 are A-isomorphic if and only if there exists an element $t \in T$ such that $I_1 = I_2 t$.*

§2. Direct Decomposition of A-Modules

The modules considered there are unitary and left modules. It will be shown in this part that every finitely generated torsion-free A-module is decomposed into the direct sum of submodules A-isomorphic to a left ideal of A.

LEMMA 2.1. *Let M be a torsion-free cyclic A-module. Then M is A-isomorphic to the module A.*

The proof is clear. Let $M = (u)$. Then every element $x \in M$ can be written as $\alpha \in M$, $\alpha \in A$, and the map

$$\varphi \colon M \to A; \varphi(\alpha u) = \alpha$$

is an A-isomorphism. □

LEMMA 2.2. *Let M be a torsion-free A-module generated by two elements. If the $R(a)$-rank of M equals 2, then M is A-isomorphic to a left ideal of A.*

PROOF. Let $M = (u_1, u_2)$. Since the $R(a)$-rank of M equals 2, the elements u_1, u_2 generate M as an $R(a)$-module too. So

$$M = \{\lambda_1 u_1 + \lambda_2 u_2 \mid \lambda_1, \lambda_2 \in R(a)\}.$$

Let L be the quotient field of $R(a)$ and form the module

$$\overline{M} = M \otimes_{R(a)} L.$$

Let $T = L(b)$ denote a skew field which is a crossed product of L with the automorphism of order 2 generated by the element b (see Lemma 3).

By the properties of tensor products the elements u_1, u_2 are generators also in the module $\overline{M}$. The L-rank of the module $\overline{M}$ is equal to 2, so it is a linear space of dimension 2 over L. Since $\overline{M}$ is, at the same time, an A module and an L-module, then it is a T-module.

But every T-module which is, at the same time, a linear space of dimension 2 over L, is isomorphic to the skew field T. Indeed, let $\bar{x} \in \overline{M}, \bar{x} \neq 0$. Then the set $T.\bar{x} \subseteq \overline{M}$ is a linear space of dimension 2 over L, hence

(2.1) $$\overline{M} = Tx\bar{x}.$$

On the other hand, the map

$$\varphi\colon T.\bar{x} \to T; \varphi(t\bar{x}) = t, t \in T$$

is an isomorphism of $T.\bar{x}$ onto T, and by (2.1) it is an isomorphism of $\overline{M}$ onto T.

Applying (2.1) to an element $x \in M$, we have

(2.2) $$x = \lambda_1 u_1 + \lambda_2 u_2 = tx(\lambda_i \in R(a), t \in T).$$

In particular,

$$u_1 = t_1\bar{x}, u_2 = t_2\bar{x}$$

holds for some $t_1, t_2 \in T$. Because of $t_i = \beta_i^{-1}t_i'(\beta_i \in R(a),\ t_i' \in A,\ i = 1, 2)$ and substituting into (2.2) we have

$$x = \lambda_1 t_1\bar{x} + \lambda_2 t_2\bar{x} = (\beta_1\beta_2)^{-1}[\lambda_1 t_1' + \lambda_2 t_2']\bar{x}.$$

Here $\beta_1, \beta_2, t_1', t_2'$ and $\bar{x}$ are fixed elements, hence the map φ defines the isomorphism

$$x \to (\lambda_1 t_1' + \lambda_2 t_2') \in A$$

of the module M into the algebra A. While x runs over M, the element $\lambda_1 t_1' + \lambda_2 t_2'$ runs over a left ideal in A. □

A more general fact is true:

LEMMA 2.3. *Let M be a finite generated torsion-free A-module. If the A-rank of M is equal to 1, then M is A-isomorphic to a left ideal of A.*

PROOF. Let us form a module

$$\overline{M} = M \bigotimes_{R(a)} L,$$

where L is the quotient field of $R(a)$. $\overline{M}$ is an A-module of A-rank 1, and is a linear space of dimension 1 over the skew field T (T is defined in Lemma 3). Consequently $\overline{M} = T.\bar{x}$ for an arbitrary fixed $\bar{x} \in \overline{M}$, $x \neq 0$, and $\varphi\colon \overline{M} - T$; $\varphi(t\bar{x}) = t$; $t \in T$ defines an A-isomorphism of $\overline{M}$ onto T. In particular $\varphi(M) \subseteq T$. Let the module M is generated by elements $u_1, \dots, u_n$. Then $u_i = t_i\bar{x} = t_i'\beta^{-1}\bar{x}$ ($\beta \in R(a), t_i' \in A, i = 1, \dots, n$) holds for some elements $t_i \in T$. Hence

$$x = (\lambda_1 t_1 + \cdots + \lambda_n t_n)\bar{x} = (\lambda_1 t_1' + \cdot + \lambda_n t_n')\beta^{-1}.\bar{x}$$

for every element $x = \lambda_1 u_1 + \cdots + \lambda_n u_n \in M$, where $t_1', \cdots, t_n', \beta, \bar{x}$ are fixed elements. Applying the isomorphism φ to the element x

$$\varphi(x) = \lambda_1 t_1' + \cdots + \lambda_n t_n' \in A.$$

While x runs over M, the element $\lambda_1 t_1' + \cdots + \lambda_n t_n'$ runs over a left ideal in A. □

The following is important:

LEMMA 2.4. *Let M be a finitely generated torsion-free A-module and $N \subseteq M$ a submodule such that the factor module M/N is A-isomorphic to a left ideal I of A. Then there exists a submodule of M A-isomorphic to I.*

PROOF. Let $\varphi\colon I \to M/N$ be an A-isomorphism mapping the left ideal I onto the M/N. If I is a principal left ideal, then it is A-isomorphic to the A and the lemma is clearly true.

If I is a non-principal left ideal, then on the basis of Lemma 1 it can be generated by elements $p,\ l+qb;\ p,q \in R(a);\ (p) = I \cap R(a)$ and it can be assumed that $\overline{p} = p$. By Lemma 2 the elements of I can be written as $ubp + v(1+qb);\ u,v \in R(a)$. Let

$$\varphi(bp) = \tilde{e}_1 \in M/N; \varphi(1+qb) = \tilde{e}_2 \in M/N,$$

where the elements $\tilde{e}_i$ are residue-classes of M/N. Let these classes be represented by fixed elements $e_1 \in \tilde{e}_1;\ e_2 \in \tilde{e}_2$, and let us consider the set

$$I_M = \{ue_1 + ve_2 \mid u,v \in R(a)\} \subseteq M. \tag{2.3}$$

Since I as an $R(a)$-module has an $R(a)$-basis $bp, l+bp$, hence I_M is an $R(a)$-submodule of the module M.

We intend to show that I_M is an A-module too. It is sufficient for this purpose to prove that the elements e_1, e_2 can be chosen such that $be_1 \in I_M, be_2 \in I_M$.

It is easy to see that the following equations hold in algebra A

$$\begin{aligned} b(bp) &= q(bp) - p(1+qb) \\ b(1+qb) &= p^{-1}(1+q\bar{q})(bp) - \bar{q}(1+qb), \end{aligned} \tag{2.4}$$

where $p^{-1}(1+q\bar{q}) \in R(a)$ because of $(p) = R(a) \cap I$. The following relations correspond to the equations (2.4) in M/N

$$\begin{aligned} b\tilde{e}_1 &= q\tilde{e}_1 - p\tilde{e}_2 \\ b\tilde{b}_2 &= p^{-1}(1+q\bar{q})\tilde{e}_1 - \bar{q}\tilde{e}_2, \end{aligned}$$

which implies the equations

$$\begin{aligned} be_1 &= qe_1 - pe_2 + n_1 \\ be_2 &= p^{-1}(1+q\bar{q})e_1 - \bar{q}e_2 + n_2 \end{aligned} \tag{2.5}$$

$(n_1, n_2 \in N)$ for the representative elements. We show that such representative elements $e'_i \in \tilde{e}_i$ $(i = 1,2)$ can be chosen that

$$\begin{aligned} be'_1 &= qe'_1 - pe'_2 \\ be'_2 &= p^{-1}(1+q\bar{q})e'_1 - \bar{q}e'_2. \end{aligned} \tag{2.5'}$$

Indeed, let $e'_1 = e_1 + x;\ e'_2 = e_2 + y$ $(x,y \in N)$. Using the equations (2.5),

$$\begin{gathered} be'_1 = b(e_1+x) = be_1 + bx = qe_1 - pe_2 + n_1 + bx \\ = q(e'_1 - x) - p(e'_2 - y) + n_1 + bx = qe'_1 - pe'_2 + (b-q)x + py + n_1, \\ be'_2 = b(e_2+y) = be_2 + by = p^{-1}(1+q\bar{q})e_1 - \bar{q}e_2 + n_2 + by \\ = p^{-1}(1+q\bar{q})(e'_1 - x) - \bar{q}(e'\_y) + n_2 + by \\ = p^{-1}(1+q\bar{q})e'_1 - \bar{q}e'_2 + [-p^{-1}(1+q\bar{q})x(\underline{+}\bar{q})y + n_2]. \end{gathered}$$

That is (2.5′) holds if and only if the system of equations

$$(2.5'') \qquad \begin{aligned} (b-q)x + py + n_1 &= 0 \\ -p^{-1}(1+q\bar{q})x + (b+\bar{q})y + n_2 &= 0 \end{aligned}$$

is solvable for $x, y \in N$. Multiplying the first equation of (2.5″) by $b+\bar{q}$ and the second one by p on the left hand, we obtain

$$(2.6) \qquad (b+\bar{q})n_1 = pn_2.$$

Since the left multiplication by b is an operator acting as $b(bz) = -z$, it follows from the first equation of (2.5) that

$$\begin{aligned} -e_1 = b(be_1) &= b(qe_1 - pe_2 + n_1) = bqe_1 - bpe_2 + bn_1 = \bar{q}be_1 - pbe_2 + bn_1 \\ &= \bar{q}(qe_1 - pe_2 + n_1) - p[p^{-1}(1+a\bar{q})e_1 - \bar{q}e_2 + n_2] + bn_1 = -e_1 \\ &\qquad + (b+\bar{q})n_1 - pn_2, \end{aligned}$$

that is $0 = (b+\bar{q})n_1 - pn_2$, consequently the equation (2.6) holds. Hence the system of equations (2.5′) is solvable. This means that M contains a submodule I_M A-isomorphism to the left ideal I. □

THEOREM 2.1. *Every finite generated torsion-free A-module is A-isomorphic to the direct sum of the left ideals of A.*

PROOF. Let M be a finite generated torsion-free A-module. We prove the theorem by induction on the A-rank of M. If the A-rank of M equals 1 then by Lemma 2.3 M is A-isomorphic to the left ideal of A.

Assume that the theorem is proved for the modules of A-rank $\leq n$ and let M be a module of A-rank $n+1$. Let $u_1, \dots, u_n, u_{n+1}$ be the A-basis of M. Let us consider the cyclic A-module N generated by the element u_{n+1}.

If the factor module M/n is a torsion-free one, then it is a finitely generated torsion-free A-module of A-rank n an by the induction hypothesis

$$M/NI_1 \dot{+} \cdots \dot{+} I_s,$$

where I_j are left ideals in A. Applying Lemma 2.4, we have

$$M \cong I_1 \dot{+} \cdots \dot{+} I_s \dot{+} N,$$

where N is A-isomorphic to A.

If M/N is not a torsion-free A-module, then

$$M/N \cong M_1/N \dot{+} P/N,$$

where M_1/N is a torsion-free A-module of A-rank n and P/N a torsion A-module. Clearly it hold that

$$(2.7) \qquad M/N/P/N \cong M_1/N,$$

and by the second theorem on isomorphisms

$$(2.8) \qquad M/N/P/N \cong M/P.$$

Using (2.7) and (2.8) and the induction hypothesis

$$M/P \cong M_1/N \cong I_1 \dot{+} \cdots \dot{+} I_s,$$

where I_j are left ideals in A. Applying Lemma 2.4 we have

$$M \cong I_1 \dot{+} \cdots + I_s \dot{+} P.$$

Here A-rank of P equals 1, hence by Lemma 2.3, it is isomorphic to a left ideal of A. □

§3. On the Isomorphism of Direct Decomposition

It was proved in the previous chapter that every finitely generated torsion-free A-module is A-isomorphic to the direct sum of left ideals in A. In this chapter we show that an arbitrary finitely generated torsion-free A-module is either A-isomorphic to a left ideal of A or is a free A-module.

LEMMA 3.1. *Let I_1, I_2, I_3, I_4 be left ideals in A. The direct sum $I_1 \dot{+} I_2$ is A-isomorphic to the direct sum $I_3 \dot{+} I_4$ if and only if there exists elements t_1, t_2, t_3, t_4 in the skew field T (defined in Lemma 2.3) such that the matrix*

$$\begin{pmatrix} t_1 & t_2 \\ t_3 & t_4 \end{pmatrix} \tag{3.1}$$

is invertible and there exists elements $i_1 \in I_1, i_2 \in I_2$ for all the elements $i_3 \in I_3, i_4 \in I_4$ such that it holds that

$$\begin{aligned} i_1t_1 + i_2t_3 &= i_3 \\ i_1t_2 + i_2t_4 &= i_4. \end{aligned} \tag{3.2}$$

PROOF. Consider the A-isomorphism

$$\varphi\colon I_1 \dot{+} I_2 \to I_3 \dot{+} I_4; \varphi[(i_1, i_2)] = (i_3, i_4); (i_j \in I_j; j = 1, \ldots, 4). \tag{3.3}$$

In particular, there exist elements $i_3' \in I_3; i_4' \in I_4$ for every fixed $i_1 \in I_1$ such that

$$\varphi[(i_1, 0)] = (i_3', i_4').$$

By Lemma 2.3 the A-isomorphism of left ideals in A is a right multiplication by an element from T. Hence there exist elements t_1, t_2 such that $i_3' = i_1t_1; i_4' = i_1t_2$, that is

$$\varphi[(i_1, 0)] = (i_1t_1, i_2t_2).$$

Similarly,

$$\varphi[(0, i_2)] = (i_2t_3, i_2t_4),$$

for some elements $t_3, t_4 \in T$, consequently

$$\varphi[(i_1, i_2)] = (i_1 + t_1 + i_2t_3, i_1t_2 + i_2t_4),$$

that is the system of equations (3.2) is solvable for $t_1, \ldots, t_4 \in T$. Since φ is an isomorphism, the matrix (3.1) is invertible. On the other hand, the system of equations (3.2) defines the A-isomorphism φ. □

LEMMA 3.2. *Let I_1 and I_2 be non-principal left ideals in A. Then the direct sum $I_1 \dotplus I_2$ is A-isomorphic to $A + \dot{A}$.*

PROOF. Since the elements $(1,0)$ and $(0,1)$ generate the direct sum $A \dotplus A$, applying Lemma 3.1, our Lemma will be proved if one can find elements $t_1, \dots, t_4 \in T$ such that the matrix of the form (3.1) is invertible and the system of equations

$$(3.4) \qquad \left.\begin{aligned} i + 1t_1 + i_2t_3 &= 1 \\ i_1t_2 + i_2t_4 &= 0 \end{aligned}\right\} \qquad \left.\begin{aligned} i'_1t_1 + i'_2t_3 &= 0 \\ i'_1t_2 + i'_2t_4 &= 1 \end{aligned}\right\}$$

is solvable. Let the left ideals I_1, I_2 be given by

$$I_1 = (p_1, l + q_1b); I_2 = (p_2, 1 + q_2b),$$

$(p_j, q_j \in R(a); (p_j = I_j \cap R(a))$, let

$$t_1 = (1+b)(p_1p_2)^{-1}; t_2 = b(p_1p_2)^{-1}; t_3 = -b(p_1p_2)^{1}; t_4 = (1-b)(p_1p_2)^{-1}.$$

Clearly, the matrix

$$\begin{pmatrix} 1 + bp_1p_2^{-1} & bp_1p_2^{-1} \\ -bp_1p_2^{-1} & 1 - bp_1p_2 \end{pmatrix}$$

is invertible. By Lemma 2

$$i_j = u_jbp_j + v_j(1 + q_jb)(u_j, v_j \in R(a); j = 1, \dots, 4).$$

Then the first system of (3.4) has a form

$$(3.5) \qquad \left.\begin{aligned} &[u_1bp_1 + v_1(1+q_1b)](1+b) + u_2bp_2 + b_2(1+q_2b)](-b) = p_1p_2 \\ &[u_1bp_1 + v_2(1+q_1b)] \cdot b + [u_2bp_2 + b_2(1+q_2b)](1-b) = 0 \end{aligned}\right\}.$$

This system implies for $u_1, v_1, u_2, v_2 \in R(a)$ the system of equations

$$(3.5') \qquad \left.\begin{aligned} -u_1p_1 + v_1(1-q_1) + u_2p_2 + v_2q_2 &= p_1p_2 \\ u_1p_1 + v_1(1+q_1) - v_2 &= 0 \\ -u_1p_1 - v_1q_1 + u_2p_2 + v_2(1+q_2) &= 0 \\ v_1 + u_2p_2 + v_2(-1+q_2) &= 0 \end{aligned}\right\}.$$

It is easy to see that the elements

$$u_1 = -p_2(1+q_1); v_1 = p_1p_2; u_2 = -p_1; v_2 = 0$$

give a solution of (3.5′).

Similarly, the second system of (3.4) implies

$$(3.6) \qquad \left.\begin{aligned} &[u'_1bp_1 + v'_1(1+q_1b)](1+b) + [u'_2bp_2 + v'_2(1+q_2b)](-b) = 0 \\ &[u'_1bp_1 + v'_1(1+q_1b)] \cdot b + [u'_2bp_2 + v'_2(1+q_2b)](1-b) = p_1p_2 \end{aligned}\right\}.$$

It is easy to show that

$$u'_1 = p_2; v'_1 = 0; u'_2 = p_1(1-q_2); v'_2 = p_1p_2$$

is a solution of (3.6). □

LEMMA 3.3. *Let I be a left ideal of AS. Then the direct sum $I\dot{+}$ is A-isomorphic to the direct sum $A\dot{+}$.*

PROOF. By Lemmas 1 and 2 the left ideal I can be given by $I = (p, 1 + qb), p, q \in R(a)$; $(p) = I \cap R(a)$ and every element of I can be expressed by $ubp + v(1 + qb)$; $u, v \in R(a)$. Let

$$t_1 = (1 + b)p^{-1}; t_2 = bp^{-1}; t_3 = (-b)p^{-1}; t_4(1 - b)p^{-1}.$$

The matrix

$$\begin{pmatrix} t_1 & t_2 \\ t_3 & t_4 \end{pmatrix}$$

is invertible. Since $A\dot{+}A$ is generated by elements $(1, 0)$ and $(0, 1)$, the lemma will be proved if the system of equations

$$(3.7)\qquad \left.\begin{aligned} [ubp + v(1 + qb)](1 + b) + (s_0 + s_1 b)(-b) &= p \\ [ubp + v(1 + qb)] \cdot b + (s_0 + s_1 b)(1 - b) &= 0 \\ [u_1 bp + v_1(1 + qb)](1 - b) + s'_0 + s'_1 b)(-b) &= 0 \\ [u_1 bp + v_1(1 + qb)] \cdot b + (s'_0 + s'_1 b)(1 - b) &= p \end{aligned}\right\}$$

is solvable with respect to $u, v, u_1, v_1 \in R(a)$; $s_0 + s_1 b, s'_0 + s'_1 b \in A$; s_Q, $s_1, s'_0, s'_1 \in R(a)$. It is easy to see that

$$u = -1 - q; v = p; s_0 = 0; s_1 = -p; u_1 = 1; v_1 = 0; s'_0 = p; s'_1 = p$$

is a solution of (3.7) . □

THEOREM 3.1. *Every finitely generated torsion-free A-module is either A-isomorphic to a left ideal of A or is a free A-module.*

PROOF. Let M be a finitely generated torsion-free A-module. If the rank of M is equal to 1, then by Lemma 2.3 it is A-isomorphic to a left ideal of A. If the rank of M is greater than 1, then by Theorem 2.1 it is A-isomorphic to the direct sum of left ideals in A

$$M \cong I_1 \dot{+} \cdots \dot{+} I_s (s < 1).$$

Applying Lemmas 3.2 and 3.3 we obtain

$$M \cong A \dot{+} \cdots \dot{+} A,$$

which proves the theorem. □

Lemmas 3.2 and 3.3 give an opportunity to determine the Grothendieck group $K_0(A)$ of algebra A.

THEOREM 3.2. *The Grothendieck group $K_0(A)$ of algebra A is isomorphic to the infinite cyclic group.*

PROOF. As usual, we define the Grothendieck group $K_0(A)$ of algebra A as a factor group S/H of a free abelian group generated by all the non-isomorphic left ideals of A by the subgroup H generated by all the relations

$$(3.8)\qquad I_j \dot{+} A \cong A \dot{+} A$$

$$(3.9)\qquad I_j \dot{+} I_k \cong A\dot{A}$$

$(j,k = 1,2\dots)$. Then $S = (A, I_1, I_2, \dots)$, where A is the left ideal A-isomorphic to all the principal left ideals, $I_1, I_2, \dots$ are all the non-isomorphic non-principal left ideals of A. Let $u_0 = A, u_i = I_i (i = 1,2,\dots)$. Then $S = (u_0, u_1, u_2, \dots)$. It follows from (3.7) and (3.9) that

$$u_j + u_0 - 2u_0 = u_j - u_0 (j = 1,2,\dots)$$
$$u_j + u_k - 2u_0 = u_j - u_0 + u_k - u_0 (j,k = 1,2,\dots).$$

Hence the subgroup is generated by all the elements $u_j - u_0 (j = 1,2,\dots)$, that is the rank $r(H)$ of H is equal to $r(H) = r(S) - 1$, consequently S/H is the infinite cyclic group.

REFERENCES

[1.] S. D. Berman and K. Buzási, On the modules over group algebras of groups containing infinite cyclic subgroup of finite index. Studia. Sci. Math. Hungarica 16 (1981), 455–470. (Russian)

[2] S. D. Berman and K. Buzási, Description of all the finite dimensional real representations of groups containing infinite cyclic subgroup of finite index. Publ. Math. Debrecen, t.31(1984), 144–144.

[3] K. Buzási, On the structure of a real crossed group algebra.

[4] K. Buzási, On isomorphisms of left ideals in a real crossed group algebra.

L. KOSSUTH UNIVERSITY
DEBRECEN
H: 4010, PF. 12
HUNGARY

Contemporary Mathematics
Volume 93 (1989)

The Minimal Number of Generators of Wreath Products of Nilpotent Groups

K. BUZÁSI AND L. G. KOVÁCS

In memory of Samuil Davidovič Berman

1. For a group G, let $d(G)$ denote the minimum of the cardinalities of the generating sets of G.

THEOREM. *Let A and B be nontrivial finitely generated nilpotent groups. Write $A \operatorname{wr} B$ for their restricted standard wreath product, and $A \times B$ for their direct product. If the commutator subgroup B' of B is finite, then* $d(A \operatorname{wr} B) = \max\{1 + d(A), d(A \times B)\}$.

Under the stronger assumption that B is either finite or abelian, this was proved by Yeo Kok Chye [7]. In the simplest case not covered by his work, B has a cyclic subgroup of finite index, and the problem can be translated to one concerning representations of such B over finite prime fields. During the last years of his life, S. D. Berman made substantial contributions to the representation theory of (not necessarily nilpotent) groups which have cyclic subgroups of finite index. He and the first author of this paper repeatedly discussed the possibility that their results could be applied in the context of wreath products. Indeed, the results which are used in the proof below and which are more recent than Yeo Kok Chye [7] come from Berman and Buzási [2].

We do not know whether the theorem remains valid if one omits the condition that B' be finite. For example, let p and q be distinct primes, A a group of order p, and B the group defined by the presentation

$$\langle x, y, z \mid [x, z] = [y, z] = [x, y]z^q = 1 \rangle :$$

1980 *Mathematics Subject Classification* (1985 *Revision*). Primary 20E22, 20C07.
The second author gratefully acknowledges the hospitality of the University of Essen and the support of the Deutsche Forschungsgemeinschaft, which he enjoyed during the preparation of this paper.

at the time of writing we have not been able to decide whether $A \operatorname{wr} B$ can be generated by 3 elements. (As usual, $[x, y]$ denotes the commutator $x^{-1}y^{-1}xy$.)

The proof of the theorem will occupy the rest of the paper.

2. Yeo Kok Chye [7] started by proving, in the last paragraph of his §1, that if A is a finitely generated nilpotent group and if B has a nontrivial finite quotient then

$$d(A \operatorname{wr} B) \geq 1 + d(A).$$

In the first paragraph of the proof of his (5), he showed that when discussing minimal generating sets of $A \operatorname{wr} B$ and $A \times B$ with nilpotent A, one may replace A by A/A'. He then noted that $d(A \operatorname{wr} B) \geq d(A \times B)$ whenever A is abelian. Thus *for proving a theorem like ours it suffices to show that* $A \operatorname{wr} B$ *can be generated by* $\max\{1 + d(A), d(A \times B)\}$ *elements whenever* A *is abelian.*

A finitely generated abelian A is the direct product of a free abelian A_0 and a finite abelian A_1, with $d(A) = d(A_0) + d(A_1)$. For finitely generated nilpotent B one has then

$$d(A \times B) = d(A \times (B/B')) = d(A_0) + d(A_1 \times (B/B')) = d(A_0) + d(A_1 \times B).$$

Consider A and B embedded in $A \operatorname{wr} B$ as usual (that is, as "first coordinate subgroup" and "top group"): then A_0, A_1, B generate $A \operatorname{wr} B$, while the subgroup generated by A_1 and B is isomorphic to $A_1 \operatorname{wr} B$. It follows that if $A_1 \operatorname{wr} B$ can be generated by $\max\{1 + d(A_1), d(A_1 \times B)\}$ elements, then $A \operatorname{wr} B$ can be generated by

$$d(A_0) + \max\{1 + d(A_1), d(A_1 \times B)\},$$

that is, by $\max\{1 + d(A), d(A \times B)\}$ elements. Thus for proving a theorem like ours it suffices to show that $A \operatorname{wr} B$ can be generated by the right number of elements *whenever* A *is abelian and finite.*

Strictly speaking, our theorem for finite B was proved in [7], as item (4), only when A is also finite: the argument above shows that no such assumption on A was needed there. For abelian B, our theorem was item (5) in [7]. Its proof involved constructing an explicit generating set of the right size for $A \operatorname{wr} B$. Beyond the initial reduction quoted above, we shall not use (5) of [7] here. Our arguments will be non-constructive in the sense that they give little help towards finding actual generating sets of minimal size in $A \operatorname{wr} B$. Even for finite A and B, we know of no constructions beyond the rather special ones given in the unpublished parts of the thesis **[6]** of Yeo Kok Chye.

3. The central step in the proof of our theorem is the following. Let F be a free group of finite rank and F/R a nilpotent quotient of F containing a finite normal subgroup Q/R such that F/Q is infinite cyclic. Choose a prime p not dividing $|Q/R|$, denote by $R/R'R^p$ the largest elementary abelian p-quotient of R, and regard $R/R'R^p$ as an $\mathbf{F}_p(F/R)$-module with $\mathbf{F}_p$ the field of p elements and the action of F/R on $R/R'R^p$ coming from conjugation in F. Since every finitely generated nilpotent group is finitely presented, R is the normal closure of finitely many elements in F, and therefore $R/R'R^p$ is a finitely generated $\mathbf{F}_p(F/R)$-module.

LEMMA. *The free $\mathbf{F}_p(F/R)$-module of rank $d(F) - 1$ is a homomorphic image of $R/R'R^p$.*

COROLLARY. *If A is a finite abelian p-group with $d(A) < d(F)$ and α is the natural homomorphism of $A \operatorname{wr} (F/R)$ onto F/R, then F has a homomorphism τ onto $A \operatorname{wr} (F/R)$ such that the composite $\tau\alpha$ is the natural homomorphism of F onto F/R.*

Deferring the proof of the lemma until later, we show here how the corollary follows from the lemma, and in the next section we derive the theorem from this corollary.

By the lemma, $R/R'R^p$ has a submodule $S/R'R^p$ such that R/S is a free $\mathbf{F}_p(F/R)$-module of rank $d(A)$. Let f be an element of F such that $F/Q = \langle fQ \rangle$: then the index $|Q/R|$ of $\langle fS \rangle(R/S)$ in F/S is prime to p, so by a theorem of Gaschütz (I.17.4a in Huppert [5]) R/S is a semidirect factor in F/S. It follows that F/S is isomorphic to $(A/A^p) \operatorname{wr} (F/R)$ with R/S matching the base group. The composite of a corresponding homomorphism σ of F onto $(A/A^p) \operatorname{wr} (F/R)$ with the natural projection of this wreath product onto F/R then has kernel R: so it differs from the natural projection $F \to F/R$ only by some automorphism of F/R. As in any wreath product, each automorphism of F/R lifts to an automorphism of $(A/A^p) \operatorname{wr} (F/R)$: hence σ can be corrected if necessary to ensure that the composite mentioned is precisely the natural map $F \to F/R$. Because F is free, the σ so chosen may be written as the composite of a $\tau \colon F \to A \operatorname{wr} (F/R)$ with the homomorphism $\pi \colon A \operatorname{wr} (F/R) \to (A/A^p) \operatorname{wr} (F/R)$ obtained from the natural projection $A \to A/A^p$. Of course then $\tau\alpha$ is the natural map $F \to F/R$.

It remains to prove that τ is surjective. Since $\tau\alpha$ is surjective, $F\tau$ supplements the kernel A^B of α. If $F\tau$ is a proper subgroup, it is contained in some maximal subgroup H of our finitely generated wreath product, and then $A^B \cap H$ is a maximal $\mathbb{Z}(F/R)$-submodule of A^B. The quotient $A^B/(A^B \cap H)$ is a simple module and so must have prime exponent: thus $A^B \cap H$ contains the kernel $(A^p)^B$ of π. It follows that $H\pi$ is still a proper subgroup, contrary to the surjectivity of σ $(= \tau\pi)$. Thus $F\tau$ cannot be proper, and the deduction of the corollary is complete.

4. We have already seen that the theorem holds when B is finite, so in the sequel B will always be assumed infinite. Then $1 + d(A) \leq d(A \times B)$, so what we have to prove is that $A \operatorname{wr} B$ can be generated by $d(A \times B)$ elements.

For an element g of a group G, let $\pi(g)$ denote the set of prime divisors of the order $|g|$ of g. When the subgroup generated by g is infinite, we say that $|g| = 0$ and $\pi(g)$ is the set of all primes. Each finitely generated abelian G has generating sets $\{g_1, \dots, g_{d(G)}\}$ such that each term of the sequence

$|g_1|, \dots, |g_{d(G)}|$ divides the next, so

$$\pi(g_1) \subseteq \cdots \subseteq \pi(g_{d(G)}).$$

Set $G = B/B'$ and choose such a generating set. Since B is nilpotent and B' is finite, in each coset g_i one can choose an element b_i such that $\pi(b_i) = \pi(g_i)$. [This is the only point where the finiteness of B' is used.] The subgroup of B generated by $b_1, \dots, b_{d(G)}$ supplements B' and so equals B: thus $d(B) = d(G)$ and $B = \langle b_1, \dots, b_{d(B)} \rangle$. The effect of the assumption that B be infinite is that $|b_{d(B)}| = 0$.

For each prime p, set

$$k(p) = \begin{cases} 0 & \text{when } p \in \pi(b_1), \\ \max\{\, i \mid p \notin \pi(b_i) \,\} & \text{when } p \notin \pi(b_1): \end{cases}$$

note that

$$k(p) < d(B) \text{ for all } p.$$

Let A be a finite abelian group and A_p the largest p-quotient of A. It is easy to see that

$$d(A_p \times B) = d(A_p \times G) = \max\{d(B), d(A_p) + d(B) - k(p)\} :$$

what we shall use is that consequently

$$d(A \times B) \geq d(A_p \times B) \geq d(A_p) + d(B) - k(p).$$

For each p, let B_p denote the subgroup generated in B by $\{\, b_i \mid 1 \leq i \leq k(p) \,\}$ and $b_{d(B)}$. Since the last set consists of elements of finite order prime to p, its normal closure in B_p is finite of order prime to p, and of course the quotient of B_p modulo this normal subgroup is infinite cyclic.

Let F be a free group of rank $d(A \times B)$, freely generated by a set $\mathbf{f} = \{f_1, \dots, f_{d(A\times B)}\}$. Consider the homomorphism ρ of F onto B defined by

$$f_i\rho = \begin{cases} b_i & \text{when } i \leq d(B), \\ 1 & \text{when } i > d(B), \end{cases}$$

and denote its kernel by R. For each p, write $\mathbf{f}$ as the disjoint union $\mathbf{f} = \mathbf{f}_p \cup \mathbf{f}_{p'}$ where

$$\mathbf{f}_p = \{\, f_i \mid 1 \leq i \leq k(p) \text{ or } d(B) \leq i \leq d(A \times B)\},$$
$$\mathbf{f}_{p'} = \{\, f_i \mid k(p) < i < d(B)\},$$

and let F_p denote the subgroup of F generated by $\mathbf{f}_p$. Note that

$$d(F_p) = k(p) + d(A \times B) - d(B) + 1,$$

so by the last line of the second last paragraph $d(F_p) - 1 \geq d(A_p)$. Write R_p for the kernel $F_p \cap R$ of the restriction $\rho \downarrow F_p$. Since ρ maps F_p onto B_p, we have $F_p/R_p \cong B_p$.

We are now ready to apply the corollary of the lemma, with F_p, R_p, A_p in place of F, R, A. The conclusion may be put as follows. Let α_p be the natural homomorphism of $A_p \operatorname{wr} B$ onto B. Consider A_p and B embedded in

$A_p \operatorname{wr} B$. The subgroup $\langle A_p, B_p \rangle$ generated by A_p and B_p is then isomorphic to $A_p \operatorname{wr} B_p$. By that corollary there is a homomorphism $\tau_p \colon F_p \to A_p \operatorname{wr} B$ such that

$$F_p \tau_p = \langle A_p, B_p \rangle \text{ and } \tau_p \alpha_p = \rho \downarrow F_p.$$

Define a homomorphism $\rho_p \colon F \to A_p \operatorname{wr} B$ by

$$f_i \rho_p = \begin{cases} f_i \tau_p & \text{when } f_i \in \mathbf{f}_p, \\ f_i \rho & \text{when } f_i \in \mathbf{f}_{p'}. \end{cases}$$

Then $F \rho_p \geq F_p \rho_p = F_p \tau_p = \langle A_p, B_p \rangle \geq B_p$ and $\langle B_p, \mathbf{f}_{p'} \rho_p \rangle = B$ together imply that ρ_p is surjective, while of course $\rho_p \alpha_p = \rho$.

For each p, let π_p be the homomorphism of $A \operatorname{wr} B$ onto $A_p \operatorname{wr} B$ corresponding to the natural homomorphism of A onto A_p, and α the natural homomorphism of $A \operatorname{wr} B$ onto B: then $\pi_p \alpha_p = \alpha$ for all p. It is easy to see that the π_p and the α_p, with p ranging through the finite set of the prime divisors of $|A|$, form a pullback. This implies that there is a homomorphism $\varphi \colon F \to A \operatorname{wr} B$ such that $\varphi \pi_p = \rho_p$ for all relevant p. We claim that this φ must be surjective. Since $\varphi \alpha$ is surjective, it suffices to show that $F\varphi$ contains the kernel of α. That kernel is the base group of $A \operatorname{wr} B$, and it is an abelian torsion group: so all we need to show is that $F\varphi$ contains each prime-power order element x of that base group. If p is the prime divisor of the order of x, use that $\varphi \pi_p = \rho_p$ and ρ_p is surjective: so there is an $f\varphi$ congruent to x modulo the kernel of π_p. Choose a multiple m of the exponent of that kernel such that $m \equiv 1 \pmod{|x|}$: then $f^m \varphi = x^m = x$. This proves that φ is surjective, so $A \operatorname{wr} B$ can be generated by $d(A \times B)$ elements, and the deduction of the theorem from the corollary of the lemma is complete.

5. It remains to prove the lemma. The notation used in the last section has served its purpose: we now return to the notation of §3. Thus F/R is a nilpotent group with a finite normal subgroup Q/R such that F/Q is infinite cyclic: $F/Q = \langle fQ \rangle$. Since Q/R is finite, so is the index of its centralizer in F/R. Since p does not divide $|Q/R|$ and F/R is nilpotent, p cannot divide that index either. Let k be the least positive integer such that f^k centralizes Q/R: by the foregoing, p does not divide k. Now $\langle f^k R \rangle$ commutes with fR and with Q/R, so it is an infinite cyclic subgroup in the centre of F/R. Choose a prime q other than p; for each nonnegative integer m, let $R_m = \langle f^{kq^m}, R \rangle$: then each R_m is a normal subgroup in F, of index $kq^m|Q/R|$, and $\bigcap_m R_m = R$.

Let $R_m/R'_m R_m^p$ denote the largest elementary abelian p-quotient of R_m, viewed as an $\mathbf{F}_p(F/R_m)$-module. This module is completely reducible by Maschke's Theorem; in fact, by a result of Gaschütz **[4]**, it is a direct sum of a 1-dimensional trivial module and a free $\mathbf{F}_p(F/R_m)$-module of rank $d(F) - 1$. Since R_m/R is cyclic and central in F/R, its largest elementary abelian p-quotient is a 1-dimensional trivial $\mathbf{F}_p(F/R_m)$-module. Obviously, that quotient is $R_m/R(R'_m R_m^p)$. It follows that $R(R'_m R_m^p)/(R'_m R_m^p)$ is a free

$\mathbf{F}_p(F/R_m)$-module of rank $d(F)-1$, and hence so is $R/(R\cap(R'_mR^p_m))$. Of course $R'_mR^p_m \geq R'R^p$, and thus our conclusion may be put as follows. For each m, the $\mathbf{F}_p(F/R)$-module $R/R'R^p$ has a quotient on which R_m/R acts trivially and which is free of rank $d(F)-1$ as $\mathbf{F}_p(F/R_m)$-module.

The group-theoretic preparations for the proof of the lemma are now complete. In what follows, the role of the R_m will be taken on by the kernels of the natural homomorphisms of $\mathbf{F}_p(F/R)$ onto the $\mathbf{F}_p(F/R_m)$.

6. The rest of the argument no longer involves groups, only rings and modules. Accordingly we abandon all previous notation and start afresh, except that we keep $\mathbf{F}_p$ as the name of the field of p elements.

Let R denote the group algebra over $\mathbf{F}_p$ of a group which has a finite normal subgroup of order prime to p such that the corresponding factor group is infinite cyclic. (We have no further use for the assumption that the group be nilpotent.) Let $I_0 \geq \cdots \geq I_m \geq \dots$ be a descending chain of ideals in R such that each quotient R/I_m is finite yet $\bigcap_m I_m = 0$. Let d be a positive integer, and U a finitely generated right R-module with submodules U_m such that, for each m, U/U_m is annihilated by I_m and is free of rank d as (R/I_m)-module. *Then the free R-module of rank d is a homomorphic image of U.* The conclusions of the previous section show that if we can prove this claim, the lemma will follow.

The proof of Theorem 1 in Berman and Buzási [2] (for some suppressed detail, see also the analogous proof of Theorem 1 in Berman and Buzási [3]) readily specializes to yield that our R is a direct sum of full matrix algebras over crossed group algebras of the infinite cyclic group over finite fields. Let $R = \bigoplus R_i$ be any (two-sided) direct decomposition of R. Then $I_0R_i \geq \cdots \geq I_mR_i \geq \dots$ is a descending chain of ideals such that each R_i/I_mR_i is finite and $\bigcap_m I_mR_i = 0$. Further, UR_i is a finitely generated R_i-module such that UR_i/U_mR_i is annihilated by I_mR_i and is free of rank d as (R_i/I_mR_i)-module. Conversely, if for each i the free R_i-module of rank d is a homomorphic image of UR_i, then the free R-module of rank d is a homomorphic image of U. This sketch shows that it suffices to prove the italicized claim above for full matrix algebras R over crossed group algebras of the kind mentioned.

Suppose then that R is the full matrix algebra of degree n over some ring S, and let e be one of the diagonal matrix units in R (that is, a matrix with one diagonal entry 1 and all the other entries 0). Then $ReR = R$, and eRe may be naturally identified with S. The functors $-\bigotimes_R Re$ and $-\bigotimes_S eR$ provide an equivalence between the categories of (right) R-modules and S-modules (see Exercise 21.6 in Anderson and Fuller [1]). Since R as right R-module is the direct sum of n copies of eR, an R-module V is free of rank d if and only if $V\otimes_R Re$ is a free S-module of rank nd. Further, $eI_0e \geq \cdots \geq eI_me \geq \dots$ is a descending chain of ideals in S with each S/eI_me finite and $\bigcap_m eI_me = 0$; also, $U\otimes_R Re$ is a finitely generated S-module, while $(U\otimes_R Re)/(U_m\otimes_R Re)$ is annihilated by eI_me and is free of rank nd as (S/eI_me)-module. Conversely, if the free S-module of rank nd is a homomorphic image of $U \otimes_R Re$, then the free R-module of rank d is a homomorphic image of U. This sketch shows that it suffices to prove the claim with S and nd in place of R and d.

Shifting notation, we assume instead that R itself is a crossed group algebra of an infinite cyclic group over a finite field. As was noted in the discussion preceding Lemma 2 in Berman and Buzási [2], now R is a (not necessarily commutative) principal ideal domain and so each finitely generated R-module U is a direct sum of cyclic modules. Moreover, all proper homomorphic images of R are now finite, so each such U is the direct sum of a free module V and a finite module W. Let c denote the rank of V. Since I_m annihilates U/U_m, we have $VI_m \leq U_m$ and so

$$|U/U_m| \leq |U/VI_m| = |V/VI_m||W| = |R/I_m|^c|W|.$$

Since U/U_m is (R/I_m)-free of rank d,

$$|U/U_m| = |R/I_m|^d.$$

Thus $|R/I_m|^{d-c} \leq |W|$. As R is infinite while each R/I_m is finite but $\bigcap_m I_m = 0$, $|R/I_m|$ tends to infinity with m. It follows that $d - c \leq 0$, and the proof is complete.

Added in Proof (9 March 1989).

(1) The death of the first author, Professor Károly Buzási, on 7 August 1988, is recorded here with deep regret.

(2) P. A. Linnell has proved that the wreath product named in the second last paragraph of §1 can be generated by 3 elements.

References

1. Frank W. Anderson and Kent R. Fuller, *Rings and categories of modules*, Graduate Texts in Mathematics 13, Springer-Verlag, New York-Heidelberg-Berlin, 1974.

2. S. D. Berman and K. Buzási, *On modules over group algebras of groups containing an infinite cyclic subgroup of finite index*, Studia Math. Sci. Hungar. **16** (1981), 455–470. (Russian)

3. ——, *On representations of groups containing an infinite cyclic group of finite index*, Publ. Math. Debrecen **29** (1982), 163–170. (Russian)

4. Wolfgang Gaschütz, *Über modulare Darstellungen endlicher Gruppen, die von freien Gruppen induziert werden*, Math. Z. **60** (1954), 274–286.

5. B. Huppert, *Endliche Gruppen* I, Springer-Verlag, Berlin-Heidelberg-New York, 1967.

6. Yeo Kok Chye, *Minimal number of generators of some classes of groups*, Ph.D. Thesis, Australian National University, 1972 [Abstract: Bull. Austral. Math. Soc. **9** (1973), 301–302].

7. ——, *Minimal generating sets for some wreath products of groups*, Bull. Austral. Math. Soc. **9** (1973), 127–136.

Kossuth Lajos Tudományegyetem Matematikai Intézete, Postafiók 12, 4010 Debrecen, Hungary

and

Mathematics, IAS, Australian National University, GPO Box 4, Canberra, ACT 2601, Australia

Contemporary Mathematics
Volume **93** (1989)

Torsion Free Space Groups and Permutation Lattices for Finite Groups

GERALD CLIFF AND ALFRED WEISS

1. Introduction. Let Γ be a torsion free space group, which means that Γ has a finitely generated free abelian normal subgroup A of finite index. If A is maximal, it is called the translation subgroup, and then Γ/A is called the point group of Γ; the rank of A is called the dimension of Γ. Such groups are of interest to differential geometers, since they arise as fundamental groups of compact flat Riemannian manifolds, and classify these manifolds up to affine equivalence, cf. **[Cha]**, **[Wol]**. We shall start by proving a result of Vasquez **[Vas]**, which says that for any finite group G, there is an integer $n(G)$ such that a torsion free space group Γ with point group G has a normal subgroup N contained in the translations such that Γ/N is still torsion free, and has dimension $\leq n(G)$. (As Vasquez put it, any compact flat manifold with holomony group G is a flat toral extension of a flat manifold of dimension $\leq n(G)$.) Vasquez said that he had no information about this function $n(G)$, except when G has prime order, in which case $n(G)$ can be taken to be 1; this was proved by using Reiner's classification **[Rei]** of indecomposable $\mathbf{Z}G$-lattices for G of prime order. Our proof, which does not use Reiner's work, shows that $n(G)$ is less than or equal to the sum of the indices of non-conjugate subgroups of G of prime orders. We do this by showing that Γ has a map, with abelian kernel, to a space group whose translation subgroup, considered as a $\mathbf{Z}G$-module, is a permutation lattice S given by the sum of inductions from non-conjugate subgroups of prime orders. Moreover, we can show that our bound is sharp if G is a finite p-group, although it is not sharp otherwise. This is done by studying $H^2(G, S)$ as a module over $\mathrm{End}_{\mathbf{Z}G}\, S$.

A related question is work of Hiller and Sah **[HiS]** on flat manifolds of (first) Betti number 0. This means that the corresponding space group Γ has finite abelianization Γ/Γ'. They proved that there is a torsion free space group of Betti number zero, with point group G, if and only if G has the property that no cyclic Sylow subgroup has a normal complement. We give

This research was supported in part by a grant from NSERC of Canada.

a conceptual proof of this result, using the same $\mathbf{Z}G$-lattice as in the study of Vasquez's number, and we also show that there is always a torsion free space group with point group G having Betti number 1 (which means that Γ/Γ' modulo its torsion subgroup has rank 1.)

2. The standard lattice. Throughout, G will denote a finite group. Let Γ be a torsion-free space group with translation subgroup A and point group G. Then A is a $\mathbf{Z}G$-lattice, that is, a $\mathbf{Z}G$-module which is finitely generated and free as a $\mathbf{Z}$-module. The extension Γ of G by A gives us an element $\alpha \in H^2(G, A)$, and since Γ is torsion-free, then for any subgroup C of G of prime order, $\operatorname{res}_C^G \alpha \in H^2(C, \operatorname{res}_C^G A)$ is not 0. Such an element α of $H^2(G, A)$ is called *special*. Conversely, a faithful ZG-lattice A together with a special element α of $H^2(G, A)$ gives us a torsion-free space group.

Let $\mathscr{X}$ be a set of representatives of the conjugacy classes of subgroups of G of prime order. Then α is special if and only if $\operatorname{res}_C^G \alpha \neq 0$ in $H^2(C, \operatorname{res}_C^G M)$ for all C in $\mathscr{X}$.

LEMMA 1. *Suppose that C is cyclic, and that $\alpha \in H^2(C, M)$, $\alpha \neq 0$. Then there exists a $\mathbf{Z}C$-homomorphism $\phi\colon M \to \mathbf{Z}$ such that $\phi_*(\alpha) \neq 0$ in $H^2(C, \mathbf{Z})$.*

PROOF. Consider the map $h\colon M \to M$ given by $h(m) = \sum_{g \in C} gm$, and let M'' denote the image of h, with M' being the kernel. Then we have the exact sequence

$$0 \to M' \to M \to M'' \to 0$$

and $M''^C = M''$, $M'^C = 0$ (where the superscript $*^C$ denotes the C-fixed points of $*$). Since C is cyclic, then $H^2(C, M') = M'^C/h(M')$, which is 0 since $M'^C = 0$. So the map $h_*\colon H^2(C, M) \to H^2(C, M'')$ is injective. Then replacing the pair (M, α) by $(M'', h_*\alpha)$, we may assume that C acts trivially on M. Choose a $\mathbf{Z}$-basis for M, thus obtaining a $\mathbf{Z}C$-isomorphism $M \cong \bigoplus_i \mathbf{Z}$, so $H^2(C, M) \cong \bigoplus_i H^2(C, \mathbf{Z})$. Since $\alpha \neq 0$ in $H^2(C, M)$, it follows that one of the projection maps $M \to \mathbf{Z}$ sends α to a non-zero element of $H^2(C, \mathbf{Z})$. This completes the proof.

Let C be a group in $\mathscr{X}$, and let T_C be a set of left coset representatives in C in G, with $1 \in T_C$. We shall write an element of $\operatorname{ind}_C^G \mathbf{Z}$ in the form $\sum_{t \in T_C} n_t tC$ where each n_t is an integer. We have the projection map

$$\pi\colon \operatorname{res}_C^G \operatorname{ind}_C^G \mathbf{Z} \to \mathbf{Z}, \ \pi\left(\sum_{t \in T_C} n_t tC\right) = n_1.$$

Define the *standard lattice* S to be the permutation $\mathbf{Z}G$-lattice given by

$$S = \bigoplus_{C \in \mathscr{X}} \operatorname{ind}_C^G \mathbf{Z}.$$

THEOREM 1. *Let $\alpha \in H^2(G, M)$ be special. Then there exists a $\mathbf{Z}G$-homomorphism $h\colon M \to S$ such that $h_*(\alpha) \in H^2(G, S)$ is special for S.*

PROOF. Since $\operatorname{res}_C^G(\alpha) \neq 0$ for $C \in \mathscr{X}$, we have by Lemma 1 a $\mathbf{Z}C$-homomorphism $\phi_C \colon \operatorname{res}_C^G M \to \mathbf{Z}$ such that $\phi_{C^*}(\operatorname{res}_C^G \alpha) \neq 0$ in $H^2(C, \mathbf{Z})$. The projection map π gives us the Frobenius reciprocity isomorphism

$$\operatorname{Hom}_{\mathbf{Z}G}(M, \operatorname{ind}_C^G \mathbf{Z}) \xrightarrow{\pi_*} \operatorname{Hom}_{\mathbf{Z}C}(\operatorname{res}_C^G M, \mathbf{Z}).$$

Let $\psi_C \in \operatorname{Hom}_{\mathbf{Z}G}(M, \operatorname{ind}_C^G \mathbf{Z})$ be the map such that $\pi_*(\psi_C) = \phi_C$. Then

$$\pi_*(\psi_{C^*}(\operatorname{res}_C^G \alpha)) = \phi_{C^*}(\operatorname{res}_C^G \alpha) \neq 0$$

so $\psi_{C^*}(\operatorname{res}_C^G \alpha) \neq 0$. Define

$$h = \bigoplus_{C \in \mathscr{X}} \psi_C \colon M \to S.$$

We must show that $h_*(\operatorname{res}_C^G \alpha) \neq 0$ in $H^2(C, S)$ for all $C \in \mathscr{X}$, since $\operatorname{res}_C^G h_*(\alpha) = h_*(\operatorname{res}_C^G \alpha)$. It clearly suffices to show that $\psi_{C^*}(\operatorname{res}_C^G \alpha) \neq 0$ in $H^2(C, \operatorname{ind}_C^G \mathbf{Z})$. But this is true, so the proof is complete.

COROLLARY. $n(G) \leq \sum_{C \in 3} |G \colon C|$.

Even though Vasquez claimed that he knew nothing about $n(G)$ except when G has prime order, his proof in fact shows that $n(G) \leq r|G|$ where r is the number of elements of $\mathscr{X}$, and moreover his proof can be modified slightly to yield the bound in the Corollary.

Let us note at this point that any normal abelian subgroup of Γ is contained in the translations. Indeed the Fitting subgroup of Γ is abelian, a fact which may be proved as follows: we may assume that Γ is nilpotent, and since G is finite, it can have no fixed points on A/A^G, since A^G is torsion free, so $A = A^G$ and then Γ is abelian, by a result of Schur, **[Gru]**, p. 190.

3. $H^2(G, S)$ as $\operatorname{End}_{\mathbf{Z}G} S$-module, and minimality of S. We continue with the same notation as in the previous section.

We want to define a special element in $H^2(G, S)$. For $C \in \mathscr{X}$, we have the projection $\pi_C \colon \operatorname{res}_C^G \operatorname{ind}_C^G \mathbf{Z} \to \mathbf{Z}$ as defined in the previous section. Then from Shapiro's Lemma (**[Gru]**, p. 91) the composite ψ

$$H^2(G, \operatorname{ind}_C^G \mathbf{Z}) \xrightarrow{\operatorname{res}_C^G} H^2(C, \operatorname{res}_C^G \operatorname{ind}_C^G \mathbf{Z}) \xrightarrow{\pi_{C^*}} H^2(C, \mathbf{Z})$$

is an isomorphism. Since C is cyclic and C acts trivially on $\mathbf{Z}$, then $H^2(C, \mathbf{Z}) = \mathbf{Z}/|C|\mathbf{Z}$. Pick a non-zero element $\gamma \in H^2(C, \mathbf{Z})$ and let $\beta_C \in H^2(G, \operatorname{ind}_C^G \mathbf{Z})$ be mapped to γ by ψ. Let

$$\beta = \bigoplus_{C \in \mathscr{X}} \beta_C \in H^2(G, S) = \bigoplus_{C \in \mathscr{X}} H^2(G, \operatorname{ind}_C^G \mathbf{Z}).$$

Then $\operatorname{res}_C^G \beta$ is not 0, since β_C is not 0, so β is special.

We shall first study the action of $\operatorname{End}_{\mathbf{Z}G} S$ on $H^2(G, S)$. We have a ring homomorphism

$$\phi \colon \operatorname{End}_{\mathbf{Z}G} S \to \operatorname{End}(H^2(G, S)), \; \phi(f) = f_*, \; f \in \operatorname{End}_{\mathbf{Z}G} S.$$

LEMMA 2. *Let C and D be two different elements of $\mathscr{X}$, and let $f \in \operatorname{End}_{\mathbf{Z}G} S$ be such that $f(\operatorname{ind}_C^G \mathbf{Z}) \subseteq \operatorname{ind}_D^G \mathbf{Z}$ while $f(\operatorname{ind}_E^G \mathbf{Z}) = 0$ for all $E \in \mathscr{X}$ other than C. Then $\phi(f) = 0$.*

PROOF. We have the commutative diagram

$$\begin{array}{ccc} H^2(G, \operatorname{ind}_C^G \mathbf{Z}) & \xrightarrow{f_*} & H^2(G, \operatorname{ind}_D^G \mathbf{Z}) \\ \big\downarrow \operatorname{res}_D^G & & \big\downarrow \operatorname{res}_D^G \\ H^2(D, \operatorname{res}_D^G \operatorname{ind}_C^G \mathbf{Z}) & \xrightarrow{\operatorname{res}_D^G f_*} & H^2(D, \operatorname{res}_D^G \operatorname{ind}_D^G \mathbf{Z}) \end{array}$$

The right vertical arrow is injective, by Shapiro's Lemma. Since C and D are different, they are not conjugate, so $\operatorname{res}_D^G \operatorname{ind}_C^G \mathbf{Z}$ is a free $\mathbf{Z}D$-module, by Mackey's formula. Therefore $\alpha \in H^2(G, \operatorname{ind}_C^G \mathbf{Z})$ gets mapped to 0 by the left vertical arrow. It follows that $f_*(\alpha) = 0$ as desired.

Our next result is the minimality of S if G is a finite p-group.

THEOREM 2. *Let G be a finite p-group. Then (S, β) is minimal, in the sense that if $e\colon S \to S'$ is any $\mathbf{Z}G$-homomorphism with non-zero kernel, then $e_*(\beta)$ is not special.*

COROLLARY. *If G is a finite p-group, then $n(G) = \sum_{C \in \mathscr{X}} |G\colon C|$.*

PROOF OF THEOREM 2. For C in $\mathscr{X}$, let $\kappa_C\colon \operatorname{ind}_C^G \mathbf{Z} \to S$ and $\lambda_C\colon S \to \operatorname{ind}_C^G \mathbf{Z}$ denote the inclusion and projection, respectively. Then $f \in \operatorname{End}_{\mathbf{Z}G} S$ can be written in the form

$$f = \sum_{C,D \in \mathscr{X}} \kappa_C f_{CD} \lambda_D \text{ for some } f_{CD} \in \operatorname{Hom}_{\mathbf{Z}G}(\operatorname{ind}_D^G \mathbf{Z}, \operatorname{ind}_C^G \mathbf{Z}).$$

The lemma tells us that $\phi(\kappa_C f_{CD} \lambda_D) = 0$ if $C \neq D$. Then set $f_C = \kappa_C f_{CC} \lambda_C \in \operatorname{End}_{\mathbf{Z}G} S$, and we have $f \equiv \sum_{C \in \mathscr{X}} f_C \pmod{\ker \phi}$, so the image of ϕ is contained in $\bigoplus_{C \in \mathscr{X}} \operatorname{End} H^2(G, \operatorname{ind}_C^G \mathbf{Z})$. By Shapiro's Lemma, $H^2(G, \operatorname{ind}_C^G \mathbf{Z})$ is cyclic of order p, and so $\operatorname{End} H^2(G, \operatorname{ind}_C^G \mathbf{Z}) \cong \mathbf{F}_p$. Therefore $H^2(G, S)$ is a semi-simple $\operatorname{End}_{\mathbf{Z}G} S$-module, and $\phi(\operatorname{End}_{\mathbf{Z}G} S)$ is isomorphic, as a ring, to a direct sum of r copies of $\mathbf{F}_p$, where r is the number of groups in $\mathscr{X}$.

Let J be the ideal of $\mathbf{Z}G$ defined by

$$J = \left\{ \sum_{g \in G} n_g g \in \mathbf{Z}G\colon \sum_{g \in G} n_g \equiv 0 \pmod{p} \right\}.$$

Let I be the ideal in $\operatorname{End}_{\mathbf{Z}G} S$ of all elements f such that $f(S) \subseteq J \cdot S$. Since G is a finite p-group, then $J^n \subseteq p\mathbf{Z}G$ for some positive integer n. We then have $I^n \subseteq p \operatorname{End}_{\mathbf{Z}G} S$, so $\phi(I)$ is a nilpotent ideal. Therefore $I \subseteq \ker \phi$, since $\operatorname{End}_{\mathbf{Z}G} S / \ker \phi$ is a direct sum of copies of $\mathbf{F}_p$. We shall show that $I = \ker \phi$.

For subgroups C and D in $\mathscr{X}$, let $[CxD]\colon \operatorname{ind}_C^G \mathbf{Z} \to \operatorname{ind}_D^G \mathbf{Z}$ be given by

$$[CxD](tC) = \sum_{u \in U} tuxD \in \operatorname{ind}_D^G \mathbf{Z}, \quad t \in T_C$$

where CxD is the disjoint union $\bigcup_{u\in U} uxD$. Then, as shown by Schur (see for instance **[Lan]**, p. 177) $\mathrm{Hom}_{\mathbf{Z}G}(\mathrm{ind}_C^G \mathbf{Z}, \mathrm{ind}_D^G \mathbf{Z})$ has a **Z**-basis given by elements $[CxD]$ corresponding to double cosets CxD, $x \in G$. We shall find it convenient to regard $[CxD]$ as an element of $\mathrm{End}_{\mathbf{Z}G} S$, by having it annihilate $\mathrm{ind}_E^G \mathbf{Z}$ for $E \in \mathscr{X}$, $E \neq C$. To prove that $I = \ker\phi$ we shall use the following result.

LEMMA 3. *The following elements of* $\mathrm{End}\, S$ *are in* I:
(i) $[CxD]$ *where* $C \neq D \in \mathscr{X}$ *or* $C = D$ *but* $x \notin N_G(C)$.
(ii) $[CxC] - [C1C]$ *where* $x \in N_G(C)$.
(iii) $p[C1C]$.

PROOF. If $C \neq D$ or if $x \notin N_G(C)$, then $[CxD]$ maps tC to $\sum_{u\in U} tuxD$ where $|U| = |C\colon C \cap xDx^{-1}| = p$ so

$$\sum_{u\in U} tuxD = pD + \sum_{u\in U}(tux - 1)D \in JS.$$

This shows that elements of type (i) are in I. For elements of type (ii),

$$([CxC] - [C1C])(tC) = txC - tC \in JS.$$

Part (iii) is clear. This completes the proof.

It follows from the lemma that I has index p^r in $\mathrm{End}_{\mathbf{Z}G} S$. Since $I \subseteq \ker\phi$ and $\ker\phi$ has index p^r in $\mathrm{End}_{\mathbf{Z}G} S$, we must have $I = \ker\phi$.

Let $e\colon S \to S'$ be a $\mathbf{Z}G$-homomorphism, having non-zero kernel, with $e_*(\beta)$ special. By Theorem 1, there is a $\mathbf{Z}G$-homomorphism $d\colon S' \to S$ so that $d_*(e_*\beta))$ is special in $H^2(G, S)$. Letting f be the composite

$$S \xrightarrow{e} S' \xrightarrow{d} S$$

we now have $f_*(\beta)$ special, and $\ker f \neq 0$. We shall show that this is impossible. We have $f \equiv \sum_{C\in\mathscr{X}} f_C \pmod{\ker\phi}$ with each f_{C_*} a non-zero element of $\mathrm{End}\, H^2(G, \mathrm{ind}_C^G \mathbf{Z})$ (since $f_*(\beta)$ is special). Since $\mathrm{End}\, H^2(G, \mathrm{ind}_C^G \mathbf{Z}) \cong \mathbf{F}_p$, it follows that for some integer m, $f^m \equiv 1 \pmod{\ker\phi}$, hence $f^m - 1 \in I$, since $\ker\phi = I$. We have $I^n \subseteq p\,\mathrm{End}_{\mathbf{Z}G} S$, so raising $f^m - 1$ to a suitable p-power, we find that $f^k \equiv 1 \pmod{p\,\mathrm{End}_{\mathbf{Z}G} S}$. But then f must be one-to-one. This completes the proof.

Now we show that this last result does not hold if G is not a p-group.

THEOREM 3. *Let G be a group whose order is divisible by at least two distinct primes. Then there exists an element* $f \in \mathrm{End}_{\mathbf{Z}G} S$ *which is not one-to-one such that* $f_*(\beta)$ *is special.*

COROLLARY. $n(G) < \sum_{C\in\mathscr{X}} |G\colon C|$ *if G is not a p-group.*

PROOF OF THEOREM 3. Let p and q be distinct primes dividing $|G|$, and let C and D be groups in $\mathscr{X}$ having orders p and q respectively. Define

$$x = \sum_{t\in T_C} tC \in \mathrm{ind}_C^G \mathbf{Z}, \quad y = \sum_{t\in T_D} tD \in \mathrm{ind}_D^G \mathbf{Z}.$$

Find integers k and l such that $kp + lq = 1$. Then define $e \in \mathrm{End}_{\mathbf{Z}G} S$ by

$$e = kp[C1C] + k[C1D] + l[D1C] + l[D1D].$$

We shall show that $e(px+qy) = px+qy$. We first claim that $[C1D](x) = qy$. Since $C \cap D = 1$, then CD is the disjoint union $\bigcup_{c\in C} cD$, so

$$[C1D](x) = \sum_{t\in T_C} [C1D]tC = \sum_{t\in T_C}\sum_{c\in C} tcD = \sum_{g\in G} gD = q\sum_{t\in T_D} tD = qy.$$

Similarly, $[D1C](y) = px$. By definition, $[C1C](y) = [C1D](y) = 0$, and $[D1D](x) = [D1C](x) = 0$. Also, $[C1C](x) = x$ and $[D1D](y) = y$. Therefore

$$\begin{aligned} e(px + qy) &= kp[C1C](px) + k[C1D](px) + l[D1C](qy) + l[D1D](qy) \\ &= kp^2x + kpqy + lpqx + lq^2y \\ &= p(kp + lq)x + q(kp + l)y = px + qy. \end{aligned}$$

From Lemma 2, $[C1D]_* = [D1C]_* = 0$. Also, $p[C1C]_* = 0$ since $H^2(G, \mathrm{ind}_C^G \mathbf{Z})$ has order p, and similarly $q[D1D]_* = 0$, so $e_* = 0$ on $H^2(G, S)$. Then $(1-e)_*(\beta)$ is special, and $1-e$ has $px+qy$ in its kernel. This completes the proof.

4. Betti numbers. In this section we shall use our standard module S and special element β of $H^2(G, S)$ to prove a result due to Hiller and Sah **[HiS]** on space groups Γ with Betti number 0. This means that $|\Gamma\colon \Gamma'|$ is finite.

LEMMA 4. *Let Γ be a space group with translations A and point group G. Then the torsion free rank of Γ/Γ' is equal to the rank of A^G.*

PROOF. Let $B = [A, \Gamma]$; then B is the smallest normal subgroup of Γ contained in A such that G acts trivially on A/B. Since Γ/B has a central subgroup of finite index, Schur tells us that Γ'/B is finite. So the torsion free ranks of Γ/Γ' and A/B are equal. The $\mathbf{Q}G$-module $\mathbf{Q}A$ is the direct sum of $\mathbf{Q}B$ and $\mathbf{Q}C$ for some non-zero $\mathbf{Z}G$-sublattice of C of A. Then the torsion free rank of A/B and the rank of A^G are both equal to the rank of C. This completes the proof.

COROLLARY. *Γ/Γ' is finite if and only if $A^G = 0$.*

LEMMA 5. *Let C be a cyclic subgroup of G. Let $a_C\colon \mathrm{ind}_C^G \mathbf{Z} \to \mathbf{Z}$ be given by*

$$a_C\left(\sum_{t\in T_C} n_t tC\right) = \sum_{t\in T_C} n_t.$$

Then $a_{C^*}\colon H^2(G, \mathrm{ind}_C^G \mathbf{Z}) \to H^2(G, \mathbf{Z})$ is 0 if and only if the transfer from G to C is trivial.

PROOF. The exact sequence

$$0 \to \mathrm{ind}_C^G \mathbf{Z} \to \mathrm{ind}_C^G \mathbf{Q} \to \mathrm{ind}_C^G \mathbf{Q}/\mathbf{Z} \to 0$$

gives us an isomorphism $\delta\colon H^1(G, \operatorname{ind}_C^G \mathbf{Q}/\mathbf{Z}) \to H^2(G, \operatorname{ind}_C^G \mathbf{Z})$, since $H^i(G, \operatorname{ind}_C^G \mathbf{Q}) = 0$. Similarly we have an isomorphism between $H^1(G, \mathbf{Q}/\mathbf{Z})$ and $H^2(G, \mathbf{Z})$. So we need only prove the same assertion about the map $a_{C^*}\colon H^1(G, \operatorname{ind}_C^G \mathbf{Q}/\mathbf{Z}) \to H^1(G, \mathbf{Q}/\mathbf{Z})$.

We exhibit elements of $H^1(G, \operatorname{ind}_C^G \mathbf{Q}/\mathbf{Z})$ via the isomorphism

$$H^1(G, \operatorname{ind}_C^G \mathbf{Q}/\mathbf{Z}) \overset{\operatorname{res}_C^G}{\to} H^1(C, \operatorname{res}_C^G \operatorname{ind}_C^G \mathbf{Q}/\mathbf{Z}) \overset{\pi_{C^*}}{\to} H^1(C, \mathbf{Q}/\mathbf{Z})$$

of Shapiro's Lemma, where π_C is the projection map as in Section 2. For $\chi \in H^1(C, \mathbf{Q}/\mathbf{Z}) = \operatorname{Hom}(C, \mathbf{Q}/\mathbf{Z})$, we construct a 1-cocycle $\hat{\chi}$ on G with values in $\operatorname{ind}_C^G \mathbf{Q}/\mathbf{Z}$; to define $\hat{\chi}(g)$ for $g \in G$, for each $t \in T_C$ write

$$gt = t_g c_{g,t} \text{ for some } t_g \in T_C \text{ and some } c_{g,t} \in C.$$

Then define

$$\hat{\chi}(g) = \sum_{t \in T_C} \chi(c_{g,t}) t_g C.$$

It is easy to check that $\hat{\chi}$ is a cocycle, so we get a class $[\hat{\chi}] \in H^1(G, \operatorname{ind}_C^G \mathbf{Q}/\mathbf{Z})$, and this class maps to χ under the isomorphism of Shapiro's Lemma. Thus $a_{C^*} = 0$ on H^1 if and only if $a_{C^*}(\hat{\chi}) = 0$ for all $\chi \in \operatorname{Hom}(C, \mathbf{Q}/\mathbf{Z})$. But

$$a_{C^*}(\hat{\chi})(g) = a_C(\hat{\chi}(g)) = a_C\left(\sum_{t \in T_C} \chi(c_{g,t}) t_g C\right) = \sum_t \chi(c_{g,t}) = \chi(V(g)),$$

where V is the transfer from G to C. Thus $a_{C^*}(\hat{\chi}) = 0$ for all χ if and only if $V(g) = 1$ for all $g \in G$, and the proof is complete.

THEOREM 4. *Let G be a finite group. Then G is the point group of a torsion free space group Γ with $|\Gamma : \Gamma'|$ finite if and only if the transfer from G to C is trivial for every subgroup C in $\mathscr{X}$.*

PROOF. Suppose that G is the point group of a space group as stated. Let A be the translation subgroup of Γ and let $\alpha \in H^2(G, A)$ be the special element arising from the extension Γ of G by A. From Theorem 1 we have a map $h\colon A \to S$ with $h_*(\alpha)$ special. Then $h_*(\alpha) = \bigoplus_{C \in \mathscr{X}} \varepsilon_C$, with ε_C a non-zero element of $H^2(G, \operatorname{ind}_C^G \mathbf{Z})$. For each $C \in \mathscr{X}$ we have $a_C\colon \operatorname{ind}_C^G \mathbf{Z} \to \mathbf{Z}$ defined above, and we set $a\colon S \to \mathbf{Z}^r = \bigoplus a_C$, giving the exact sequence

$$0 \to \ker a \overset{i}{\to} S \overset{a}{\to} \mathbf{Z}^r \to 0$$

with G acting trivially on $\mathbf{Z}^r$ and $(\ker a)^G = 0$. From the corollary to Lemma 4, $A^G = 0$, so the composite

$$A \overset{h}{\to} S \overset{a}{\to} \mathbf{Z}^r$$

is 0. It follows that $a_{C^*}(\varepsilon_C) = 0$, so we deduce from Lemma 5 that the transfer is trivial.

Conversely, suppose that the transfer is trivial for every $C \in \mathscr{X}$. Let $\beta = \bigoplus \beta_C$ be our special element in $H^2(G,S)$. By the lemma, $a_{C^*}(\beta_C) = 0$ for each C, so $a_*(\beta) = 0$ in $H^2(G, \mathbf{Z}^r)$. We have the exact sequence

$$H^2(G, \ker a) \xrightarrow{i_*} H^2(G,S) \xrightarrow{a_*} H^2(G,\mathbf{Z}^r)$$

so $\beta = i_*(\alpha)$ for some $\alpha \in H^2(G, \ker a)$, and α must be special since β is. Moreover $(\ker a)^G = 0$, so α gives us the space group we want, except that G might not act faithfully on $\ker a$ so the point group of our extension of G by $\ker a$ may not be G. Take a $\mathbf{Z}G$-lattice L on which G acts faithfully, and for which $L^G = 0$; let $A = \ker a \oplus L$, and let $\gamma \in H^2(G,A)$ be given by $\alpha \oplus 0$. Then the extension of G by A defined by γ gives us the desired space group.

To get the Hiller-Sah result from this, we need the following transfer theorem (which may well be known).

PROPOSITION. *Let G be a finite group, and let C be a subgroup of G of prime order p. Then the following are equivalent:*

(a) *A Sylow p-subgroup of G is cyclic and is contained in the centre of its normalizer.*

(b) *A Sylow p-subgroup of G is cyclic and has a normal complement.*

(c) *The transfer from G to C is onto.*

PROOF. That (a) implies (b) follows from Burnside's transfer theorem, **[Gor]**, p. 252. Assume (b). Let P be a cyclic Sylow p-subgroup of G containing C and let H be its normal complement. Then the transfer from G to C is the composite of the transfer from G to P with the transfer from P to C; this latter transfer is onto since P is cyclic. The transfer from G to P is onto, since one may compute it using H as the set of coset representatives of P in G. This proves (c).

Now assume (c). Let P be a p-subgroup of G maximal with respect to having C in its centre. The transfer from G to C is the composite of the transfers from G to K and from K to C, for any subgroup K of G containing C, and so the transfer from K to C cannot be trivial for any such K. We shall use the following lemma.

LEMMA. *Let C be a normal subgroup of a group K with C having prime order p. Then the transfer from K to C is trivial if one of the following holds:*

(i) *K is non-cyclic of order p^2.*

(ii) *K is the quaternion group of order* 8.

(iii) *K/C is cyclic of order prime to p and C is not central in K.*

PROOF. This is an easy calculation.

Returning to the proof of the proposition, if P is not cyclic, then either it contains a non-cyclic subgroup H of order p^2, or else it is a generalized quaternion 2-group. In the first instance, let K be generated by C and one of the subgroups of H which is not C, and apply the lemma. In the second instance, P has a unique central subgroup of order 2, namely C, and a quaternion subgroup K of order 8 containing C, and we apply the lemma again.

Thus P is cyclic. If P is not a Sylow p-subgroup of G, then it is not a Sylow p-subgroup of $N = N_G(P)$. Since p does not divide the order of the automorphism group of C, every p-subgroup of N centralizes C, hence is contained in P by maximality. So P is a cyclic Sylow p-subgroup of G.

Finally, if $g \in N$ and $g \notin C_G(P)$, then conjugation by g induces a non-trivial automorphism of C, of order prime to p. Replace g by a suitable p-power so that it has order prime to p, and apply part (iii) of the lemma to $K = \langle C, g\rangle$. This completes the proof.

COROLLARY (Hiller-Sah). *A finite group G is the point group of a torsion free space group Γ with $|\Gamma : \Gamma'|$ finite if and only if no cyclic Sylow subgroup of G has a normal complement.*

We next show that for any finite group G, there is a torsion free space group of Betti number 1 with point group G.

THEOREM 5. *Suppose that G is any finite group. Then there exists a torsion free space group Γ with point group G such that Γ/Γ' modulo its torsion subgroup has rank* 1.

PROOF. If G does not have a cyclic Sylow subgroup with a normal complement, let Γ_0 be a torsion free space group of Betti number 0, with point group G, and let $\Gamma = \Gamma_0 \times \mathbf{Z}$, which has Betti number 1.

Suppose that G has a cyclic Sylow subgroup with a normal complement. Partition $\mathscr{X}$ into the union of two sets, $\mathscr{Y}$ and $\mathscr{Z}$, such that the transfer from G to C is trivial if $C \in \mathscr{Y}$ but not if $C \in \mathscr{Z}$. Each C in $\mathscr{Z}$ is contained in a cyclic Sylow subgroup P_C, and the transfer from G to P_C is onto. Let $\phi \colon G \to \prod_{C\in\mathscr{Z}} P_C = K$ be the direct product of the transfers from G to P_C for $C \in \mathscr{Z}$. Then ϕ is onto, and K is cyclic, since no two elements of $\mathscr{Z}$ have the same order. Letting H be the kernel of ϕ, it follows that G is the split extension HK and we regard K as a subgroup of G, and each $C \in \mathscr{Z}$ as a subgroup of K.

For each $C \in \mathscr{Z}$, let β_C be a non-zero element of $H^2(G, \operatorname{ind}_C^G \mathbf{Z})$ as before. Since the transfer from G to C is trivial, then denoting by i_C the inclusion from $\ker a_C$ to $\operatorname{ind}_C^G \mathbf{Z}$, it follows that $\beta_C = i_{C*}(\alpha_C)$ for some $\alpha_C \in H^2(G, \ker a_C)$. Moreover, $\operatorname{res}_C^G \alpha_C \neq 0$. From Shapiro's Lemma $H^2(G, \operatorname{ind}_K^G \mathbf{Z})$ is cyclic of order K; let α be a generator, so $\operatorname{res}_C^G \alpha \neq 0$ for each $C \in \mathscr{Z}$ since each $C \in \mathscr{Z}$ is a subgroup of K. Then set

$$M = \operatorname{ind}_K^G \mathbf{Z} \oplus \left(\bigoplus_{C\in\mathscr{Y}} \ker a_C \right), \quad \gamma = \alpha \oplus \left(\bigoplus_{C\in\mathscr{Y}} \alpha_C \right)$$

and it follows that γ is a special element of $H^2(G, M)$. Since $(\ker a_C)^G = 0$ for each $C \in \mathscr{Y}$, and $(\operatorname{ind}_K^G \mathbf{Z})^G$ has rank 1, then M^G has rank 1. In order to have G act faithfully, we may need to form the direct sum A of M with a faithful $\mathbf{Z}G$-lattice without fixed points. Then the extension of G by A, using $\alpha \oplus 0 \in H^2(G, A)$, has Betti number 1 (by Lemma 4) and point group G. This completes the proof.

References

[Cha] L. S. Charlap, "Bieberbach groups and flat manifolds," Springer-Verlag, New York, 1986.

[Gor] D. Gorenstein, "Finite groups," Harper & Row, New York, 1968.

[Gru] K. W. Gruenberg, "Cohomological topics in group theory," Lecture Notes in Math. **143**, Springer-Verlag, Berlin, 1970.

[HiS] H. Hiller and C.-H. Sah, *Holonomy of flat manifolds with* $b_1 = 0$, *I*, Quarterly J. Math. Oxford **37** (1986), 177–187.

[Lan] P. Landrock, "Finite group algebras and their modules," London Math. Soc. Lecture Note Series **84**, Cambridge University Press, Cambridge, 1983.

[Rei] I. Reiner, *Integral representations of cyclic groups of prime order*, Proc. Amer. Math. Soc. **8** (1957), 142–146.

[Vas] A. T. Vasquez, *Flat Riemannian manifolds*, J. Differential Geometry **4** (1970), 367–382.

[Wol] J. A. Wolf, "Spaces of constant curvature," McGraw-Hill, New York, 1968.

University of Alberta
Department of Mathematics
Edmonton, Alberta
Canada T6G 2G1

Contemporary Mathematics
Volume **93** (1989)

Clifford Theory and Induction from Subgroups

EVERETT C. DADE[1]

Dedicated to the memory of S. D. Berman

ABSTRACT. We shall show that Clifford theory is compatible with induction of modules. The proof of this result is complicated by the fact that we work over arbitrary group-graded rings, for which induction is not just a tensor product. On the other hand, it is helped considerably by our use of direct, 'one-step' Clifford theory in place of the original, 'two-step' theory. For the benefit of those using the latter theory, we translate our result into its language at the end of this note.

1. Induction of Modules.

We adopt the notation and assumptions of **[1]**, with minor modifications. Thus we fix a multiplicative group G (with identity $1 = 1_G$) and a *G-graded ring* R. So R is an associative ring (with identity $1 = 1_R$) having a direct sum decomposition:

$$R = \dot{\sum_{\sigma \in G}} R_\sigma \qquad \textit{(as additive groups)}, \tag{1.1a}$$

where the additive subgroups R_σ of R satisfy:

$$R_\sigma R_\tau \subseteq R_{\sigma\tau}, \textit{ for all } \sigma, \tau \in G. \tag{1.1b}$$

The *1_G-component* R_1 is a subring of R containing 1_R, and each *σ-component* R_σ, for $\sigma \in G$, is a two-sided R_1-submodule of R. Hence the *G-grading* (1.1a) is a direct sum of two-sided R_1-modules.

For any subset C of G we define R_C to be the two-sided R_1-submodule:

$$R_C = \dot{\sum_{\sigma \in C}} R_\sigma \qquad \textit{(as two-sided } R_1\textit{-modules)} \tag{1.2}$$

of R. If C is empty, then $R_C = 0$.

Fix a subgroup H of G. Then we clearly have:

(1.3) *R_H-is a subring of R containing 1_R. Furthermore, R_H is an H-graded ring with the σ-component $(R_H)_\sigma = R_\sigma$, for each $\sigma \in H$.*

[1] This research was supported by a grant from the National Science Foundation.

Let $H\backslash G$ be the right G-set formed by the left cosets $H\sigma$ of H in G. An *$H\backslash G$-graded R-module* M is a right, unitary R-module M, together with an *$H\backslash G$-grading*:

$$M = \dot{\sum_{C \in H\backslash G}} M_C \qquad \textit{(as } R_1\textit{-modules)}, \tag{1.4a}$$

whose *C-components* M_C are R_1-submodules of M satisfying:

$$M_C R_\sigma \subseteq M_{C\sigma}, \textit{ for all } C \in H\backslash G \textit{ and } \sigma \in G. \tag{1.4b}$$

Such M are the objects in an abelian category $\operatorname{Gr Mod}(H\backslash G, R)$. The morphisms $\phi\colon M \to N$ in $\operatorname{Gr Mod}(H\backslash G, R)$ are those homomorphisms of R-modules which are *$H\backslash G$-grade-preserving*, in the sense that:

$$\phi(M_C) \subseteq N_C, \textit{ for all } C \in H\backslash G. \tag{1.5}$$

The H-null socle of any module $M \in \operatorname{Gr Mod}(H\backslash G, R)$ was denoted by $S_H(M)$ in **[1]**. Here we prefer to write $\operatorname{Soc}_H(M)$ for this socle, to avoid confusion with the module S_H which will later become ubiquitous. It and the related H-null objects of $\operatorname{Gr Mod}(H\backslash G, R)$ are defined by:

(1.6a) An $H\backslash G$-graded R-module M is *H-null* if its H-component M_H is zero.

(1.6b) The *H-null socle* $\operatorname{Soc}_H(M)$ of any $M \in \operatorname{Gr Mod}(H\backslash G, R)$ is the largest (under inclusion) H-null $H\backslash G$-graded R-submodule of M.

Notice that any sum of H-null $H\backslash G$-graded R-submodules of M is H-null. So $\operatorname{Soc}_H(M)$ is well-defined. Clearly any morphism $\phi\colon M \to N$ in $\operatorname{Gr Mod}(H\backslash G, R)$ must carry $\operatorname{Soc}_H(M)$ into $\operatorname{Soc}_H(N)$, and hence restrict to a morphism $\operatorname{Soc}_H(\phi)\colon \operatorname{Soc}_H(M) \to \operatorname{Soc}_H(N)$ in $\operatorname{Gr Mod}(H\backslash G, R)$. In this way Soc_H becomes an additive functor from $\operatorname{Gr Mod}(H\backslash G, R)$ into itself.

Tensor products define an additive functor $\cdot \otimes R = \cdot \otimes_{R_H} R$ from the category $\operatorname{Mod}(R_H)$ of all right, unitary R_H-modules and their R_H-homomorphisms to the corresponding category $\operatorname{Mod}(R)$ for the ring R. In fact, we may regard $\cdot \otimes R$ has an additive functor from $\operatorname{Mod}(R_H)$ to $\operatorname{Gr Mod}(H\backslash G, R)$. To see this, notice that R_C is an R_H, R_1-sub-bimodule of R, for any $C = H\sigma \in H\backslash G$. Since G is the disjoint union of the cosets $C \in H\backslash G$, it follows from (1.1a) and (1.2) that:

$$R = \dot{\sum_{C \in H\backslash G}} R_C \qquad \textit{(as } R_H, R_1\textit{-bimodules)}.$$

Hence we may identify each $K \otimes R_C = K \otimes_{R_H} R_C$ naturally with its image in $K \otimes R$, for any R_H-module K and coset $C \in H\backslash G$. Then we have:

$$K \otimes R = \dot{\sum_{C \in H\backslash G}} (K \otimes R_C) \qquad \textit{(as } R_1\textit{-modules)}. \tag{1.7}$$

One easily verifies that the $H\backslash G$-grading (1.7) makes $K \otimes R$ an $H\backslash G$-graded R-module. Furthermore, the R-homomorphism $\phi \otimes R\colon K \otimes R \to L \otimes R$ determined by any morphism $\phi\colon K \to L$ in $\mathrm{Mod}(R_H)$ preserves the $H\backslash G$-gradings (1.7). Thus $\cdot \otimes R$ becomes an additive functor from $\mathrm{Mod}(R_H)$ to $\mathrm{Gr\,Mod}(H\backslash G, R)$.

Fix a module $K \in \mathrm{Mod}(R_H)$. By (1.6) the $H\backslash G$-graded R-module $K \otimes R$ has the $H\backslash G$-graded R-submodule $\mathrm{Soc}_H(K \otimes R)$. As in Section 5 of **[1]**, the module $K \overline{\otimes} R = K \overline{\otimes}_{R_H} R$ *induced* by K is the $H\backslash G$-graded R-factor module:

(1.8) $$K \overline{\otimes} R = (K \otimes R)/\mathrm{Soc}_H(K \otimes R) \qquad (\textit{in } \mathrm{Gr\,Mod}(H\backslash G, R)).$$

If $k \in K$ and $r \in R$, then $k \overline{\otimes} r \in K \overline{\otimes} R$ denotes the image of the element $k \otimes r \in K \otimes R$. From (1.7) and (1.8) we get:

(1.9) *For any $C \in H\backslash G$, the C-component $(K \overline{\otimes} R)_C$ of $K \overline{\otimes} R$ is the R_1-submodule $K \overline{\otimes} R_C$ generated by all $k \overline{\otimes} r$ with $k \in K$ and $r \in R_C$.*

Since $\mathrm{Soc}_H(K \otimes R)$ is H-null by (1.6), while $(K \otimes R)_H = K \otimes_{R_H} R_H$ is naturally isomorphic to K, we have:

(1.10) *There is a natural R_H-isomorphism $\cdot \overline{\otimes} 1\colon K \xrightarrow{\sim} K \overline{\otimes} R_H = (K \overline{\otimes} R)_H$ sending any $k \in K$ to $k \overline{\otimes} (1_R)$.*

Since both $\cdot \otimes R\colon \mathrm{Mod}(R_H) \to \mathrm{Gr\,Mod}(H\backslash G, R)$ and $\mathrm{Soc}_H\colon \mathrm{Gr\,Mod}(H\backslash G, R) \to \mathrm{Gr\,Mod}(H\backslash G, R)$ are additive functors, it is easy to make $\cdot \overline{\otimes} R$ into an additive functor from $\mathrm{Mod}(R_H)$ to $\mathrm{Gr\,Mod}(H\backslash G, R)$ sending any morphism $\phi\colon K \to L$ in $\mathrm{Mod}(R_H)$ into the unique morphism $\phi \overline{\otimes} R\colon K \overline{\otimes} R \to L \overline{\otimes} R$ in $\mathrm{Gr\,Mod}(H\backslash G, R)$ determined by:

(1.11) $$(\phi \overline{\otimes} R)(k \overline{\otimes} r) = \phi(k) \overline{\otimes} r, \textit{ for all } k \in K \textit{ and } r \in R.$$

We can also forget $H\backslash G$-gradings and consider $\cdot \overline{\otimes} R$ to be an additive functor from $\mathrm{Mod}(R_H)$ to $\mathrm{Mod}(R)$. Then Theorem 5.15 of **[1]** tells us the induction is transitive, in the sense that:

If I is a subgroup of H and J is any module in $\mathrm{Mod}(R_I)$, *then there is a unique isomorphism of R-modules*:

(1.12) $$\mu\colon (J \overline{\otimes}_{R_I} R_H) \overline{\otimes}_{R_H} R \xrightarrow{\sim} J \overline{\otimes}_{R_I} R \qquad (\textit{in } \mathrm{Mod}(R))$$

sending $(j \overline{\otimes} t) \overline{\otimes} r$ to $j \overline{\otimes} (tr)$, for all $j \in J$, $t \in R_H$ and $r \in R$.

2. Direct Clifford theory.

When the above subgroup H is 1, the set $1\backslash G$ of its cosets can be identified naturally with G. Then $\mathrm{Gr\,Mod}(1\backslash G, R)$ is just the abelian category $\mathrm{Gr\,Mod}(R)$ of all R-graded R-modules and their grade-preserving R-homomorphisms.

From now on we fix U and S satisfying:

(2.1a) *U is a simple right R_1-module.*

(2.1b) *$S = U \,\overline{\otimes}\, R = U \,\overline{\otimes}_{R_1} R$ is the G-graded R-module induced by U.*

From (1.8) and (1.10) we get:

(2.2) *$S_1 = U \,\overline{\otimes}\, R_1 \simeq U$ is a simple R_1-submodule generating the R-module $S = S_1 R$.*

Lemma 6.3 of **[1]** tells us that:

(2.3) *S is a simple G-graded R-module, i.e., a simple object in* $\operatorname{Gr Mod}(R)$.

The *support* $\operatorname{Supp}(S)$ of S is the set of all $\sigma \in G$ for which the σ-component S_σ is non-zero. We can reconstruct S for any such S_σ by (7.4) in **[1]**, which says that:

(2.4) *If $\sigma \in \operatorname{Supp}(S)$, then S_σ is a simple R_1-submodule of S, and there is a unique isomorphism $\mu_\sigma : S_\sigma \,\overline{\otimes}\, R \xrightarrow{\sim} S$ of R-modules sending $s_\sigma \,\overline{\otimes}\, r$ to $s_\sigma r$ for any $s_\sigma \in S_\sigma$ and $r \in R$. If $\tau \in G$, then the restriction of μ_σ is an R_1-isomorphism of $S_\sigma \,\overline{\otimes}\, R_\tau$ onto $S_{\sigma\tau}$.*

It follows from (2.2) that $S = s_1 R$ for some non-zero $s_1 \in S_1$. As in Proposition 2.4 of **[2]**, this implies that:

(2.5) *The R-endomorphism ring $E(S) = \operatorname{End}_R(S)$ of S is a G-graded ring. For any $\rho \in G$, its ρ-component is:*

$$E(S)_\rho = \{\phi \in E(S) | \phi(S_\sigma) \subseteq S_{\rho\sigma}, \text{ for all } \sigma \in G\}.$$

Restriction to S_σ is a homomorphism of $E(S)_\rho$ into the additive group $\operatorname{Hom}_{R_1}(S_\sigma, S_{\rho\sigma})$ of all R_1-homomorphisms of S_σ into $S_{\rho\sigma}$, whenever $\rho, \sigma \in G$. In fact, (2.4) implies (see Proposition 2.5 in **[2]**) that:

(2.6) *If $\sigma \in \operatorname{Supp}(S)$ and $\rho \in G$, then restriction to S_σ is an isomorphism of $E(S)_\rho$ onto $\operatorname{Hom}_{R_1}(S_\sigma, S_{\rho\sigma})$ as additive groups.*

From this and Schur's Lemma for the simple R_1-submodules S_σ of S, we conclude (as in Proposition 9.4 of **[1]**) that:

(2.7) *The support $\operatorname{Supp}(E(S))$ of $E(S)$ is a subgroup $G\{S\}$ of G consisting of all $\sigma \in G$ such that S_σ is R_1-isomorphic to U. The $G\{U\}$-graded subring $E(S)_{G\{U\}}$ is equal to $E(S)$ as a ring, and is a crossed product of $G\{U\}$ over the division ring $E(S)_1 \simeq \operatorname{End}_{R_1}(U)$.*

The category $\operatorname{Mod}(R|U)$ of all R-modules *lying over U* is best defined by Theorem 8.8 of **[1]**, which says that:

(2.8) *$\operatorname{Mod}(R|U)$ is the full, additive subcategory of $\operatorname{Mod}(R)$ generated by S, i.e., whose objects are the R-epimorphic images of direct sums $S^{(J)}$ of copies of S.*

It follows from (1.4) and (2.4) that:

(2.9) *S is the direct sum of its simple R_1-submodules S_σ for $\sigma \in \operatorname{Supp}(S)$.*

This and (2.8) imply immediately that:

(2.10) *Any module $M \in \operatorname{Mod}(R|U)$ is semi-simple as an R_1-module, and any simple R_1-submodule V of M is isomorphic to S_σ for some $\sigma \in \operatorname{Supp}(S)$.*

An important consequence of this and (2.6) (see Lemma 10.4 in [**1**] or Lemma 3.3 in [**2**]) is:

(2.11) *Suppose that $\sigma \in \operatorname{Supp}(S)$ and $M \in \operatorname{Mod}(R|U)$. Then any R_1-homomorphism $\phi\colon S_\sigma \to M$ has a unique extension to an R-homomorphism $\phi^e\colon S \to M$.*

From (2.10) and (2.11) we easily deduce (see Propositions 3.4 and 3.5 in [**2**]) that:

(2.12) $\operatorname{Mod}(R|U)$ *is an abelian subcategory of* $\operatorname{Mod}(R)$, *and is closed under arbitrary direct sums. The object S is a finitely-generated projective generator of* $\operatorname{Mod}(R|U)$.

Now a Theorem of Gabriel and Mitchell (see Corollary IV.15.28 in [**3**]) gives us the equivalence $\operatorname{Mod}(R|U) \approx \operatorname{Mod}(E(S))$ which is direct Clifford theory (see Theorems 10.6 in [**1**] or 3.7 in [**2**]):

(2.13) *The restrictions of the additive functors*

$$\cdot\,\overline{\otimes}_{E(S)}S\colon \operatorname{Mod}(E(S)) \to \operatorname{Mod}(R)$$

and

$$\operatorname{Hom}_R(S,\cdot)\colon \operatorname{Mod}(R) \to \operatorname{Mod}(E(S))$$

form an equivalence between the abelian category $\operatorname{Mod}(E(S))$ *and the abelian subcategory* $\operatorname{Mod}(R|U)$ *of* $\operatorname{Mod}(R)$.

3. Their relationship.

We return to an arbitrary subgroup H of G, and to the H-graded subring $R_H = \dot{\sum}_{\sigma\in H} R_\sigma$ of R. The G-graded R-module S in (2.1) has an H-graded R_H-submodule:

(3.1a) $$S_H = \dot{\sum_{\sigma\in H}} S_\sigma \qquad \textit{(as } R_1\textit{-modules)}$$

with the σ-components:

(3.1b) $$(S_H)_\sigma = S_\sigma, \textit{ for all } \sigma \in H.$$

In fact, this is naturally isomorphic to the simple H-graded R_H-module $U \mathbin{\overline{\otimes}} R_H = U \mathbin{\overline{\otimes}}_{R_1} R_H$ given by (2.1) and (2.3) with H and R_H in place of G and R.

PROPOSITION 3.2. *In* $\operatorname{Gr}\operatorname{Mod}(R_H)$ *there is a unique isomorphism of $U \mathbin{\overline{\otimes}} R_H$ onto S_H sending $u \mathbin{\overline{\otimes}} t \in U \mathbin{\overline{\otimes}} R_H$ to $u \mathbin{\overline{\otimes}} t \in U \mathbin{\overline{\otimes}} R = S$, for any $u \in U$ and $t \in R_H$.*

PROOF. We apply the transitivity (1.12) of induction with $I = 1$ and $J = U$. It gives us an isomorphism:

$$\mu\colon (U \mathbin{\overline{\otimes}} R_H) \mathbin{\overline{\otimes}}_{R_H} R \xrightarrow{\sim} U \mathbin{\overline{\otimes}} R = S \text{ (as } R\text{-modules)}$$

sending $(u \mathbin{\overline{\otimes}} t) \mathbin{\overline{\otimes}} r$ to $u \mathbin{\overline{\otimes}} (tr)$, for all $u \in U$, $t \in R_H$ and $r \in R$. From (1.10) for $K = U \mathbin{\overline{\otimes}} R_H$ we obtain an R_H-monomorphism $\cdot \mathbin{\overline{\otimes}} 1$ of $U \mathbin{\overline{\otimes}} R_H$ into $(U \mathbin{\overline{\otimes}} R_H) \mathbin{\overline{\otimes}} R$ sending $u \mathbin{\overline{\otimes}} t$ to $(u \mathbin{\overline{\otimes}} t) \mathbin{\overline{\otimes}} 1_R$, for all $u \in U$ and $t \in R_H$. The composition $u(\cdot \mathbin{\overline{\otimes}} 1)$ is an R_H-monomorphism of $U \mathbin{\overline{\otimes}} R_H$ into S sending $u \mathbin{\overline{\otimes}} t \in U \mathbin{\overline{\otimes}} R_H$ to $u \mathbin{\overline{\otimes}} t \in U \mathbin{\overline{\otimes}} R$, for all $u \in U$ and $t \in R_H$. It is clear from (1.9) that $u(\cdot \mathbin{\overline{\otimes}} 1)$ sends $(U \mathbin{\overline{\otimes}} R_H)_\sigma = U \mathbin{\overline{\otimes}} R_\sigma$ onto $S_\sigma = U \mathbin{\overline{\otimes}} R_\sigma$, for all $\sigma \in H$. So it is the H-grade-preserving R_H-isomorphism $U \mathbin{\overline{\otimes}} R_H \xrightarrow{\sim} S_H$ of the proposition.

From now on:

(3.3) *We identify $U \mathbin{\overline{\otimes}} R_H$ with S_H as H-graded R_H-modules via the isomorphism of Proposition 3.2. So the expression $u \mathbin{\overline{\otimes}} t$, for $u \in U$ and $t \in R_H$, has the same value, whether computed in $U \mathbin{\overline{\otimes}} R_H$ or in $U \mathbin{\overline{\otimes}} R$.*

Then we have:

COROLLARY 3.4. *There is a unique isomorphism $\mu\colon S_H \mathbin{\overline{\otimes}} R \xrightarrow{\sim} S$ in* $\operatorname{Gr Mod}(H\backslash G, R)$ *sending $s \mathbin{\overline{\otimes}} r$ to sr, for all $s \in S_H$ and $r \in R$.*

PROOF. After the identification (3.3), the map μ in the proof of Proposition 3.2 is an R-isomorphism of $S_H \mathbin{\overline{\otimes}} R = S_H \mathbin{\overline{\otimes}}_{R_H} R$ onto S sending $s \mathbin{\overline{\otimes}} r$ to sr for all $s \in S_H$ and $r \in R$. By (1.9) we have:

$$\mu((S_H \mathbin{\overline{\otimes}} R)_C) = \mu(S_H \mathbin{\overline{\otimes}} R_C) = S_H R_C \subseteq S_{HC} = S_C, \text{ for all } C \in H\backslash G.$$

So μ preserves $H\backslash G$-gradings, and the corollary is proved.

The G-graded ring $E(S)$ of (2.5) has an H-graded subring $E(S)_H$ as in (1.3). We may also use (2.5), with H and R_H in place of G and R, to construct an H-graded ring $E(S_H) = \operatorname{End}_{R_H}(S_H)$.

PROPOSITION 3.5. *Restriction to S_H is an isomorphism* $\operatorname{Res} = \operatorname{Res}_{S_H}$ *of $E(S)_H$ onto $E(S_H)$ as H-graded rings.*

PROOF. It follows from (2.5) that Res is an H-grade-preserving homomorphism of the ring $E(S)_H$ into $E(S_H)$. We know from (2.2) that $S_1 \neq 0$. Hence 1_G lies in both $\operatorname{Supp}(S)$ and $\operatorname{Supp}(S_H)$. So (2.6), for both G, R, S and H, R_H, S_H, tells us that restriction to S_1 sends both $E(S)_\rho$ and $E(S_H)_\rho$ isomorphically onto $\operatorname{Hom}_{R_1}(S_1, S_\rho)$, for any $\rho \in H$. We conclude that Res_{S_H} maps $E(S)_\rho$ isomorphically onto $E(S_H)_\rho$, for all $\rho \in H$. So it is an isomorphism of H-graded rings, and the proposition is proved.

From now on:

(3.6) *We identify $E(S)_H$ with $E(S_H)$ as H-graded rings via the restriction isomorphism* Res_{S_H} *of Proposition 3.5. So the value $\phi(s)$ of any ϕ in these rings at any $s \in S_H$ is the same, whether ϕ is considered as an element of $E(S)$ or of $E(S_H)$.*

After this identification, induction in the G-graded ring $E(S)$ becomes an additive functor:

$$\cdot \overline{\otimes} E(S) = \cdot \overline{\otimes}_{E(S)_H} E(S) = \cdot \overline{\otimes}_{E(S_H)} E(S)$$

from $\mathrm{Mod}(E(S_H)) = \mathrm{Mod}(E(S)_H)$ to $\mathrm{Gr\,Mod}(H\backslash G, E(S))$ or to $\mathrm{Mod}(E(S))$. Direct Clifford theory (2.13) for G, R and U is an equivalence between this last category and the subcategory $\mathrm{Mod}(R|U)$ of $\mathrm{Mod}(R)$. That same theory for H, R_H and U is an equivalence between $\mathrm{Mod}(E(S_H))$ and the subcategory $\mathrm{Mod}(R_H|U)$ of $\mathrm{Mod}(R_H)$. Thus we have the two rows and the right column in the following diagram:

$$\text{(3.7)}\qquad \begin{array}{ccc} \mathrm{Mod}(R|U) & \underset{\cdot\otimes_{E(S)}S}{\overset{\mathrm{Hom}_R(S,\cdot)}{\rightleftarrows}} & \mathrm{Mod}(E(S)) \\ \uparrow{\scriptstyle \cdot\,\overline{\otimes}_{R_H} R} & & \uparrow{\scriptstyle \cdot\,\overline{\otimes}_{E(S_H)} E(S)} \\ \mathrm{Mod}(R_H|U) & \underset{\cdot\otimes_{E(S_H)}S_H}{\overset{\mathrm{Hom}_{R_H}(S_H,\cdot)}{\rightleftarrows}} & \mathrm{Mod}(E(S_H)) \end{array}$$

In order to complete this diagram with its left column, we need:

PROPOSITION 3.8. *Induction* $\cdot\,\overline{\otimes}\, R\colon \mathrm{Mod}(R_H) \to \mathrm{Mod}(R)$ *sends the subcategory* $\mathrm{Mod}(R_H|U)$ *of* $\mathrm{Mod}(R_H)$ *into the subcategory* $\mathrm{Mod}(R|U)$ *of* $\mathrm{Mod}(R)$.

PROOF. We need only show that $K \,\overline{\otimes}\, R \in \mathrm{Mod}(R|U)$ whenever K is an R_H-module in $\mathrm{Mod}(R_H|U)$. By (2.8) our assumption about K gives us an R_H-epimorphism $\phi\colon S_H^{(J)} \twoheadrightarrow K$, where $S_H^{(J)}$ is a direct sum of copies of S_H indexed by some set J. Then $\phi \otimes R$ is an R-epimorphism of $S_H^{(J)} \otimes R$ onto $K \otimes R$, and determines an R-epimorphism $\phi \,\overline{\otimes}\, R$ of $S_H^{(J)} \,\overline{\otimes}\, R$ onto $K \,\overline{\otimes}\, R$ (see (1.8)).

The tensor product $\cdot \otimes R$ preserves direct sums. It is clear from (1.6) that Soc_H preserves direct sums. It follows from this and (1.8) that $\cdot\,\overline{\otimes}\, R$ preserves direct sums. In particular, we have:

$$\begin{aligned} S_H^{(J)} \,\overline{\otimes}\, R &\simeq (S_H \,\overline{\otimes}\, R)^{(J)} && (\textit{in } \mathrm{Gr\,Mod}(H\backslash G, R)). \\ &\simeq S^{(J)}, && (\textit{by Corollary } 3.4). \end{aligned}$$

Thus the above epimorphism $\phi \,\overline{\otimes}\, R$ gives an R-epimorphism of $S^{(J)}$ onto $K \,\overline{\otimes}\, R$. So $K \,\overline{\otimes}\, R \in \mathrm{Mod}(R|U)$ by (2.8), and the proposition is proved.

The compatibility of Clifford theory with induction is expressed by the commutativity of the diagram (3.7) to within natural equivalences of functors. The main obstacle in the proof of this commutativity will be the fact that our induction functors are not necessarily tensor products. The following result removes part of that obstacle.

LEMMA 3.9. *If K is any right $E(S_H)$-module, then the $H\backslash G$-graded $E(S)$-module $K \otimes E(S) = K \otimes_{E(S_H)} E(S)$ satisfies:*

$$\mathrm{Soc}_H(K \otimes E(S)) = 0.$$

Hence $K \otimes E(S)$ *is naturally isomorphic to its factor module* $K \overline{\otimes} E(S)$ *modulo* $\mathrm{Soc}_H(K \otimes E(S))$. *So the two functors* $\cdot \otimes E(S)$ *and* $\cdot \overline{\otimes} E(S)$ *from* $\mathrm{Mod}(E(S_H))$ *to* $\mathrm{Mod}(E(S))$ *are naturally equivalent.*

PROOF. We know from (2.7) that $E(S)$ is equal to its subring $E(S)_{G\{U\}}$, which is a crossed product of the subgroup $G\{U\} = \mathrm{Supp}(E(S))$ of G with the subring $E(S)_1$ of $E(S)$. Let C be any coset in $H\backslash G$. If $C \cap G\{U\}$ is empty, then $E(S)_C = 0$. Hence:

$$\mathrm{Soc}_H(K \otimes E(S))_C \subseteq K \otimes E(S)_C = 0$$

in this case. On the other hand, if $C \cap G\{U\}$ contains some element σ, then C is $H\sigma$ and $E(S)_\sigma$ contains a unit e_σ whose inverse e_σ^{-1} in $E(S)$ lies in $E(S)_{\sigma^{-1}}$. So (1.4b) and (1.6b) give:

$$\begin{aligned}\mathrm{Soc}_H(K \otimes E(S))_C e_\sigma^{-1} &\subseteq \mathrm{Soc}_H(K \otimes E(S))_C E(S)_{\sigma^{-1}} \\ &\subseteq \mathrm{Soc}_H(K \otimes E(S))_{C\sigma^{-1}} = \mathrm{Soc}_H(K \otimes E(S))_H = 0.\end{aligned}$$

Because e_σ is a unit in $E(S)$, this implies that $\mathrm{Soc}_H(K \otimes E(S))_C$ is zero in this case too.

We have shown that every C-component of the $H\backslash G$-graded module $\mathrm{Soc}_H(K \otimes E(S))$ is zero. So the first statement of the lemma holds. The rest of the lemma follows immediately from this and (1.8).

In some special cases the other induction functor $\cdot \overline{\otimes} R$ in (3.8) is also a tensor product. This happens when R is fully G-graded, i.e., when $R_\sigma R_\tau = R_{\sigma\tau}$ for all $\sigma, \tau \in G$. Of course, this includes the case when R is a crossed product of G over R. As in Lemma 3.9, we can even show the natural equivalence of $\cdot \overline{\otimes} R$ and $\cdot \otimes R$ when $\mathrm{Supp}(R)$ is a subgroup I of G and $R = R_I$ is a fully I-graded ring. However, in general there is no natural equivalence between $\cdot \overline{\otimes} R$ and $\cdot \otimes R$. We must use more subtle arguments to establish:

THEOREM 3.10. *The diagram* (3.7) *commutes to within natural equivalences of functors.*

PROOF. Since the rows of (3.7) are equivalences (2.13) of categories, it will suffice to provide a natural equivalence between the composite functors $(\cdot \overline{\otimes} E(S)) \otimes S$ and $(\cdot \otimes S_H) \overline{\otimes} R$ from $\mathrm{Mod}(E(S_H))$ to $\mathrm{Mod}(R|U)$ in that diagram. Thus we must construct, for any right $E(S_H)$-module K, a natural R-isomorphism:

$$\lambda_K : (K \overline{\otimes}_{E(S_H)} E(S)) \otimes_{E(S)} S \xrightarrow{\sim} (K \otimes_{E(S_H)} S_H) \overline{\otimes}_{R_H} R. \tag{3.11}$$

In view of Lemma 3.9 and Corollary 3.4 we already have natural R-isomorphisms:

$$\begin{aligned}(K \overline{\otimes}_{E(S_H)} E(S)) \otimes_{E(S)} S &\simeq (K \otimes_{E(S_H)} E(S)) \otimes_{E(S)} S \\ &\simeq K \otimes_{E(S_H)} (E(S) \otimes_{E(S)} S) \\ &\simeq K \otimes_{E(S_H)} S \\ &\simeq K \otimes_{E(S_H)} (S_H \overline{\otimes}_{R_H} R).\end{aligned} \tag{3.12}$$

So the only problem is to construct an R-isomorphism from $K \otimes (S_H \mathbin{\overline{\otimes}} R)$ to $(K \otimes S_H) \mathbin{\overline{\otimes}} R$. If induction were tensor products we could use the natural R-isomorphism:

$$K \otimes_{E(S_H)} (S_H \otimes_{R_H} R) \xrightarrow{\sim} (K \otimes_{E(S_H)} S_H) \otimes_{R_H} R. \tag{3.13}$$

In any event, the two R-modules in (3.13) are $H\backslash G$-graded with the C-components $K \otimes (S_H \otimes R_C)$ and $(K \otimes S_H) \otimes R_C$, respectively, for any $C \in H\backslash G$. So (3.13) is an isomorphism in $\operatorname{Gr Mod}(H\backslash G, R)$, and hence determines an isomorphism:

$$[K \otimes (S_H \otimes R)]/\operatorname{Soc}_H(K \otimes (S_H \otimes R)) \xrightarrow{\sim} (K \otimes S_H) \mathbin{\overline{\otimes}} R \tag{3.14}$$

in $\operatorname{Gr Mod}(H\backslash G, R)$ by (1.8).

Left multiplication by an element $\phi \in E(S_H)$ is an endomorphism $\phi \otimes R$ of the $H\backslash G$-graded R-module $S_H \otimes R$, and thus sends $\operatorname{Soc}_H(S_H \otimes R)$ into itself. So $S_H \otimes R$, its submodule $\operatorname{Soc}_H(S_H \otimes R)$ and their factor module $S_H \mathbin{\overline{\otimes}} R$ are all '$H\backslash G$-graded $E(S_H), R$-bimodules'. We conclude that the natural exact sequence:

$$0 \to \operatorname{Soc}_H(S_H \otimes R) \to S_H \otimes R \to S_H \mathbin{\overline{\otimes}} R \to 0$$

leads to an exact sequence:

$$\begin{aligned} K \otimes_{E(S_H)} \operatorname{Soc}_H(S_H \otimes R) \xrightarrow{\psi} K \otimes_{E(S_H)} (S_H \otimes R) \xrightarrow{\theta} \\ K \otimes_{E(S_H)} (S_H \mathbin{\overline{\otimes}} R) \to 0 \end{aligned} \tag{3.15}$$

in $\operatorname{Gr Mod}(H\backslash G, R)$.

The image of ψ in (3.15) is an $H\backslash G$-graded R-submodule N of $K \otimes (S_H \otimes R)$ such that θ gives an isomorphism:

$$\overline{\theta}\colon [K \otimes (S_H \otimes R)]/N \xrightarrow{\sim} K \otimes (S_H \mathbin{\overline{\otimes}} R) \qquad (in \operatorname{Gr Mod}(H\backslash G, R)).$$

Since $\operatorname{Soc}_H(S_H \otimes R)$ is H-null (see (1.6)), so are $K \otimes \operatorname{Soc}_H(S_H \otimes R)$ and its image N. It follows that N is contained in $\operatorname{Soc}_H(K \otimes (S_H \otimes R))$ and that:

$$\operatorname{Soc}_H([K \otimes (S_H \otimes R)]/N) = \operatorname{Soc}_H(K \otimes (S_H \otimes R))/N.$$

We conclude that $\overline{\theta}$ induces an isomorphism:

$$\begin{aligned} [K \otimes (S_H \otimes R)]/\operatorname{Soc}_H(K \otimes (S_H \otimes R)) \xrightarrow{\sim} \\ [K \otimes (S_H \mathbin{\overline{\otimes}} R)]/\operatorname{Soc}_H(K \otimes (S_H \mathbin{\overline{\otimes}} R)) \end{aligned} \tag{3.16}$$

in $\operatorname{Gr Mod}(H\backslash, R)$.

If we compare (3.16) with (3.14), we see that we can construct the desired isomorphism of $K \otimes (S_1 \mathbin{\overline{\otimes}} R)$ with $(K \otimes S_H) \mathbin{\overline{\otimes}} R$ provided that:

$$\operatorname{Soc}_H(K \otimes_{E(S_H)} (S_H \mathbin{\overline{\otimes}}_{R_H} R)) = 0. \tag{3.17}$$

So the proof of the entire theorem comes down to that of (3.17).

Let M be the $H\backslash G$-graded R-module:

$$M = K \otimes S = K \otimes_{E(S_H)} S, \tag{3.18a}$$

whose C-component, for any $C \in H\backslash G$, is:

(3.18b) $$M_C = K \otimes S_C = K \otimes_{E(S_H)} S_C,$$

where S_C is the C-component:

(3.18c) $$S_C = \dot{\sum}_{\sigma \in C} S_\sigma \qquad (as\ R_1\text{-}modules)$$

in the natural $H\backslash G$-grading:

(3.18d) $$S = \dot{\sum}_{C \in H\backslash G} S_C \qquad (as\ R_1\text{-}modules)$$

of S. Notice that each S_C is an $E(S_H)$, R_1-sub-bimodule of S by (3.6) and (2.5). So each tensor product $K \otimes S_C$ in (3.18b) can be identified with its image in the tensor product $K \otimes S$ in (3.18a).

It is clear from (3.18a) that M is an R-epimorphic image of a direct sum $S^{(K)}$ of copies of S. So (2.8) says that:

(3.19) $$M \in \mathrm{Mod}(R|U).$$

If $C \in H\backslash G$, then (3.18b) implies that M_C is an R_1-epimorphic image of a direct sum $S_C^{(K)}$ of copies of S_C. But (3.18c) tells us that S_C is the direct sum of its R_1-submodules S_σ, for $\sigma \in C \cap \mathrm{Supp}(S)$. Furthermore, each such S_σ is R_1-simple by (2.8). We conclude that:

(3.20) *For any $C \in H\backslash G$, the R_1-submodule M_C of M is semi-simple, and any simple R_1-submodule V of M_C is isomorphic to S_σ for some $\sigma \in C \cap \mathrm{Supp}(S)$.*

Suppose that (3.17) is false. The R-isomorphism $S_H \mathbin{\overline{\otimes}} R \xrightarrow{\sim} S$ of Corollary 3.4 preserves $H\backslash G$-gradings. Hence so does the R-isomorphism $K \otimes (S_H \mathbin{\overline{\otimes}} R) \simeq K \otimes S = M$ of (3.12). So $\mathrm{Soc}_H(M) \neq 0$. In particular, there is some $C \in H\backslash G$ such that:

$$\mathrm{Soc}_H(M)_C \neq 0.$$

In view of (3.20) the non-zero R_1-submodule $\mathrm{Soc}_H(M)_C$ of M_C must contain a simple R_1-submodule V. Furthermore, there is some R_1-isomorphism ϕ of S_σ onto V, where σ is some element of $C \cap \mathrm{Supp}(S)$. Now (3.19) and (2.11) give us an extension of ϕ to an R-homomorphism ϕ^e of S into M. Since C is $H\sigma$ we have:

$$\begin{aligned}\phi^e(S_\sigma R_{\sigma^{-1}}) = \phi(S_\sigma)R_{\sigma^{-1}} &\subseteq \mathrm{Soc}_H(M)_C R_{\sigma^{-1}} \\ &\subseteq \mathrm{Soc}_H(M)_{C\sigma^{-1}} = \mathrm{Soc}_H(M)_H = 0.\end{aligned}$$

Therefore the R_1-submodule $S_\sigma R_{\sigma^{-1}} R_\sigma$ of S_σ satisfies:

$$\phi(S_\sigma R_{\sigma^{-1}} R_\sigma) = \phi^e(S_\sigma R_{\sigma^{-1}})R_\sigma = 0.$$

Because ϕ is an isomorphism of S_σ onto V, this implies that:

$$S_\sigma R_{\sigma^{-1}} R_\sigma = 0.$$

But $\sigma \in \mathrm{Supp}(S)$. So (2.4) easily implies (see Theorem 8.6 in [**1**]) that:

$$S_\sigma R_{\sigma^{-1}} R_\sigma = S_1 R_\sigma = S_\sigma \neq 0.$$

This contradiction proves that (3.17) must hold.

In view of (3.17) the isomorphism (3.16) is really an isomorphism:

$$(3.21) \qquad [K \otimes (S_H \otimes R)]/\mathrm{Soc}_H(K \otimes (S_H \otimes R)) \xrightarrow{\sim} K \otimes (S_H \overline{\otimes} R)$$

in $\mathrm{Gr\,Mod}(H\backslash G, R)$. The composition of the isomorphisms (3.12), the inverse of the isomorphism (3.21) and the isomorphism (3.14) is the desired isomorphism λ_K in (3.11). One easily computes that the inverse λ_K^{-1} of λ_K is determined by:

$$(3.22) \qquad \lambda_K^{-1}((k \otimes s) \overline{\otimes} r) = (k \overline{\otimes} 1_{E(S)}) \otimes sr, \ \textit{for all } k \in K,\ s \in S_H \textit{ and } r \in R.$$

It follows that the λ_K form a natural equivalence λ between the functors $(\cdot \overline{\otimes} E(S)) \otimes S$ and $(\cdot \otimes S_H) \overline{\otimes} R$ in (3.7). So Theorem 3.10 is proved.

4. Two-step results.

If H is the stabilizer $G\{U\}$ of U in G, then (3.6) and (2.7) give:

$$(4.1) \qquad E(S_{G\{U\}}) = E(S)_{G\{U\}} = E(S) \qquad (\textit{as rings}).$$

It follows that:

$$(4.2) \qquad \mathrm{Mod}(E(S_{G\{U\}})) = \mathrm{Mod}(E(S)) \qquad (\textit{as abelian categories}).$$

We know from Lemma 3.9 that $\cdot \overline{\otimes} E(S)$ and $\cdot \otimes E(S)$ are equivalent functors from $\mathrm{Mod}(E(S_{G\{U\}}))$ to $\mathrm{Mod}(E(S))$. In view of (4.1) the latter functor is actually $\cdot \otimes_{E(S_{G\{U\}})} E(S_{G\{U\}})$, and thus is equivalent to the identity functor on $\mathrm{Mod}(E(S_{G\{U\}}))$. So we have:

(4.3) *Induction* $\cdot \overline{\otimes} E(S)\colon \mathrm{Mod}(E(S_{G\{U\}})) \to \mathrm{Mod}(E(S))$ *is naturally equivalent to the identity functor on* $\mathrm{Mod}(E(S_{G\{U\}})) = \mathrm{Mod}(E(S))$. *In particular, it is an equivalence of abelian categories.*

This and Theorem 3.10 for $H = G\{U\}$ imply that:

(4.4a) *Induction* $\cdot \overline{\otimes} R\colon \mathrm{Mod}(R_{G\{U\}}|U) \to \mathrm{Mod}(R|U)$ *is an equivalence of abelian categories.*

(4.4b) *The composite equivalence:*

$$\mathrm{Mod}(E(S_{G\{U\}})) \xrightarrow[\cdot \otimes S_{G\{U\}}]{\approx} \mathrm{Mod}(R_{G\{U\}}|U) \xrightarrow[\cdot\, \overline{\otimes}\, R]{\approx} \mathrm{Mod}(R|U)$$

of two-step Clifford theory is naturally equivalent to the direct equivalence:

$$\mathrm{Mod}(E(S_{G\{U\}})) = \mathrm{Mod}(E(S)) \xrightarrow[\cdot \otimes S]{\approx} \mathrm{Mod}(R|U)$$

of (4.2) *and* (2.13).

The above results should be compared with Theorem 11.18 and its Corollary 11.19 in **[1]**, which give further information about the inverse to the equivalence in (4.4a).

We return once more to an arbitrary subgroup H of G. Our hypotheses on G, R and U are also satisfied by H, R_H and U (see (1.3) and (2.1a)). In view of (2.1b) and (3.3), the equivalent of $S = U \overline{\otimes} R$ in this new situation is $S_H = U \overline{\otimes} R_H$. Furthermore, (3.6) says that the equivalent $E(S_H)$ of $E(S)$ is just $E(S)_H$. Since $G\{U\}$ is the support of $E(S)$ (see (2.7)), this implies that:

$$(4.5) \qquad H\{U\} = G\{U\} \cap H.$$

Now our hypotheses are satisfied with $G\{U\}$, $R_{G\{U\}}$, $U, S_{G\{U\}}$, $E(S_{G\{U\}})$ and $H\{U\}$ in place of G, R, U, S, $E(S)$ and H, respectively. The identification (3.3) of $U \overline{\otimes} R_{H\{U\}}$ with $S_{H\{U\}}$ is compatible with its similar identification with $(U \overline{\otimes} R_{G\{U\}})_{H\{U\}} = (S_{G\{U\}})_{H\{U\}}$. So the equivalent of (3.7) in this new situation is the left square in the following diagram:

$$(4.6) \qquad \begin{array}{ccccc}
\mathrm{Mod}(E(S_{G\{U\}})) & \xrightarrow[\cdot \otimes S_{G\{U\}}]{\approx} & \mathrm{Mod}(R_{G\{U\}}|U) & \xrightarrow[\cdot \overline{\otimes} R]{\approx} & \mathrm{Mod}(R|U) \\
\uparrow \cdot \overline{\otimes}\, E(S_{G\{U\}}) & & \uparrow \cdot \overline{\otimes}\, R_{G\{U\}} & & \uparrow \cdot \overline{\otimes}\, R \\
\mathrm{Mod}(E(S_{H\{U\}})) & \xrightarrow[\cdot \otimes S_{H\{U\}}]{\approx} & \mathrm{Mod}(R_{H\{U\}}|U) & \xrightarrow[\cdot \overline{\otimes} R_H]{\approx} & \mathrm{Mod}(R_H|U).
\end{array}$$

That square commutes to within natural equivalences of functors by Theorem 3.10 for our new situation.

The top row in (4.6) is the two-step Clifford equivalence of (4.4b). The bottom row is that same equivalence with H, R_H and U in place of G, R and U. The right column of (4.6) is the induction functor in Proposition 3.8. The entire right square in (4.6) commutes to within natural equivalences of functors by the transitivity (1.12) of induction. Thus we have proven:

THEOREM 4.7. *The diagram* (4.6) *commutes to within natural equivalences of functors, i.e., induction carries the two-step Clifford theory for* H, R_H *and* U *into the two-step Clifford theory for* G, R *and* U.

We should remark that the commutativity of the outer rectangle in (4.6) is equivalent to that of (3.7), once one notices that $\cdot \overline{\otimes}\, E(S_{G\{U\}})$ in (4.6) is naturally equivalent to $\cdot \overline{\otimes}\, E(S)$ from $\mathrm{Mod}(E(S_{H\{U\}})) = \mathrm{Mod}(E(S_H))$ to $\mathrm{Mod}(E(S_{G\{H\}})) = \mathrm{Mod}(E(S))$.

REFERENCES

[1] E. C. Dade, *Clifford theory for group-graded rings*, J. Reine Angew. Math. **369** (1986), 40–86.

[2] ——, *Clifford theory for group-graded rings, II*, J. Reine Angew. Math. **387** (1988), 148–181.

[3] C. Faith, "Algebra: Rings, Modules and Categories. I", Springer-Verlag, Berlin-Heidelberg-New York, 1973.

Contemporary Mathematics
Volume **93** (1989)

The Isomorphism Problem for Modular Group Algebras of Groups with Large Centres

VESSELIN DRENSKY[1)]

Dedicated to the memory of S. D. Berman

ABSTRACT. We give an affirmative answer to the isomorphism problem for modular group algebras for a class of finite p-groups which are close to the abelian groups. The following theorem is established:

Let K be a field of positive chartacteristic p and let G be a finite p-group such that the centre $Z(G)$ is of index p^2 in G. Then for a group $\tilde{G}$ the K-algebra isomorphism $K[G] \cong K[\tilde{G}]$ implies $G \cong \tilde{G}$.

0. Introduction.

One of the well known problems in the theory of group algebras is the following (see [**17**, Problem 6.6], [**18**, Conjecture 9.4]):

Let K be a field of positive characteristic p and let G be a finite p-group. Does the group algebra $K[G]$ determine the group G up to isomorphism? More precisely, let $\tilde{G}$ be another group such that $K[G]$ and $K[\tilde{G}]$ are isomorphic as K-algebras. Does it hold that G and $\tilde{G}$ are also isomorphic?

Deskins [**5**] solved positively this problem for abelian groups; an alternate proof of this result was given by Berman [**4**]. Passman [**15**] showed that the group algebra over $K = GF(p)$ determines the group G when $|G| = p^4$. Makasikis [**8**] studied the case $|G| = 2^5$ and $K = GF(2)$; as the referee has mentioned the work was done again more efficiently in [**9**]. Recently Bagiński has established similar results fore some Miller-Moreno groups and $K = GF(p)$ [**1**] as well as for G metacyclic and $K = GF(p)$, $p > 3$ [**2**]. For other results concerning the isomorphism problem for modular group algebras we refer to the surveys of Zalesskii and Mikhalev [**18**] and Sandling [**17**] and to the book by Passman [**14**].

1980 *Mathematics Subject Classification* (1985 *Revision*). 20C05; Secondary 16A26, 20D15.
Key words and phrases. group ring, p-group, isomorphism problem.
1) Partially supported by the Committee of Science (Bulgaria) Contract No. 62/March 25, 1987.

It is natural to generalize the result of Deskins **[5]** for groups which are close to abelian and especially for groups G with a centre $Z(G)$ of index p^2. The case $p = 2$ was handled by Nachev and Mollov **[12]**. They proved that if K is a finite ring of characteristic 2 and without nilpotent elements, then $K[G] \cong K[\tilde{G}]$ and $|G/Z(G)| = 4$ yield $G \cong \tilde{G}$ under the additional assumption $Z(G) = C_{2^n} \times C_2 \times \cdots \times C_2$ or $Z(G) = C_4 \times \cdots \times C_4 \times C_2 \times \cdots \times C_2$.

When $K = GF(p)$ is the prime field Sandling **[17**, Theorem 6.25**]** established the following: If $(\gamma_2(G))^p = \gamma_3(G) = \langle 1 \rangle$ then the group algebra determines G. Here $\gamma_i(G)$ is the ith element of the lower central series. In particullar, the groups G with $|G/Z(G)| = p^2$ have such a property. Unfortunately, the method of Sandling is essentially based on the primality of the base field and does not work over an arbitrary field of characteristic p.

The main purpose of the present paper is to establish the following result:

THEOREM. *Let p be a prime integer, let K be a field of characteristic p and let G be a finite p-group such that $|G/Z(G)| = p^2$. Then for a group $\tilde{G}$ the K-algebra isomorphism $K[G] \cong K[\tilde{G}]$ implies $G \cong \tilde{G}$.*

As an immediate consequence of this result we obtain:

COROLLARY. *Let p be prime, let K be a commutative ring with* 1 *such that p is not invertible element in K and let G be a finite p-group with $|G/Z(G)| = p^2$. Then the isomorphism of K-algebras $K[G] \cong K[\tilde{G}]$ yields $G \cong \tilde{G}$.*

In the proof of the theorem we do not work with the multiplicative group of $K[G]$ as in **[12]**. We linearize the problem and apply the structure properties of the group algebra established by Jennings [7]. Additionally, we make use of some ideas of Passman **[15]**, Nachev and Mollov **[12]** and of homological methods.

1. Preliminaries.

We fix some standard notation: p is a prime integer; G–*a* finite p-group; $Z = Z(G)$–the centre of G; $|G|$ and $|g|$–the orders of G and $g \in G$, respectively; $(g,h) = g^{-1}h^{-1}gh$–the commutator of $g,h \in G$; (H_1,H_2)–the subgroup generated by (h_1,h_2), where $h_1 \in H_1$, $h_2 \in H_2$ for $H_1, H_2 \subset G$; in particular $G' = (G,G)$ is the commutator subgroup of G; $\langle \dots \rangle$ is the subgroup of G generated by $\dots$; if $H \triangleleft G$ then $\overline{G} = G/H$ and $\overline{g} = gH$, $g \in G$. By C_n we denote the cyclic group of order n; $G_1 \times G_2$ is the direct product of the groups G_1 and G_2.

Besides, K is a fixed field of characteristic p, as we shall see, without loss of generality we can consider K algebraically closed; $R = K[G]$ is the group algebra of G; $N = N(R) = N(K[G])$-the Jacobson radical of R, N coincides with the augmentation ideal of R; $R' = [R,R]$-the commutator ideal of R generated by $[a,b] = ab - ba$, where $a,b \in R$. Moreover, if S is a subset of R, then $sp(S)$ is the vector space spanned by S; $S^{(n)} = \{a^n | a \in S\}$ and $S^n = sp(a_1 \dots a_n | a_i \in S)$. As a rule we denote the elements of R by small italics and use small greeks for the coefficients (from K) of linear combinations of elements from R.

When we consider another group $\tilde{G}$ we denote by $\tilde{Z}$, $\tilde{R}$ etc. respectively $Z(\tilde{G})$, $K[\tilde{G}]$ etc.

If L is an extension of the field K, then $L[G] \cong L \otimes_K K[G]$ (and we suppose that $K[G]$ is canonically embedded into $L[G]$). Hence the isomorphism $K[G] \cong K[\tilde{G}]$ implies $L[G] \cong L[\tilde{G}]$ and in the sequel we assume that K is an algebraically closed field. Moreover if $GF(p)$ is the prime field with p elements and the nth power $N^n(GF(p)[G])$ of the radical of $GF(p)[G]$ has a basis $a_1, \dots, a_k$ modulo $N^{n+1}(GF(p)[G])$, then the same elements form a basis of $N^n(K[G])$ modulo $N^{n+1}(K[G])$. Hence the results on the radical of $GF(p)[G]$ established in [7] are transfered without any restrictions to the case of an arbitrary base field K of characteristic p. In particular, R is obtained from N by a formal adjunction of 1, i.e. $R = K + N$.

In the sequel, an essential role will be played by the Brauer-Jennings-Zassenhaus M-series [7], [**14**, p. 670] defined by $M_1 = M_1(G) = G$ and $M_k = M_k(G) = \langle (M_{k-1}, G), M_i^{(p)} \rangle$ for $k > 1$, where i is the smallest integer satisfying $ip \geq k$.

LEMMA 1.1. [7, *Theorems 2.3 and 3.6*] (i) $M_k = M_k(G) = \{g \in G | g - 1 \in N^k\}$ *and* M_k/M_{k+1} *is an elementary abelian p-group.*

(ii) *Let* $M_k/M_{k+1} = \langle a_{k1}M_{k+1} \rangle \times \cdots \times \langle a_{kd_k}M_{k+1} \rangle$. *Then* N^n *has a basis modulo* N^{n+1} *consisting of all products* $\prod_{k \geq 1} \prod_{i=1}^{d_k} (a_{ki} - 1)^{\alpha_{ki}}$ *such that* $0 \leq \alpha_{ki} < p$ *and* $\sum k\alpha_{ki} = n$.

LEMMA 1.2. [**14**, *Lemma 2.7, p. 670*] *Assume that* $K[G] \cong K[\tilde{G}]$. *Then*:
(i) $M_k(G)/M_{k+1}(G) \cong M_k(\tilde{G})/M_{k+1}(\tilde{G})$;
(ii) $G/G' \cong \tilde{G}/\tilde{G}'$;
(iii) $Z(G) \cong Z(\tilde{G})$.

We start on the description of the p-groups G such that $|G/Z(G)| = p^2$. When $p = 2$ these groups are described by Nachev [**11**, Theorem 7].

LEMMA 1.3. *Let* $A = \langle a_1 \rangle \times \langle a_2 \rangle$ *be a p-group and let k be a nonnegative integer such that* $|a_1| \leq p^k |a_2|$. *Then* $A = \langle a_1 \rangle \times \langle a_1^{p^k} a_2 \rangle$.

PROOF. Let $|a_1| = p^s$, $|a_2| = p^t$. Then $s \leq k + t$, $(a_1^{p^k} a_2)^{p^{t-1}} = a_1^{p^{k+t-1}} a_2^{p^{t-1}} \notin \langle a_1 \rangle$ and $(a_1^{p^k} a_2)^{p^t} = 1$. Hence $|a_2| = |a_1^{p^k} a_2|$ and $\langle a_1 \rangle \cap \langle a_1^{p^k} a_2 \rangle = \langle 1 \rangle$. Obviously $A = \langle a_1, a_1^{p^k} a_2 \rangle$ and, therefore, $A = \langle a_1 \rangle \times \langle a_1^{p^k} a_2 \rangle$.

LEMMA 1.4. *Let A be a finite abelian p-group and let b be an element of A of order p. Then there exists a decomposition* $A = \langle a_1 \rangle \times \cdots \times \langle a_k \rangle$ *such that* $b \in \langle a_1 \rangle$.

PROOF. Let $A = \langle a_1 \rangle \times \cdots \times \langle a_k \rangle$ be a decomposition of A as a direct product of cyclic subgroups, $|a_i| = p^{s_i}$. Then b is expressed as $b = a_1^{u_1} \dots a_k^{u_k}$, where $0 \leq u_i < p^{s_i}$. Then $b^p = 1$ implies that $p^{s_i - 1}$ divides u_i for $i =$

$1,\dots,k$. Without loss of generality we can assume that $b = a_1^{p^{s_1-1}} \dots a_m^{p^{s_m-1}}$ with $m \le k$, $s_1 \le \dots \le s_m$. Then $|a_i| = p^{s_i - s_1}|a_1|$ for $i \le m$ and by Lemma 1.3 $A = \langle a_1' \rangle \times \langle a_2 \rangle \times \dots \times \langle a_k \rangle$, where $a_1' = a_1 a_2^{p^{s_2-s_1}} \dots a_m^{p^{s_m-s_1}}$. Therefore $b = (a_1')^{p^{s_1-1}}$.

LEMMA 1.5. *Let* $|G/Z(G)| = p^2$. *Then*:
(i) $|G'| = p$;
(ii) $\overline{G} = G/G' = \langle \bar{a} \rangle \times \langle \bar{b} \rangle \times \overline{C}$, *where* $(a,b) \neq 1$ *and* $C < Z(G)$.

PROOF. (i) Since $|G/Z| = p^2$, then $G/Z = \langle aZ \rangle \times \langle bZ \rangle$, where a, b are arbitrary elements of G such that $d = (a,b) \neq 1$. Clearly, $G' = \langle d \rangle < Z$. But $b^p \in Z$ and $a = b^{-p}ab^p = ad^p$, hence $d^p = 1$.

(ii) Assume that $\overline{G} = G/G' = \langle \bar{a} \rangle \times \langle \bar{b} \rangle \times \langle \bar{c}_1 \rangle \times \dots \times \langle \bar{c}_l \rangle$, where $(a,b) \neq 1$. Suppose $c_1 \notin Z$, then $c_1 = a^\alpha b^\beta z$, $z \in Z$ and $p \nmid \alpha$ or $p \nmid \beta$. Let for example $p \nmid \alpha$. We can assume that $|\bar{a}| \le |\bar{c}_1|$. (If $|\bar{a}| > |\bar{c}_1|$, then $\langle \bar{a}, \bar{b}, \bar{c}_1 \rangle = \langle \bar{a}_0 \rangle \times \langle \bar{b}_0 \rangle \times \langle \bar{c}_{10} \rangle$, where $a_0 = c_1$, $b_0 = b$, $c_{10} = a$, $(a_0, b_0) \neq 1$ and again $|\bar{a}_0| \le |\bar{c}_{10}|$.) We substitute $\bar{c}_1' = \bar{a}^{-\alpha}\bar{c}_1$, then $\langle \bar{a} \rangle = \langle \bar{a}^{-\alpha} \rangle$ and by Lemma 1.3 $\langle \bar{a} \rangle \times \langle \bar{c}_1 \rangle = \langle \bar{a}^{-\alpha} \rangle \times \langle \bar{c}_1 \rangle = \langle \bar{a}^{-\alpha} \rangle \times \langle \bar{c}_1' \rangle = \langle \bar{a} \rangle \times \langle \bar{c}_1' \rangle$; hence c_1' does not depend on a. Applying a similar procedure for $\bar{b}$ and $\bar{c}_1'$ we obtain that $\overline{G} = \langle \bar{a} \rangle \times \langle \bar{b} \rangle \times \langle \bar{z}_1 \rangle \times \langle \bar{c}_2 \rangle \times \dots \times \langle \bar{c}_l \rangle$, where $z_1 \in Z$. Using an obvious induction on l, we obtain $\overline{G} = \langle \bar{a} \rangle \times \langle \bar{b} \rangle \times (\langle \bar{z}_1 \rangle \times \dots \times \langle \bar{z}_l \rangle)$, where $z_i \in Z$, i.e. $\overline{G} = \langle \bar{a} \rangle \times \langle \bar{b} \rangle \times \overline{C}$ with $C = \langle z_1, \dots, z_l \rangle$.

PROPOSITION 1.6. *Let G be a finite p-group such that* $|G/Z(G)| = p^2$. *Then* $G = H \times A$, *where A is an abelian group and H belongs to one of the following families of groups*:

$H_1 = H_1(m,n,s) = \langle a,b,c \mid a^{p^m} = b^{p^n} = c^{p^s} = (a,c) = (b,c) = 1, (a,b) = c^{p^{s-1}} \rangle$, $m \ge n \ge 1$, $s \ge 1$, *and when $p = 2$ we require additionally* $m + n + s \ge 4$;
$H_2 = H_2(m,n) = \langle a,b \mid a^{p^m} = b^{p^n} = 1,\ (a,b) = a^{p^{m-1}} \rangle$, $m \ge 2$, $n \ge 1$;
$H_3 = Q$, *the quaternion group of order* 8.

PROOF. In virtue of Lemma 1.5 (ii) $G = \langle a, b, C \rangle$, where the commutator $d = (a,b)$ is of order p and generates G', $C < Z$ and $\overline{G} = G/G' = \langle \bar{a} \rangle \times \langle \bar{b} \rangle \times \overline{C}$. Therefore $\langle a,b \rangle \cap C < G'$. If $C \cap G' = \langle 1 \rangle$, then $G = \langle a,b \rangle \times A$ with $A = C$. Now, let $C \cap G' = G' = \langle d \rangle$. In virtue of Lemma 1.4 there exists a decomposition $C = \langle c_1 \rangle \times A$ such that $d \in \langle c_1 \rangle$. Then $\langle a,b,c_1 \rangle \cap A = \langle 1 \rangle$, i.e. $G = \langle a,b,c_1 \rangle \times A$. Therefore we can assume that $G = \langle a,b,c \rangle$, $(a,b) = d \neq 1$, $c \in Z$, $\overline{G} = G/G' = \langle \bar{a} \rangle \times \langle \bar{b} \rangle \times \langle \bar{c} \rangle$ (in some cases $c = 1$). Since $a^p, b^p \in Z$, all maximal subgroups of the group $F = \langle a,b \rangle$ are abelian and F is a p-group of Miller-Moreno **[10]**. By **[16**, Theorem 444, p. 736] $F = \langle a,b \rangle$ has one of the following presentations:

$F_1 = \langle a,b,d \mid a^{p^m} = b^{p^n} = d^p = (a,d) = (b,d) = 1, (a,b) = d \rangle$, $m \ge n \ge 1$;

$F_2 = \langle a,b \mid a^{p^m} = b^{p^n} = 1, (a,b) = a^{p^{m-1}} \rangle$, $m \ge 2$, $n \ge 1$, $p^{m+n} > 8$;

$F_3 = \langle a,b \mid a^4 = 1, (a,b) = a^2 = b^2 \rangle$-the quaternion group of order 8.

When $p = 2$, $m = n = 1$, $F_1 \cong D$ is the dihedral group of order 8 which is isomorphic to F_2 for $p^{m+n} = 8$. We attach D to the class F_2 and we assume for F_1 that $m + n \geq 3$ when $p = 2$.

Let $|c| = p^s$. If $c^{p^{s-1}} \notin G'$, then $G \cong \langle a, b\rangle \times \langle c\rangle$ and $\langle a, b\rangle \cong F_i$ for some $i = 1, 2, 3$. Let us assume that $c^{p^{s-1}} = (a, b)^\delta$, $p \nmid \delta$. Without loss of generality we can consider the case $\delta = 1$ only, i.e. $c^{p^{s-1}} = (a, b) = d$. If $\langle a, b\rangle \cong F_1$, then $\langle a, b, c\rangle \cong H_1$. Now, let $\langle a, b\rangle \cong F_2$, $|a| = p^m$, $|b| = p^n$, $|c| = p^s$, $|G| = p^{m+n+s-1}$. For $m > s$ we substitute $c_1 = ca^{-p^{m-s}}$. Then $|c_1| = p^{s-1}$ and $c_1 \leq Z$ because $a^p \in Z$. Therefore $G \cong F_2 \times \langle c_1\rangle$. When $m < s$ we substitute $a_1 = ac^{-p^{s-m}}$, $|a_1| = p^{m-1}$, $(a_1, b) = (a, b)$ and $\langle a_1, b\rangle \cong F_1$; hence $G \cong H_1$. When $p = 2$, $\langle a, b\rangle \cong Q$, $c^{2^{s-1}} = a^2$, we have $s \geq 2$ and replacing a by $a_1 = ac^{2^{s-1}}$ we obtain $G = \langle b, a_1, c | b^4 = a_1^2 = c^{2^s} = (b, c) = (a_1, c) = 1, (b, a_1) = c^{2^{s-1}}\rangle \cong H_1$.

REMARK 1.7. We are able to show that the groups of Proposition 1.6 are pairwise non-isomorphic. We do not submit the proof because in the sequel we shall establish that these groups have nonisomorphic group algebras.

2. Main results.

We fix the finite p-group G such that $|G/Z(G)| = p^2$. Hence $G \cong H \times A$, where $H = H_1 = H_1(m, n, s)$, $H = H_2 = H_2(m, n)$ or $H = H_3 = Q$ is a group of the list from Proposition 1.6 and A is an abelian group, $A = \langle e_1\rangle \times \cdots \times \langle e_t\rangle$, $|e_j| = p^{u_j}$. The following two lemmas give the description of the M-series of G. The proofs are obtained by direct calculations.

LEMMA 2.1. (i) *$M_k(G) = M_k(H) \times M_k(A)$; $M_k(A) = \langle e_1^k\rangle \times \cdots \times \langle e_t^k\rangle$ if k is a power of p and $M_k(A) = M_{k+1}(A)$ in the other case;*

(ii) *$M_1(H_i) = H_i$; $M_k(H_1) = \langle a^k\rangle \times \langle b^k\rangle \times \langle c^k\rangle$, $M_k(H_2) = \langle a^k\rangle \times \langle b^k\rangle$, when $k > 1$ is a power of p, with the exception of the case $p = 2$, $H = H_1$, $k = 2$, $s = 1$; $M_2(H_1) = \langle a^p\rangle \times \langle b^p\rangle \times \langle c\rangle$, when $s = 1$; $M_2(Q) = \langle a^2\rangle$, $M_3(Q) = \langle 1\rangle$; $M_k(H_i) = M_{k+1}(H_i)$, $i = 1, 2, 3$, in all other cases.*

LEMMA 2.2. (i) *$M_k(G)/M_{k+1}(G) = M_k(H)/M_{k+1}(H) \times M_k(A)/M_{k+1}(A)$, $M_k(A)/M_{k+1}(A) = \langle \bar{e}_1^k\rangle \times \cdots \times \langle \bar{e}_t^k\rangle$, when k is a power of p (and $\langle \bar{e}_j^k\rangle$ participates as a direct factor if and only if $|e_j| > k$), $M_k(A)/M_{k+1}(A) = \langle 1\rangle$ in the other case;*

(ii) *$M_k(H_1)/M_{k+1}(H_1) = \langle \bar{a}^k\rangle \times \langle \bar{b}^k\rangle \times \langle \bar{c}^k\rangle$, if k is a power of p, with the exception of the case $k = s = 1$, when $M_1(H_1)/M_2(H_1) = \langle \bar{a}\rangle \times \langle \bar{b}\rangle$; $M_2(H_1)/M_3(H_1) = \langle \bar{c}\rangle$, if $s = 1$, $p > 2$; $M_k(H_2)/M_{k+1}(H_2) = \langle \bar{a}^k\rangle \times \langle \bar{b}^k\rangle$, if k is a power of p; $M_1(Q)/M_2(Q) = \langle \bar{a}\rangle \times \langle \bar{b}\rangle$, $M_2(Q)/M_3(Q) = \langle \bar{a}^2\rangle$; $M_k(H_i)/M_{k+1}(H_i) = \langle 1\rangle$, $i = 1, 2, 3$, in all other cases.*

PROPOSITION 2.3. *Let k be the greatest integer such that the commutator ideal R' is contained in N^k and let r be an arbitrary element of R' for which $r \not\equiv 0 \pmod{N^{k+1}}$. The following table gives the values of k and the existence* *of solutions of the congruence* $x^k \equiv r \pmod{N^{k+1}}$ *in* N *and* $Z(N)$.

H	p	Constraints	k	Solutions in N	Solutions in $Z(N)$
$H_1(m,n,s)$	odd	$s>1$			
			$p^{s-1}>2$		yes
	2	$s>2$			
$H_2(m,n)$	odd				
			$p^{m-1}>2$		no
	2	$m>2$			
$H_1(m,n,1)$	odd		2		
$H_1(m,n,1)$				no	
$H_1(m,n,2)$					yes
	2		2		
$H_2(2,n)$					
				yes	no
$H_3=Q$					

PROOF. Using the equality $[g_1,g_2] = g_1g_2 - g_2g_1 = g_2g_1(g_1^{-1}g_2^{-1}g_1g_2 - 1) = g_2g_1((g_1,g_2) - 1)$, $g_1,g_2 \in G$, we obtain that R' is generated by $g-1$, where g runs on $G' = \langle (a,b) \rangle$. Then, by Lemma 1.1 (i) k is determined by the property $(a,b) \in M_k$, $(a,b) \notin M_{k+1}$. The assertion follows immediately from Lemma 2.1 (ii): If $G = H_1 \times A$, then $(a,b) = c^{p^{s-1}}$; hence $k=2$ for $s=1$ and $k=p^{s-1}$ for $s>1$. If $G=H_2\times A$, then $(a,b) = a^{p^{m-1}}$ and $k=p^{m-1}$; if $G = Q\times A$, then $(a,b)=a^2$ and $k=2$.

Let $k>2$. First, let $G = H_1(m,n,s)\times A$; then by Lemma 2.3 $s>1$ and $k=p^{s-1}$, $(a,b) = c^{p^{s-1}} = d$, i.e. $r \equiv \delta(d-1) \pmod{N^{k+1}}$ and $0 \neq \delta \in K$. But $c-1 \in Z(N)$, the field K is algebraically closed and

$$(\delta^{1/k}(c-1))^k = \delta(c^{p^{s-1}} - 1) \equiv r \pmod{N^{k+1}},$$

i.e. the congruence has a solution in $Z(N)$.

Now, let $G = H_2(m,n)\times A$; then $k=p^{m-1}$, $(a,b) = a^{p^{m-1}}$ and by Lemma 1.1 (ii) and Lemma 2.2 N has a basis $a-1, b-1, e_1-1, \dots, e_t-1$ modulo N^2. Let $x = \alpha(a-1)+\beta(b-1)+\sum \varepsilon_j(e_j-1)+n$, $n\in N^2$. Since $x\in Z(N)$ then $0 = [x,b-1] \equiv \alpha[a,b] \equiv \alpha(a^{p^{m-1}}-1) \pmod{N^{k+1}}$, i.e. $\alpha=0$. Similarly, $\beta=0$ and $x^k = x^{p^{m-1}} \equiv \sum(\varepsilon_j^{p^{m-1}}-1) \not\equiv \delta(a^{p^{m-1}}-1) \pmod{N^{k+1}}$, i.e. the congruence has no solution in $Z(N)$.

Let $p=k=2$.

1) Let $G = H_1(m,n,1)\times A$. Then N has a basis $a+1, b+1, e_1+1, \dots, e_t+1$ modulo N^2. Let $x = \alpha(a+1)+\beta(b+1)+\sum\varepsilon_j(e_j+1)+n$, $n\in N^2$, $r\equiv \delta(c+1) \pmod{N^3}$, $0\neq\delta\in K$. Direct calculations show that $ab+ba\equiv c+1 \pmod{N^3}$, $x^2 \equiv \alpha^2(a^2+1)+\beta^2(b^2+1)+\alpha\beta(c+1)+\sum\varepsilon_j^2(e_j^2+1) \equiv \delta(c+1) \pmod{N^3}$. Since a^2+1, b^2+1, $c+1$ and those e_j^2+1 for which $e_j^2\neq 1$ form a subset of the basis of N^2 modulo N^3, then $\alpha=\beta=0$, i.e. $x^2\equiv r \pmod{N^3}$ has no solution in N.

2) Let $G = H_1(m,n,2) \times A$, then $r \equiv \delta(c^2+1) \pmod{N^3}$, $0 \neq \delta \in K$ and for $x = \delta^{1/2}(c+1) \in Z(N)$ it holds $x^2 \equiv r \pmod{N^3}$.

3) Let $G = H_2(2,n) \times A$. Then $r \equiv \delta(a^2+1) \pmod{N^3}$, $0 \neq \delta \in K$ and $(\delta^{1/2}(a+1))^2 \equiv r \pmod{N^3}$. On the other hand, if $x = \alpha(a+1) + \beta(b+1) + \sum \varepsilon_j(e_j+1) + n \in Z(N)$, $n \in N^2$, then $\alpha = \beta = 0$ and $x^2 \equiv \sum \varepsilon_j^2(e_j^2+1) \not\equiv r \pmod{N^3}$, i.e. $x^2 \equiv r \pmod{N^3}$ has no solution in $Z(N)$.

4) The case $G = Q \times A$ is handled similarly to 3).

LEMMA 2.4. *Let $p = 2$ and let D and Q be the dihedral and the quaternion groups of order 8, respectively, and let A be a finite 2-group. Then the group algebras $K[D \times A]$ and $K[Q \times A]$ are non-isomorphic.*

PROOF. As a background of homological methods we refer to [**3**]. Since $H^n(G,K) \cong \mathrm{Ext}^n_{K[G]}(K,K)$, the Poincaré series $\xi_K(G,t) = \sum t^n \dim_K H^n(G,K)$ of a group G is an invariant of $K[G]$. By [**3**, p. 185–186] $\xi_K(D,t) = 1/(1-t)^2$, $\xi_K(Q,t) = (1+t+t^2)/(1-t)(1+t^2)$. Since $\xi_K(G \times A, t) = \xi_K(G,t)\xi_K(A,t)$, the Poincaré series $\xi_K(D \times A, t)$ and $\xi_K(Q \times A, t)$ are different. This immediately yields that the group algebras $K[D \times A]$ and $K[Q \times A]$ are non-isomorphic.

REMARK 2.5. If the base field $K = GF(2)$ is the prime field with two elements and A is abelian, the assertion of Proposition 2.4 follows immediately from [**13**, Theorem 3] which states that $GF(p)[G] \cong GF(p)[\tilde{G}]$ for some p-groups G and $\tilde{G}$ implies $M_i(G)/M_{i+2}(G) \cong M_i(\tilde{G})/M_{i+2}(\tilde{G})$. Supposing $K[D \times A] \cong K[Q \times A]$ we obtain

$$D \times A/A^{(4)} \cong M_1(D \times A)/M_3(D \times A) \cong M_1(Q \times A)/M_3(Q \times A) \cong Q \times A/A^{(4)},$$

which is a contradiction. But in our case the field K is arbitrary and the arguments from [**13**] do not hold. When A is an abelian group and K is algebraically closed, one can give an alternate proof of the proposition using an idea of Holvoet [**6**]. Bearing in mind that $R_D = K[D \times A] \cong K[D] \otimes_K K[A]$ and $R_Q = K[Q \times A] \cong K[Q] \otimes_K K[A]$, direct computations show that the vector spaces $X_D = sp[x \in N(R_D) | x^2 \equiv 0 \pmod{N^5(R_D)})$ and $X_Q = sp(x \in N(R_Q) | x^2 \equiv 0 \pmod{N^t(R_Q)})$ have different dimensions. Therefore the algebras R_D and R_Q are non-isomorphic.

2.6. PROOF OF THE THEOREM. Let $|G/Z(G)| = p^2$ and let $\tilde{G}$ be a group such that $R = K[G] \cong K[\tilde{G}] = \tilde{R}$. It follows from Lemma 1.2 (iii) that $Z \cong \tilde{Z}$ and as an immediate consequence of $|G| = |\tilde{G}|$ we obtain that $|G/Z| = |\tilde{G}/\tilde{Z}| = p^2$. Hence $\tilde{G}$ is a group of the same class. Assume that $\tilde{G} = \tilde{H} \times \tilde{A}$, where $\tilde{H} = H_1(\tilde{m},\tilde{n},\tilde{s})$, $\tilde{H} = H_2(\tilde{m},\tilde{n})$ or $\tilde{H} = Q$, $\tilde{A} = \langle \tilde{e}_1 \rangle \times \cdots \times \langle \tilde{e}_{\tilde{t}} \rangle$. We shall prove the theorem in several steps.

Step 1. Let $G = H \times A$, let k be the greatest integer such that $R' \subset N^k$. Since k is an invariant of R, then k has the same property with respect to $\tilde{R}$. Similarly, an invariant of R is the existence of solutions in N and $Z(N)$ of the congruence $x^k \equiv r \pmod{N^{k+1}}$ for any $r \in R'$, $r \not\equiv 0 \pmod{N^{k+1}}$.

Using Proposition 2.3 and comparing these two invariants, we establish the following:

1) If $H = H_1(m, n, s)$, then $\tilde{H} = H_1(\tilde{m}, \tilde{n}, s)$;

2) If $H = H_2(m, n)$, $m > 2$, then $\tilde{H} = H_2(m, \tilde{n})$;

3) If $H = H_2(2, n)$, then $\tilde{H} = H_2(2, \tilde{n})$ or $\tilde{H} = Q$;

4) If $H = Q$, then $\tilde{H} = H_2(2, \tilde{n})$ or $\tilde{H} = Q$.

Step 2. Let $G = H_1(m, n, s) \times \langle e_1 \rangle \times \cdots \times \langle e_t \rangle$, $|e_j| = p^{u_j}$, $\tilde{G} = H_1(\tilde{m}, \tilde{n}, \tilde{s}) \times \langle \tilde{e}_1 \rangle \times \cdots \times \langle \tilde{e}_{\tilde{t}} \rangle$, $|\tilde{e}_j| = p^{\tilde{u}_j}$ and by Step 1 $s = \tilde{s}$. Clearly, $Z(G) = \langle a^p \rangle \times \langle b^p \rangle \times \langle c \rangle \times \langle e_1 \rangle \times \cdots \times \langle e_t \rangle$, $G/G' = \langle \bar{a} \rangle \times \langle \bar{b} \rangle \times \langle \bar{c} \rangle \times \langle \bar{e}_1 \rangle \times \cdots \times \langle \bar{e}_t \rangle$ and $|a^p| = p^{m-1}$, $|b^p| = p^{n-1}$, $|c| = p^s$, $|\bar{a}| = p^m$, $|\bar{b}| = p^n$, $|\bar{c}| = p^{s-1}$, $|e_j| = |\bar{e}_j| = p^{u_j}$. Similar formulas hold for $Z(\tilde{G})$ and $\tilde{G}/\tilde{G}'$. We present the invariants of the abelian groups $Z(G)$ and G/G' respectively as vectors $z(G) = (m-1, n-1, s, u_1, \ldots, u_t)$, $f(G) = (m, n, s-1, u_1, \ldots, u_t)$, analogously $z(\tilde{G}) = (\tilde{m}-1, \tilde{n}-1, s, \tilde{u}_1, \ldots, \tilde{u}_{\tilde{t}})$, $f(\tilde{G}) = (\tilde{m}, \tilde{n}, s-1, \tilde{u}_1, \ldots, \tilde{u}_{\tilde{t}})$. By Lemma 1.2 (ii), (iii) the vector $z(\tilde{G})$ is obtained from the vector $z(G)$ by permuting of the coordinates, similarly for $f(\tilde{G})$ and $f(G)$. The comparison of $z(G)$ and $z(\tilde{G})$, $f(G)$ and $f(\tilde{G})$ shows that $m = \tilde{m}$, $n = \tilde{n}$ and hence $H \cong \tilde{H}$, $A \cong \tilde{A}$, i.e. $G \cong \tilde{G}$.

Step 3. Let $G = H_2(m, n) \times \langle e_1 \rangle \times \cdots \times \langle e_t \rangle$ and let $m > 2$ when $p = 2$. By Step 1 $\tilde{G} = H_2(\tilde{m}, \tilde{n}) \times \langle \tilde{e}_1 \rangle \times \cdots \times \langle \tilde{e}_{\tilde{t}} \rangle$ and $m = \tilde{m}$. Again the vectors $z(G) = (m-1, n-1, u_1, \ldots, u_t)$ and $f(G) = (m-1, n, u_1, \ldots, u_t)$ define uniquely $Z(G)$ and G/G'. As in Step 2 we see that $n = \tilde{n}$, $H \cong \tilde{H}$, $A \cong \tilde{A}$. Hence $G \cong \tilde{G}$. In particular, this completes the proof for $p > 2$.

Step 4. Let $p = 2$. As in Step 2 we handle the cases $H = H_2(2, n)$, $\tilde{H} = H_2(2, \tilde{n})$ and $H \cong \tilde{H} \cong Q$. The only case to be considered is $H = H_2(2, n)$, $\tilde{H} = Q$. Again, the comparison of the invariants of $Z(G)$ and $Z(\tilde{G})$, G/G' and $\tilde{G}/\tilde{G}'$ gives that $n = 1$ and $A \cong \tilde{A}$. Therefore $H = D$ is the dihedral group of order 8. But the group algebras $R = K[D \times A]$ and $\tilde{R} = K[Q \times A]$ are non-isomorphic in virtue of Lemma 2.4. This completes the proof of the theorem.

2.7. Proof of the corollary. Let K be a commutative ring with 1 such that p is not invertible in K and let G be a group of the investigated type. There exists a maximal ideal I of K such that $p \in I$. Hence, if $\tilde{G}$ is a group with $K[G] \cong K[\tilde{G}]$, then $K/I[G] \cong K/I[\tilde{G}]$. Since K/I is a field of characteristic p, the corollary follows immediately from the theorem.

Acknowledgments

The author is very grateful to: A. S. Kondrat′ev and K. Tchakerian for help with group theory; T. Mollov and P. Siderov for discussions on group algebras; L. Avramov and K. W. Roggenkamp for help with homological algebra; C. Genov for general discussions and the referee for numerous remarks and suggestions for improving the exposition.

REFERENCES

1. C. Baginski, *Groups of units of modular group algebras*, Proc. Amer. Math. Soc., **101** (1987), 619–624.

2. ——, *The isomorphism question for modular group algebras of metacyclic p-groups*, preprint.

3. D. Benson, *Modular representation theory: New trends and methods*, Lecture Notes in Math., 1081, Springer Verlag, Berlin, Heidelberg, New York, Tokyo, 1984.

4. S. D. Berman, *Group algebras of countable abelian p-groups*, Publ. Math. Debrecen, **14** (1967), 365–405 (Russian).

5. E. Deskins, *Finite abelian groups with isomorphic group algebras*, Duke Math. J., **23** (1956), 35–40.

6. R. Holvoet, *Sur l'isomorphie d'algèbres de groupes*, Bull. Soc. Math. Belg., **20** (1968), 264–282.

7. S. A. Jennings, *The structure of the group ring of a p-group over a modular field*, Trans. Amer. Math. Soc., **50** (1941), 175–185.

8. A. Makasikis, *Sur l'isomorphie d'algèbres de groupes sur un champ modulaire*, Bull. Soc. Math. Belg., **28** (1976), No. 2, 91–109.

9. G. O. Michler, M. F. Newman, E. A. O'Brien, *Modular group algebras*, preprint.

10. G. A. Miller, H. C. Moreno, *Non-abelian groups in which every subgroup is abelian*, Trans. Amer. Math. Soc., **4** (1903), 398–404.

11. N. Nachev, *Finite 2-groups with a factor-group by centre of order four*, Université de Plovdiv "Paissi Hilendarski", Travaux Scientifiques, mathématiques, **19** (1981), No. 1, 27–48 (Bulgarian).

12. N. A. Nachev, T. Zh. Mollov, *An isomorphism of modular group algebras of finite 2-groups for which the order of the factor group with respect to the centre equals four*, Pliska, Studia Math. Bulg., **8** (1986), 3–20 (Russian).

13. I. B. S. Passi, S. K. Sehgal, *Isomorphism of modular group algebras*, Math. Z., **129** (1972), 65–73.

14. D. S. Passman, *The algebraic structure of group rings*, Interscience, New York, 1977.

15. ——, *The group algebras of groups of order p^4 over a modular field*, Michigan Math. J., **12** (1965), 405–415.

16. L. Rédei, *Algebra*, vol. 1, Akadémiai Kiadó, Budapest, 1967.

17. R. Sandling, *The isomorphism problem for group rings: a survey, in "Orders and their Applications,"* Lecture Notes in Math., **1142**, 256–288, Springer Verlag, Berlin, Heidelberg, New York, Tokyo, 1985.

18. A. E. Zalesskii, A. V. Mikhalev, *Group rings, in "Itogi Nauki i Tekhniki (Sovremennye Problemy Matematiki),"* **2** (1973), 5–118 (Russian); English translation: J. Soviet Math., **4** (1975), 1–78.

INSTITUTE OF MATHEMATICS,
BULGARIAN ACADEMY OF SCIENCES,
1090 SOFIA,
P. O. BOX 373,
BULGARIA

Contemporary Mathematics
Volume **93** (1989)

Zero Divisors in Group Rings: Something Old, Something New[1]

DANIEL R. FARKAS[2] AND PETER A. LINNELL

One of the oldest and most frustrating problems in the subject of infinite group rings is the zero divisor conjecture. If G is a torsion free group and k is a field, the presumption is that the group algebra $k[G]$ has no zero divisors.

How can one prove that a noncommutative ring is a domain? We can fall back on our experience with commutative rings for some strategies.

(i) Imbed the ring in a field (i.e. localize).
(ii) Find a degree-like map that makes the ring resemble a polynomial ring.
(iii) Use homological techniques as in the proof that a regular local Noetherian ring is a domain.

This essay reports on some insightful ways to pursue these methods. The first section is an exposition of part of J. A. Moody's thesis ([**8**]). His very clever ideas lead to strategy (iii), as introduced in [**4**]. His result implies that the zero divisor conjecture is valid for all torsion free solvable groups. The second section unearths an overlooked theorem of M. Lazard ([**5**]), which sets up a deep version of method (ii). One consequence is that the zero divisor conjecture holds for congruence subgroups, namely the kernels of $\mathrm{GL}(n, \mathbb{Z}) \to \mathrm{GL}(n, \mathbb{Z}/(p))$ for odd primes p.

1. Moody's Theorem.

To get started we need a whirlwind review of the structure theory for noncommutative Noetherian rings. Recall that a ring is prime provided that the product of two of its non-zero two-sided ideals always remains non-zero.

GOLDIE'S THEOREM ([**2**]): *A prime Noetherian ring has a classical ring of fractions which is isomorphic to the ring of all $n \times n$ matrices over some division ring.*

Recall that if $\mathrm{Fract}(R)$ is the classical ring of fractions for R then each non-zero-divisor in R becomes a unit in $\mathrm{Fract}(R)$ and every member of $\mathrm{Fract}(R)$

[1] Everything borrowed, everything true.

[2] Partially supported by the National Science Foundation.

can be written in the form $a^{-1}b$ for $a, b \in R$, with a not a zero-divisor. It is also possible to "localize" at smaller sets of non-zero-divisors.

What sort of invariant is the n in Goldie's Theorem? It is certainly the composition length of Fract(R). More generally, if M is a finitely generated module over the prime Noetherian ring R then the *rank* of M is defined to be the length of $\mathrm{Fract}(R) \otimes_R M$ as a module over the ring of fractions. The rank turns out to have the following nice property: If $0 \to A \to B \to C \to 0$ is a short exact sequence of finitely generated R-modules then $rk(B) = rk(A) + rk(C)$. The formula can be extended in the obvious way to longer exact sequences ([**2**]).

The essence of the third strategy is embodied in

WALKER'S THEOREM ([**2**]): *A prime regular local Noetherian ring is a domain.*

PROOF: If R is the ring, take a minimal left ideal of Fract(R) and let L be its intersection with R. Then $rk(L) = 1$. Since R has finite global dimension and is Noetherian, L has a finite resolution by finitely generated projective modules. Furthermore, projectives over a local ring are free. Thus we have

$$0 \to R^{n(t)} \to \cdots \to R^{n(2)} \to R^{n(1)} \to L \to 0.$$

In particular, $rk(L)$ is an integer multiple of $rk(\mathrm{Fract}(R))$. It follows that $rk(\mathrm{Fract}(R)) = 1$.

Let's look at the instant replay. The key ingredients are finite projective resolutions and the fact that finitely generated projective modules are stably free. This brings to mind one of the fundamental results in algebraic K-theory which relates the Grothendieck group on finitely generated projective modules to the Grothendieck group on all finitely generated modules.

RESOLUTION THEOREM ([**1**]): *If R is a regular Noetherian ring then $K_0(R) \simeq G_0(R)$.*

The point for us, of course, is that $G_0(R)$ is the universal source of all "ranks".

WALKER REVISITED: *If R is a prime Noetherian ring such that $G_0(R)$ is generated by $[R]$ then R is a domain.*

The original problem has shifted to proving that $G_0(R)$ is trivial. This is of interest for the case that R is a group ring to both algebraists and topologists. For the latter, the vanishing of $\tilde{K}_0$ implies the triviality of a geometric obstruction.

The first serious triviality theorem is Grothendieck's result that $K_0(k[X_1, X_2, \ldots, X_n]) \simeq K_0(k) = \mathbb{Z} \cdot [k]$ for a field k. This has far-reaching generalizations due to Serre and Quillen; we will need the following special case. (An algebraic proof can be found in the notes [**3**] of Cliff and Weiss).

GROTHENDIECK-SERRE-QUILLEN THEOREM ([**10**]): *Let $R = R_0 \oplus R_1 \oplus R_2 \oplus \ldots$ be a left Noetherian graded ring such that R has finite projective dimension over R_0 and R_0 has finite projective dimension over R. Then $G_0(R_0) \simeq G_0(R)$.*

Although it is not obvious why it's an isomorphism, the map is given by $[M] \to [R \otimes_{R_0} M]$ when R is projective over R_0.

We will require some easier properties of G_0 of a Noetherian ring. An important "induction" occurs when $R \subseteq S$ and S is a flat R-module. (In most applications S will be a free R-module). In this case $S \otimes_R$ -preserves exact sequences of R-modules so it yields a map $G_0(R) \to G_0(S)$.

Another pair of rings which comes up consists of R and the ring of matrices $M_n(R)$. Left R-modules are sent to left $M_n(R)$-modules by

$$M \to R^n \otimes_R M = \begin{pmatrix} M \\ M \\ \vdots \\ M \end{pmatrix}$$

and $M_n(R)$-modules are sent to R-modules by

$$N \to R^n \otimes_{M_n(R)} N = e_{11}N.$$

It follows that $G_0(M_n(R)) \simeq G_0(R)$.

Finally, we will need something about localization. If R is a left Noetherian ring and T is a multiplicative semigroup of R at which one can localize then tensoring induces a surjection $G_0(R) \to G_0(T^{-1}R)$.

In order to state Moody's Theorem, we must introduce crossed products. Suppose R is an arbitrary ring and Γ is an arbitrary group. The *group ring* $R[\Gamma]$ is the free R-module with distinguished basis Γ whose multiplication is inherited from the product on Γ via the distributive laws. This definition is too rigid for arguing by mathematical induction on Γ. For example, if $N \lhd \Gamma$ then $R[\Gamma]$ is not generally the group ring of Γ/N over the base ring $R[N]$ even though every element of $R[\Gamma]$ can be written uniquely in the form $\sum s_x x$ where x runs over a collection of coset representatives of Γ/N and s_x is an R-linear combination of elements in N. The difficulty is that the product of coset representatives is not necessarily a coset representative and that members of $R[N]$ don't always commute with coset representatives.

Let R be a ring and let Γ be a group. The *crossed product* $R * \Gamma$ consists of all formal finite sums $\sum a_u \overline{u}$ with $a_u \in R$ and $u \in \Gamma$ together with data consisting of a "twist" $f\colon \Gamma \times \Gamma \to \text{Units}(R)$ and an "action" $^{\overline{u}}(\ ) \in \text{Aut}(R)$. Multiplication in $R*\Gamma$ is given by $(a_u\overline{u})(b_w\overline{w}) = a_u{}^{\overline{u}}(b_w)f(u,w)\overline{uw}$. In order to make this gadget a ring, we require that

$$f(u,v)f(uv,w) = {}^{\overline{u}}f(v,w)f(u,vw)$$

and that

$$\overline{u} \cdot \overline{v} \text{ and } ad(f(u,v))\overline{uv} \text{ have the same action on } R.$$

Notice that if Γ_1 is a subgroup of Γ_2 then $R * \Gamma_1$ inherits the structure of a crossed product from $R * \Gamma_2$ and, in this way, $R * \Gamma_2$ becomes a free left $R * \Gamma_1$-module. Indeed, if T is a right transversal for Γ_1 in Γ_2 then

$$R * \Gamma_2 = \bigoplus \sum_{t \in T} (R * \Gamma_1)t.$$

As observed before, tensoring now gives rise to a map $G_0(R{*}\Gamma_1) \to G_0(R{*}\Gamma_2)$.

Recall that a group is said to be *abelian-by-finite* if it has an abelian normal subgroup whose factor group is finite.

MOODY'S FUNDAMENTAL THEOREM: *Let S be a left Noetherian ring and let Γ be a finitely generated abelian-by-finite group. Then for any crossed product $S * \Gamma$, the induced map*

$$\bigoplus \sum_{H} G_0(S * H) \to G_0(S * \Gamma)$$

is surjective, where H runs over a finite collection of finite subgroups of Γ.

As a quick application, if Γ is torsion free and S is a field k then Γ has no finite subgroups except $\{1\}$ and $G_0(k) = \mathbb{Z} \cdot [k]$. Hence $G_0(k[\Gamma])$ is generated by $k[\Gamma]$. We conclude that $k[\Gamma]$ is a domain.

It is also immediate that Moody's Theorem holds when Γ is a polycyclic-by-finite group. Indeed, we may assume Γ is infinite and contains a normal subgroup N such that Γ/N is abelian-by-finite and the Hirsch length of N is smaller than that of Γ. View $S * \Gamma$ as a crossed product $(S * N) * \Gamma/N$. Since $S * N$ is left Noetherian, the result follows by an induction on Hirsch length.

The Proof

Most of the proof consists of a formal analysis of crossed products. We will get the deep application of G_0 out of the way at once.

GSQ Reduction: *Let $F = A \rtimes G$ where A is a finitely generated torsion free abelian group, G is a finite group, and A is a free $\mathbb{Z}[G]$-module. If R is a left Noetherian ring and $R * F$ is a crossed product then*

$$G_0(R * G) \to G_0(R * F)$$

is surjective.

PROOF: By the free action of G on A, we can pick a basis $\mathbf{X}$ for A which is permuted by G. Let B be the subsemigroup of A generated by $\mathbf{X}$. Accepting the obvious abuse of notation, $R * B$ looks like an iterated skew polynomial ring. Versions of the Hilbert Basis Theorem and the Koszul Complex say that $R * B$ is left Noetherian and R has finite projective dimension as an $R * B$-module.

Since $G\mathbf{X} = \mathbf{X}G$, it makes sense to talk about the semigroup BG. Clearly $R * (BG) = \bigoplus \sum_{g \in G} (R * B)g$, a finitely generated free module over $R * B$, and $R * G \simeq R \otimes_{R*B} R * (BG)$. Hence $R * (BG)$ is Noetherian and $R * G$ has finite projective dimension as an $R * (BG)$-module. Moreover, $R * (BG)$ is a free $R * G$-module with basis $\overline{B}$: it is easily seen to be a ring graded by the total degree on $\mathbf{X} = \{X_1, \dots, X_n\}$, with the degree zero component $R * G$.

It follows from the Grothendieck-Serre-Quillen Theorem that

$$G_0(R * G) \to G_0(R * (BG))$$

is an isomorphism. Now $R * F$ is just a localization of $R * (BG)$.

Most of our work will be devoted to describing crossed-products.

LEMMA [9]: *Let W be a ring that contains a set of elements*

$$\{e_{ij} | i, j = 1, 2, \ldots, n\}$$

satisfying the multiplication rule for matrix units and suppose $1 = e_{11} + \cdots + e_{nn}$. *If C is the centralizer in W of all these elements then* $W \simeq M_n(C)$ *and* $C \simeq e_{11} W e_{11}$.

PROOF: If $w \in W$ set $w_{ij} = \sum_k e_{ki} w e_{jk}$. Then $w_{ij} \in C$ and $\sum_{i,j} w_{ij} e_{ij} = w$. The proof is now straightforward.

THEOREM (Clifford): *Let W be a ring and let $1 = e_1 + e_2 + \cdots + e_n$ be a decomposition of* 1 *into a sum of orthogonal idempotents. Let G be a subgroup of units of W such that G permutes the set $\{e_1, \ldots, e_n\}$ transitively by conjugation. Then $W \simeq M_n(T)$ where T is the ring $e_1 W e_1$.*

PROOF: Choose $g_i \in G$ such that $g_i e_1 g_i^{-1} = e_i$ and then set $e_{ij} = g_i e_1 g_j^{-1}$. Calculation shows that we have produced "matrix units".

CORESTRICTION THEOREM FOR CROSSED PRODUCTS: *Let S be a ring and let $E \subseteq F$ be groups with $|F : E| = n < \infty$. If $S * E$ is a given crossed product then there exists a crossed product $S^n * F$ such that*

(i) $S^n * F \simeq M_n(S * E)$

(ii) *"conjugation" by F transitively permutes the components of S^n and the stabilizer of some component is E.*

PROOF: Consider a coset decomposition $F = Eb_1 \cup Eb_2 \cup \cdots \cup Eb_n$ with $b_1 = 1$. Right multiplication by $y \in F$ permutes these cosets, inducing an element $\pi y \in \mathscr{S}_n$. Notice that $\pi(yz) = \pi(z)\pi(y)$ and $b_i y = c b_{\pi y(i)}$ for some $c \in E$. We will write $\langle i, y \rangle$ for this element c. This gives us the nice and useful formula

$$\langle i, x \rangle \langle \pi x(i), y \rangle = \langle i, xy \rangle.$$

Next define an "action" of F on S^n by

$$[{}^{\overline{x}}\sigma]_i = {}^{\overline{\langle i, x \rangle}}[\sigma]_{\pi x(i)}.$$

(Thus $\overline{x}$ permutes the coordinates of σ and acts on the ith entry by its ith monomial contribution.) If f is the twist for $S * E$ define a new twist $f^+ : F \times F \to \text{Units}(S^n)$ by

$$[f^+(x, y)]_i = f(\langle i, x \rangle, \langle \pi x(i), y \rangle).$$

It is an exercise to verify that this data defines a crossed product $S^n * F$.

By construction, conjugation by an element of $\overline{F}$ permutes the components of S^n. Indeed, if $\varepsilon_1, \ldots, \varepsilon_n$ are the canonical idempotents of S^n then ${}^{\overline{x}^{-1}}(\varepsilon_i) = \varepsilon_{\pi x(i)}$. Since F acts transitively on the cosets of E, $\{\overline{x}^{-1} | \overline{x} \in \overline{F}\}$ acts transitively on these idempotents. Also,

$$\text{Stab}_F(\varepsilon_i) = \text{Stab}_F(Eb_i) = b_i^{-1} E b_i.$$

Therefore (ii) holds.

Notice that

$$\varepsilon_1\overline{x}\varepsilon_1 = \overline{x}({}^{\overline{x}^{-1}}\varepsilon_1)\varepsilon_1 = \begin{cases} \overline{x}\varepsilon_1 & \text{if } \pi x(1) = 1 \\ 0 & \text{otherwise} \end{cases} \quad \text{and}$$

$\pi x(1) = 1$ if and only if $x \in E$. As a consequence $\varepsilon_1(S^n * F)\varepsilon_1 \simeq S * E$. Apply Clifford's Theorem to obtain (i).

STABILIZER THEOREM: *Let S be a ring and let G be a group. Assume that $S^n * G$ is a crossed product and that conjugation by elements of $\overline{G}$ permutes the canonical idempotents of S^n. Then*

$$S^n * G \simeq \prod_{\eta} M_{d(\eta)}(S * \operatorname{Stab}_G(\eta))$$

where one η is taken from each orbit of canonical idempotents under the action induced by G.

PROOF: Partition the set of idempotents into orbits under the action of G and for each orbit Θ let e_Θ denote the sum of idempotents in that orbit. Then e_Θ is a central idempotent of $S^n * G$. Moreover, if $\Theta \neq \Theta'$ then

$$e_\Theta(S^n * G)e_{\Theta'} = e_\Theta e_{\Theta'}(S^n * G) = 0.$$

Hence

$$S^b * G = \prod_{\Theta} e_\Theta(S^n * G)e_\Theta$$

and $e_\Theta(S^n * G)e_\Theta \simeq (e_\Theta S^n e_\Theta) * G$. (If τ is the original twist for $S^n * G$ then $e_\Theta\tau$ is the new twist.)

Notice that $\overline{G}$ acts transitively on the canonical idempotents $\eta_1, \dots, \eta_d$ of $e_\Theta S^n e_\Theta$, so Clifford's Theorem yields

$$e_\Theta(S^n * G)e_\Theta \simeq M_d(\eta_1(S^n * G)\eta_1).$$

Furthermore, if $x \in G$ then $\eta_1\overline{x}\eta_1 = \eta_1({}^{\overline{x}}\eta_1)\overline{x}$, which is zero unless $x \in \operatorname{Stab}_G(\eta_1)$. Hence

$$\eta_1(S^n * G)\eta_1 \simeq S * \operatorname{Stab}_G(\eta_1)$$

where the action on S is the induced action on $\eta_1 S^n$ and where the twist is given by $\eta_1\tau$.

COROLLARY: *Assume the hypothesis and notation of the Corestriction Theorem. If G is a finite subgroup of F then*

$$S^n * G \simeq \prod_{H} M_{\ell(H)}(S * H)$$

where H runs over a finite collection of finite subgroups of E.

PROOF: Recall that, from the Corestriction Theorem, if $F = Eb_1 \cup \dots \cup Eb_n$ then $\operatorname{Stab}_F(\varepsilon_i) = b_i^{-1}Eb_i$ where $\varepsilon_1, \dots, \varepsilon_n$ are the canonical idempotents of S^n. A direct calculation now shows that the crossed product $S * \operatorname{Stab}_G(\varepsilon_i)$ with the data described in the Stabilizer Theorem is isomorphic to $S * (b_iGb_i^{-1} \cap E)$ with the original twist and action inherited from $S * E$.

We must borrow another construction from the cohomology of groups. Let $1 \to B \to E \to \Gamma \to 1$ be a short exact sequence of groups. If $S * \Gamma$ is a crossed product then the action and twist can be pulled back to E. (The new twist is essentially the "inflation" of the old.) Thus we obtain a new crossed product $S * E$. There is an obvious homomorphism $S * E \to S * \Gamma$ and the inherited crossed product $S * B$ is just the ordinary group ring $S[B]$.

LEMMA (Tor): *Let E be a group with a normal subgroup B and suppose that $S * E$ is the inflation of some crossed product $S * (E/B)$. If Y is any subgroup of E such that $Y \cap B = 1$ and M is an $S * Y$-module then*

$$\mathrm{Tor}_1^{S[B]}(S, S * E \otimes_{S*Y} M) = 0.$$

PROOF: Since $Y \cap B = 1$, we can identify $S*Y \subseteq S*E$ with $S*Y \subseteq S*E/B$. Let

$$(1) \cdots \to F_2 \to F_1 \to F_0 \to M \to 0$$

be a free resolution of M by $S * Y$-modules. Then

$$(2) \cdots \to (S * E) \otimes_{S*Y} F_2 \to (S * E) \otimes_{S*Y} F_1 \to (S * E) \otimes_{S*Y} F_0 \to (S * E) \otimes_{S*Y} M \to 0$$

is a free resolution of $(S * E) \otimes M$ as an $S * E$-module because $S * E$ is a free $S * Y$-module. For a similar reason, (2) is also a resolution of $(S * E) \otimes M$ by free $S[B]$-modules. Now if N is any $S * Y$-module,

$$S \otimes_{S[B]} ((S * E) \otimes_{S*Y} N) \simeq (S * E/B) \otimes_{S*Y} N.$$

Thus tensoring the resolution (2) with $S \otimes_{S[B]} -$ can naturally be identified with tensoring the resolution (1) with $(S * E/B) \otimes_{S*Y} -$. But the latter functor is exact.

To complete Moody's Theorem we need two well known observations about lattices (i.e., finitely generated free $\mathbb{Z}$-modules).

LEMMA 1: *Assume G is a finite group. Let M be a $\mathbb{Z}[G]$-module which is a $\mathbb{Z}$-lattice. If $\mathbb{Q} \otimes_{\mathbb{Z}} M$ is a free $\mathbb{Q}[G]$-module then there exists a free $\mathbb{Z}[G]$-module N, containing a copy of M, such that N/M is finite.*

PROOF: Pick a basis $x_1, \ldots, x_n$ for $\mathbb{Q} \otimes_{\mathbb{Z}} M$ as a free $\mathbb{Q}[G]$-module. Let L be the free $\mathbb{Z}$-module with basis $\{gx_j | g \in G,\ 1 \le j \le n\}$. Since this set is also a basis for the rational vector space $\mathbb{Q} \otimes M$, we can find a "common denominator" such that $M \subset \frac{1}{d}L$. Let $N = \frac{1}{d}L$. Since M and N have the same rank as abelian groups, N/M is finite.

LEMMA 2: *Assume G is a finite group. If V is a finite dimensional $\mathbb{Q}[G]$-module then there is a G-invariant lattice $M \subseteq V$ with $\mathbb{Q} \otimes_{\mathbb{Z}} M \simeq V$ as $\mathbb{Q}[G]$-modules.*

PROOF: If we regard V as a rational vector space, we can find a $\mathbb{Z}$-lattice $\Lambda \subseteq V$ such that $\mathbb{Q} \otimes_{\mathbb{Z}} \Lambda = V$. Let $M = \sum_{g \in G} g\Lambda$.

PROOF OF MOODY'S THEOREM: Recall that S is a left Noetherian ring and Γ is a finitely generated abelian-by-finite group. We first do the special case in which Γ is replaced by E where

$$1 \to C \to E \to G \to 1$$

is a short exact sequence of groups with G finite, C a finitely generated free abelian group, and $\mathbb{Q} \otimes_{\mathbb{Z}} C$ is a free $\mathbb{Q}[G]$-module under the induced action.

According to Lemma 1, we can expand C by finite index to produce a free $\mathbb{Z}[G]$-module D. We obtain a new extension

$$1 \to D \to F \to G \to 1$$

where we may assume that $E \subseteq F$ and $|F : E| = n < \infty$. By the Corestriction Theorem we have

$$G_0(S^n * F) \stackrel{\simeq}{\to} G_0(M_n(S * E)) \stackrel{\simeq}{\to} G_0(S * E).$$

Now Shapiro's Lemma implies that the extension describing F splits. Thus the hypotheses of the GSQ Reduction hold, so

$$G_0(S^n * G) \stackrel{\text{onto}}{\to} G_0 n * F).$$

Finally, we apply the corollary to the Stabilizer Theorem:

$$\bigoplus \sum_H G_0(S * H) \stackrel{\simeq}{\to} G_0 \left(\prod_H M_{\ell(H)}(S * H) \right) \stackrel{\simeq}{\to} G_0(S^n * G)$$

where H runs over a finite collection of finite subgroups of E and ℓ is borrowed from the corollary. The theorem follows for E. It is a thankless but routine task to check that the composition of all of these maps is just the expected induced map.

Now consider the general case of an arbitrary finitely generated abelian-by-finite group Γ, say

$$1 \to A \to \Gamma \to G \to 1$$

with A a finitely generated free abelian group and G a finite group. It is easy to find a finite dimensional $\mathbb{Q}[G]$-module V such that $(\mathbb{Q} \otimes_{\mathbb{Z}} A) \oplus V$ is a free $\mathbb{Q}[G]$-module. Use Lemma 2 to find a $\mathbb{Z}[G]$-lattice B inside V such that $\mathbb{Q} \otimes_{\mathbb{Z}} B \simeq V$ as $\mathbb{Q}[G]$-modules. Then Γ acts on B by making A act trivially on B. Let $E = B \rtimes \Gamma$. We have an extension

$$1 \to A \times B \to E \to G \to 1$$

which satisfies the requirements of our special case. Notice that, since B is torsion free, a finite subgroup of H of E is isomorphic to its image $\overline{H}$ in Γ.

$S * \Gamma$ is given as a crossed product; let $S * E$ be its inflation. If H is a finite subgroup of E then $S * H$ (a subring of $S * E$) is isomorphic to $S * \overline{H}$ (a subring of $S * \Gamma$). From now on we will make no distinction between crossed products of finite subgroups of E and Γ.

Let $\mathscr{C}$ be the full subcategory of finitely generated $S * E$-modules whose objects are modules M such that $\operatorname{Tor}_1^{S[B]}(S, M) = 0$. There is an obvious map $G_0(\mathscr{C}) \to G_0(S * E)$. We are going to argue that this is an isomorphism by producing the map in the opposite direction. Since B is a finitely generated free abelian group, the Koszul complex yields a resolution of S by d free $S[B]$-modules. A dimension shifting argument then implies that for every $S[B]$-module N and every resolution by d free $S[B]$-modules,

$$0 \to K \to \text{Free} \to \text{Free} \to \cdots \to \text{Free} \to N \to 0,$$
$$|\longleftarrow d \longrightarrow|$$

the kernel K satisfies $\operatorname{Tor}_1^{S[B]}(S, K) = 0$. Now if N happens to be a finitely generated $S * E$-module, we can choose the free $S[B]$-modules to be finitely generated free $S * E$-modules. In particular, each module in the resolution of N lies in $\mathscr{C}$. Schanuel's Lemma shows that all such "almost free" resolutions of N determine a single element of $G_0(\mathscr{C})$. The Horseshoe Lemma implies that this map from $S*E$-modules to $G_0(\mathscr{C})$ gives rise to the map $G_0(S*E) \to G_0(\mathscr{C})$.

There are several advantages in replacing the category of finitely generated $S * E$-modules with $\mathscr{C}$. One is that $S \bigotimes_{S[B]} -$ is an exact functor from $\mathscr{C}$ to the category of finitely generated $S * \Gamma$-modules. Thus there is a map

$$G_0(\mathscr{C}) \to G_0(S * \Gamma).$$

Secondly, the Tor Lemma for $Y = \Gamma$ says that if N is a finitely generated $S * \Gamma$-module then $(S * E) \otimes_{S * \Gamma} N \in \mathscr{C}$ and

$$S \otimes_{S[B]} ((S * E) \otimes N) \simeq N.$$

Thus the map, above, on G_0 is surjective. We have

$$\bigoplus \sum_H G_0(S * H) \overset{\text{onto}}{\to} G_0(S * E) \overset{\simeq}{\to} G_0(\mathscr{C}) \overset{\text{onto}}{\to} G_0(S * \Gamma).$$

The only remaining question is whether the composition is correct. However, another application of the Tor Lemma (for $Y = H$) shows that a finitely generated $S * H$-module is induced into $\mathscr{C}$, so the composition "avoids resolutions" in the middle isomorphism.

2. Lazard's Theorem. Moody's remarkable results complete the chapter on zero divisors for the group algebras of solvable groups. Which class of torsion free groups should be analyzed next? Some progress has already been made for groups which are almost free ([7]); perhaps small cancellation groups are a suitable class. However, if one prefers to work with strong finiteness restrictions, linear groups are the candidates of choice. In particular, the principal congruence subgroups for odd primes are well known to be torsion free.

The conventional wisdom has been that the zero divisor problem is intractable for congruence subgroups. The discovery we will discuss in this section is a bit of "library research": the problem was solved in 1965! The lost technique is the p-analogue to a celebrated observation of Lazard ([**6**]), Quillen ([**11**]), and others. The familiar result is that if G is a finitely generated torsion free nilpotent group and Γ is its associated Lie ring then the completion of the group algebra $\mathbb{C}[G]$ at its augmentation ideal is isomorphic to the completion of the universal enveloping algebra $U(\mathbb{C}\otimes\Gamma)$ at its augmentation ideal. This yields a convoluted proof of the zero divisor problem for $\mathbb{C}[G]$ based on the consequence of the Poincaré-Birkhoff-Witt Theorem that the universal enveloping algebra is a domain.

Lazard, in the monograph [**5**], observes that the filtration given by the descending central series for a torsion free nilpotent group has a counterpart for congruence subgroups. We begin with the observation that the filtration can be described by its "degree function" or "valuation".

Let p be a prime. A *p-valuation* on a group G is a function $w\colon G\to\mathbb{Z}^+\cup\{\infty\}$ such that for all $x, y, \varepsilon G$,

(i) $w(xy^{-1})\geq\min\{w(x),w(y)\}$
(ii) $w(x^{-1}y^{-1}xy)\geq w(x)+w(y)$
(iii) $(p-1)^{-1}<w(x)<\infty$ for all $x\neq 1$
(iv) $w(1)=\infty$
(v) $w(x^p)=w(x)+1$.

To make these axioms appear more natural, set $G_a=\{x\in G|w(x)\geq a\}$. Then (i) says G_a is a subgroup of G, giving rise to a descending filtration. Condition (ii) means $[G_a,G_b]\subseteq G_{a+b}$; in particular, $G_a\triangleleft G$. It follows from (iv) that $\bigcap_{a>0}G_a=\{1\}$. At first glance, (v) just seems to imply that G_a/G_{a+1} is an elementary abelian p-group. However, the strict equality in (v) requires G to be torsion free: since $\bigcap G_a=\{1\}$, any element of finite order has p-power order. But if $x^{p^n}=1$ for $x\neq 1$ then

$$\infty=w(1)=w(x^{p^n})=w(x)+n<\infty.$$

The examples which interest us are the principal congruence subgroups. Recall that the p-adic valuation v on $\mathbb{Z}$ (or, for that matter, on the p-adic integers $\mathbb{Z}_p$) is defined so that $v(p^m b)=m$ when $(p,b)=1$ and $v(0)=\infty$. This can be extended to $\overline{v}$ on matrices $M_n(\mathbb{Z})$ so that if $A=(a_{ij})$ then $\overline{v}(A)=\min_{i,j}v(a_{ij})$. It's easy to check that

$$\overline{v}(A-B)\geq\min\{\overline{v}(A),\overline{v}(B)\}$$
$$\overline{v}(AB)\geq\overline{v}(A)+\overline{v}(B)$$
$$\text{and }\overline{v}(0)=\infty.$$

Define $C_n(p)=\{A\in GL(n,\mathbb{Z})|\overline{v}(A-I)>(p-1)^{-1}\}$. (Here the $(p-1)^{-1}$ is a red herring. For every prime except 2 it can obviously be replaced by 0).

First notice that $AB^{-1}-I=(A-I)-(B-I)-(AB^{-1}-I)(B-I)$. Hence not only is $C_n(p)$ a group, but if we define $w\colon C_n(p)\to\mathbb{Z}^+\cup\{\infty\}$ by $w(A)=\overline{v}(A-I)$, then w satisfies (i). Inequality (ii) is a consequence of

$$A^{-1}B^{-1}AB-I=A^{-1}B^{-1}[(A-I)(B-I)-(B-I)(A-I)].$$

We will have shown that w is a p-valuation on $C_n(p)$ if we can prove that $w(A^p) = w(A) + 1$. Now

$$A^p = ((A-I)+I)^p = I + p(A-I) + \binom{p}{2}(A-I)^2 + \cdots + (A-I)^p.$$

Since $\overline{v}(p \cdot (A-I)) = 1 + w(A)$ we will be done provided that $\overline{v}$ on each remaining term $\binom{p}{2}(A-I)^2, \ldots, (A-I)^p$ is at least $2 + w(A)$. This amounts to checking that $\overline{v}(p \cdot (A-I)^2) > 1 + w(A)$ and that $\overline{v}((A-I)^p) > 1 + w(A)$. Indeed,

$$\overline{v}(p \cdot (A-I)^2) \geq 1 + 2w(A) > 1 + w(A)$$

since $w(A) > (p-1)^{-1} > 0$ and

$$\overline{v}((A-I)^p) \geq pw(A) > 1 + w(A)$$

again because $w(A) > (p-1)^{-1}$.

We turn to the Lie theory next. Assume that G is a group with p-valuation w. The corresponding filtration by normal subgroups induces a graded Lie ring $\mathscr{G}$ over $\mathbb{F}_p$, the field with p elements. The function $g \to g^p$ on the group G induces a graded linear transformation $\pi\colon \mathscr{G} \to \mathscr{G}$ with the properties that $[\pi(x), y] = \pi[x, y]$ and that π raises the degree by 1. (This is the consequence of an intricate commutator calculation which we shall omit.) At this point, axiom (v) plays a crucial role: it says that if x is homogeneous then $x \neq 0$ implies $\pi(x) \neq 0$. In other words, if we designate an indeterminate T then $\mathscr{G}$ is a *torsion free* graded $\mathbb{F}_p[T]$-module when T acts like π. A well known proposition about graded modules says that such modules are actually free.

Knowing that $\mathscr{G}$ has a basis as an $\mathbb{F}_p[T]$-module allows one to follow the constructions of the Poincaré-Birkhoff-Witt Theorem and conclude that $U(\mathscr{G})$, the universal enveloping algebra over $\mathbb{F}_p[T]$, is a domain. Moreover, if $\mathscr{G}$ is finitely generated as a module over $\mathbb{F}_p[T]$, then $U(\mathscr{G})$ is a Noetherian ring. For example, the associated graded Lie ring for $C_n(p)$ is finitely generated. Each of its homogeneous components is isomorphic to the same $(\mathbb{F}_p)^n$ and so powers of T shift the first component to all others.

The second side of Lazard's isomorphism involves the group ring with coefficients in the p-adic integers. The required filtration of ideals is concocted from the prime p and the augmentation ideal. In $\mathbb{Z}_p[G]$ let I_m be the $\mathbb{Z}_p$-module generated by

$$p^h(g_1 - 1)(g_2 - 1)\ldots(g_d - 1)$$

where $h + w(g_1) + \cdots + w(g_d) \geq m$. This filtration on $\mathbb{Z}_p[G]$ yields an associated graded ring. We can now state the expected isomorphism.

THEOREM: ([**5**], 2.3.3) $U(\mathscr{G}) \simeq \operatorname{gr} \mathbb{Z}_p[G]$.

COROLLARY: ([**5**], 2.3.4) *If G has a p-valuation then $\mathbb{Z}[G]$ is a domain. In particular, $\mathbb{Z}[C_n(p)]$ is a domain.*

PROOF: We have already observed that $U(\mathscr{G})$ is a domain. Hence $\operatorname{gr} \mathbb{Z}_p[G]$ is a domain. A "leading term" argument implies that $\mathbb{Z}_p[G]$ is a domain.

COROLLARY: *If $\mathcal{G}$ is finitely generated as an $\mathbb{F}_p[T]$-module then the completion of $\mathbb{Z}_p[G]$ with respect to the filtration is a Noetherian local ring.*

PROOF: Standard completion nonsense.

Lazard also shows that, in the case of the second corollary, the completion of $\mathbb{Z}_p[G]$ is isomorphic to $\varprojlim \mathbb{Z}_p[G/G_a]$.

We pause to outline a proof of the theorem. First use the filtration $\{I_n\}$ of $\mathbb{Z}_p[G]$ to define a new valuation on G by thinking of it as a "congruence subgroup", i.e., $w^*(g)$ is the largest index m such that $g-1 \in I_m$. It's not difficult to see that this induced valuation coincides with the original one w. Viewing w in this way, a general argument produces a graded Lie monomorphism $\mathcal{G} \to \operatorname{gr} \mathbb{Z}_p[G]$: the image of x in $\mathcal{G}$ is sent to the image of $x-1$ in the graded ring. In addition, $\operatorname{gr} \mathbb{Z}_p[G]$ is an $\mathbb{F}_p[T]$-algebra. (Roughly speaking, T behaves on the $\mathbb{F}_p$-part like multiplying by p and acts on the augmentation part by raising to the pth power.) Thus we obtain a graded $\mathbb{F}_p[T]$-algebra map $U(\mathcal{G}) \to \operatorname{gr} \mathbb{Z}_p[G]$ which is surjective.

Choose $\{x_j | j \in J\} \subseteq G$ so that its image in $\mathcal{G}$ is a basis for $\mathcal{G}$ as a free $\mathbb{F}_p[T]$-module. Order the set J and let $z_j = x_j - 1$. If $\alpha_{j(1)}, \alpha_{j(2)}, \ldots, \alpha_{j(d)}$ are natural numbers with $j(1) < j(2) < \cdots < j(d)$ in J, define

$$z^\alpha = z_{j(1)}^{\alpha_{j(1)}} z_{j(2)}^{\alpha_{j(2)}} \cdots z_{j(d)}^{\alpha_{j(d)}}.$$

Surjectivity of the map above means that for each $n > 0$, the $\mathbb{Z}_p$-module I_n is generated modulo I_{n+1} by $\{p^h z^\alpha | h + \sum_t \alpha_{j(t)} = n\}$. These representatives can be used to successively approximate an element of $\mathbb{Z}_p[G]$ going down one layer of the filtration at a time. The result is that every element of $\mathbb{Z}_p[G]$ can be interpreted to have the form $\sum_\alpha \lambda_\alpha z^\alpha$ where λ_α is a p-adic integer and $v(\lambda_\alpha) + \sum \alpha_{j(t)} \to \infty$. The form is seen to be unique by studying the images of a fixed element of $\mathbb{Z}[G]$ in each $\mathbb{Z}_p[G/G_a]$. At this point the desired isomorphism can be exhibited rather explicitly.

REFERENCES

1. H. Bass, *Algebraic K-Theory*, W. A. Benjamin, New York, 1968.
2. A. W. Chatters and C. R. Hajarnavis, *Rings with chain conditions*, Res. Notes in Math. **44**, Pitman, Boston, 1980.
3. G. Cliff and A. Weiss, *Moody's induction theorem*, Illinois J. Math. **32** (1988), 489–500.
4. D. R. Farkas and R. L. Snider, *K_0 and Noetherian group rings*, J. Alg. **42** (1976), 192–198.
5. M. Lazard, *Groupes analytiques p-adiques*, I.H.E.S. Pub. Math. **26**, 1965.
6. M. Lazard, *Sur les groupes nilpotents et les anneaux de Lie*, Annales E.N.S., **71** (1954), 101–190.
7. J. Lewin and T. Lewin, *An embedding of the group algebra of a torsion-free one-relator group in a field*, J. Alg. **52** (1978), 39–74.
8. J. A. Moody, *Induction theorems for infinite groups*, Bull. Amer. Math. Soc. **17** (1987), 113–116.
9. D. S. Passman, *The algebraic structure of group rings*, Wiley-Interscience, New York, 1977.
10. D. G. Quillen, *Higher algebraic K-theory I, Algebraic K-theory, I*, Springer Lect. Notes in Math. **341**, Springer, Berlin, 1973.
11. ——, *On the associated graded ring of a group ring*, J. Alg. **10** (1968), 411–418.

Contemporary Mathematics
Volume **93** (1989)

The Augmentation Quotients of Finite Abelian p-Groups

A. W. HALES AND I. B. S. PASSI

1. Introduction. Let G be a finite Abelian p-group, $\mathbf{Z}G$ its integral group ring and Δ the augmentation ideal of $\mathbf{Z}G$. The purpose of this paper is to present a proof of the following known result.

THEOREM 1.1. *For sufficiently large n, the structure of the Abelian group Δ^n/Δ^{n+1} is independent of n and the order of the terminal quotient is p^{d-1}, where d is the number of cyclic subgroups of G.*

In [**1**] Bachmann and Grünenfelder showed that, for large n, the structure of Δ^n/Δ^{n+1} as an Abelian group does not depend on n. Singer ([**10**], [**11**]) determined the order of the terminal quotient as an application of his earlier results ([**7**], [**8**], [**9**]) on invertible powers of fractional ideals in algebras. The importance of these results in the study of integral group rings, especially the applications in [**3**], has motivated us to give a relatively straightforward and self-contained proof (modeled after Singer's approach) of these results. While an exposition of the result of Bachmann-Grünenfelder is available in [**5**], we hope that the present exposition will, in particular, make Singer's important result on the order of the terminal quotient more accessible.

2. Extensions and localizations. Let $G = \prod_{i=1}^{r}\langle g_i\rangle$, where $\langle g_i\rangle$ is a cyclic group of order p^{α_i} generated by g_i, with p prime and $\alpha_i \geq 1$. The rational group algebra $\mathbf{Q}G$ splits as a direct sum

$$\mathbf{Q}G \cong \sum_{j=1}^{d} k_j$$

where d is the number of cyclic subgroups of G (see [**4**], Theorem 2.9, page 56). Here each k_j is a field extension $k_j = \mathbf{Q}(\zeta_j)$, where ζ_j is a primitive p^{β_j}th root of unity, and the projection $\pi_j\colon \mathbf{Q}G \to k_j$ is induced by a character $\chi_j\colon G \twoheadrightarrow \langle\zeta_j\rangle$. We assume that $k_1 = \mathbf{Q}$ with χ_1 the trivial character. Recall that if $\mathscr{O}_j$ denotes the ring of integers in k_j, then $\mathscr{O}_j = \mathbf{Q}[\zeta_j]$ and the prime p ramifies totally in $\mathscr{O}_j$. More precisely, $p\mathscr{O}_j = \mathscr{P}_j^{\varphi(p^{\beta_j})}$ where φ is the Euler totient and $\mathscr{P}_j = (\zeta_j - 1)\mathscr{O}_j$ is the unique prime ideal of $\mathscr{O}_j$ over p. The maximal order $\mathscr{B}$ of $\mathbf{Q}G$ is clearly just

$$\mathscr{B} \cong \sum_{j=1}^{d} \mathscr{O}_j.$$

As general references on algebraic number theory and orders in algebras the reader may consult **[2]** and **[6]** respectively.

If $G = \langle g \rangle$ were cyclic, the element $u = g - 1$ would satisfy $\Delta^2 = u\Delta$. In the general case we aim at finding such an element u; however, we need to suitably enlarge our base ring.

Choose any monic irreducible polynomial $f(x) = x^r + a_{r-1}x^{r-1} + \cdots + a_0$ of degree r, the rank of G, over the p-element field $\mathbf{F}_p$. Let $F(x) = x^r + A_{r-1}x^{r-1} + \cdots + A_0 \in \mathbf{Z}[x]$ lie over $f(x)$, i.e. $\overline{A}_i = a_i$ in $\mathbf{F}_p$ for all i. Then $F(x)$ is irreducible over $\mathbf{Z}$ and hence over $\mathbf{Q}$. Let θ be a root of $F(x)$ and let R be the ring of integers in $\mathbf{Q}(\theta)$. The irreducibility of $f(x)$ over $\mathbf{F}_p$ guarantees that $R/pR \cong \mathbf{F}_p(\bar{\theta})$, where $\bar{\theta}$ (the image of θ mod pR) is a root of $f(x)$ over $\mathbf{F}_p$, and hence that the prime ideal (p) of $\mathbf{Z}$ stays prime in R. Then $R/pR \cong \mathbf{F}_{p^r}$. Note that pR/p^2R also has order p^r with the images of $p, p\theta, \ldots, p\theta^{r-1}$ as a basis over $\mathbf{F}_p$. Since p ramifies totally in $\mathscr{O}_j$ for each j but stays prime in R, the minimal polynomial of ζ_j over $\mathbf{Q}$ stays irreducible over $\mathbf{Q}(\theta)$, as may be seen by examining the splitting of (p) in $\mathbf{Q}(\theta, \zeta_j)$. Hence $\mathbf{Q}(\theta) \otimes_{\mathbf{Q}} k_j$ is a field of degree $\varphi(p^{\beta_j})$ over $\mathbf{Q}(\theta)$, isomorphic to $\mathbf{Q}(\theta, \zeta_j)$. Consequently

$$\mathbf{Q}(\theta)(G) \cong \mathbf{Q}(\theta) \otimes_{\mathbf{Q}} \mathbf{Q}G \cong \sum_{i=1}^{d} k_j'$$

with $k_j' \cong \mathbf{Q}(\theta, \zeta_j)$ a field of degree $\varphi(p^{\beta_j})$ over $\mathbf{Q}(\theta)$, $1 \le j \le d$. Furthermore the maximal order of $\mathbf{Q}(\theta)(G)$ over R is just

$$\mathscr{B}' \cong \sum_{j=1}^{d} \mathscr{O}_j'$$

where $\mathscr{O}_j'$ is the ring of integers of k_j' and (though we shall not use this) $\mathscr{O}_j' = R[\zeta_j]$. Note that the prime p factors in $\mathscr{O}_j'$ as follows:

$$p\mathscr{O}_j' = (\mathscr{P}_j')^{\varphi(p^{\beta_j})}$$

where $\mathscr{P}_j' = (\zeta_j - 1)\mathscr{O}_j'$ and $\mathscr{O}_j'/\mathscr{P}_j' \cong R/pR \cong \mathbf{F}_{p^r}$. The quotient $\mathscr{P}_j'/\mathscr{P}_j'^2$ also has order p^r with the images of $(\zeta_j - 1), (\zeta_j - 1)\theta, \ldots, (\zeta_j - 1)\theta^{r-1}$ as a basis over $\mathbf{F}_p$.

We next localize at the prime ideal $(p) = pR$ of R. Tensoring (over R) each component of the decomposition

$$\mathscr{B}' \cong \sum_{j=1}^{d} \mathscr{O}_j'$$

with $R_{(p)}$ gives a decomposition for the maximal order $\mathscr{B}'_{(p)}$ of $\mathbf{Q}(\theta)(G)$ over $R_{(p)}$

$$\mathscr{B}'_{(p)} \cong \sum_{j=1}^{d} \mathscr{O}'_{j(p)}$$

where $\mathscr{O}'_{j(p)}$ denotes the localization of $\mathscr{O}'_j$ at $\mathscr{P}'_j$. Then we still have $\mathscr{O}'_{j(p)}/\mathscr{P}'_{j(p)} \cong R/pR \cong \mathbf{F}_{p^r}$ and $\mathscr{P}'_{j(p)}/\mathscr{P}'^2_{j(p)}$ still has order p^r with the images of $(\zeta_j - 1), (\zeta_j - 1)\theta, \dots, (\zeta_j - 1)\theta^{r-1}$ as a basis over $\mathbf{F}_p$. It may be noted that we could equally well have first localized at the prime ideal (p) of $\mathbf{Z}$ and then "extended by θ".

Let $e = |G|^{-1} \sum_{g \in G} g$ be the idempotent which is the "identity" of k_1. Let $\Delta'_{(p)} = \Delta \otimes_{\mathbf{Z}} R_{(p)}$ denote the augmentation ideal of $R_{(p)}(G)$. Define

$$u = e + 1(g_1 - 1) + \theta(g_2 - 1) + \cdots + \theta^{r-1}(g_r - 1).$$

Then $u \in R_{(p)}e + \Delta'_{(p)}$ and we have

LEMMA. *The element u is invertible in $\mathbf{Q}(\theta)G$ and $R_{(p)}e + \Delta'_{(p)} \subseteq u\mathscr{B}'_{(p)}$.*

PROOF. For $j \geq 2$ we know that the jth component $\pi_j(x)$ of an element x in $R_{(p)}e + \Delta'_{(p)}$ lies in $\mathscr{P}'_{j(p)}$. Since $u\mathscr{B}'_{(p)} = \sum_{j=1}^{d} \pi_j(u)\mathscr{O}'_{j(p)}$ it will suffice (for both parts of the lemma) to show that, for $j \geq 2$, $\pi_j(u)$ lies in $\mathscr{P}'_{j(p)}$ but not $\mathscr{P}'^2_{j(p)}$.

Fix $j \geq 2$ and let $\chi_j(g_i) = \zeta_j^{m_i}$. Since χ_j is surjective, for some i_0 we have $p \nmid m_{i_0}$. Now

$$\pi_j(u) = (\zeta_j^{m_1} - 1) + \theta(\zeta_j^{m_2} - 1) + \cdots + \theta^{r-1}(\zeta_j^{m_r} - 1)$$

clearly lies in $\mathscr{P}'_{j(p)}$. However

$$\zeta_j^m - 1 \equiv m(\zeta_j - 1) \bmod \mathscr{P}'^2_{j(p)}.$$

Therefore we have

$$\pi_j(u) \equiv (m_1 + m_2\theta + \cdots + m_r\theta^{r-1})(\zeta_j - 1) \bmod \mathscr{P}'^2_{j(p)}.$$

Since the images of the elements $(\zeta_j - 1), \theta(\zeta_j - 1), \dots, \theta^{r-1}(\zeta_j - 1)$ form a basis for $\mathscr{P}'_{j(p)}/\mathscr{P}'^2_{j(p)}$ over $\mathbf{F}_p$ and $p \nmid m_{i_0}$, it follows that $\pi_j(u) \notin \mathscr{P}'^2_{j(p)}$.

We are now ready to prove the main result.

3. Augmentation Quotients.

3.1 PROOF OF THEOREM 1.1. Let $M = \mathbf{Z}e + \Delta$ and $M'_{(p)} = M \otimes_{\mathbf{Z}} R_{(p)} = R_{(p)}e + \Delta'_{(p)}$. Then clearly

$$\Delta^n/\Delta^{n+1} \cong M^n/M^{n+1}.$$

Write $N = M'_{(p)}$ and observe that N^n/N^{n+1} is isomorphic to the direct sum of r copies of M^n/M^{n+1}. (This uses the fact that $R_{(p)}$ is the direct sum of r copies of $\mathbf{Z}_{(p)}$ and that Δ^n/Δ^{n+1} is a p-group.) By Lemma 2.1 there is an invertible element u with $u \in N$, $N \subseteq u\mathscr{B}'_{(p)}$. Hence $1 \in u^{-1}N \subseteq \mathscr{B}'_{(p)}$. Therefore the sequence

$$u^{-1}N, (u^{-1}N)^2, \dots, (u^{-1}N)^n, \dots$$

is ascending and must eventually stabilize with $(u^{-1}N)^{n_0} = (u^{-1}N)^{n_0+1} = \dots$ for some n_0. This implies that $uN^{n_0} = N^{n_0+1}$ and so $N^{n_0} \cong N^{n_0+1}$ as $R_{(p)}G$ modules. Hence $N^n/N^{n+1} \cong N^{n_0}/N^{n_0+1}$ for all $n \geq n_0$. But N^n/N^{n+1} is isomorphic to the direct sum of r copies of M^n/M^{n+1}, so we conclude that the structure of the finite Abelian p-group Δ^n/Δ^{n+1} is independent of n for $n \geq n_0$. Further $|N^n/N^{n+1}| = |N^n/uN^n| = |\mathscr{B}'_{(p)}/u\mathscr{B}'_{(p)}|$, since $\mathscr{B}'_{(p)}/N^n \cong u\mathscr{B}'_{(p)}/uN^n$. However,

$$|\mathscr{B}'_{(p)}/u\mathscr{B}'_{(p)}| = \prod_{j=2}^{d} |\mathscr{O}'_{j(p)}/\mathscr{P}'_{j(p)}| = p^{r(d-1)}.$$

It follows that $|\Delta^n/\Delta^{n+1}| = p^{d-1}$ for all $n \geq n_0$ and the proof is complete.

3.2 *Remark.* For the precise stage from which the augmentation quotients stabilize and for the structure of the augmentation terminal, the reader may refer to [**3**].

This work was completed during the visit of the second author to the University of California, Los Angeles, under the Indo-American Fellowship Program. He would like to thank the University for its warm hospitality.

The first author would like to acknowledge the support of NSA Grant No. MDA904-85-H-0014.

REFERENCES

1. F. Bachmann and L. Grünenfelder, *The periodicity in the graded ring associated with an integral group ring*, J. Pure Appl. Algebra **5** (1974), 253–264.

2. J. W. S. Cassels and A. Frohlich, *Algebraic number theory*, Academic Press, London, 1967.

3. A. W. Hales, *Stable augmentation quotients of abelian groups*, Pacific J. Math. **118** (1985), 405–410.

4. G. Karpilovsky, *Commutative group algebras*, Marcel Dekker, New York, 1983.

5. I. B. S. Passi, *Group rings and their augmentation ideals*, LNM vol. 715, Springer Verlag, 1979.

6. K. W. Roggenkamp and V. Huber-Dyson, *Lattices over orders I*, LNM 115, Springer-Verlag, 1970.

7. M. Singer, *Invertible powers of ideals over orders in commutative separable algebras*, Proc. Camb. Phil. Soc. **67** (1970), 237–242.

8. ——, *A product theorem for lattices over orders*, Math. Scand. **29** (1971), 50–54.

9. ——, *An elementary proof of the invertible powers theorem*, Proc. Camb. Phil. Soc. **73** (1973), 289–291.

10. ——, *On the augmentation terminal of a finite abelian group*, J. Algebra **41** (1976), 196–201.

11. ——, *Determination of the augmentation terminal for finite Abelian groups*, Bull. Amer. Math. Soc. **83** (1977), 1321–1322.

DEPARTMENT OF MATHEMATICS
UNIVERSITY OF CALIFORNIA
LOS ANGELES, CA 90024
U.S.A.

CENTRE FOR ADVANCED STUDY IN MATHEMATICS
PANJAB UNIVERSITY
CHANDIGARH 160014
INDIA

Contemporary Mathematics
Volume 93 (1989)

Free Groups in Normal Subgroups of Unit Groups and Arithmetic Groups

B. HARTLEY

1. Introduction. Let G be a finite group, and $\mathbf{Z}G$ its integral group ring. A special case of the main result of **[4]** states that the unit group $U(\mathbf{Z}G)$ of $\mathbf{Z}G$ contains a non-abelian free group unless G is abelian or a Hamiltonian 2-group; that is, the direct product of an elementary abelian 2-group and a quaternion group of order 8. The original purpose of the present work was to consider free subgroups of normal subgroups of $U(\mathbf{Z}G) = U$. Roughly speaking, one might hope that a normal subgroup of U should either contain a non-abelian free subgroup, or be central in U. The well known fact that $U = \pm Q_8$ when $G = Q_8$, the quaternion group of order 8, shows that this is not quite true. We do, however, have the following, in which, following Gonçalves, Ritter and Sehgal **[3]**, we consider more general subgroups than normal ones.

DEFINITION. A subgroup X of a group Y is called *almost subnormal*, if there is a finite chain of subgroups $X = X_0 \leq X_1 \leq \cdots \leq X_n = Y$, such that X_i is either normal or of finite index in X_{i+1} $(0 \leq i \leq n-1)$.

THEOREM 1. *Let G and U be as above. Then U contains a normal subgroup U_0 of finite index such that if N is any almost subnormal subgroup of U, then either N contains a non-abelian free group or $N \cap U_0$ lies in the centre $Z(U)$ of U. Further, $U_0 = U$ unless some non-abelian quotient of G is embeddable in the group of real unit quaternions.*

The theorem is a quite straightforward consequence of the well known fact that U can be identified with an arithmetic group, together with a density theorem of Borel for arithmetic groups. In fact we prove the following result on arithmetic groups, from which we deduce a result like Theorem 1 for orders in more general semisimple $\mathbf{Q}$-algebras.

THEOREM 2. *Let H be a connected reductive algebraic $\mathbf{Q}$-group and Γ an arithmetic subgroup of H. Then Γ contains a normal subgroup Γ_0 of finite*

0271-4132/89 $1.00 + $.25 per page

index such that if Δ is almost subnormal in Γ, then either Δ contains a non-abelian free group or $\Delta \cap \Gamma_0 \leq Z(\Gamma)$.

2. Arithmetic groups. We recall that an algebraic **Q**-group H is a subgroup of some $\mathrm{GL}(n, \mathbf{C})$ that is the set of zeros of some set of polynomials over **Q** in n^2 indeterminates X_{ij} together with $1/\det(X_{ij})$. We write $H_{\mathbf{R}} = H \cap \mathrm{GL}(n, R)$ and interpret $H_{\mathbf{Q}}$ and $H_{\mathbf{Z}}$ similarly. An arithmetic subgroup of H is a subgroup Γ of $H_{\mathbf{Q}}$ that is commensurable with $H_{\mathbf{Z}}$, in the sense that the indices $|\Gamma : \Gamma \cap H_{\mathbf{Z}}|$ and $|H_{\mathbf{Z}} : \Gamma \cap H_{\mathbf{Z}}|$ are both finite. For completeness, we indicate the proof of

LEMMA 2.1. *Let A be any finite dimensional **Q**-algebra and Λ any **Z**-order in A. Then the group of units $U = U(A)$ can be identified with the group $H_{\mathbf{Q}}$ of rational points in an algebraic **Q**-group H, and $U(\Lambda)$ with an arithmetic subgroup of H. If A is semisimple, then H is connected and reductive.*

PROOF. Let $a_1, \ldots, a_n$ be any **Z**-basis of Λ, and hence also a **Q**-basis of A. For $a \in A$, let $\rho(a)$ be the matrix with respect to this basis of the operation of right multiplication by a. Since the endomorphisms of the left A-module A are precisely the right multiplications by elements of A, $\rho(A)$ consists precisely of all $n \times n$ matrices over **Q** that commute with all matrices of left multiplications of A. Thus, an $n \times n$ matrix over **Q** belongs to $\rho(A)$ if and only if its entries satisfy certain linear conditions over **Q**. The elements of $\mathrm{GL}(n, \mathbf{C})$ satisfying these conditions constitute H. Clearly $H \cap \mathrm{GL}(n, \mathbf{Q})$ corresponds to $U(A)$ and $H \cap \mathrm{GL}(n, \mathbf{Z})$ to $U(\Lambda)$. Note that we can use the same basis to identify H with $U(A \otimes_{\mathbf{Q}} \mathbf{C})$. If A is semisimple, then $A \otimes_{\mathbf{Q}} \mathbf{C}$ is a direct sum of full matrix rings over **C**, and its unit group is a direct product of various groups $\mathrm{GL}(r, \mathbf{C})$.

PROOF OF THEOREM 2. Let $S = H' = [H, H]$ be the semisimple part of H, and let $S_1, \ldots, S_n$ be the minimal non-trivial closed connected normal subgroups of S. Then $S = S_1 \ldots S_n$, the S_i commute elementwise in pairs, and each pair has finite intersection [**5**, p. 167]. The Galois group $\mathcal{G}$ of **C** over **Q** clearly leaves S invariant, so S is an algebraic **Q**-group. Then $\mathcal{G}$ permutes the S_i, and the product of those in an orbit is a minimal non-trivial connected normal **Q**-subgroup of S. Every non-trivial connected normal **Q**-subgroup of S is a product of such subgroups formed from various $\mathcal{G}$-orbits. Collecting those whose real points form a compact group (in the usual metric topology of complex matrices) we can write $S = MT$, where $[M, T] = 1$, $M \cap T$ is finite, M_R is compact, and no non-trivial connected normal **Q**-subgroup of T has the property that its real points form a compact group.

Let Z be the connected component of the identity of the centre of H. Clearly Z is an algebraic **Q**-group. Also $H = SZ = TMZ$ and $|S \cap Z| < \infty$ [**5**, p. 168]. Let φ be a **Q**-epimorphism of H onto a **Q**-group K, with kernel TZ. Then φ maps M surjectively, and it follows from [**1**, 7.4] that $K_{\mathbf{R}}$ is compact.

Now let Γ be any arithmetic subgroup of H. Then $\varphi(\Gamma)$ is an arithmetic subgroup of K [**1**, 8.11]. As $K_{\mathbf{Z}}$ is a discrete subgroup of the compact group $K_{\mathbf{R}}$, it is finite, and as $\varphi(\Gamma)$ is commensurable with $K_{\mathbf{Z}}$, it is also finite. Therefore, putting $\Gamma_0 = \Gamma \cap TZ$, we have $|\Gamma : \Gamma_0| < \infty$.

Next let ψ be a $\mathbf{Q}$-epimorphism of TZ onto a $\mathbf{Q}$-group L, with kernel Z. Then ψ maps T onto L, and it is easy to deduce that if L^* is any normal $\mathbf{Q}$-subgroup of L such that $L^*_{\mathbf{R}}$ is compact, then L^* is finite. As above, $\Gamma_1 = \varphi(\Gamma_0)$ is arithmetic in L, and so Borel's density theorem [**1**, 15.12] tells us that Γ_1 is Zariski-dense in L.

Let Δ be almost subnormal in Γ, and suppose Δ contains no non-abelian free group. Let bars denote closures in the Zariski topology. By Tits' Theorem [**7**, **8**], Δ is soluble-by-finite, and so is $\Delta_1 = \psi(\Delta \cap TZ)$. Also Δ_1 is almost subnormal in Γ_1. By well known properties of the Zariski topology [**8**, Chapter 5], $\overline{\Delta}_1$ is soluble-by-finite and almost subnormal in $\overline{\Gamma}_1 = L$. Now $\mathbf{Z}(L)$ is finite and $L/\mathbf{Z}(L)$ is a direct product of infinite simple groups, and so every almost subnormal subgroup of $L/\mathbf{Z}(L)$ has the same structure. It follows that $\overline{\Delta}_1 \leq \mathbf{Z}(L)$, and so, if $\Delta_0 = \Delta \cap TZ = \Delta \cap \Gamma_0$, then $[T, \Delta_0] \leq T \cap Z$. Hence $[\Delta_0, T, T] = 1$ and since $T = T'$, we have $[\Delta_0, T] = 1$. Therefore Δ_0 is central in TZ and hence, since $K = TZM$ and $[M, TZ] = 1$, we have that Δ_0 is central in H, and *a fortiori* in Γ. This establishes Theorem 2.

3. Unit groups. By Theorem 2 and Lemma 1, we now have the subgroup U_0 of finite index of $U = U(\mathbf{Z}G)$, required in Theorem 1. To obtain conditions under which $U = U_0$, we require now to identify the subgroup M entering into the proof of Theorem 2. We do this in a more general context. Let A be any finite-dimensional semisimple $\mathbf{Q}$-algebra and Λ be any $\mathbf{Z}$-order in A. Write

$$A = A_1 \oplus \cdots \oplus A_n \tag{3.1}$$

where each A_i is a simple $\mathbf{Q}$-algebra and so is isomorphic to a full matrix ring over some division ring. We may identify $U = U(\Lambda)$ with an arithmetic group as in the proof of Lemma 2.1, but since the property of being arithmetic is invariant under rational changes of basis, it is permissible to do so with respect to a basis adapted to the decomposition (3.1). This is more convenient. Then U is identified with an arithmetic subgroup of

$$H = H_1 \times \cdots \times H_n, \tag{3.2}$$

where $H_i \cong U(A_i \otimes_{\mathbf{Q}} \mathbf{C})$, and the decomposition (3.2) is reflected by a block diagonal form for the matrices. We have $U(A_i) \cong (H_i)_{\mathbf{Q}}$.

Let B denote one of the simple algebras A_i, let C be its centre, and let $r = \dim_{\mathbf{Q}} C$. Then $C \otimes_{\mathbf{Q}} \mathbf{C} \cong \mathbf{C}^r$. If $\varphi_1, \ldots, \varphi_r$ are the distinct embeddings of C in $\mathbf{C}$, an isomorphism is given by

$$\alpha \otimes \lambda \to (\varphi_1(\alpha)\lambda, \ldots, \varphi_r(\alpha)\lambda) \ (\alpha \in C, \lambda \in \mathbf{C}).$$

We have

$$
(3.3)\qquad
\begin{aligned}
B \otimes_{\mathbf{Q}} \mathbf{C} &\cong B \otimes_C (C \otimes_{\mathbf{Q}} \mathbf{C}) \\
&= \bigoplus_{j=1}^{r} B \otimes_{C,\varphi_j} \mathbf{C}
\end{aligned}
$$

where we have written $B \otimes_{C,\varphi_j} \mathbf{C}$ to emphasize that in forming the tensor product, $\mathbf{C}$ is to be viewed as a C-module via φ_j. If $\dim_C B = t^2$, then each of these summands is isomorphic to $\mathbf{M}_t(\mathbf{C})$. Now the Galois group $\mathscr{G}$ of $\mathbf{C}$ acts on $B \otimes_{\mathbf{Q}} \mathbf{C}$ via its action on $\mathbf{C}$. Since $\mathscr{G}$ permutes the embeddings φ_j transitively, it permutes the summands (3.3) transitively. Correspondingly, the derived group of $B \otimes_{\mathbf{Q}} \mathbf{C}$ is the direct product of r copies of $\mathrm{SL}(t, \mathbf{C})$, permuted transitively by $\mathscr{G}$. Returning to the notation (3.2), we see from this discussion that the groups $T_i = H_i'$ are the minimal non-trivial connected normal $\mathbf{Q}$-subgroups of H, and M is the product of those T_i such that $(T_i)_{\mathbf{R}}$ is compact.

Now if A_i is non-commutative, then $A_i \otimes_{\mathbf{Q}} \mathbf{R}$ is the direct sum of a number of simple $\mathbf{R}$-algebras, and each of these has the form $\mathbf{M}_m(E)$, where E is either $\mathbf{R}$, $\mathbf{C}$ or the real quaternions $\mathbf{H}$, and $m > 1$ in the first two cases. If $m > 1$, then $\mathrm{GL}(m, E)$ contains the group D generated by the elementary matrices $1 + \lambda e_{ij}$ $(1 \leq i \neq j \leq m;\ \lambda \in E)$. This is well known to be perfect [**2**], and hence, up to conjugacy by a real invertible matrix, it is contained in $(\mathbf{T}_i)_{\mathbf{R}}$. Since D is not bounded, $(T_i)_{\mathbf{R}}$ cannot be compact. It follows that M is the product of those T_i such that $A_i \otimes_{\mathbf{Q}} \mathbf{R}$ is a direct sum of copies of H. This amounts to saying that A_i is a totally definite quaternion algebra [**6**, p. 293]. If $M \neq 1$ and $A = \mathbf{Q}G$ is the group ring of a finite group, then $RG = \mathbf{H} \oplus \dots$, and we complete the proof of Theorem 1 by considering the projection of G into H. More generally we have

THEOREM 3. *Let A be a finite dimensional semisimple $\mathbf{Q}$-algebra and Λ a $\mathbf{Z}$-order in A. Then $U = U(\Lambda)$ contains a normal subgroup U_0 of finite index such that if N is almost subnormal in U, then either N contains a non-abelian free group or $N \cap U_0 \leq \mathbf{Z}(U)$. Further, $U = U_0$ unless some simple component of A is a totally definite quaternion algebra.*

Finally we discuss the relation between our work and that of Gonçalves, Ritter and Sehgal [**3**]. They study subgroups N of $U = U(\mathbf{Z}G)$ slightly less general than our almost subnormal ones, and prove by completely different methods that if N contains G, then N contains a non-abelian free group, unless G is abelian or a Hamiltonian 2-group. This is not a direct consequence of Theorem 1, but is easily proved by the same methods. Suppose N contains G but contains no non-abelian free group. We apply the proof of Theorem 2 to the projection of U into $U(A_i \otimes_{\mathbf{Q}} \mathbf{C})$. By the discussion above, we have $M = 1$ or $T = 1$ here. If $M = 1$, then the projection of N must be central in U_i. Since this projection spans A_i, we find that A_i is commutative. Therefore every simple component of A_i is either a commutative field or a totally definite quaternion algebra. Hence RG is a direct sum of copies of R, $\mathbf{C}$ and H, and this yields that G is abelian or a Hamiltonian 2-group.

References

1. Armand Borel, *Introduction aux groupes arithmetiques*, (Publ. de l'Institut de Mathématiques de l'Université de Strasbourg, Vol. XV, Hermann, Paris, 1969).

2. J. Dieudonné, *La géometrie des groupes classiques* (Springer-Verlag, Berlin-Heidelberg-New York, 1955).

3. J. Gonçalves, J. Ritter and S. K. Sehgal, *Subnormal subgroups in* $U(\mathbf{Z}G)$, Proc. Amer. Math. Soc. (to appear)

4. B. Hartley and P. F. Pickel, *Free subgroups in the unit groups of integral group rings*, J. Canad. Math. Soc. **32** (1980), 1342–1352.

5. J. E. Humphreys, *Linear algebraic groups*, (Springer-Verlag, Berlin-Heidelberg-New York, 1975).

6. I. Reiner, *Maximal orders* (London Mathematical Society Monographs, Vol. 5, Academic Press, London, 1975).

7. J. Tits, *Free subgroups in linear groups*, J. Algebra **20** (1972), 250–270.

8. B. A. F. Wehfritz, *Infinite linear groups* (Springer-Verlag, Berlin-Heidelberg-New York, 1973).

Contemporary Mathematics
Volume 93 (1989)

On the Bass-Milnor Index of Abelian p-Groups

KLAUS HOECHSMANN

0. Introduction. According to a theorem of Bass and Milnor (cf. [**1**], Theorem 2), the unit group $U(\mathbf{Z}A)$ in the integral group ring $\mathbf{Z}A$ of a finite abelian group A is generated *up to finite index* by the subgroups $U\mathbf{Z}C$, as C ranges over all cycles in A. Actually their theorem concerns $K_1(\mathbf{Z}A)$ for *any* finite group A, not necessarily abelian. For the abelian case (and U instead of K_1), a simplified version of Bass's proof is included at the end of this paper, both to serve as reference and to underline that it gives no clue as to the size or nature of the finite index mentioned above.

Letting $\dot{U}(\mathbf{Z}A)$ denote the unit group modulo its torsion subgroup, consider the obvious product of inclusions

$$\alpha_A\colon \prod_C \dot{U}(\mathbf{Z}C) \to \dot{U}(\mathbf{Z}A),$$

C running over all cyclic subgroups of A. The order $\mathrm{ind}(\alpha_A)$ of its cokernel will be called the *Bass-Milnor index* of A. The aim of this paper is to prove the following assertion for any natural prime p.

THEOREM. *If the order of A is a p-power, then so is* $\mathrm{ind}(\alpha_A)$.

In view of the results of [**6**], this entails the strange conclusion that for odd *regular* p the map α_A is surjective. In the case of an *elementary* abelian A, both the theorem and this consequence have been proved previously (cf. [**4**] and [**5**]), and the proof given below is a rather convoluted variation on [**4**]. It does not use the finiteness established by Bass-Milnor, but depends instead on an "elementary" lemma, whose terms will now be explained.

Fix a prime power $p^m > 1$ and a non-trivial multiplicative character $\chi\colon U(R_m) \to U(\mathbf{C})$ from the units of the ring $R_m = \mathbf{Z}/p^m\mathbf{Z}$ to complex numbers. Define the *level* $l(\chi)$ of χ to be the smallest integer k such that $\chi(1 + p^k\mathbf{Z}) = 1$, i.e. such that χ can be pulled back from a character on $U(R_k)$. The level of χ is a measure of its non-triviality, but the relation between it and the (multiplicative) order of χ is somewhat messy, especially for $p = 2$. To smoothen the terminology, a level will also be assigned to elements x of any R_m-module X: $l(x)$ is the smallest k such that $(1+p^k\mathbf{Z})x = x$, i.e., it is just the p-logarithm of the (additive) order of x. In particular, for $0 \neq r \in R$, the quotient $r/p^{m-l(r)}$ is a unit in $R_{l(r)}$. Thus it makes sense to define a function $u_\chi \colon R_m \to \mathbf{C}$ by setting

$$u_\chi(r) = \chi(r/p^{m-l(r)})$$

if $l(r) = l(\chi)$, and 0 otherwise.

Now choose another integer $n \geq 1$, and consider the module X of $(n+1)$-tuples $x = (x_0, \ldots, x_n)$ with $x_i \in R_m$. For every cyclic subgroup $C \subset X$ pick a generator, and collect these in a set P. Writing $P_\chi = \{x \in P | l(x) \geq l(\chi)\}$, form a $P_\chi \times P_\chi$-matrix $M_{P,\chi}$ by putting

$$M_{P,\chi}(x, y) = u_\chi(x \cdot y),$$

where $x \cdot y = \sum_{i=0}^{n} x_i y_i$ is the usual "dot-product" of the $(n+1)$-tuples $x, y \in P$. Note: replacing any element $x \in P$ by an equivalent generator vx, with $v \in U(R_m)$, has the effect of multiplying the x-th row and column of $M_{P,\chi}$ by the unimodular complex number $\chi(v)$; hence it does not affect

$$D(\chi, n) = |\det M_{P,\chi}|^2,$$

which clearly is an algebraic integer depending only on χ and n. Here, then, comes the crux of our theorem.

MAIN LEMMA. *For every non-trivial χ and every n, $D(\chi, n)$ is a p-power depending only on $l(\chi)$ and n.*

After some easy preliminaries, paragraph 2 deduces the theorem from this lemma, the proof of which occupies paragraphs 3 to 6.

Added in proof: Actually the theorem is proved in a more general form (cf. Proposition 1.2) which can be applied, for instance, to certain subfunctors of $\dot{U}$. The lemma, in turn, is derived from a detailed eigenvalue calculation for the product of $M_{P,\chi}$ with its complex conjugate.

1. Generalities. Keeping p^m fixed, let ζ_k stand for the primitive p^k-th root of unity $\zeta_k = \exp(2\pi i p^{-k})$ for $k = 0, 1, \ldots, m$. Let $\mathscr{C}_{p^m}$ denote the full subcategory of finite dimensional $\mathbf{Q}$-algebras whose simple components are isomorphic to some of the cyclotomic fields $\mathbf{Q}[\zeta_k]$. Note that $G = U(R_m)$ acts naturally on all objects of $\mathscr{C}_{p^m}$, and that $\mathbf{Q}[\zeta_k] = \mathbf{Q}[\zeta_l]^{G_k}$, where $k \leq l$ and $G_k = (1 + p^k\mathbf{Z}) \cap G$. We shall soon have occasion to refer to the inclusions $\iota_{k,l} \colon \mathbf{Q}[\zeta_k] \to \mathbf{Q}[\zeta_l]$ by name. By a *lattice* we shall mean a finitely generated $\mathbf{Z}$-module.

DEFINITION 1.1. A lattice-valued functor V on $\mathscr{C}_{p^m}$ is said to be *admissible*, if

(i) $V(\mathbf{Q}) = \{0\}$, and V commutes with direct products;

(ii) $V(\iota_{k,l})$ bijects $V(\mathbf{Q}[\zeta_k])$ onto $V(\mathbf{Q}[\zeta_l])^{G_k}$ for all $0 \leq k \leq l \leq m$.

The second property ensures in particular that the image-lattices never have any (non-trivial) G-invariant elements, no non-zero submodules on which $G = G_0$ acts via the trivial character. This is related to the proviso in the

Main Lemma that χ be non-trivial. In fact, the next paragraph will be devoted to deducing the following assertion from that Lemma.

PROPOSITION 1.2. *Let A be a finitely generated R_m-module and V be an admissible functor on $\mathscr{C}_{p^m}$. Then the cokernel of the map*

$$\alpha_{V,A}\colon \prod_C V(\mathbf{Q}C) \to V(\mathbf{Q}A),$$

with C running over all cyclic subgroups of A, is a finite p-group.

Because of property (i), an admissible functor preserves surjections. In particular, any epimorphism $A \to B$ of finite groups leads to a surjection $\operatorname{coker}(\alpha_{V,A}) \to \operatorname{coker}(\alpha_{V,B})$. This has a noteworthy consequence.

REMARK 1.3 It suffices to prove Proposition 1.2 for free R_m-modules A.

Trusting that this will be accomplished in due course, the Proposition must still be connected to the Theorem; i.e. $\dot{U}(\mathbf{Z}\text{—})$ must be related to $V(\mathbf{Q}\text{—})$ for some admissible functor V. This can be done as follows.

Let $\mathscr{M}(\mathbf{Q}A)$ denote the integral closure of $\mathbf{Z}A$ in $\mathbf{Q}A$, and $W(\mathbf{Q}A)$ be its unit-group. Under the Wedderburn decomposition $\mathbf{Q}A \simeq \prod \mathbf{Q}[\zeta_k]$, the ring $\mathscr{M}(\mathbf{Q}A)$ corresponds to $\prod \mathbf{Z}[\zeta_k]$. Hence $W(\mathbf{Q}A) \simeq \prod U(\mathbf{Z}[\zeta_k])$ is well on its way to being an admissible functor (just shed the torsion...), but unfortunately it is too big. To cut it down to size, we take only the subgroup $W_1(\mathbf{Q}A)$ of 1-units, i.e. those which are congruent to 1 modulo the ideal $\Delta\mathscr{M}(\mathbf{Q}A)$ generated by the differences $\{a - 1 | a \in A\}$. It is important to note that this is *smaller* than the kernel of the so-called augmentation map $\kappa\colon \mathbf{Q}A \to \mathbf{Q}$ (sum of the coefficients) restricted to $\mathscr{M}(\mathbf{Q}A)$; in fact it is the direct product of ideals $I_k = (\zeta_k - 1)\mathbf{Z}[\zeta_k]$. Of course, in $\mathbf{Z}A$ the ideal $\Delta(\mathbf{Z}A)$ generated by the same differences *is* the kernel of κ, and hence $U(\mathbf{Z}A) = \{\pm 1\} \times U_1(\mathbf{Z}A)$, where the subscript 1 again means $\equiv 1$ modulo $\Delta(\mathbf{Z}A)$.

Thus, after factoring out the torsion everywhere, $\dot{U}(\mathbf{Z}A) = \dot{U}_1(\mathbf{Z}A)$ can be seen as a subgadget of $\dot{W}_1(\mathbf{Q}A)$, and the latter finally is the desired $V(\mathbf{Q}A)$. For fields this means that $V(\mathbf{Q}[\zeta_k])$ is the group $U(\mathbf{Z}[\zeta_k]) \cap (1 + I_k)$ taken modulo its torsion subgroup. Property (ii) holds by Galois theory and the fact that $I_k = I_l \cap \mathbf{Q}[\zeta_k]$ for $k \leq l$. The gap between Proposition 1.2 and the Theorem is now bridged by the next little lemma.

LEMMA 1.4. *$U_1(\mathbf{Z}A)$ has p-power index in $W_1(\mathbf{Q}A)$.*

PROOF. Let $u \in W_1(\mathbf{Q}A) \subset W_1(\mathbf{Q}_pA)$. By general principles (cf. [7], proof of II.2.9, pp. 49–50), $U_1(\mathbf{Z}_pA)$ is of finite index in $W_1(\mathbf{Q}_pA)$, but since the latter is a $\mathbf{Z}_p$-module (cf. [3], II.15.5), this index must be a p-power. Thus $v = u^N \in \mathbf{Z}_pA$ for some p-power N. But $v \in \mathscr{M}(\mathbf{Q}A) \subset \frac{1}{|A|}\mathbf{Z}A$, hence $v \in \mathbf{Z}A$.

As a consequence, $\dot{U}(\mathbf{Z}A)$ has p-power index in $V(\mathbf{Q}A)$, as does the composite

$$\prod_C \dot{U}(\mathbf{Z}C) \to \prod_C V(\mathbf{Q}C) \to V(\mathbf{Q}A),$$

by Proposition 1.2. Factorization via $\dot{U}(\mathbf{Z}A)$, however, shows this to be divisible by the Bass-Milnor index, completing our bridge.

To change the topic somewhat, let Γ be a finite group and K be a commutative ring in which the order $|\Gamma|$ is invertible. Then the group ring $K\Gamma$ contains the obvious idempotents $|\Gamma|^{-1}T$ and $1-|\Gamma|^{-1}T$, where $T=\sum_{g\in\Gamma} g$ is the sum over all group elements; these decompose any $K\Gamma$-module X into $X^\Gamma \oplus \ker T$. Applying this recursively to $\Gamma = G_{k-1}$ and $X = Y^{G_k}$, for some KG-module Y and $k = m, \dots, 2$, we obtain the decomposition

$$Y = Y_1 \oplus \cdots \oplus Y_m, \text{ with } Y_k = Y^{G_k} \cap \ker T_{k-1} \text{ and } Y_1 = Y^{G_1}.$$

Intuitively, Y_k is the part of Y which is fixed under G_k but, for $k > 1$, is moved full circle by the larger group G_{k-1}. The reason for stopping the process at $k = 2$, i.e. $\Gamma = G_1$, is that $[G\colon G_1]$ is prime to p and the K to be chosen later might not allow division by this index. Of course, this means that there is nothing to decompose if $m = 1$, which is just as well.

Another way to look at the meaning of Y_k is contained in the following observation, whose details are left to the reader.

REMARK 1.5. If $K = \mathbf{C}$ and if $Y = \prod_\chi Y_\chi$ is the decomposition according to characters $\chi\colon G \to U(\mathbf{C})$, then $Y_k = \prod_{l(\chi)=k} Y_\chi$.

Since these components are defined in terms of the G-action, which is respected by every morphism in $\mathscr{C}_{p^m}$, the next assertion is not surprising either.

LEMMA 1.6. *Let p be invertible in the commutative ring K, and let V be an admissible functor. Then there is a functorial decomposition $\mathscr{V} = \mathscr{V}_1 \oplus \cdots \oplus \mathscr{V}_m$, with $\mathscr{V} = K \otimes V$, $\mathscr{V}_1 = \mathscr{V}^{G_1}$, and $\mathscr{V}_k = \mathscr{V}^{G_k} \cap \ker T_{k-1}$ for $k > 1$. For $h \le k$ the maps $\mathscr{V}_h(\iota_{k,l})$ are isomorphisms.*

The proof is immediate from the discussion above. In fact, the statement would hardly be worth making, except to avoid blurring the next paragraph with these notational niceties.

There is a similar motive (applying to paragraph 5) for recording the last of these preparatory lemmas, which concerns a subgroup chain $\{1\} = \Gamma_h \subset \Gamma_{h-1} \subset \cdots \subset \Gamma_0 = \Gamma$ and an element $s = \sum_{g\in\Gamma} c(w(g))g$ of the ring $K\Gamma$ whose coefficient for any $g \in \Gamma$ depends only on $w(g) = \max\{i | g \in \Gamma_i\}$, i.e. on the location of g in the chain. Assume that K is a field.

LEMMA 1.7. *In the preceding context, let $\chi\colon K\Gamma \to K$ be a homomorphism which is non-trivial on Γ_{h-1}. Then $\chi(s) = c(h) - c(h-1)$.*

PROOF. Defining $c(-1) = 0$ and $t_k = \sum_{w(g)=k} g$, for $0 \le k \le h$, we have

$$s = \sum_{k=0}^{h} c(k) t_k = \sum_{k=0}^{h} t_k \sum_{j=0}^{k} (c(j) - c(j-1)) = \sum_{j=0}^{h} (c(j) - c(j-1)) \sum_{k=j}^{h} t_k.$$

Now, $\chi(\sum_{k=j}^h t_k) = \sum_{w(g)\le j} \chi(g) = \sum_{g\in\Gamma_j} \chi(g)$ is the sum over a non-trivial subgroup of $U(K)$, hence $= 0$, unless $j = h$. This finishes the proof.

2. Reduction to Main Lemma. In preparation for the announced reduction, we need to describe the Wedderburn decomposition of $\mathbf{Q}A$ more concretely than usual, not just "up to isomorphism" but in terms of nuts and bolts. Any functional $f: A \to R_m$ yields a $\mathbf{Q}$-algebra homomorphism $\mathbf{Q}A \to \mathbf{Q}[\zeta_{l(f)}]$ by $a \mapsto \zeta_m^{f(a)}$, and every G-orbit in $\operatorname{Hom}(A, R_m)$ constitutes a set of $\mathbf{Q}$-isomorphic irreducible representations, with $G = U(R_m)$ as before. Letting $F \subset \operatorname{Hom}(A, R_m)$ be a set of representatives of G-orbits, we get an algebra *isomorphism*:

$$\mathbf{Q}A \to \prod_{f \in F} \mathbf{Q}[\zeta_{l(f)}],$$

induced by $a \mapsto (\ldots, \zeta_m^{f(a)}, \ldots)$. This is the "Wedderburn decomposition" with respect to F. In case $A = C_x$ is cyclic with a fixed generator x, we can even be more explicit, inducing the map by $x \mapsto (\ldots, \zeta_k, \ldots)$, with $0 \le k \le l(x)$. It is crucial to relate these two methods of bookkeeping for elements $x \in A$. Indeed, for a single $x \in A$ and a single $f: A \to R_m$, they fit into a commutative square

$$\text{(1)} \qquad \begin{array}{ccc} \mathbf{Q}C_x & \longrightarrow & \prod_{k \le l(x)} \mathbf{Q}[\zeta_k] \\ \downarrow & & \downarrow \\ \mathbf{Q}A & \longrightarrow & \mathbf{Q}[\zeta_{l(f)}] \end{array}$$

with the top arrow representing the Wedderburn decomposition. The point is to describe the right vertical arrow, and here is what it does:

1. Project $\prod_{k \le l(x)} \mathbf{Q}[\zeta_k] \to \mathbf{Q}[\zeta_{l'}]$ onto the l'-th component, where $l' = l(f(x))$.
2. Apply the automorphism of $\mathbf{Q}[\zeta_{l'}]$ which changes $\zeta_{l'} \mapsto \zeta_m^{f(x)}$.
3. Include $\mathbf{Q}[\zeta_{l'}] \to \mathbf{Q}[\zeta_{l(f)}]$ using $\iota_{l', l(f)}$.

The stage is now clear for deducing Proposition 1.2 from the Main Lemma. This will be achieved by an ascending chain of "claims", each of which will imply the previous one. To lighten the notation, we agree to omit reference to $\mathbf{Q}$ from our functor V, writing simply $V(A)$, $V(\zeta_k)$, etc. Moreover, we take advantage of Remark 1.3, and let $A = R_m^{n+1}$ with $n \ge 1$. Any $f: A \to R_m$ can then be obtained in the form $y^*: x \mapsto x \cdot y$, so that the set P of representatives of G-orbits in A (cf. Introduction) can also serve as the set F mentioned at the beginning of this paragraph. All in all, we have a commutative square

$$\text{(2)} \qquad \begin{array}{ccc} \prod_{x \in P} V(C_x) & \longrightarrow & \prod_{x \in P} \prod_{k \le l(x)} V(\zeta_k) \\ \downarrow & & \downarrow \\ V(A) & \longrightarrow & \prod_{y \in P} V(\zeta_{l(y)}) \end{array}$$

whose horizontal arrows are lattice-isomorphisms and whose left vertical is our deal old α. All these maps are compatible with the G-action. Referring to the right vertical as γ, Proposition 1.2 boils down to

Claim 1. The cokernel of γ is a finite p-group.

Tensoring all these lattices by the ring of quotients $K = \mathbf{Z}[p^{-1}]$, and abbreviating $\mathscr{V} = K \otimes V$ as well as $\delta = K \otimes \gamma$, we must show that δ is surjective. This can be accomplished by looking separately at each component of the decomposition $\mathscr{V} = \mathscr{V}_1 \oplus \cdots \oplus \mathscr{V}_m$ which exists by virtue of Lemma 1.6. So, we are down to

Claim 2. The map δ_h is surjective, for $h = 1, \ldots, m$.

Aforesaid map is depicted as the left vertical arrow in our third and final commutative square, whose horizontal arrows are made from the routine identifications $\mathscr{V}_h(\zeta_k) \simeq \mathscr{V}_h(\zeta_m)$ via $\iota_{k,m}$ compounded, in the case of the top arrow, by a drastic thinning out of the Wedderburn components belonging to each C_x: of all the components corresponding to indices k with $h \leq k \leq l(x)$ only the one belonging to $k = h$ is kept, the others are mapped to zero. Here is the diagram:

$$\begin{array}{ccc} \prod_{x \in P_h} \prod_{h \leq k \leq l(x)} \mathscr{V}_h(\zeta_k) & \longrightarrow & \prod_{x \in P_h} \mathscr{V}_h(\zeta_m) \\ \downarrow & & \downarrow \\ \prod_{y \in P_h} \mathscr{V}_h(\zeta_{l(y)}) & \longrightarrow & \prod_{y \in P_h} \mathscr{V}_h(\zeta_m) \end{array} \tag{3}$$

The bottom arrow represents a straightforward isomorphism and the top one is a projection onto a direct summand. Therefore, letting ε_h stand for the morphism on the right of this square, it would certainly suffice to prove

Claim 3. The map ε_h is an isomorphism, for $h = 1, \ldots, m$.

For fixed x, y the map δ_h is still described fairly well by items 1. to 3. following diagram (1). Translating from that description, here is what happens under the y-th component of our ε_h to the x-th component of the product $\prod_x \mathscr{V}_h(\zeta_m)$:

1′. Annihilate if $l' = l(x \cdot y)$ does not equal h.

2′. Otherwise apply the automorphism $\zeta_h \mapsto \zeta_m^{x \cdot y}$.

Note that the third stage is now subsumed in the $\iota_{k,m}$ of the bottom arrow. The main advantage of our new arrangement is that the map ε_h can be described as a *matrix* M_h with entries in the ring-endomorphisms of $\mathbf{Q}[\zeta_h]$, namely $M_h(x, y) = u_h(x \cdot y)$ where, for $r \in R_m$,

$$u_h(r) = \begin{cases} r/p^{m-h}, & \text{if } l(r) = h; \\ 0, & \text{otherwise.} \end{cases}$$

As written, these entries are 0 or in the group $U(R_h)$, so they act on $X_h = \mathscr{V}_h(\zeta_m)$ as elements of the group ring K_h of that group over $\mathbf{Z}[p^{-1}]$. Therefore the matrix M_h defines an isomorphism, if and only if

Claim 4. $d_h = \det M_h \in K_h$ acts on X_h as an isomorphism.

To show this, it would obviously suffice to find that $d_h c$ acts invertibly on X_h for a suitable $c \in K_h$. Indeed, let the matrix M_h^- be defined by $M_h^-(x, y) = M_h(x, y)^{-1}$ wherever that makes sense (and 0 elsewhere), and let d_h^- be its determinant. Since p is a unit in K_h, it is enough to establish

Claim 5. $d_h d_h^-$ acts on X_h as a p-power.

Now we are very close to the Main Lemma. In fact, a character $\chi\colon U(R_h) \to U(\mathbf{C})$ yields a ring homomorphism $\chi\colon K_h \to \mathbf{C}$, and if χ is of level h the definitions given here obviously match those of the Introduction, so that $\chi(d_h d_h^-) = D(\chi, n)$. But by Remark 1.5 all characters occurring in the $\mathbf{C}[G]$-module $\mathbf{C} \otimes X_h$ *are* of level h.

3. Hermitian Product. Just as every divisor of p^m occurs both as the order of a subgroup and of a factor group of R_m, the quantity $m - l(x)$ inevitably crops up alongside the level $l(x)$ of any element x of an R_m-module. Instead of complicating several formulas with too many minus signs, let us agree to use the notation $v(x) = m - l(x)$. For the character $\chi\colon G \to U(\mathbf{C})$ of the Main Lemma, let us set the level $l(\chi) = h$ and put $m - h = \kappa$. Note that $l(x) > l(\chi)$ is exactly the same condition as $v(x) \le \kappa$.

Since the basic set-up $P \subset A = R_m^{n+1}$, as described in the Introduction, will be kept fixed, it need not always be referred to by the notation for the matrix M_χ whose determinant is the subject of the Main Lemma. Actually, that lemma is about the determinant of the Hermitian matrix $S_\chi = M_\chi M_{\bar{\chi}}$ whose entries are given by

$$S_\chi(x,y) = \sum_{z \in P} u_\chi(x \cdot z)\bar{u}_\chi(z \cdot y),$$

for $x, y \in P_\chi$, i.e. $v(x), v(y) \le \kappa$. Note that the restriction $l(z) \ge l(\chi)$ is automatically enforced by the function u_χ. Still, the first problem with this formula is the unknown and possibly chaotic nature of the summation index $z \in P$. This is easily straightened out: imagine $|G|$ such formulas, each obtained from the one displayed by writing gz instead of z everywhere on the right ($g \in G$), and then letting the $\chi(g)$ be cancelled by the $\bar{\chi}(g)$ in each summand; finally add together all these equations. The index of the resulting sum will now run over complete G-orbits, covering each non-trivial orbit Gz exactly $p^{v(z)}$ times, as this is the order of the fix-group $G_{l(z)} = \{g | gz = z\}$ of any $z \neq 0$. Here is the result of this symmetrizing summation:

$$|G| S_\chi(x,y) = \sum_{z \in A} p^{v(z)} u_\chi(x \cdot z)\bar{u}_\chi(z \cdot y). \tag{4}$$

CONVENTION 3.1: In the subsequent formulas, it will be supposed that $l(x) \le l(y)$, in other words that $v(y) \le v(x)$. Since S_χ is Hermitian, this is a harmless constraint.

The relationship between the indices y and x can be made more visible by transforming x, y to a kind of standard form by a matrix $L \in \mathrm{GL}(n+1, R_m)$. We picture a $2 \times (n+1)$-matrix in which y forms the first, x the second row. Applying elementary column operations, we can reduce all but one of the entries in the first row to zero, and so on. The outcome of this exercise is

$$\begin{aligned} yL &= [p^\mu, 0, \ldots, 0] \\ xL &= [r, p^\lambda, \ldots, 0] \end{aligned} \tag{5}$$

where the off-diagonal term r can be made zero unless $v(r) < \lambda$, in which case $v(r) = v(x) \geq \mu$. In all instances we can therefore write $r = tp^\mu$, with $t \in R_m$ determined modulo $p^{\lambda-\mu}$. Incidentally, $\lambda \geq \mu$, because $\lambda < \mu$ would imply $v(x) = \lambda$ and $v(x) < \mu = v(y)$, contrary to our convention.

The next step toward understanding the sum $S_\chi(x, y)$ comes from the observation that (4) is not affected if we substitute zN for z in each summand, where N denotes a fixed element of $\mathrm{GL}(n+1, R_m)$, say the transpose of the L used above. The individual summands now appear as

$$p^{v(z)} u_\chi(z_0 tp^\mu + z_1 p^\lambda)\bar{u}_\chi(z_0 p^\mu),$$

where $z = [z_0, \ldots, z_n]$. The second factor in this is $\neq 0$ if and only if $l(z_0 p^\mu) = l(\chi)$, i.e. if and only if $v(z_0) = \kappa - \mu$, i.e. if and only if $z_0 = gp^{\kappa-\mu}$ with $0 \leq \kappa - \mu < m$ and $g \in G$. Since the fix-group of such a z_0 has order $p^{\kappa-\mu}$, the number of possible values for z_0 is $p^{\mu-\kappa}|G|$. Therefore, if we take only *one* of these, that is, if we fix $z_0 = p^{\kappa-\mu}$ and restrict the summation to $z' = [z_1, \ldots, z_n] \in A' = R_m^n$, formula (4) mutates to

$$p^{\kappa-\mu} S_\chi(x, y) = \sum_{z' \in A'} p^{(\kappa-\mu)\cap v(z')} u_\chi(tp^\kappa + z_1 p^\lambda), \tag{6}$$

where the notation $a \cap b$ is used instead of the lengthier $\min\{a, b\}$.

DEFINITION 3.2. A pair $x, y \in X$ from a free R_m-module X is *k-congruent* if $v(x - y) \geq k$; is *k-similar* if gx, y is k-congruent for some $g \in G$; and is *k-dependent* if tx, y or x, ty is k-congruent for some $t \in R_m$. It will be called *quasi-congruent* (*quasi-similar*) if it is k-congruent (k-similar) for some $k > v(x), v(y)$.

LEMMA 3.3. *With $v(y) \leq v(x)$, let λ, μ, t be the parameters appearing in* (5) *and* (6). *Then*

a) *$\mu = v(y)$, and λ is the largest integer such that x, y is λ-dependent.*
b) *x, sy is λ-congruent if and only if $s \equiv t$ modulo $p^{\lambda-\mu}$.*
c) *x, y is quasi-similar if and only if $\mu < \lambda$ and t is a unit.*

PROOF. (a) is clear from (5); (b) amounts to the observation that $v((s-t)y) \geq \lambda$ if and only if $v(s-t) \geq \lambda - \mu$. The if-part of (c) is clear; the converse follows from (b).

Having thus obtained an intrinsic characterization of the terms of our formula for $S_\chi(x, y)$, we turn back to it with a view to "divide" the expression $tp^\kappa + z_1 p^\lambda$ by p^κ. Since $v(z_1 p^\lambda) < \kappa$ implies that $u_\chi(tp^\kappa + z_1 p^\lambda) = 0$, we might as well constrain z_1 by the condition $v(z_1) + \lambda \geq \kappa$, that is $z_1 p^\lambda = sp^\kappa$ for suitable $s \in R_m$. Actually, as $\kappa + h = m$, this element s is only determined modulo p^h. It is systematically obtained from z_1 by first applying the surjection $R_m \to p^\lambda R_m$ (multiplication) and then the isomorphism $p^\kappa R_m \to R_h$ (division). What happens here is best understood by visualizing R_m as the set of m-digit p-adic integers, and the maps as a λ-fold forward shift (overflow allowed but lost) and a κ-fold backward shift (no underflow allowed), respectively. The composite of these maps will be denoted by $\phi_\kappa^\lambda(z_1)$ and is defined only if $v(z_1) \geq \kappa - \lambda$ (no condition if $\kappa < \lambda$). Similarly, first multiplying and then dividing the parameter $t \in R_m$ by p^κ we get a certain element $t_h = \phi_\kappa^\kappa(t) \in R_h$, which is none other than the image of t under the canonical map $R_m \to R_h$, and like t is unique modulo $p^{\lambda-\mu}$. The formula now looks like:

$$p^{\kappa-\mu} S_\chi(x,y) = \sum_{z' \in A', v(z_1) \geq \kappa-\lambda} p^{(\kappa-\mu) \cap v(z')} \chi(t_h + \phi_\kappa^\lambda(z_1)), \tag{7}$$

with the understanding that $\chi\colon R_h \to \mathbf{C}$ takes non-units to zero.

We now leave aside S_χ for a while, in order to make an adjustment on the index set P.

DEFINITION 3.4. A subset $F \subset A$ is *fully transversal* if k-similarity implies k-congruence for any pair x, y from F and any $k = 1, \ldots, m$.

LEMMA 3.5. *The set of $P \subset A$ of G-representatives may be chosen to be fully transversal.*

PROOF. For every $x = [x_0, \ldots, x_n] \in A$, let $i(x) = \min\{i | v(x_i) = v(x)\}$. Take P be the set of $(n+1)$-tuples $x \neq 0$ whose $i(x)$-th component is a power of p (possibly 1). Clearly this catches exactly one element in every non-trivial G-orbit.

Now take $x, y \in P$ with $v(y) \leq v(x)$. From $v(x - gy) \geq k$ we must deduce that $v(x-y) \geq k$. To make the task non-trivial, we may assume that $v(y) < k$. Then $v(gy) = v(y) < v(x)$ would imply $v(gy - x) = v(gy) < k$, and not $\geq k$; hence we must have $v(x) = v(y)$. Moreover, $i(x) \leq i(gy)$ or else the difference would not get shorter at the $i(x)$-th component. Hence $i(x) = i(gy) = i(y)$ by symmetry.

Now let i be that common index $i(x) = i(y)$. By definition of P, we know that $x_i = y_i = p^\mu$ where $\mu = v(x) = v(y) < k$. Since $p^\mu - gp^\mu \in p^k R_m$, it follows that $1 - g \in p^{k-\mu} R_m$. Hence $x \equiv gy \equiv y$ modulo p^k, as was to be shown.

To sum up the results of this paragraph, the following lemma can now be stated without further proof.

SUMMARY 3.6. Let $P \subset A$ be a fully transversal set of G-representatives, and let $x, y \in P_\chi$ with $v(x) \geq v(y) = \mu$, and λ be the largest integer such that x, y is λ-dependent. Then there is a $t_h \in R_h$ such that (7) holds; with (i) $\mu < \lambda$ and $t_h = 1$ if x, y is quasi-similar (hence quasi-congruent), and (ii) $t_h \equiv 0 \pmod p$ if not.

It is worth noting that t_h is *unique* if $\lambda - \mu \geq h$. At the other extreme, if $\lambda = \mu$ we can take $t_h = 0$; so this falls under case (ii). For $0 < \lambda - \mu < h$, there is some vagueness $\pmod{p^{\lambda-\mu}}$ to the value of t_h, but its status as unit or non-unit is fixed; if it is a unit, we can and will put $t_h = 1$.

4. Block Decomposition. In this short paragraph, $S_\chi(x, y)$ will be computed in two extreme situations, namely x, y not being quasi-similar and x, y being equal, and a certain organizational matter will be attended to. We keep the notation and terminology of the last paragraph, but need not yet assume that P is fully transversal.

LEMMA 4.1. *If $x, y \in P_\chi$ are not quasi-similar, then $S_\chi(x, y) = 0$.*

PROOF. Since S_χ is Hermitian, we may assume $v(x) \geq v(y) = \mu$ and apply formula (7) with $t_h \equiv 0 \pmod{p}$. Then $\chi(t_h + \phi_\kappa^\lambda(z_1)) = 0$ unless $\phi_\kappa^\lambda(z_1)$ is a unit, which happens if and only if $v(p^\lambda z_1) = \kappa$, i.e. $v(z_1) = \kappa - \lambda$ (remember that $\kappa < m$). But there are no such z_1, and hence everything is zero, unless $\kappa \geq \lambda$, in which case the homomorphism $\phi_\kappa^\lambda \colon B = \{b \in R_m | v(b) \geq \kappa - \lambda\} \to p^\kappa R_m \to R_h$ is surjective with kernel of order p^λ, so that every $w \in U(R_h)$ occurs exactly p^λ times in the image of ϕ_κ^λ. Letting $z'' = [z_2, \ldots, z_n]$ run over $A'' = R_m^{n-1}$ (possibly the trivial group), we get

$$\begin{aligned} p^{\kappa-\mu} S_\chi(x, y) &= \sum_{z''} \sum_{b \in B} p^{(\kappa-\mu) \cap v(b) \cap v(z'')} \chi(t_h + \phi_\kappa^\lambda(z_1)) \\ &= \sum_{z''} p^{(k-\lambda) \cap v(z'') + \lambda} \sum_{w \in W} \chi(t_h + w), \end{aligned}$$

with $W = U(R_h)$. The first line is just a literal translation of (7), the second uses the fact that $v(b) = \kappa - \lambda \leq \kappa - \mu$, etc. Since $w \mapsto w + t_h$ yields a permutation of W, the term t_h can be omitted from the second line, and since W is non-trivial the final sum is zero (remember that $h \geq 1$), as was to be shown.

For any subset $Q \subset P_\chi$ let $(S_\chi | Q)$ denote the $Q \times Q$-matrix obtained by restriction to $x, y \in Q$. Since P_χ is the disjoint union of quasi-similarity classes, the preceding lemma can also be stated as follows.

COROLLARY 4.2. *S_χ is the direct sum of $(S_\chi | Q)$ as Q ranges over quasi-similarity classes in P_χ.*

As the computation of the diagonal entries does not fit well into the context of the next paragraph, it will be used to fill out the present one. The following notation will be used repeatedly.

NOTATION 4.3. If T is an indeterminate, define the polynomials $\pi_k(T)$ by

$$\pi_k(T) = \begin{cases} T^m & \text{if } k = m; \\ (p^{m-k} - p^{m-k-1}) T^k, & \text{if } k < m. \end{cases}$$

LEMMA 4.4. *For $v(x) = \mu \leq \kappa$ we have*

$$p^{\kappa-\mu} S_\chi(x, x) = \sum_{k=0}^{\kappa-\mu} \pi_{m-k}(p^n).$$

PROOF. Let us put $q = p^n$. Then q^{m-k} is the number of $z' \in A'$ such that $v(z') \geq k$, for $0 \leq k \leq m$. Writing $\kappa - \mu = s$, we have by (7):

$$p^s S_\chi(x,x) = \sum_{z' \in A'} p^{s \cap v(z')} = q^{m-s} p^s + \sum_{v(z')<s} p^{v(z')} = q^{m-s} p^s + \sum_{0 \leq k < s} (q^{m-k} - q^{m-k-1}) p^k,$$

whence the result by rearrangement of summands (summation by parts).

If $m = 1$, the two extreme cases treated here are the only ones possible, and the matrix S_χ is diagonal. Since $\kappa < m$, we must have $\kappa = \mu = 0$, and the diagonal terms are $S_\chi(x,x) = q = p^n$ in happy agreement with the Lemma of **[4]**.

5. Positive Entries. Before getting into the actual agenda of this paragraph we will present its crucial computation in abstract form, out here where the plot is not so thick.

LEMMA 5.1. *Let B be cyclic of order p^l, and ϕ: $B \to R_h$ be a homomorphism whose kernel is of order p^λ, with $0 < \lambda < l$. Let f: $\{0, \ldots, l\} \to \mathbf{C}$ be a function which is constant up to $\nu > 0$, i.e. such that $f(k) = f(\nu)$ for $k \leq \nu$. If χ: $U(R_h) \to \mathbf{C}$ is a character of level h, we have*

$$\sum_{b \in B} f(l(b)) \chi(1 + \phi(b)) = \sum_{\nu \leq k \leq \lambda} p^k (f(k) - f(k+1)),$$

it being understood that χ takes non-units to zero, and that the second sum is trivial if $\lambda < \nu$.

PROOF. The sum to be computed is obviously equal to $\sum_{a \in D} F(a) \chi(1+a)$, where $D = \{a \in R_h | 1 + a \notin pR_h\}$ and $F(a) = \sum_{\phi(b)=a} f(l(b))$. We claim that

(8)
$$F(a) = \begin{cases} p^\lambda f(\lambda+1) + \sum_{\nu \leq k \leq \lambda} p^k (f(k) - f(k+1)), & \text{if } a = 0; \\ p^\lambda f(l(a) + \lambda), & \text{if } 1 \leq l(a) \leq l - \lambda; \\ 0, & \text{if } l - \lambda < l(a). \end{cases}$$

First, let $a \neq 0$. Then $\phi(b) = a$ means $l(b) = l(a) + \lambda$; moreover there are exactly p^λ such b if $l(a) + \lambda \leq l$ and none otherwise—whence the last two cases of the formula. In particular, $F(a) = p^\lambda(1+\lambda)$ for $l(a) = 1$, for instance $a = p^{h-1}$.

Next, $F(0) = \sum_{l(b) \leq \lambda} f(l(b))$ is easy if $\lambda \leq \nu$, since then all summands equal $f(\nu) = f(\lambda)$, and their number is again p^λ. This agrees with the first line of (8), where the second term is trivial for $\lambda < \nu$.

Finally, suppose that $\nu < \lambda$. Then

$$F(0) = \sum_{l(b)<\nu} f(\nu) + \sum_{\nu \leq l(b) \leq \lambda} f(l(b)) = p^{\nu-1} f(\nu) + \sum_{\nu \leq j \leq \lambda} (p^j - p^{j+1}) f(j),$$

which gives (8) after summation by parts.

To finish the proof we call on Lemma 1.7, which we apply to $\Gamma = U(R_h)$ with the usual filtration of subgroups $\Gamma_k = \{1 + p^k R_h\} \cap U(R_h)$. Since $F(a)$ depends only on $l(a)$, this lemma says that the sum to be computed equals $F(0) - F(p^{h-1}) = F(0) - p^\lambda f(\lambda+1)$. This easily yields the desired conclusion.

LEMMA 5.2. *Let $0 < \nu \leq m$ be integers, and $d\colon \{0,\dots,m\} \to \mathbf{Z}_{\geq 0}$ be a function which is constant up to ν and such that $d(k+1) = d(k) - 1$ if $\nu \leq k < m$. Let I be a finite set, and $e\colon I \to \mathbf{Z}_{\geq 0}$ be any function. Putting*

$$f(k) = \sum_{i \in I} p^{e(i) \cap d(k)} \quad \textit{we have}$$

$$f(k) - f(k+1) = \begin{cases} 0, & \text{for } 0 \leq k < \nu; \\ N_k(p^{d(k)} - p^{d(k+1)}), & \text{if } \nu \leq k < m, \end{cases}$$

where N_k denotes the number of $i \in I$ such that $e(i) \geq d(k)$.

PROOF. Obviously $f(k)$ is constant for $k \leq \nu$, because $d(k)$ is. For the rest it suffices to subtract the identities

$$f(k) = \sum_{e(i) < d(k)} p^{e(i)} + \sum_{e(i) \geq d(k)} p^{d(k)},$$

$$f(k+1) = \sum_{e(i) \leq d(k+1)} p^{e(i)} + \sum_{e(i) > d(k+1)} p^{d(k)},$$

whose summation indices match because, for $\nu \leq k < m$, the condition $< d(k)$ is the same as $\leq d(k+1)$, as $d(k) - d(k+1) = 1$.
We now turn back to the context and notation of the previous two paragraphs.

PROPOSITION 5.3. *Let $x, y \in P_\chi$ be quasi-congruent, with $v(x) = v(y) = \mu$, and let $h = l(\chi)$. Then*

$$p^m S_\chi(x,y) = p^{h+\mu} \sum_{h+\mu \leq k \leq \lambda} \pi_k(p^n),$$

where λ is the largest integer $\leq m$ for which x, y are λ-congruent.

PROOF. Let us first agree that the case $\lambda = m$, i.e. $x = y$, has been dealt with in Lemma 4.4. To recognize the two results as identical, we have to recall the relation $\kappa - \mu = m - (h + \mu)$ and perform the change of variable $k \mapsto m - k$. Henceforth we may assume that $\mu < \lambda < m$.

Setting $B = \{b \in R_m | v(b) \geq \kappa - \lambda\}$, we apply formula (7) with $t_h = 1$ to get

$$p^{\kappa-\mu} S_\chi(x,y) = \sum_{b \in B} s(b)\chi(1 + \phi_\kappa^\lambda(b)) \text{ with } s(b) = \sum_{z'' \in A''} p^{v(z'') \cap v(b) \cap (\kappa-\mu)},$$

just as in the proof of Lemma 4.1. This can be written as $s(b) = f(l(b))$, where f has the shape shown in Lemma 5.2, with $I = A''$, $e = v$, and $d(k) = (m-k) \cap (\kappa - \mu)$. The set-up now obviously calls for the application of Lemmas 5.1 and 5.2, but we must first check out their parameters and premises.

B is either all of R_m (if $\kappa - \lambda < 0$) or $p^{\kappa-\lambda}R_m$, hence its order is p^l, with $l = m \cap (h+\lambda)$. Since $h \geq 1$, we have $\lambda < h+\lambda$; together with the assumption that $\lambda < m$, this assures $0 < \lambda < l$, as is required by the first sentence of 5.1. The role of ν will be played by $h + \mu$, which is positive because h is. The function f needs to be defined only on $\{0, \ldots, l\}$, so the possibly larger domain of definition of d is ample. Constancy up to ν is inherited from d, which is stationary until $(m - k)$ reaches $(\kappa - \mu)$, i.e. $k = m - \kappa + \mu = \nu$. All the premises of 5.1 and 5.2 are now verified.

We are almost ready to pull the switch. The $d(k)$ appearing in the conclusion of 5.2 is just $m - k$ for the range in question, but what is N_k? It is the number of $z'' \in A'' = R_m^{n-1}$ for which $v(z'') \geq m - k$, namely $(p^{n-1})^k$. Altogether we get

$$\sum_{h+\mu\leq k\leq\lambda} p^k (p^{n-1})^k (p^{m-k} - p^{m-k-1}),$$

as was to be shown.

The final result depends only on $h = l(\chi)$ (fixed throughout), $\mu = v(x) = v(y)$ (fixed within each quasi-similarity class), and λ which can range from $\mu+1$ to m. It is 0 for λ less than $\mu+h$ and takes on strictly increasing integer values beyond that point. Its triviality for small λ leads to a refinement of the block decomposition of 4.2: for $S_\chi(x, y)$ to be non-zero, it is necessary (and sufficient) that x, y be $(h+\mu)$-congruent; let us call them *h-strictly* congruent if $v(x) = v(y) = \mu$, and $v(x - y) \geq h + \mu$.

COROLLARY 5.4. *If P is fully transversal, S_χ is the direct sum of $(S_\chi|Q)$ as Q ranges over $l(\chi)$-strict congruence classes of P_χ.*

The full transversality (cf. 3.4) is necessary to force quasi-similar pairs (cf. 4.2) to be quasi-congruent, so that 5.3 can be applied.

6. Eigenvalues. Again we begin by presenting the main computations *in abstracto*. Let $q \geq 2$, $s \geq 0$ be integers, and $a_0, \ldots, a_s$ be a finite sequence of elements of some field F. The $q^s \times q^s$-matrix $L_q(a_0, \ldots, a_s)$ is defined recursively as follows. $L_q(a_0) = [a_0]$, and

$$L_q(a_0, \ldots, a_s) = a_s I_{q^s} + E_q \otimes L_q(a_0, \ldots, a_{s-1}),$$

where E_q is the $q \times q$-matrix with all entries $= 1$, and $\otimes$ denotes the Kronecker product (cf. **[2]**, p. 69). A more visual description would be: take $L_q(a_0, \ldots, a_{s-1})$, blow up each entry to a $q \times q$-block, and add a_s to the main diagonal of the result.

REMARK 6.1. We need to formalize this description somewhat. For every natural number $0 < t \leq s$, let $\rho\colon Q_t \to Q_{t-1}$ be a q-to -1 surjection of finite sets, and let Q_0 have cardinality 1. All these maps get the same name ρ, as there is no danger of confusion. (For instance, ρ might be the dropping of the last digit, if Q_s is the set of s-digit q-adic numerals.) Such a structure clearly amounts to the same as the imposition on Q_s of a nested sequence of equivalence relations $\{\sim_k \,|k = 0, \dots, s\}$ such that each class defined by $\sim_k$ breaks up into exactly q classes given by $\sim_{k+1}$.

If we imagine our matrix as a function on the Cartesian product $Q_s \times Q_s$, the recursion process consists of pulling back along $\rho \times \rho$ and then adding a_s to the diagonal. A simple induction now proves that the i, j-th entry of $L_q(a_0, \dots, a_s)$ equals $a_0 + \cdots + a_{s-k}$, where k is the smallest integer such that $\rho^k(i) = \rho^k(j)$; or in the other language, $a_0 + \cdots + a_t$, where t is the largest integer such that $i \sim_t j$.

LEMMA 6.2. *The eigenvalues of $L_q(a_0, \dots, a_s)$ are $\varepsilon_k = a_s + qa_{s-1} + \cdots + q^k a_{s-k}$, with $k = 0, \dots, s$. Their respective multiplicities are $m_s = 1$ and $m_k = q^{s-k} - q^{s-k-1}$, for $k < s$.*

PROOF. Clear for $s = 0$. For $s > 0$, we have

$$\mathscr{E}(L_q(a_0, \dots, a_s)) = a_s + \mathscr{E}(E_q \otimes L_q(a_0, \dots, a_{s-1})) = a_s + \mathscr{E}(E_q) \times \mathscr{E}(L_q(a_0, \dots, a_{s-1})),$$

where $\mathscr{E}$ denotes the generic eigenvalue. Now, E_q has eigenvalues 0 (multiplicity $q - 1$) and q (multiplicity 1). Thus, for $L_q(a_0, \dots, a_s)$ we get a_s with multiplicity $(q - 1)q^{s-1}$, and $a_s + q\mathscr{E}(L_q(a_0, \dots, a_{s-1}))$ with the same multiplicity as before—whence the result.

COROLLARY 6.3.. *Let $b_0, \dots, b_s$ be a finite sequence in F, and put $a_t = q^t(b_t - b_{t+1})$ for $t < s$, setting $a_s = q^s b_s$. Then the eigenvalues of $L_q(a_0, \dots, a_s)$ are $\varepsilon_k = q^s b_{s-k}$ with multiplicity m_k as in 6.2. In particular, if $b_t = p^{s-t}$, we get $\varepsilon_k = q^s p^k$.*

The point of this corollary, whose proof is a trivial verification, is to narrow the gap between the matrices considered here and those of the last paragraph. In fact, with $q = p^n$ and $s = m - \nu = m - (h + \mu) = \kappa - \mu$ as in the proof of 5.3, we shall presently see that

$$p^s(S_\chi|Q) = L_q(\pi_\nu(q), \dots, \pi_{\nu+s}(q)), \tag{9}$$

where $Q \subset P_\chi$ is an h-strict congruence class. This will lead, via 6.3, to the following conclusion.

PROPOSITION 6.4. *In the situation of 5.4, each submatrix $(S_\chi|Q)$ has eigenvalues $\sigma_k = q^m p^{k-s}$ with multiplicities m_k as in 6.2, for $0 \le k \le s = \kappa - \nu$.*

PROOF. A typical Q looks like the set of all $n + 1$-tuples of the form $[p^\mu, x_1, \dots, x_n]$ with $l(x_i) \le s$, and therefore has the right size. If we mimic 6.1 by writing $x \sim_t y$ to mean $v(x-y) \ge \nu + t$, the second part of 6.1 translates 5.3 into (9). On the other hand, if we construct the sequence a_t by the prescription of 6.3, setting $b_t = p^{s-t}$, we easily check that $a_t = q^{-\nu}\pi_{\nu+t}(q)$. By 6.3 and formula (9), the eigenvalues of $p^s(S_\chi|Q)$ are therefore $q^\nu \varepsilon_k$, whence the proposition.

Since $q = p^n$ with $n \geq 2$, all eigenvalues of S_χ are integral p-powers—provided that P was chosen fully transversal. This proves the Main Lemma.

7. Appendix: Finiteness A Priori. The following is a proof, in the abelian case, of a general theorem by Bass and Milnor about K_1 of integral group rings. Apart from avoiding stabilization, reduced norms, etc., the main simplification here consists in sidestepping induced characters (cf. Lemma 7.1). The rest is essentially copied from Bass's paper **[1]**. We shall completely abandon our previous notational conventions—they have used up too many letters.

1. *Rationality.* Let Q be a field. By an *algebra* over Q we shall always mean a Q-algebra A which is finite dimensional and separable. A set $\mathscr{S}$ of subalgebras is called a *saturated family* if, for each simple component A_i of A, the obvious map $B \to A \to A_i$ is surjective for suitable $B \in \mathscr{S}$. Example: $A = Q\pi$, where π is a finite abelian group whose order does not vanish in Q, and $\mathscr{S} = \{Q\pi'\}$ with π' running over all cyclic subgroups.

Now let R/Q be an arbitrary field extension. By an *R-character* of A we shall mean an element of $X_R(A) = R \otimes_{\mathbf{Z}} K_0(R \otimes A)$, where $K_0(R \otimes A)$ denotes the Grothendieck group of representations of the R-algebra $R \otimes A$. An R-character will be said to be *Q-rational* if it lies in the obvious subspace $R \otimes_Q X_Q(A)$. The following easy rationality criterion will turn out to be useful.

LEMMA 7.1. *An R-character ϕ of A is Q-rational if and only if this is so for all its restrictions $\phi|B$ to members B of a saturated family of subalgebras of A.*

PROOF. If $A = \prod_i A_i$ over Q, then $X_F(A) = \prod_i X_F(A_i)$ for both $F = Q$ and $F = R$. Hence rationality can be tested on the components ϕ_i of ϕ corresponding to this decomposition. But each ϕ_i also occurs somewhere among the analogous decompositions of the $\phi|B$.

2. *Units.* From now on we take Q and R to be the fields of rationals and reals, respectively.

First let $A = K$ be a field and D be its ring of integers. $R \otimes_Q K = \prod_j L_j$, where $L_j \simeq \mathbf{R}$ for $j = 1, \ldots, r_1$ and $L_j \simeq \mathbf{C}$ for $j = r_1 + 1, \ldots, r_1 + r_2$ in the standard notation. Let $V = \mathbf{R}^{r_1+r_2}$ and let λ: $\prod_j L_j^\times \to V$ be defined componentwise by $\lambda_j(x) = \log|x|$ for $j \leq r_1$ and by $\lambda_j(x) = \log|x|^2$ for the rest. Then Dirichlet's Unit Theorem says that λ maps the units of D onto a lattice of corank 1, trivializing only torsion units. We now present Bass's reformulation of this result.

First we note that, according to Dirichlet, λ yields an R-linear surjection $\operatorname{Hom}_R(V, R) \to \operatorname{Hom}_{\mathbf{Z}}(D^\times, R)$ with 1-dimensional kernel. Secondly, mapping the j-th irreducible representation of $R \otimes_Q K$ to the j-th coordinate function of V obviously produces an isomorphism $X_R(K) \to \operatorname{Hom}_R(V, R)$.

The composite of these two maps is a surjection $X_R(K) \to \mathrm{Hom}_{\mathbf{Z}}(D^\times, R)$ whose kernel, since it is 1-dimensional, must be $R \otimes_{\mathbf{Z}} X_Q(K)$. In fact, this composite maps a representation ψ of $R \otimes K$ to the homomorphism $u \mapsto \log|\det \psi(u)|$, which is zero for rational ψ, since such ψ is a multiple of the regular representation of K, and $\det \psi(u)$ is a unit in $\mathbf{Z}$.

LEMMA 7.2. *Let D be an order in a commutative algebra A over Q. Then*

$$(*) \qquad 0 \to R \otimes X_Q(A) \to X_R(A) \to \mathrm{Hom}_{\mathbf{Z}}(D^\times, R) \to 0$$

is exact.

PROOF. For $A = K$ and D as above, this is just a summary of the preceding arguments. For general A and maximal order D, it results from splitting everything according to the decomposition $A = \prod_i A_i$ into simple components. Finally, if D is allowed to be an arbitrary order, the term $\mathrm{Hom}_{\mathbf{Z}}(D^\times, R)$ does not notice the difference, since $D^\times$ is of finite index in the units of the maximal order.

THEOREM 7.3. *Let D be an order in a commutative Q-algebra A, and let $\mathscr{T}$ be a set of subrings of D such that $\{Q \otimes_{\mathbf{Z}} C | C \in \mathscr{T}\}$ form a saturated family of subalgebras of A. Then $\{C^\times | C \in \mathscr{T}\}$ generates a subgroup of finite index in $D^\times$.*

PROOF. We shall prove that $\mathrm{Hom}_{\mathbf{Z}}(D^\times, R) \to \prod_C \mathrm{Hom}_{\mathbf{Z}}(C^\times, R)$, as C ranges over $\mathscr{T}$, is injective. Indeed, let $h \in \mathrm{Hom}_{\mathbf{Z}}(D^\times, R)$ and $\phi \in X_R(A)$ be a pre-image of h in the middle term of (*). Then the vanishing of $h|C^\times$ in $\mathrm{Hom}_{\mathbf{Z}}(C^\times, R)$ means that $\phi | R \otimes_{\mathbf{Z}} C$ vanishes in $X_R(Q \otimes_{\mathbf{Z}} C)$ and therefore lies in $R \otimes_{\mathbf{Z}} X_Q(Q \otimes_{\mathbf{Z}} C)$, by Lemma 7.2. In other words, ϕ is Q-rational when restricted to every $R \otimes_{\mathbf{Z}} C$, hence is itself Q-rational, by Lemma 7.1; and thus gives to a trivial h. In short, $h|C^\times$ is trivial for all C only if it was trivial to begin with, Q.E.D.

REMARK. This theorem can be applied to $D = \mathbf{Z}\pi$ and $C = \mathbf{Z}\pi'$, where π is a finite abelian group, and π' runs over all cyclic subgroups. It implies that the units of the latter generate a subgroup of finite index in the units of the former. However, this quasi-surjectivity of lattices is deduced from the surjectivity of the Euclidean spaces they generate. The crucial intervention of transcendental tools (logarithms), which cannot be arithmetized in an obvious way (say, by p-adic analysis), destroys all arithmetical information about the index.

REFERENCES

1. H. Bass, *The Dirichlet unit theorem, Induced characters, and Whitehead groups of finite groups*, Topology **4** (1966), 391–410.

2. C. Curtis and I. Reiner, *Representation theory of finite groups and associative algebras*, J. Wiley, New York (1962).

3. H. Hasse, *Zahlentheorie*, Akademie Verlag, Berlin (1963).

4. K. Hoechsmann, S. K. Sehgal, and A. Weiss, *Cyclotomic units and the unit group of an elementary abelian group ring*, Arch. d. Math. **45** (1985), 5–7.

5. K. Hoechsmann and S. K. Sehgal, *Units in regular elementary abelian group rings*, Arch. d. Math. **47** (1986), 413–417.

6. ——, *Units in regular abelian p-group rings*, J. of Number Theory **30**, 375–381.

7. S. K. Sehgal, *Topics in group rings*, Marcel Dekker, New York (1978).

Department of Mathematics
University of British Columbia
Vancouver, B. C., Canada

Contemporary Mathematics
Volume **93** (1989)

A Remark on Algebraic Geometric Codes

G. L. KATSMAN AND M. A. TSFASMAN

In memory of S. D. Berman

ABSTRACT. In [**1**] V. D. Goppa mentioned that for algebraic geometric codes restricted to a subfield the parameters sometimes can be in fact better than the trivial estimate (just as for BCH-codes). We prove a respective theorem, and apply it to the study of polynomial asymptotic bounds. The bound established in [**2**] is ameliorated so that for $\delta \to 0$ the asymptotic behaviour of the obtained bound coincides (for $q = 2$) with that of the Gilbert-Varshamov bound.

BCH-codes can be obtained as a restriction to a subfield of Reed-Solomon codes (i.e. of algebraic geometric codes of genus 0). Their redundancy sometimes appears to be $(q-1)/q$ times better than the trivial estimate for a restriction of a code. The reason is that functions f and f^q have equal values on the elements of $\mathbf{F}_q$. Here is a natural generalization. A restriction to $\mathbf{F}_q$ of an algebraic geometric $\mathbf{F}_{q^m}$-code is called a BCH-code of arbitrary genus.

Let an algebraic geometric $[n, a-g+1, \geq n-a]_{q^m}$-code C_1 be constructed with the help of a curve X of genus g over $\mathbf{F}_{q^m}$, a subset $\mathscr{P} = \{P_1, \dots, P_n\} \subseteq X(\mathbf{F}_{q^m})$, and a positive divisor $D_1 \geq 0$ of degree $\deg D_1 = a$, $2g-2 < a < n$, such that $(\operatorname{Supp} D) \cap \mathscr{P} = \varnothing$. The dual code $C_0 = C_1^{\perp}$ can be also obtained from X, $\mathscr{P}$, and another divisor D_0, $\deg D_0 = n-a+2g-2$. Its parameters are

$$[n, n-a+g-1, \geq a-2g+2]_{q^m}.$$

Consider $C = C_0 \cap \mathbf{F}_q^n$. Its parameters are

$$[n, \geq n-m(a-g+1), \geq a-2g+2]_q.$$

Just as for BCH-codes this estimate can be ameliorated.

Suppose there exists a divisor $D = \sum a_i Q_i$ such that $0 \leq a_i \leq q-1$, and $qD \leq D_1$, $b = \deg D \leq a/q$. By Riemann-Roch $1(D) = \dim_{\mathbf{F}_{q^m}} L(D) \geq b-g+1$.

Let us choose $b-g+1$ independent vectors $f_0 = 1, f_1, \dots, f_{b-g} \in L(D)$. We claim that $1, f_1, \dots, f_{b-g}, f_1^q, \dots, f_{b-g}^q \in L(D_1)$ are also $\mathbf{F}_{q^m}$-independent.

In fact, suppose $x_0 + \sum x_i f_i + \sum y_i f_i^q = 0$. The order of any pole of $\sum y_i f_i^q$ is divided by q, and the order of any pole of $x_0 + \sum x_i f_i \in L(D)$ cannot be greater than $q-1$, hence $\sum y_i f_i^q = y_0 = \text{const}$. There exist z_i such that $z_i^q = y_i$, thus $(\sum z_i f_i)^q = \sum y_i f_i^q = z_0^q$, i.e. $z_0 = \sum z_i f_i$ which contradicts the independence of $1, f_1, \ldots, f_{b-g} \in L(D)$.

Extend the set $1, f_1, \ldots, f_{b-g}, f_1^q, \ldots, f_{b-g}^q$ to a $\mathbf{F}_q$-basis of $L(D_1)$. The images of the elements of this basic under the values-in-P_i map are the $(a-g+1)$ rows of the parity check matrix of the code C_0. For $v \in \mathbf{F}_q^n$ the condition $(v, f_i) = 0$ is equivalent to the condition $(v, f_i^q) = 0$. Therefore the number of independent checks is at least by $(b-g)$ less, i.e. $\operatorname{codim}_{\mathbf{F}_q} C \le m(a-b+1)$.

Let D be as above. Setting $D_1 = qD$ we get

THEOREM 1. *Let X be a smooth curve of genus g over $\mathbf{F}_{q^m}$, let $\mathscr{P} = \{P_1, \ldots, P_n\} \subseteq X(\mathbf{F}_{q^m})$, $D = \sum a_i Q_i$, $0 \le a_i \le q-1$, $(\operatorname{Supp} D) \cap \mathscr{P} = \varnothing$, $\frac{2g-2}{q} < \deg D < \frac{n}{q}$. Then there exists a $[n, \ge n - m((q-1)\cdot \deg D + 1), \ge q \cdot \deg D - 2g + 2]_q$-code.*

Now let us consider a family of curves, e.g. modular, over $\mathbf{F}_{q^m}$ (if q is not a square, let m be even) such that the ratio of the number of $\mathbf{F}_{q^m}$-points to the genus is asymptotically (for $g \to \infty$) the best possible: $N = X(\mathbf{F}_{q^m}) \sim (q^{m/2} - 1)g$. Let $\gamma = (q^{m/2} - 1)^{-1}$.

The number N_e of the $\mathbf{F}_{q^{me}}$-points on X satisfies

$$q^{me} + 1 + 2gq^{me/2} \ge N_e \ge q^{me} + 1 - 2gq^{me/2}.$$

By brute force one checks that starting from some e the number N_e increases monotonously, and in particular for $b > g$ there always exists a point defined exactly over $\mathbf{F}_{q^{mb}}$. Such a point being fixed let $Q_1, \ldots, Q_b$ be its conjugates over $\mathbf{F}_{q^m}$. Let $D = \sum Q_i$ and $\mathscr{P} = X(F_{q^m})$, of course $(\operatorname{Supp} D) \cap \mathscr{P} = \varnothing$.

From Theorem 1 it follows that for $\beta = b/n > \gamma$ there exist asymptotic families of codes with parameters

$$R \sim 1 - m(q-1)\beta, \qquad \delta \sim q\beta - 2\gamma,$$

i.e. for

$$\frac{q-2}{q^{m/2}-1} \le \delta \le \frac{q}{m(q-1)} - \frac{2}{q^{m/2}-1}$$

(these inequalities correspond to $b \ge g$ and $R \ge 0$) we get a segment of the line

$$R = 1 - \frac{2m(q-1)}{q(q^{m/2}-1)} - \frac{(q-1)m}{q}\delta.$$

Note that for $m = 1, 2$ the conditions on δ are contradictory, and that for any q and $m \ge 6$ the segment is nonempty. The construction of C from C_1 is clearly polynomial. In [3] the polynomiality was proved for algebraic geometric codes on modular curves with some special choice of the divisor. S. C. Vlăduţ informed us that mutatis mutandis this proof is also valid for our choice of D. We get

THEOREM 2. *For any m (such that q^m is a square) the segment*

$$R_m(\delta) = 1 - \frac{2m(q-1)}{q(q^{m/2}-1)} - \frac{(q-1)m}{q}\delta,$$
$$\frac{q-2}{q^{m/2}-1} \le \delta \le \frac{q}{(q-1)m} - \frac{2}{q^{m/2}-1}$$

is a lower asymptotic bound for linear q-ary codes with polynomial construction.

In terms of [2], where asymptotic problems are accurately stated, it means

$$\alpha_q^{\mathrm{pol,lin}}(\delta) \ge R_{**}(\delta) = \max_m R_m(\delta)$$

(here m is even, if q is not a square). Let us recall that the bound $R_*(\delta)$ obtained in [2] behaves as $R_*(\delta) \sim 1 + 2\delta\log_q\delta$ when $\delta \to 0$, and the Gilbert-Varshamov bound behaves as $R_{GV}(\delta) \sim 1 + \delta\log_q\delta$.

PROPOSITION 3. *The bound $R_{**}(\delta)$ for $\delta \to 0$ is asymptotically*

$$R_{**}(\delta) \sim 1 + \frac{2(q-1)}{q}\delta\log_q\delta.$$

For $q = 2$ the bounds R_{**} and R_{GV} are asymptotically equal (for $\delta \to 0$).

Sketch of proof. We follow the same lines as in [2].

Find the nodes of the polygon R_{**}. Direct calculation shows that for $m \to \infty$ their coordinates are

$$\delta_m \sim \frac{m(q-1)}{q}q^{-m/2}, \qquad 1 - R_m \sim \frac{m^2(q-1)^2}{q^2}q^{-m/2} \sim$$
$$m\frac{q-1}{q}\delta_m \sim (-2\log_q\delta_m)\frac{q-1}{q}\delta_m,$$

and for m sufficiently large δ_m satisfies the inequalities of Theorem 2.

Acknowledgment. We are deeply grateful to S. G. Vlăduţ for his valuable comment.

REFERENCES

1. V. D. Goppa, *Algebraico-geometric codes*, Izv. Akad. Nauk, SSSR, 46:4 (1982), 762–781 (=Math. USSR Izvestija **21** (1983), 75–91).

2. S. G. Vlăduţ, G. L. Katsman and M. A. Tsfasman, *Modular curves and codes with polynomial construction complexity*, Probl. Peredachi Info., 20:1 (1984), 47–55 (=Probl. Info. Trans., **20** (1984), 35–42).

3. Yu. I. Manin and S. G. Vlăduţ, *Linear codes and modular curves*, Sovrem. Probl. Math., VINITI **25** (1984), 209–257 (=J. Soviet Math., **30** (1985), 2611–2643).

Contemporary Mathematics
Volume **93** (1989)

On Decomposition of a Binomial in a Skew Polynomial Ring

E. SH. KERER

Dedicated to the memory of S. D. Berman

A skew polynomial ring $R = K[x, \sigma]$ over a field K with an automorphism $\sigma \colon K \to K$ is an associative ring, consisting of polynomials over the field K in a variable x, where a multiplication is defined by $ax = x\sigma(a)$ $(a \in K)$.

A ring R is noncommutative ring of principle left and right ideals. An element $A \in R$ is called an irreducible one when it is not divisor of unity and has not nontrivial noninvertible divisors. The elements $A \in R$ and $A_1 \in R$ are called similar when for some $B, B_1 \in R$, $AB = BA_1$ and elements A and B have no common noninvertible left divisors, and A_1 and B_1 have no common noninvertible right divisors. It is known that noninvertible element $A \in R$ could be factored into a product of irreducible factors, determined uniquely up to invertible factors, the order of consiquence and similarity. (See **[1]**–**[3]**).

A significant example of skew polynomial ring is the one $S = K[x, \sigma_\lambda]$, where $K = C(y)$ is the field of rational functions of a variable y over complex field C and automorphism σ_λ is defined by a formula $\sigma_\lambda(\gamma y) = \lambda^{-1}\gamma y$ where λ is a fixed element of the field C. This ring S appears in a natural way on considering a group of algebras of nilpotent torsion free groups. To descript irreducible representations of such groups one should consider maximal one-sided ideals of group rings, and that is connected with considering maximal one-sided ideals of the ring S. The mentioned ideals are generated by irreducible elements. The most interest to the theory of representations is a case, when λ is not a root of unity. (This case corresponds to nonmonomial irreducible representations of nilpotent groups.) (See **[4]**, **[5]**).

In this study the decomposition of a binomial $x^n - f \in S = K[x, \sigma_\lambda]$ into product of irreducible factors, where λ is not a root of unity, is investigated. All irreducible divisors of this binomial are shown not to be similar in pairs, to have equal degrees and to be binomials. Moreover, the way to find those irreducible divisors effectively is shown.

For a rational function $g \in C(y)$ let us denote $g^{(n)} = x^n y x^{-n}$. It's easy to show that $(g_1 g_2)^{(n)} = g_1^{(n)} g_2^{(n)}$, $(g_1/g_2)^{(n)} = g_1^{(n)}/g_2^{(n)}$. If $g = (y-\alpha_1)(y-\alpha_2)\dots(y-\alpha_m) \in C[y]$ then $g^{(n)} = \lambda^{mn}(y-\alpha_1\lambda^{-n})\dots(y-\alpha_m\lambda^{-n})$.

THEOREM 1. *A binomial $F = X^n - f \in S$ has a binomial $H = x^k - h$ among its right divisors if and only if k divides n and h satisfies an equation*

$$hh^{(k)}h^{(2k)}\dots h^{(n-k)} = f. \tag{1}$$

The equation (1) defines h uniquely up to a factor $\varepsilon = \sqrt[l]{1}$ where $l = n/k$.

PROOF. It is easily seen that

$$x^n - f = (x^{n-k} + h^{(n-k)}x^{n-2k} + \dots + \\ + h^{(n-(l-2)^k)}\dots h^{(n-k)}x^{n-(l-1)k})(x^k - h) + P,$$

where $P = h^{(n-(l-1)^k)}\dots h^{(n-k)}x^{n-lk} - f$. Obviously, $P = 0$ if and only if $n = lk$ and the equality (1) is true.

Assume that

$$hh^{(k)}\dots h^{(n-k)} = tt^{(k)}\dots t^{(n-k)} = f.$$

Let $h = h_0/h_1$, $t = t_0/t_1$, where $h_0, h_1, t_0, t_1 \in C[y]$ and let $u = h_0 t_1$, $v = t_0, h_1$. Then

$$uu^{(k)}\dots u^{(n-k)} = vv^{(k)}\dots v^{(n-k)}. \tag{2}$$

Let us show that $u = \varepsilon v$, where $\varepsilon^l = 1$. If $d = \deg u = \deg v = 0$ then, obviously, $u = \varepsilon v$. Assume this equality has been proved for every degree d less than d_1 and let us prove it for $d = d_1$. Assume α is a root of the polynomial u that $\lambda^m\alpha$ $(m > 0)$ is not a root of u and hence it could not be a root of the polynomial $uu^{(k)}\dots u^{(n-k)}$. Hence α could not be a root of the polynomial $v^{(k)}\dots v^{(n-k)}$ and, consequently α is a root of the polynomial v. Dividing both parts of equality (2) on $(y-\alpha)\dots(y-\alpha)^{n-k}$ and taking in account the induction hypothesis we receive $u = \varepsilon v$ or $h_0 t_1 = \varepsilon t_1 h_1$, from/where $h = \varepsilon t$. That proves the theorem.

REMARK. From Theorem 1 it follows that if k is a minimal number for which the equation (1) is resolved and rational function h satisfies the equation (1) with this k, then the binomial $x^k - h$ has no binomial among is own divisors.

THEOREM 2. *Let $f = f_0/f_1$ where $f_0, f_1 \in C[y]$. The equation* (1) *has solution if and only if the set of all roots of the polynomial $f_1 f_0^{(k)} f_0^{(2k)}\dots f_0^{(n-k)}$ break down into disjoint subsets of power n/k, in each of them all roots are either zeroes, or congruent modulo λ^k and non-congruent in pairs modulo λ^n in multiplicative group of the field C.*

PROOF. Assume the equation (1) has solution $h = h_0/h_1$ where $h_0, h_1 \in C[y]$. Then the equality (1) has the form

$$h_0 h_0^{(k)}\dots h_0^{(n-k)} f_1 = h_1 h_1^{(k)}\dots h_1^{(n-k)} f_0. \tag{1'}$$

Let us multiply the both parts of equality $(1')$ by $f_0^{(k)}\dots f_0^{(n-k)}$. Obviously the roots of the right part of the result and of polynomial $h_0h_0^{(k)}\dots h_0^{(n-k)}$ satisfy the theorem condition. Hence the roots of polynomial $t = f_1f_0^{(k)}\dots f_0^{(n-k)}$ satisfy this condition.

Now assume that the roots of $t = f_1f_0^{(k)}\dots f_0^{(n-k)}$ satisfy the Theorem 2 condition, p is a number of subsets, and the roots of ith subset are α, $\alpha\lambda^{k+m_1n}$, $\alpha\lambda^{2k+m_2n},\dots,\alpha\lambda^{(l-1)k+m_{l-1}n}$ $(m_1,\dots,m_{l-1}>0)$. Set for $i=1,\dots,p$

$$\begin{aligned} v_i &= (y-\alpha\lambda^n)(y-\alpha\lambda^{2n})\dots(y-\alpha\lambda^{m_1n})\\ &\quad\times(y-\alpha\lambda^{k+n})(y-\alpha\lambda^{k_2n})\dots(y-\alpha\lambda^{k+m_2n)}\dots\\ &\quad\times(y-\alpha\lambda^{(l-2)k+n})(y-\alpha\lambda^{(l-2)k+2n})\dots(y-\alpha\lambda^{(l-2)k+m_{l-1}n}),\\ u_i &= \lambda^{-k-\dots-(n-k)}(y-\alpha\lambda^{n-k})\\ &\quad\times(y-\alpha\lambda^{k+n})(y-\alpha\lambda^{k+2n})\dots(y-\alpha\lambda^{k+m_1n})\dots\\ &\quad\times(y-\alpha\lambda^{(l-1)k+n})(y-\alpha\lambda^{(l-1)k+2n})\dots(y-\alpha\lambda^{(l-1)k+m_{l-1}n}),\\ t_i &= (y-\alpha)(y-\alpha\lambda^{k+m_1n})\dots(y-\alpha\lambda^{(l-1)k+m_{l-1}n}) \end{aligned}$$

($t=\gamma\prod_{i=1}^p t_i$, where γ is a leading coefficient of polynomial t).

It is easy to verify that

$$v_iv_i^{(k)}\dots v_i^{(n-k)}t_i = u_iu_i^{(k)}\dots u_i^{(n-k)}. \tag{3}$$

Let us multiply equations (3) for all $i=1,\dots,p$. Assuming $h_0 = f_0\prod_{i=1}^p v_i$, $h_1=\beta\prod_{i=1}^p u_i$ $(\beta\in C)$ and having chosen properly the coefficient β, we have equality $(1')$. The theorem is proved.

REMARK. In Theorem 2 the polynomial $f_1f_0^{(k)}f_0^{(2k)}\dots f_0^{(n-k)}$ may be changed to the polynomial $f_0f_1^{(k)}f_1^{(2k)}\dots f_1^{(n-k)}$.

THEOREM 3. *A polynomial $F = X^n - f\in S$ is irreducible if and only if the equation* (1) *has no solution for all k-divisors n.*

To prove this we need the following lemmas:

LEMMA 1. *Assume the complex polynomial $\varphi = x^k+\alpha_1x^{k-1}+\dots+\alpha_k\in C[x]$ divides a binomial $x^n-\beta^n\in C[x]$ and let $\alpha_1=\dots=\alpha_{m-1}=0$, $\alpha_m\neq 0$. Then m divides n.*

PROOF. Without loss of generality we can set $\beta=1$, because this case may be received from general one by replace $x_1\beta x$.

Let $\xi_1,\dots,\xi_k$ ($\xi_j=\xi^{i_j}$ where ξ is a primitive root of 1 of degree n)-are the roots of polynomial φ and $s_i=\xi_1^i+\dots+\xi_k^i$. Accordingly with Newton formula for $i\le n$ $s_i-s_{i-1}\alpha_1+s_{i-2}\alpha_2-\dots+(-1)^{i-1}s_1\alpha_{i-1}+(-1)^ik\alpha_i=0$. Hence $s_1=\dots=s_{m-1}=0$, $s_m\neq 0$.

Assume m is not divisor of n and $n=p_1^{n_1}\dots p_l^{n_l}$, $m=p_1^{m_1}\dots p_i^{m_i}\dots p_l^{m_l}d$, where $p_1,\dots,p_l$ prime numbers, $m_j\le n_j$ for $j<i$, $m_j>n_j$ for $i\le j\le l$ and $(d,n)=1$. Set $u_1=p_i^{m_i-n_i}\dots p_l^{m_l-n_l}d$, $u_2=p_1^{\tau_1}\dots p_{i-1}^{\tau_{i-1}}$, where $\tau_j>n_j-m_j$, $u=u_1+u_2$ and $t=p_1^{m_1}\dots p_{i-1}^{m_{i-1}}p_i^{n_i}\dots p_l^{n_l}$. Obviously $tu\equiv m \pmod n$.

Since every prime divisor of n either divides u_1 and does not divide u_2, or divides u_2 and does not divide u_1 then $(u, n) = 1$, and a map $\xi \to \xi^u$ defines automorphism of a field $Q(\xi)$. (Q is a rational numbers field.) This automorphism transforms s_t into s_m. But $s_t = 0$, from where $s_m = 0$, which is impossible. Lemma is proved.

LEMMA 2. *Assume a polynomial* $H = x^k + h_1x^{k-1} + \cdots + h_k \in s$ *is a right divisor of a binomial* $F = x^n - f \in s$. *Then*

$$(4)\qquad \left.\begin{array}{c} h_1^{(n)} f^{(k-1)} = h_1 f^{(k)} \\ \cdots\cdots\cdots\cdots\cdots\cdots \\ h_i^{(n)} f^{(k-i)} = h_i f^{(k)} \\ \cdots\cdots\cdots\cdots\cdots\cdots \\ h_k^{(n)} f = h_k f^{(k)} \end{array}\right\}$$

PROOF. Assume $h_{m_1}, h_{m_2}, \ldots, h_k$ are all non-zero coefficients of polynomial H. The following sequence of implications takes place: H divides $F \Rightarrow H$ divides $T_1 = x^{n-k}H - F = h_{m_1}^{(n-k)}x^{n-m_1} + h_{m_2}^{(n-k)}x^{n-m_2} + \cdots + h_k^{(n-k)}x^{n-k} + f \Rightarrow H$ divides $T_2 = x^{m_1}T_1 - h_{m_1}^{(n-k+m_1)}F = h_{m_2}^{(n-k+m_1)}x^{n-m_2+m_1} + h_{m_3}^{(h-k+m_1)}x^{n-m_3+m_1} + \cdots + h_k^{(n-k+m_1)}f^{(m_1)}x^{m_1} + h_{m_1}^{(n-k+m_1)}f \Rightarrow \cdots \Rightarrow H$ divides $f^{(k)}x^k + h_{m_1}^{(n)}f^{(k-m_1)}x^{k-m_1} + \cdots + h_k^{(n)}f$. Hence the system of equalities (4) follows. Lemma is proved.

For $f = f_1/f_2 \in C(y)$, where $f_1, f_2 \in C[y]$, $(f_1, f_2) = 1$, define $\deg f = \deg f_1 - \deg f_2$, where $\deg f_1$, $\deg f_2$ are minimal degrees of terms appearing in polynomials $f_1, \ldots, f_2$ respectively.

LEMMA 3. *Let H and F be as in Lemma 2. Then* $\deg h_i = i \cdot \deg f/n$.

PROOF. Having cleared the denominators in the ith equality of the system (4) and having compared the coefficients for minimal degrees of the variable y in the both parts we obtain

$$\lambda^{n \deg h_i + (k-i)\deg f} = \lambda^{k \deg f}.$$

Since λ is not a root of unity, it follows that $n \deg h_i = i \deg f$. Lemma is proved.

REMARK. Evidently Lemma 3 is correct also for left divisors of binomials in the ring S.

If H and F are as in Lemma 2 then $F = (x^{n-k} + t_1x^{n-k-1} + \cdots + t_{n-k})H$ and the following system of equations takes place:

$$(5)\qquad \left.\begin{array}{l} h_1^{(n-k)} + t_1 = 0 \\ h_2^{(n-k)} + t_1h_1^{(n-k-1)} + t_2 = 0 \\ \cdots\cdots\cdots\cdots\cdots\cdots\cdots\cdots\cdots\cdots\cdots\cdots \\ \qquad t_{n-k-2}h_k^{(2)} + t_{n-k-1}h_{k-1}^{(1)} + t_{n-k}h_{k-2} = 0 \\ \qquad\qquad t_{n-k-1}h_k^{(1)} + t_{n-k}h_{k-1} = 0 \\ f - t_{n-k}h_k = 0 \end{array}\right\}$$

LEMMA 4. *Let H and F be as in Lemma 2 and assume* $h_1 = \cdots = h_{m-1} = 0$, $h_m \neq 0$. *Then m divides n.*

PROOF. Without loss of generality we may suppose that $0 \leq \deg f < n$ (the general case is reduced to this by substitution $x_1 = xy^l$ where $l = [\deg f/n]$).

For $n = 1, 2$ lemma is evidently true. Assume that the lemma statement is true for all $n < n_0$ for any $\lambda(\lambda^\tau \neq 1)$ and prove it for $n = n_0$.

1. Let $0 < \deg f < n$. Then $(\deg f, n) = n/d$, where $d \neq 1$. From Lemma 3 it follows that if i is not multiply to d then $h_i = 0$. Hence if we set $x_1 = x^d$ and use the induction hypothesis, the lemma would be proved.

2. Let $\deg f = 0$. From the Lemma 3 and its remark it follows that $h_i \neq 0$ implies $\deg h_i = 0$ and $t_i \neq 0$ implies $\deg t_i = 0$. We may assume that denominators of rational function f and all non-zero h_i, t_i have free terms equal to 1. Let α_i, β_i, γ be free terms of numerators of h_i, t_i, f respectively. From (5) it follows that

$$\begin{aligned}
&\alpha_1 + \beta_1 = 0\\
&\alpha_2 + \beta_1\alpha_1 + \beta_2 = 0\\
&\cdots\cdots\cdots\cdots\cdots\cdots\cdots\cdots\cdots\cdots\\
&\quad\beta_{n-k-2}\alpha_k + \beta_{n-k-1}\alpha_{k-1} + \beta_{n-k}\alpha_{k-2} = 0\\
&\qquad\qquad\beta_{n-k-1}\alpha_k + \beta_{n-k}\alpha_{k-1} = 0\\
&\gamma - \beta_{n-k}\alpha_k = 0
\end{aligned}$$

from where $x^n - \gamma = (x^{n-k} + \beta_1 x^{n-k-1} + \cdots + \beta_{n-k})(x^k + \alpha_1 x^{k-1} + \cdots + \alpha_k)$. Since $\alpha_1 = \cdots = \alpha_{m-1} = 0$, $\alpha_m \neq 0$, then accordingly to Lemma 1 it follows that m divides n. Lemma is proved.

LEMMA 5. *Let H and F be as Lemma 2. Assume the coefficients* $h_{i_1}, \ldots, h_{i_p}$, $h_{j_1}, \ldots, h_{j_q}$ *are non-zeroes and* $i_1 + \cdots + i_p = j_1 + \cdots + j_q = i$. *Then the following equation is true*:

$$h_{i_1}^{(m_1)} \ldots h_{i_p}^{(m_p)} = \varepsilon h_{j_1}^{(l_1)} \ldots h_{j_q}^{(l_q)} \tag{6}$$

where $m_1 = l_1 = n-k$, $m_\tau = n-k-(i_1+\cdots+i_{\tau-1})$, $l_\tau = n-k-(j_1+\cdots+j_{\tau-1})$, $\varepsilon \in C$.

PROOF. Let

$$\begin{aligned}
h_{i_1}^{(m_1)} \ldots h_{i_p}^{(m_p)} &= u = u_1/u_2, h_{j_1}^{(l_1)} \ldots h_{j_q}^{(l_q)}\\
&= v = v_1/v_2 (u_1, u_2, v_1, v_2 \in C[y]).
\end{aligned}$$

From Lemma 2

$$h_i^{(n)}/h_i = f^{(k)}/f^{(k-i)} = t^{(k-1)}t^{(k-2)} \ldots t^{(k-i)},$$

where $t = f^{(1)}/f$. Hence

$$u^{(n)}/u + t^{(n-1)} \ldots t^{(n-i_1)}t^{(n-i_1-1)} \ldots t^{(n-i_1-i_2)} \ldots t^{(n-i)},$$

and similarly $v^{(n)}/v = t^{(n-1)} \ldots t^{(n-i)}$. Consequently $u^{(n)}/u = v^{(n)}/v$ or

$$(u_1v_2)^{(n)}(v_1u_2) = (u_1v_2)(v_1u_2)^{(n)}. \tag{7}$$

We assume equality $u_1v_2 = \varepsilon v_1u_2$ ($\varepsilon \in C$) has been proved for a case $\deg u_1v_2 < d$ (for a case $\deg u_1v_2 = 0$ it is true) and we shall prove it for $\deg u_1v_2 = d$.

Let α be such root of the polynomial u_1v_2 that $\lambda^n\alpha$ is not a root of u_1v_2 and hence α is not a root of the polynomial $(u_1v_2)^{(n)}$. Then α is a root of the polynomial v_1u_2. Dividing the both parts of the equality (7) on $(y-\alpha)(y-\alpha)^n$ and using the induction hypothesis we receive $u_1v_2 = \varepsilon v_1u_2$ or $u = \varepsilon v$. Lemma is proved.

PROOF OF THEOREM 3. The necessity of Theorem 3 condition is follows from Theorem 1. We shall prove that it is sufficient.

Let us denote the rational function (6) as $\gamma_i w_i$, where $\gamma_i \in C$, and leading coefficients in the numerator and denominator of w_i are equal 1. Let a polynomial $H = x^k + h_1x^{k-1} + \cdots + h_k$ be a right divisor of the binomial $F = x^n - f$. From Lemma 5 it follows that $h_i \neq 0$ implies $h_i^{(n-k)} = \gamma_i w_i$. Let $h_1 = \cdots = h_{m-1} = 0$, $h_m \neq 0$. Then from the system (5) it follows that $t_1 = \cdots = t_{m-1} = 0$,

$$t_m = h_m^{(n-k)} = \varepsilon_m W_m \; (\varepsilon_m = \gamma_m).$$

Assume, that for all $i < i_0$ the implication $t_i \neq 0 \Rightarrow t_i = \varepsilon_i W_i$ had been proved. Let us prove it for $i = i_0$.

For $t_i = 0$ it is evidently true. Let $t_i \neq 0$. In corresponding with the induction hypothesis every item of the left part of the ith equality of system (5), which has the form $t_jh_{i-j}^{(n-k-j)}$, is equal $\varepsilon_j W_j h_{i-j}^{(n-k-j)}$. From Lemma 5 it follows that it equal $\delta h_{j_1}^{(n-k)}h_{j_2}^{(n-k-j_1)}\ldots h_{j_p}^{(n-k(j-j_p))}h_{i-j}^{(n-k-j)} = \gamma W_i$ ($\delta, \gamma \in C, j_1 + \cdots + j_p = j$). From (5) we receive $t_i = \varepsilon_i W_i$. Consequently $t_{n-k}h_k = \varepsilon_{n-k}W_{n-k}h_k = \gamma W_n$ ($\gamma \in C$) and from Lemma 4 m divides n, and hence $t_{n-k}h_k = \mu h_m^{(n-k)}\ldots h_m^{(n-k-(l-1)m)}$, where $\mu \in C$, $l = m/n$. By substitution this expression into the last equality of the system (5) we receive $f = gg^{(m)}\ldots g^{(n-m)}$, where $g = \sqrt[l]{\mu}h_m^{(i)}$. It means, that the equation (1) is resolved for $k = m$. Theorem 3 is proved.

PROPOSITION. *Binomials $H = x^{k_1} - h \in s$ and $T = x^{k_2} - t \in s$ are similar, if and only if $k_1 = k_2 = k$ and such n and such rational function u exist, that*

$$h^{(n)}u^{(k)} = tu. \tag{8}$$

PROOF. Let H and T be similar. Then (see [2]) $k_1 = k_2 = k$ and we may choose such $U, V \in S$ that $\deg U = \deg V < k$ and $UH = TV$. Let n be a such number, that the coefficient u at x^n in the polynomial U is differ from zero. Then the coefficient v at x^n in the polynomial V is also differ from zero and $ux^n(x^k - h) = (x^k - t)vx^n$. Hence $u = v^{(k)}$, $uh^{(n)} = tv$, and the equality (8) is true.

Suppose that $k_1 = k_2 = k$ and for any $u \in C(y)$ and n the equality (8) is true. Then $ux^nH = Tvx^n$, where $V^{(k)} = u$, and hence the elements H and T are similar.

COROLLARY. *All binomial's right divisors of any binomial $F \in S$ are nonsimilar pairwise.*

Let us assume that $H = x^k - h$ and $T = x^k - t$ are right divisors of F, and H and T are similar. Then from Theorem 1 $t = \xi h$, where ξ is the root of unity. Comparing leading coefficients of left and right parts of (8), we obtain $\lambda^{n \cdot \deg h + k \cdot \deg u} = \xi$, that is not possible, because ξ is not a root from 1.

THEOREM 4. *All right irreducible divisors of the binomial $F = x^n - f \in S$ are binomials.*

PROOF. Assume k is the minimal natural number, for which the equation (1) is resolved, and let h be its solution for this k. From Theorem 1 $H_i = x^k - \xi^i h$ for $i = 1, \dots, l = n/k$ are right divisors of F, where ξ is a primitive root of 1. By Theorem 2 polynomials $H_1, \dots, H_l$ are irreducible. Polynomials $H_1, \dots, H_l$ are nonsimilar in paris, and the sum of their degrees is equal to degree of the polynomial F. Hence, if $F = A_1 \dots A_{l_1}$ is decomposition of the polynomial F into irreducible factors, then $l_1 = l$ and after the corresponding renumbering of polynomials $H_1, \dots, H_l$, polynomials H_i and A_i become similar.

It is well known ([**1**]) the following fact: if in the skew polynomial ring a polynomial C is a right divisor of a polynomial $A \cdot B$ and is not divisor of a polynomial B, then the polynomial A has similar to C polynomial among its right divisors. Hence follows that, if polynomial $A_l \neq H_l$, then polynomial $F_1 = A_1 \dots A_{l-1}$ has among its right divisors a polynomial similar to H_l. It is a contradiction, because in the decomposition of polynomial F_1 into irreducible factors are no factors similar to H_l. Consequently $H_1, \dots, H_l$ are all right irreducible divisors of polynomial F. The theorem is proved.

This research has been carried in the 1976 year under supervision of S. D. Berman.

REFERENCES

1. Ore, *On non-commutative polynomials*, Ann. of Math., 1933, v. 34, 480–508.
2. N. Jacobson, *Theory of rings*, 1943, Math. Surveys No. 11.
3. P. M. Cohn, *Free rings and their relations*, Academic Press, London, New York, 1971.
4. S. D. Berman and E. Sh. Kerer, *On representations of the free tortion nilpotent group of the class two with two generalisators. Functional analysis and its application*, 1977, v. 11, No. 4, 70–71. (in Russian).
5. E. Sh. Kerer, *On ideals of group algebra of a torsion free nilpotent groups of class two*, Dep. to "ВИНИТИ", No. 1523-79. (in Russian).

Contemporary Mathematics
Volume **93** (1989)

The Associated Graded Algebra of a Group Algebra

L. KROP

Dedicated to the memory of S. D. Berman

§0. Introduction

We shall be concerned here with the algebras that admit action of the full semigroup (group) of endomorphisms of some vector space by the algebra endomorphisms. One source of such algebras is various quotients of free (or tensor) algebra by ideals invariant with respect to the linear substitutions of the free generators. The process is similar to that of forming a relatively free P. I. Algebra ([**OS**]) or a relatively free group ([**H.N.**]). Before pursuing this subject any further we fix some notation.

In what follows $X = \{x_1, x_2, \dots\}$ will be a countable set and K will be a field of characteristic $p \geq 0$. For every subring with $1L$ of K we designate LX for the free L-module generated by X. Similarly if Z is a subset of a K-space V, then LZ or $\langle Z \rangle_L$ denotes the L-envelope of Z. The semigroup of L endomorphisms of LX will be denoted by $E(LX)$ or simply $E(L)$ if X is kept fixed in the discussion.

The algebras considered in this paper fall within the following class introduced by M. Schooneveldt ([**SCH**]). Let us say that K-algebra Φ is L-weakly free for a subring L of K if Φ has a generating set X such that every function $X \to LX$ extends to an algebra endomorphism of Φ. Below we shall omit L if it coincides with the ground field K. Furthermore, let $\mathcal{F}(X)$ be the tensor algebra over KX. An element $f(x_1, \dots, x_n)$ of $\mathcal{F}(X)$ is said to be an L-weak identity of an algebra A with a distinguished set of generators Y if $f(u_1, \dots, u_n) = 0$ for all substitutions of $x_1, \dots, x_n$ by $u_1, \dots, u_n$ from LY. Of course, every L-weakly free algebra $\Phi(X)$ is an $E(LX)$-algebra meaning that $E(LX)$ acts on $\Phi(X)$ by the algebra endomorphisms.

Throughout the paper we adopt notation Z_p for the smallest subring of K with 1, i.e., Z_p is the prime subfield of K if $p > 0$, and $Z_0 = Z$, the ring of rational integers. A class of Z_p-weakly free algebras arises naturally from the group algebras of relatively free groups as follows. Let $F = F(X)$

0271-4132/89 $1.00 + $.25 per page

be a relatively free group on a set of generators X ([**H.N.**]). Let KF be the corresponding group algebra and $I = I(KF)$ be the augmentation ideal of KF. Let

$$\operatorname{gr}(KF) = \bigoplus_{n\geq 0} I^n/I^{n+1}$$

denote the associated graded algebra of KF for the I-adic filtration. We refer to $\operatorname{gr}_n(KF) := I^n/I^{n+1}$ as the nth homogeneous component of $\operatorname{gr}(KF)$. We denote by $\overline{X} \subset KF$ the set of all $\bar{x}_i = x_i - 1$, $x_i \in X$ and moreover abusing notations we also use $\bar{x}_i$ for the image of $x_i - 1$ in I/I^2, that is $(x_i - 1) + I^2$. Then it can easily be seen that $\operatorname{gr}(KF)$ is Z_p-weakly free on $\overline{X}$. Consequently, $\operatorname{gr}_n(KF)$ becomes $E(Z_pX)$-module for every $n \geq 1$. It is our goal in this paper to study $E(Z_pX)$-submodules of $\operatorname{gr}_n(KF)$ and more generally $\operatorname{gr}(KF)$. In one instance, viz. F is a free nilpotent group of class 2 and $\operatorname{char} K = 0$, our results are complete.

Let us give a brief description of the paper. Besides introduction, it falls into three parts. For at least two of them the connection between $\operatorname{gr}(KF)$ and the dimension subgroups of F over K is essential.

Let K be a field and $p = \operatorname{char} K \geq 0$. We denote by $D_{m,p}(F)$ the mth dimension subgroup of F, i.e., $D_{m,p}(F) = (1 + I^m(KF)) \cap F$. We form further the graded Lie algebra ([**Hart 1**]).

$$D_p(F) = \bigoplus_{m\geq 1} D_{m,p}(F)/D_{m+1,p}(F) = \bigoplus_{m\geq 1} D_p^m(F)$$

where $D_p^m(F) = D_{m,p}(F)/D_{m+1,p}(F)$. A link between $\operatorname{gr}(KF)$ and $D_p(F)$ is provided by the fact established by D. Quillen ([**Q**]) that the universal envelope of $K \otimes D_0(F)$ if $p = 0$ or the restricted universal envelope of $K \otimes D_p(F)$ if $p > 0$, suitably graded, are isomorphic with $\operatorname{gr}(KF)$.

In part one we establish necessary and sufficient conditions for $\operatorname{gr}(KF)$ to be weakly free for every extension field K of Z_p.

Part two contains the construction of a $E(Z_pX)$-filtration in $\operatorname{gr}_n(KF)$, called canonical, whose quotients are the tensor products of the plethysms $S_p^{\lambda_i}(K \otimes D_{i,p}(F))$, namely, $\bigotimes_{i=1}^n S_p^{\lambda_i}(K \otimes D_{i,p})$, where $\sum i\lambda_i = n$ and $S_0^{\lambda_i}(V)$ is just the λ_ith symmetric power of a vector space V, while $S_p^{\lambda_i}(V)$, $p > 0$, is the certain quotient of $S_0^{\lambda_i}(V)$. We remark briefly that if K is an infinite field and F a free abelian group of finite rank, then $\operatorname{gr}(KF) \simeq S(KX)$, the symmetric algebra over KX. Furthermore, $\operatorname{gr}_n(KF) \simeq S^n(KX)$, the nth symmetric power of KX, and the canonical filtration forms actually a composition series for $S^n(KX)$ ([**K**] and [**KO**]; see also [**D**]).

The last part is taken up by the investigation of $\operatorname{gr}(KF)$, when $F = F(X)$ is a free nilpotent group of class 2 and $\operatorname{char} K = 0$. We aim at proving that $\operatorname{gr}(KF)$ satisfies the maximal condition for its $E(KX)$-ideals. There are several reasons for looking for such result. On the one hand $\operatorname{gr}(KF)$ is weakly free subject to the weak identity $[x, y, z] = 0$. Here $[x_1, x_2]$) denotes $x_1x_2 - x_2x_1$ and moreover we set $[x_1, \dots, x_c] = [[x_1, \dots, x_{c-1}], x_c]$. In more detail the weak freeness of $\mathrm{gr}(KF)$ means that for every K-homomorphism $\phi\colon K\overline{X} \to A$ of $K\overline{X}$ into an algebra A, satisfying $[\phi(x), \phi(y), \phi(z)] = 0$ for all $x, y, z \in K\overline{X}$ there exists a homomorphism $\psi\colon \mathrm{gr}(KF) \to A$ such that the diagram

$$\begin{array}{ccc} K\overline{X} & \xrightarrow{\varphi} & A \\ \searrow & & \nearrow \psi \\ & \mathrm{gr}(KF) & \end{array}$$

commutes.

On the other hand algebras satisfying weak identity $[x, y, z] = 0$ plus some additional identities have been investigated on the maximal condition in the literature. The first example is that of the exterior algebra on $V = K\overline{X}$, $\Lambda(V)$. It is well known that $\Lambda(V)$ is weakly free and defined by the weak identity $v^2 = 0$, $v \in V$ (**[GRE]**). Moreover $\Lambda(V)$ satisfies the polynomial identity $[x, y, z] = 0$, i.e., $\Lambda(V)$ is a P.I. algebra (**[JA]**). It is also known (**[LA]**) that the variety generated by $\Lambda(V)$ is Specht, that is to say that $\mathscr{F}(X)/\mathscr{J}$, where $\mathscr{J}$ is the ideal of all identities of $\Lambda(V)$ satisfies the maximal condition on the ideals invariant under the endomorphisms of $\mathscr{F}(X)/\mathscr{J}$. Clearly $\mathscr{F}(X)/\mathscr{J}$ is an image of $\mathrm{gr}(KF)$ whence the results of (**[LA]**) is obtained.

A relatively free P.I. algebra $R(X)$, like $\mathscr{F}(X)/\mathscr{J}$ above, is defined by the property that every mapping $X \to R(X)$ extends to an algebra endomorphism of $R(X)$. This is the strongest possible case. By selecting various families of mappings $X \to R(X)$ that are to be extendable other kinds of relatively free algebras ensue. For example L-weakly free algebras correspond to the family of mappings $X \to LX \subset R(X)$. In the same spirit two more classes of free algebras can be obtained and, indeed, appeared in the literature (see [**R1**, R2], [**P**], respectively). Let $L(X)$ be a relatively free Lie algebra on set X and $U(L)$ be its universal enveloping algebra. Then every mapping $X \to L$ extends to an algebra endomorphism of $U(L)$. An ideal J of $U(L)$ is called verbal if it is invariant under all endomorphisms of $U(L)$ induced by the functions $X \to L$. Likewise, if $F(X)$ is a relatively free group on X, then every function $X \to F(X)$ extends to an algebra endomorphism of $KF(X)$. Again an ideal J of $KF(X)$ is called verbal if J is invariant under all endomorphisms induced by the functions $X \to F(X)$.

It was proved recently in (**[V]**) that if L is a free nilpotent algebra of class 2 and $\mathrm{char}\, K = 0$, then $U(L)$ satisfies maximal condition for the verbal ideals. As $U(L) \simeq \mathrm{gr}(KF)$, F nilpotent of class 2, this result is also a particular case of our theorem.

The last result of this paper we want to comment on is the computation of the so-called axiomatic rank of the ideals $I(KF)$, where F is nilpotent of class 2. Here under the axiomatic rank of a verbal ideal J of $KF(X)$ we understand, like in universal algebra (**[CO]**), the smallest ordinal $n \leq \omega$ for which the elements of J depending on $\leq n$ letters from X generate J. Let J_n denote the verbal ideal generated by elements of J depending on $\leq n$ letters.

For $J = I^d$ we show that the chain

$$I_1^d < I_2^d < \cdots < I_d^d = I^d$$

is proper throughout.

The notations defined in this section will be kept throughout the paper. The end of a proof or the absence of it is signaled by a triangle ($\triangleleft$).

§1. Preliminary Results

First of all we fix some more notation and recall some standard facts on the quotients of the powers of augmentation ideals. For a group G and a natural n, G^n denotes the subgroup of G generated by all g^n, $g \in G$. The exponent of G, $\exp(G)$, is the smallest natural e such that $G^e = 1$. Following ([**H**]) we put $\gamma_n(G)$ for the nth term of the lower central series in G. The isolator of $\gamma_n(G)$, that is the set of all g such that $g^m \in \gamma_n(G)$ for some natural m, is denoted by $\bar{\gamma}_n(G)$. Further we recall the structure of the dimension subgroups $D_{n,p}(G)$ of a group G over K, $\operatorname{char} K = p$. If $p > 0$, then by Lazard's formula ([**L**]) $D_{n,p}(G) = \prod_{ip^j \geq n} \gamma_i(G)^{p^j}$ and for $p = 0$ it is known (see [**H**]) that $D_{n,0}(G) = \bar{\gamma}_n(G)$.

Before embarking on the proofs of our results we describe in detail the isomorphism between $U(K \otimes D_p(F))$ and $\operatorname{gr}(KF)$.

(a) Let G be a group and $p = 0$. We select a maximal Z-independent set $B = \{b_a / \alpha \in A\}$ in $D_0(G)$ by taking the union of the maximal Z-independent sets in every $D_0^i(G)$, $i \geq 1$. We define the weight function on B by saying that $\operatorname{wt}(b_a) = i$ if $b_a \in D_0^i(G)$. Further we arrange B in order of increasing weight, but otherwise arbitrarily, i.e., $\operatorname{wt}(b_a) < \operatorname{wt}(b_\beta)$ implies $b_\alpha < b_\beta$. Finally we endow A with the induced order, i.e., $\alpha \leq \beta \Leftrightarrow b_\alpha \leq b_\beta$. We turn now to $U(K \otimes D_0(G))$. Firstly it is elementary to show that the set $[1 \otimes b_\alpha | \alpha \in A]$ is a K-basis for $K \otimes D_0(G)$. Abusing notations we identify b_α with $1 \otimes b_\alpha$ and also set $\operatorname{wt}(1 \otimes b_\alpha) = \operatorname{wt}(b_\alpha)$. We extend the function 'wt' on the set of the products of $b_\alpha' s$, namely we put $\operatorname{wt}(b_{\alpha_1}^{n_1} \dots b_{\alpha_m}^{n_m}) = \sum_{i=1}^n n_i \operatorname{wt}(b_{\alpha_i})$. Lastly we define the grading on $U(K \otimes D_0)$ by setting the nth homogeneous component $U^n = U^n(k \otimes D_0)$ to be the K-span of the standard products $b_{\alpha_1}^{n_1} \dots b_{\alpha_m}^{n_m}$, $\alpha_1 < \cdots < \alpha_m$, for which $\sum n_i \operatorname{wt}(b_{\alpha_i}) = n$. Using the usual 'straightening' out process (see §2) it is easy to see that $U^n U^m \subseteq U^{n+m}$, so $\{U^n\}$ forms, indeed, a grading for $U(K \otimes D_0)$. Likewise it is easy to show that U^n contains every product of the form $b_{\gamma_1} \dots b_{\gamma_t}$, satisfying $\sum \operatorname{wt}(b_{\gamma_i}) = n$.

Next we take up the algebra $\operatorname{gr}(KG)$. We recall the Hall-Hartley approach to bases for $\operatorname{gr}_n(KG)$ ([**H**, Ch. 7], [**HART2**]). One starts with an N-series in G ([**L**])

$$(\mathcal{S}) \qquad G = H_1 \supset H_2 \supset \cdots \supset H_n > \dots$$

The series ($\mathcal{S}$) induces a weight function on G:

$$\operatorname{wt}(g) = \begin{cases} t, & \text{if } g \in H_t - H_{t+1}; \\ \infty, & \text{otherwise.} \end{cases}$$

Extend the weight function on the products $(g_1-1)\dots(g_s-1)$, $g_i \in G$, by formula $\operatorname{wt}(\prod_{i=1}^{s}(g_i-1)) = \sum \operatorname{wt}(g_i)$. Finally, define filtration $\{\Phi_m(\mathscr{S})\}$ of $I(G)$ induced by $(\mathscr{S})$ by setting up Φ_m as the K-span of all products $(g_1-1)\dots(g_s-1)$ with $\sum \operatorname{wt}(g_i) \geq m$. Note that $\Phi_m(\mathscr{S}) \supset I^m$ for all m and every $(\mathscr{S})$.

Let $(\mathscr{S})$ be the K-dimension series in G:

$$(D) \qquad G = D_{0,1}(G) \supset D_{0,2}(G) \supset \dots$$

Then $\Phi_m(D) = I^m$ for all m. Let $R = \{g_\alpha | \alpha \in A\}$ be the set of representatives of the cosets b_α of $D_0(G)$. Clearly one has $\operatorname{wt}(g_\alpha) = \operatorname{wt}(b_\alpha)$ for all α. Further for every natural valued function $\phi\colon A \to N \cup \{0\}$ define a standard product $\pi(\phi)$ as $(g_{\alpha_1}-1)^{n_1}\dots(g_{\alpha_m}-1)^{n_m}$, where $\alpha_1 < \dots < \alpha_m$, $\phi(\alpha_i) = n_i$ and $\phi(\beta) = 0$ for $\beta \notin \{\alpha_1,\dots,\alpha_m\}$. Now it follows from the proof of ([**H**, Thm. 7.1]) that I^m/I^{m+1} is K-freely generated by the cosets of $\pi(\phi)$ with $\operatorname{wt}(\pi(\phi)) = m$.

Actually Hall's statement includes the provision that G is finitely generated. However, Hall's proof carries through in the general case with some changes. We indicate these briefly below (cf. [**HART1**], [**PA**]).

(i) Let B be the set of cosets $\pi(\phi) + I^{m+1}$ with $\operatorname{wt}(\pi(\phi)) = m$. We claim that B spans I^m/I^{m+1} over K. For let $\rho = (g_1-1)\dots(g_m-1) + I^{m+1}$ be a generator of I^m/I^{m+1}, whence $\operatorname{wt}(g_i) = 1$. Let t be a natural number such that $g_i^t = g_{\alpha_1}^{s_1}\dots g_{\alpha_m}^{s_m}$ modulo $D_{0,2}(G)$. Then it is easy to see that $(g_i - 1) = \frac{1}{t}\sum s_i(g_{\alpha_1}-1)$ modulo I^2. Hence $\rho = \sum k_\psi (g_{\alpha_1}-1)\dots(g_{\alpha_m}-1) + I^{m+1}$, $\psi = (\alpha_1,\dots,\alpha_m)$, $k_\psi \in K$. ψ is not in general 'standard', that is, it is not nondecreasing.

A typical step in 'straightening' out $(g_{\alpha_1}-1)\dots(g_{\alpha_m}-1) + I^{m+1}$ is to replace $(g_\beta-1)(g_\gamma-1) + I^{v+1}$ by $(g_\gamma-1)(g_\beta-1) + ((g,h)-1) + I^{v+1}$, where $\operatorname{wt}(g_\beta) \geq \operatorname{wt}(g_\gamma)$ and $v = \operatorname{wt}(g_\beta) + \operatorname{wt}(g_\gamma)$. If $\operatorname{wt}(g,h) > v$, then it can be dropped out. Otherwise, using our opening argument we obtain $(g,h) - 1 = \frac{1}{t}(\sum s_i(g_{\alpha_1}-1)) + I^{v+1}$ for some integers t, s_i and base elements $g_{\alpha_i} \in R$ of weight v. The assertion follows:

(ii) B is linearly independent. Any linear relation of the form

$$(*) \qquad \sum_{i=1}^{t} k_i \pi(\phi_i) \equiv 0 \text{ modulo } I^{m+1}$$

involves only a finite number of g_α and all of them of weight $\leq m$. Hence the subgroup T generated by these g_α plus m arbitrary elements $h_i, h_i \in D_{0,i} - D_{0,i+1}$, $i = 1,2,\dots,m$ is finitely generated. The induced N-series of T, $\{T \cap D_{0,i}\}$, has cosets of the g_α appearing in the bases for $T \cap D_{0,i}/T \cap D_{0,i+1}$, $i = 1,2,\dots,m$, and with the same weights. Let $\{\Phi_m\}$ be the filtration of $I(T)$ induced by N-series $\{T \cap D_{0,i}\}$. Then by ([**HART2**, Ch. 3]) the set of $\pi(\phi_i)$, $\operatorname{wt}\pi(\phi_i) = m$, is linearly independent modulo Φ_{m+1}. Since $I^{m+1}(T) \subset \Phi_{m+1}$ no relation of the form $(*)$ is possible.

Of course, sending b_α, $\mathrm{wt}(b_\alpha) = i$, into $(g_\alpha - 1) + I^{i+1}$ induces Lie homomorphism of $D_0(G) \to \mathrm{gr}(KG)$ which extends to a homomorphism $U(K \otimes D_0(G)) \to \mathrm{gr}(KG)$. Under the latter every $b_{\alpha_1}^{n_1} \dots b_{\alpha_t}^{n_t}$ is mapped into $(g_{\alpha_1}-1)^{n_1} \dots (g_{\alpha_t}-1)^{n_t} + I^{m+1}$, where $\sum_{i=1}^{t} n_i \,\mathrm{wt}(b_{\alpha_i}) = m$ and the graded isomorphism between $U(K \otimes D_0(G))$ and $\mathrm{gr}(KG)$ ensues. We remark finally that our argument gives characteristic zero part of Quillen's theorem ([**Q**]). Indeed, in spite of the fact that Quillen starts from the graded Lie ring $L(G)$ associated with the lower central filtration in G using the rather simple fact that $L(G)/T$ is a dense subalgebra in $D_0(G)$ where T is the torsion ideal of $L(G)$, it follows that $K \otimes L(G) \simeq K \otimes L(G)/T \simeq K \otimes D_0(G)$ for every field of characteristic zero. Hence $U(K \otimes L(G)) \simeq U(K \otimes D_0(G))$.

(b) Let $p = \mathrm{char}\, K > 0$. In the foregoing notations adapted to the p-dimension series of G the restricted universal envelope of $K \otimes D_p(G)$, $U^r(K \otimes D_p(G))$, is a K-span of the products $b_{\alpha_1}^{r_1} \dots b_{\alpha_m}^{r_m}$, $\alpha_1 < \dots < \alpha_m$ and $0 \le r_i < p$. The nth homogeneous component ${}^nU^r(K \otimes D_p)$ of $U^r(K \otimes D_p)$ is defined as the k-span of all $b_{\alpha_1}^{r_1} \dots b_{\alpha_m}^{r_m}$ such that $\sum_{i=1}^{m} r_i \,\mathrm{wt}(b_{\alpha_i}) = n$. As for the $\mathrm{gr}(KG)$ its standard basis is formed by all $(g_{\alpha_1} - 1)^{r_1} \dots (g_{\alpha_m} - 1)^{r_m} + I^{n+1}$, $\alpha_1 < \dots < \alpha_m$ and $\sum n_i \,\mathrm{wt}(g_{\alpha_1}) = n$. This fact comes directly from Jenning's paper ([**JE**]). As before the correspondence $b_{\alpha_1}^{r_1} \dots b_{\alpha_m}^{r_m} \leftrightarrow (g_{\alpha_1} - 1)^{r_1} \dots (g_{\alpha_m} - 1)^{r_m}$ induces the graded isomorphism between $U^r(K \otimes D_p(G))$ and $\mathrm{gr}(KG)$. ◁

We now state the

PROPOSITION 1.1. *Let $F = F(X)$ be a relatively free group on set X. Then*

(i) *If $p = \mathrm{char}\, K = 0$, then $\mathrm{gr}(KR)$ has only a finite number of nonzero components if and only if $\mathrm{gr}_n(KF) = 0$ for all $n \ge 1$ which occurs if and only if $\exp(F) \ne 0$. Moreover, if $\exp(F) = 0$, then $\mathrm{gr}_1(KF)$ is freely generated by $\overline{X}$.*

(ii) *If $\mathrm{char}\, K = p > 0$, then $\mathrm{gr}(KF)$ has only a finite number of nonzero component if and only if $\dim_{Z_p} D_p(F) < \infty$. Furthermore, $\mathrm{gr}_n(KF) = 0$ for all $n \ge 1$ if and only if p does not divide $\exp(F)$. If p divides $\exp(F)$, then $\mathrm{gr}_1(KF)$ is freely generated by $\overline{X}$.*

PROOF. (i) Let G be an arbitrary group and $p = \mathrm{char}\, K = 0$. We know from the previous discussion that $\mathrm{gr}_n(KG) \simeq U^n(K \otimes D_0(G))$ possesses a basis of the products $b_{\alpha_1}^{n_1} \dots b_{\alpha_m}^{n_m}$: $\sum n_i \,\mathrm{wt}(b_{\alpha_i}) = n$. It becomes evident that $U^n(K \otimes D_0(G)) = 0$ if and only if $K \otimes D_0(G) = 0$. Since $D_0(G)$ is torsion free the latter implies that $D_0(G) = 0$, whence $D_0^1(G) = 0$. Conversely, we recall that $K \otimes D_0(G)$ is generated as Lie algebra by $K \otimes D_0^1(G)$. For, let again $L(G)$ be the graded Lie ring associated with the lower central filtration $\{\gamma_i(G) \colon i = 1, 2, \dots\}$. Then $K \otimes L(G)$ is clearly generated by its first component $K \otimes L^1(G)$, where $L^1(G) = G/\gamma_2(G)$. Since $K \otimes D_0(G) \simeq K \otimes L(G)$ the same is true for $K \otimes D_0(G)$. Hence $D_0^1(G) = 0$ implies $D_0(G) = 0$ which gives the first part of (i). Next we remark that if $G = F$, then $D_0^1(F) = 0$ if and only if $x_1^m = 1$ is an identity on F for some m, that is $\exp(F) \ne 0$. Finally, if $\exp(F) = 0$, then $\bar{\gamma}_2(F) = \gamma_2(F)$ and $F/\gamma_2(F)$ is free on $X\gamma_2(F)$ ([**H.N.**, Ch. 1]). As $I/I^2 \simeq K \otimes F/\gamma_2(F)$, $\mathrm{gr}_1(KF)$ is K-free on $\overline{X}$.

(ii) Now let $p > 0$. If $\dim z_p D_p(G) < \infty$, then $U^r(K \otimes D_p(G))$ is finite dimensional and so $\mathrm{gr}(KG)$ has only a finite number of components. Conversely, if $\dim_{Z_p} D_p(G) = \infty$ then either $\dim_{Z_p} D^i(G) = \infty$ for some i or there are infinitely many i such that $\dim_{Z_p} D^i(G) \neq 0$. Evidently in both cases there are infinitely many nonzero components of $U^r(K \otimes D_p(G))$. In addition to that $\mathrm{gr}_n(KG) = 0$ for all $n \geq 1$ if and only if $D_p(G) = 0$. As $D_p(G)$ is generated by $D_p^1(G)$ as p-Lie algebra the latter condition is equivalent to $D_p^1(G) = 0$. Now let $G = F$. Since $D_{2,p}(F) = F^pF'$ it is easy to see that $D_p^1(F) = 0$ if and only if $x_1 = x_1^{pm}$ or $x_1^{pm-1} = 1$ is an identity on F for some m. This means that p does not divide $\exp(F)$. If p divides $\exp(F)$, then F/F^pF' is free on the image of X. As before this implies that I/I^2 is freely generated by $\overline{X}$. ◁

In what follows we assume that p divides $\exp(F)$, if $p > 0$ and $\exp(F) = 0$, otherwise. We explore now the question of what kind of freeness does $\mathrm{gr}(KF)$ have. The natural one for $\mathrm{gr}(KF)$ is described in

Proposition 1.2. *$\mathrm{gr}(KF)$ is Z_p-weakly free on $\overline{X}$.*

Proof. We recall that $\overline{X}$ generates I/I^2 over K and hence it generates $\mathrm{gr}(KF)$ as K-algebra. Let E be the semigroup of all endomorphisms of $F/D_{2,p}(F)$. It is immediate that every $\gamma \in \Gamma = \mathscr{E}nd(F)$, the semigroup of all endomorphisms of F, induces an endomorphism $\bar{\gamma}$, say, in $F/D_{2,p}(F)$. Clearly the mapping $\gamma \mapsto \bar{\gamma}$ gives a homomorphism of Γ into E and since F is relatively free it is easy to check that '$-$' is onto. Therefore, we can define an action of E on $\mathrm{gr}_n(KF) = I^n/I^{n+1}$ by setting $\bar{u}\varepsilon = \overline{u\gamma}$, where $\varepsilon = \bar{\gamma} \in E$ and $\bar{u}$ denotes $u + I^{n+1}$ for $u \in I^n$. An easy verification shows that the action of E so defined does not depend on the choice of u and γ. Thus $\mathrm{gr}_n(KF)$ gets an E-module structure. Moreover, since γ induces an algebra endomorphism in KF, so does ε in $\mathrm{gr}(KF)$. This is exactly what we need because E induces the semigroup of substitutions $\bar{x}_\alpha \mapsto \sum n_{\alpha\beta}\bar{x}_\beta$, $n_{\alpha\beta} \in Z_p$, in I/I^2. ◁

In the next two propositions we shall look into the conditions under which $\mathrm{gr}(KF)$ is actually weakly free for every ground field K. If $p = \mathrm{char}\, K = 0$, then it turns out that no conditions are necessary.

Proposition 1.3. *Let K be a field of characteristic zero. Then* $\mathrm{gr}(KF)$ *is weakly free on $\overline{X}$.*

Proof. We seek to show that every function $\overline{X} \to K\overline{X}$ extends to an endomorphism of $\mathrm{gr}(KF)$. Let $E(K)$ be the semigroup $E(K\overline{X})$ and $E = E(Z\overline{X})$ be the subsemigroup of points with integer coordinates. Then E is Zariski dense in $E(K)$, whence every polynomial function on E has a unique extension on $E(K)$. Since the representation of E afforded by $\mathrm{gr}_n(KF)$ is polynomial, homogeneous of degree n it can be uniquely lifted to $E(K)$. Moreover, E acts in $\mathrm{gr}(KF)$ by the algebra endomorphisms. The identity

$(\overline{uv})\varepsilon = (\bar{u}\varepsilon)(\bar{v}\varepsilon)$, $\bar{u}, \bar{v} \in \mathrm{gr}(KF)$ and $\varepsilon \in E$, leads up to a set of polynomial conditions on the coordinates of ε. These conditions remain valid for $E(K)$, meaning that $E(K)$ acts on $\mathrm{gr}(KF)$ by the graded endomorphisms. ◁

We proceed to the case of positive characteristic. We need some preliminary remarks. Let us fix a ground field $K, p = \mathrm{char}\, K > 0$, and a countable set X. We simplify our notation as follows: Write $D_i(G)$ for $D_{i,p}(G)$, $D^i(G)$ for $D_i(G)/D_{i+1}(G)$, $D(G)$ for $\bigoplus_{i\geq 1} D^i(G)$ and G_i for $\gamma_i(G)$. All relatively free groups are assumed to have X as a set of free generators. Let $\mathscr{F}$ be absolutely free group and F be the given relatively free group. There exists epimorphism $\pi : \mathscr{F} \to F$ induced by the identity map $X \to X$. π induces the epimorphisms $D_i(\mathscr{F}) \to D_i(F)$ for every i and hence epimorphisms $\pi_*^i : D^i(\mathscr{F}) \to D^i(F)$. The family of these comprise graded epimorphism $\pi_* : D(\mathscr{F}) \to D(F)$ and hence the graded epimorphism $U^r(\pi_*): U^r(K \otimes D(\mathscr{F}) \to U^r(K \otimes D(F))$. On the other hand π induces epimorphism $K\mathscr{F} \to KF : \sum k_f f \mapsto \sum k_f \pi(f)$, $f \in \mathscr{F}$ and $k_f \in K$, under which $I^n(\mathscr{F})$ is mapped onto $I^n(F)$. Hence we obtain the induced mappings $\mathrm{gr}_n(\pi) : \mathrm{gr}_n(K\mathscr{F}) \to \mathrm{gr}_n(KF)$ that give rise to epimorphism $\mathrm{gr}(\pi) : \mathrm{gr}(K\mathscr{F}) \to \mathrm{gr}(KF)$. Identifying $U^r(K \otimes D(G))$ with $\mathrm{gr}(KG)$ via isomorphism described in the opening remarks to this section we can assume $U^r(\pi_*) = \mathrm{gr}(\pi)$.

We need some facts on the structure of $D(G)$. Let $B^1 = \{b_\gamma\}$ be a basis for $D^1(G)$. It is relatively easy to deduce from Lazard's formula for $D_n(G)$, $D_n(G) = \prod_{ip^j \geq n} G_i^{p^j}$, that $D^n(G)$ is spanned by the set $[b_{\gamma_1}, \ldots, b_{\gamma_m}]^{p^t}$, where $[b_{\gamma_1}, \ldots, b_{\gamma_m}]$ denotes an m-fold left-normed commutator, satisfying condition $mp^t = n$ (cf. **[P-S]**). Hence $D(G)$ is generated by $D^1(G)$ as p-Lie algebra over Z_p.

Let us specialize to $G = \mathscr{F}$. The isomorphism $D^1(\mathscr{F}) \simeq I(Z_p\mathscr{F})/I^2(Z_p\mathscr{F}) \simeq Z_p\overline{X}$ (see Proposition 1.1), where $\overline{X} = \{\bar{x}_j = x_j - 1 | x_j \in X\}$ makes it clear that the images of $\bar{x}_j$ in $I(Z_p\mathscr{F})/I^2(Z_p\mathscr{F})$, which we will denote by $\bar{x}_j$, play the role of the base elements b_γ above. Hence $D^n(\mathscr{F})$ is spanned by the set of elements $[\bar{x}_{j_1}, \ldots, \bar{x}_{j_m}]^{p^t}$, $mp^t = n$, over Z_p. Algebra $\mathrm{gr}(K\mathscr{F})$ is in fact the tensor algebra over $K\overline{X}$ (e.g., **[G1**, Ch. 3]). In particular $\mathrm{gr}_n(K\mathscr{F}) \simeq K\overline{X} \otimes_K K\overline{X} \otimes \cdots \otimes_K K\overline{X}$ and it becomes evident that $E(K)$ acts on $\mathrm{gr}(K\mathscr{F})$ by the graded endomorphisms. The description of $D^n(\mathscr{F})$ given above shows that $K \otimes D^n(\mathscr{F})$ as well as $K \otimes D(\mathscr{F})$ are $E(K)$-submodules of $\mathrm{gr}(K\mathscr{F})$. Turning to F we observe firstly that $\mathrm{gr}_1(KF) \simeq K\overline{X}$ is the same as $\mathrm{gr}_1(K\mathscr{F})$. $E(K)$ acts naturally on $\mathrm{gr}_1(KF)$ by means of linear substitutions $\bar{x}_j \mapsto \sum k_{j_m}\bar{x}_m$, $k_{j_m} \in K$, and we want to find out when this action can be extended to one by the graded endomorphisms of $\mathrm{gr}(KF)$. We remark that if such extension exists, then $\mathrm{gr}(\pi) : \mathrm{gr}(K\mathscr{F}) \to \mathrm{gr}(KF)$ is an $E(K)$-epimorphism and therefore $\mathrm{Ker}\, \mathrm{gr}(\pi)$ is $E(K)$-ideal of $\mathrm{gr}(K\mathscr{F})$. Conversely, if $\mathrm{Ker}\, \mathrm{gr}(\pi)$ is an $E(K)$-ideal in $\mathrm{gr}(K\mathscr{F})$, then $\mathrm{gr}(KF)$ clearly has the desired $E(K)$-structure.

To compute $\operatorname{Ker}\operatorname{gr}(\pi)$ we view $\operatorname{gr}(K\mathscr{F})$ as $U^r(K\otimes D(\mathscr{F}))$. Let $V = \operatorname{Ker}\pi$. For every i the kernel of $\pi_*^i = D^i(\mathscr{F}) \to D^i(F)$ equals to $(V\cap D_i(\mathscr{F})D_{i+1}(\mathscr{F}))$. We let $V^i = \operatorname{Ker}\pi_*^i$ and we further form $V_* = \bigoplus V^i$. Of course $V_* = \operatorname{Ker}\pi_*$ and it clearly is $E(Z_p)$-ideal of p-Lie ring $D(\mathscr{F})$, where $E(Z_p) = \operatorname{End}(Z_p\overline{X})$. We define $\mathscr{J}_V$ to be the ideal of $U^r(K\otimes D(\mathscr{F}))$ generated by V_*. Then it is easy to see that $\mathscr{J}_V = \operatorname{Ker} U^r(\pi_*)$ or what is the same $\mathscr{J}_V = \operatorname{Ker}\operatorname{gr}(\pi)$.

Let D_K be the semigroup of the diagonal endomorphisms of $K\overline{X}$, that is, the mappings $\delta(\lambda_i)\colon \bar{x}_i \to \lambda_i\bar{x}_i$, $\bar{x}_i \in \overline{X}$, $\lambda_i \in K$. We say that an element $u \in D^n(\mathscr{F})$ is K-homogeneous if $u\delta(\lambda_i) = \chi(\delta(\lambda_i))u$ holds in $K\otimes D^n(\mathscr{F})$ for a suitable polynomial character x of D_K. We can now state the

THEOREM 1.4. *$E(K)$ acts on $\operatorname{gr}(KF)$ by the graded endomorphisms if and only if V_* has a Z_p-spanning set made up of K-homogeneous elements.*

PROOF. Sufficiency is almost immediate. For ideal $\mathscr{J}_V$ has the form $\operatorname{gr}(K\mathscr{F})V_*$ and V_* is an $E(Z_p)$-invariant subspace, since V is a verbal subgroup of $\mathscr{F}$. Furthermore, the K-envelope of V_*, KV_*, is invariant under D_K since V_* is spanned by its K-homogeneous elements. It remains to notice that these two invariances imply the invariance of KV_* and hence $\mathscr{J}_V$ under $E(K)$.

To this end we remark that if $|X| = \infty$ we can replace $E(K)$ by its subsemigroup of a finitary endomorphisms. Thus it will suffice to consider the case of finite X. Now $E(K)$ is generated by the transvections $\tau_{ij}(K)\colon \bar{x}_i \mapsto \bar{x}_i + k\bar{x}_j$, $x_m \mapsto x_m$, $m \neq i$, and D_K. Further, for every $k \neq 0$ in K it is easy to verify that $\tau_{ij}(k) = \delta_j(k^{-1})\tau_{ij}(1)\delta_j(k)$, where $\delta_j(k)\colon x_j \mapsto kx_j$, $x_i \mapsto x_i$, $i \neq j$. The assertion follows.

Conversely, suppose that $\mathscr{J}_V$ is an $E(K)$-ideal. Then so is $\mathscr{J}_V \cap KD(\mathscr{F})$. We claim that $\mathscr{J}_V \cap KD(\mathscr{F}) = KV_*$. For we can pick a basis for $D(\mathscr{F})$, say $\{b_\alpha/\alpha \in A\}$ in such a way that $V_* \cap \{b_\alpha/\alpha \in A\}$ spans V_*. In this case the standard monomials of the form $b_{\alpha_1}^{n_1}\dots b_{\alpha_m}^{n_m}$ with at least one b_{α_j} from V_* form a K-basis for $\mathscr{J}_V$. Hence $\sum k_i b_{\alpha_i} \in \mathscr{J}_V$ if and only if $b_{\alpha_1} \in \mathscr{J}_V$ for all i. Finally, if $u \in V_*$, then we can assume that u is homogeneous, i.e., $u \in V^n$ for some n. Then it follows from our remarks on $D^n(\mathscr{F})$ above that $u = \sum k_\xi \xi$, where ξ has the form $[\bar{x}_{i_1},\dots,\bar{x}_{i_m}]^{p^j}$. Since every such ξ is an eigenvector for D_K it is K-homogeneous. Collecting all ξ in u having the same associated character we deduce that $u = \sum_{q=1} u_q$ when every u_q is a K-homogeneous element. Let χ_q be the character associated with u_q. Then $u\delta(\lambda_i) = \sum_{q=1}^m \chi_q(\delta(\lambda_i))u_q$ lies in KV_*. Since the set of distinct characters $\{\chi_q\}$ is linearly independent over K it follows that $u_q \in V_*$ for every q. Thus V_* possesses a K-homogeneous spanning set, as claimed. ◁

PROPOSITION 1.5. *Let F be a free nilpotent group of class c on set X. Then* $\operatorname{gr}(KF)$ *is weakly free algebra on $\overline{X}$. Moreover* $\operatorname{gr}(KF)$ *is defined by a unique weak identity* $[\bar{x}_1,\dots,\bar{x}_{c+1}] = 0$.

PROOF. The kernel of $\mathcal{F}(X) \to F(X)$ is, of course, $\mathcal{F}_{c+1}$ the $c+1$st term of the lower central series for $\mathcal{F}$. In view of the preceding result we must show that $\mathcal{F}^n_{c+1}$ is spanned by a K-homogeneous set for every n. We recall that $\mathcal{F}^n_{c+1} = (\mathcal{F}_{c+1} \cap D_n)D_{n+1}/D_{n+1}$. In order to find a basis for $\mathcal{F}^n_{c+1}$ we revert to Lazard's formula for D_n, $D_n = \prod_{ip^j \geq n} \mathcal{F}_i^{p^j}$.

Passing over to $D^n = D_n/D_{n+1}$ we see from the formula that only $\mathcal{F}_i^{p^j}$ such that $ip^j = n$ contribute to D^n. Let $n = p^\alpha n'$, $(p, n') = 1$. To every h, $0 \leq h \leq a$, we associate the subspace D^n_h of D^n by the rule: $D^n_h = \prod_{j \leq h} \mathcal{F}^{p^j}_{p^{\alpha - j} n'} D_{n+1}/D_{n+1}$. Then $D^n_0 \subseteq D^n_1 \subseteq \cdots \subseteq D^n_\alpha = D^n$ is a finite filtration in D^n starting with $D^n_0 = \mathcal{F}_n D_{n+1}/D_{n+1}$.

Let $c_1, c_2, \ldots$ be a basic sequence of commutators in $\mathcal{F}$ in symbols of X in the sense of P. Hall ([**H**, Ch. 5]). Since $\mathcal{F}_n^p \mathcal{F}_{n+1}$ lies in D_{n+1} it is immediate that D^n_0 is spanned by the basic commutators of weight n over Z_p. In general for $m = p^{\alpha-h} n'$ every group commutator $(x_{i_1}, \ldots, x_{i_m})$ of weight m is a product of basic commutators of the same weight modulo $\mathcal{F}_{m+1}$, i.e., $(x_{i_1}, \ldots, x_{i_m}) \equiv c^{\nu_1}_{j_1} \ldots c^{\nu_s}_{j_s} \bmod \mathcal{F}_{m+1}$. Applying Hall-Petresco formulas ([**H**, Ch. 6]) to $(x_{i_1}, \ldots, x_{i_m})^{p^h}$ it is a rather straightforward matter to show that modulo D^n_{h-1} one has $(x_{i_1}, \ldots, x_i)^{p^h} \equiv (c^{\nu_1}_{j_1}, \ldots, c^{\nu_s}_{j_s})^{p^h} \equiv (c^{p^h}_{j_1})^{\nu_1} \ldots (c^{p^h}_{j_s})^{\nu_s}$ mod. D^n_{h-1}. Hence D^n_h / D^n_{h-1} is spanned by the p^hth power of $p^{a-h}n'$-fold basic commutators. In all D^n is a Z_p-span of the p powers of basic commutators that belong to D^n. Moreover the set of these $B^n_p = \{c_i^{p^j} \mid ip^j = n\}$ is linearly independent. For the images of $c_i^{p^j}$ in I^n/I^{n+1} under the canonical embedding $D_n/D_{n+1} \to I^n/I^{n+1}$, $xD_{n+1} \to (x-1) + I^{n+1}$, are of the form $\bar{c}_i^{p^j}$, where $\bar{c}_i$ is the ring commutator associated with c_i, i.e., $\bar{c}_i$ is obtained by replacing each x_i by $\bar{x}_i$ and each group commutator $(u, v) = u^{-1}v^{-1}uv$ by the ring commutator $[\bar{u}, \bar{v}]$. As $I^n/I^{n+1} \simeq K\overline{X} \otimes \cdots \otimes K\overline{X}$ (n times) can be viewed as the nth component of free associative K-algebra on $\overline{X}$ it follows from ([**H**, 5.3]) that $\bar{c}_i^{p^j}$ are linearly Independent in I^n/I^{n+1} and our claim follows. Let us order each B^n_p somehow and let B_p be the ordered union of B^n_p, meaning that $c_i^{p^j} < c_k^{p^l}$ in B_p if $ip^j < kp^l$ and otherwise if it is so in B^n_p, $n = ip^j$. Then it follows from the above remarks that for every $f \in \mathcal{F}$ and every $n \in Nf$ has a unique representation $f \equiv u_1^{r_1} \ldots u_s^{r_s}$ modulo D_n, where $u_1 < \cdots < u_s$ and $0 \leq r_i < p$ for all i. Turning to $\mathcal{F}_{c+1}$ we notice that every $f \in \mathcal{F}_{c+1}$ can be written down in the form $f \equiv c^{\nu_1}_{i_1} \ldots c^{\nu_m}_{i_m}$ modulo $\mathcal{F}_n$ with the c_i being commutators of weight $\geq c+1$ and $\nu_i \in Z$. Replacing every ν, by its p-adic expansion, actually taking modulo p^j, where p^j is the smallest p-power $> n$, we employ the collecting process to expand $c^{\nu_1}_{i_1} \ldots c^{\nu_m}_{i_m}$ along the base B_p modulo D_n. Since the commutators appearing in this process have weight no less than c_{i_j} we conclude that every $f \in \mathcal{F}_{c+1}$ is expressible in the form $f \equiv v_1^{s_1} \ldots v_m^{s_m}$, where each v_i is some p-power of a commutator of weight $\geq c+1$. It becomes apparent that $\mathcal{F}^n_{c+1} = (\mathcal{F}_{c+1} \cap D_n)D_{n+1}/D_{n+1}$ is spanned by the p-powers of the basic commutators of weight $\geq c+1$. In particular $\mathscr{F}_{c+1}^{n}$ is K-homogeneous for every extension field K of Z_p.

Finally, let $\pi\colon \mathscr{F}(X) \to F(X)$ be the canonical epimorphism of $\mathscr{F}(X)$ on $F(X)$. By the foregoing $\mathscr{N} = \operatorname{Ker} \operatorname{gr}(\pi)$ is an $E(K)$-ideal of $\operatorname{gr}(K\mathscr{F})$. Hence the weak identities of $\operatorname{gr}(KF)$ coincide with elements of $\mathscr{N}$. Then the above characterization of $(\mathscr{F}_{c+1})_* = \bigoplus_{n\geq 1} \mathscr{F}_{c+1}^{n}$ as Z_p-span of p-powers of the basic commutators of weight $\geq c+1$ makes it clear that every element of $\operatorname{gr}(K\mathscr{F})(\mathscr{F}_{c+1})_* = \mathscr{N}$ is the weak consequence of $[\bar{x}_1, \dots \bar{x}_{c+1}]$. ◁

§2. The Canonical Filtration in $\operatorname{gr}_n(KF)$

Below p denotes a prime number or 0 and Z_p stands for the prime field of characteristic $p > 0$ or Z, if $p = 0$. Let $F = F(X)$ be a relatively free group on a finite set X. As before $E(K)$ denotes $\operatorname{End}_K(KX)$ for every commutative ring K. In preparation to the statement below we introduce one operation on the representations of $E(Z_p)$.

First, if $p = \operatorname{char} K = 0$ and V is an $E(Z)$-module, then $S_0(V)$ denotes the symmetric algebra over $K \otimes V$ (**[GRE]**). By definition $S_0(V)$ is graded algebra and its dth homogeneous component is denoted by $S_0^d(V)$, $d \geq 0$. Secondly, if $p = \operatorname{char} K > 0$ we form again $S_0(V)$, the symmetric algebra over $K \otimes_{Z_p} V$, and then define $S_p(V)$ as the quotient algebra of $S_0(V)$ obtained by factoring out the homogeneous ideal generated by all v^p, $v \in V$. Of course, this ideal is a graded $E(Z_p)$-submodule of $S_0(V)$ which shows that $S_p(V)$ inherits the structure of graded K-algebra with $E(Z_p)$ acting on it by graded endomorphisms. The dth homogeneous component of $S_p(V)$ is denoted by $S_p^d(V)$ and is referred to as the p-truncated dth symmetric power of V.

We recall the structure of a canonical basis for $U(K \otimes D_p(G))$ (see §1). Select a basis in every nonzero $K \otimes D_p^i(F)$, $i = 1, 2, \dots$, and order them arbitrarily. The union of these bases $B = \{b_1, b_2, \dots\}$ is endowed with a weight function: $\operatorname{wt}(b_m) = i$ if and only if $b_m \in K \otimes D_p^i(F)$. Furthermore B is ordered by the rule: If $b_m, b_n \in B$, then $b_m < b_n$ if $\operatorname{wt}(b_m) < \operatorname{wt}(b_n)$ and if $\operatorname{wt}(b_m) = \operatorname{wt}(b_n) = i$, then $b_m < b_n$ if it is so in $K \otimes D_p^i(F)$.

Let N be the set of natural numbers and let $\mathbf{\Phi}$ be the collection of all functions $\phi\colon N \to N \cup \{0\}$ if $p = 0$, or $\phi\colon N \to \{0, 1, \dots, p-1\}$, if $p > 0$, with finite support. To each ϕ we associate the standard monomial $b(\phi) = \prod_{i\in N} b_i^{\phi(i)}$. Then the set of $b(\phi)$, $\phi \in \mathbf{\Phi}$, forms a basis for $U(K \otimes D_p(F)) \simeq \operatorname{gr}(KF)$. Moreover, setting $\operatorname{wt}(b(\phi)) = \sum \phi(i) \operatorname{wt}(b_i)$, it turns out that the subset $\mathscr{B}_n = \{b(\phi) | \operatorname{wt}(b(\phi)) = n\}$ forms a basis for $\operatorname{gr}_n(KF)$.

Let us fix our n and define n functions on $\mathbf{\Phi} = \{\phi\}$. Namely, for every $i = 1, 2, \dots, n$ we let $\partial_i\colon \mathbf{\Phi} \to N \cup \{0\}$ be given by the formula

$$\partial_i(\phi) = \sum_{\operatorname{wt}(b_k)=i} \phi(k).$$

Let $\mathbf{\Phi}_n$ be the set of those $\phi \in \mathbf{\Phi}$ that label $b(\phi)$ from $\mathscr{B}_n$. It is immediate from the definition of weight that for every $\phi \in \mathbf{\Phi}_n$ the equality

$$\partial_1(\phi) + 2\partial_2(\phi) + \cdots + n\partial_n(\phi) = n$$

holds. We want to clarify the standard monomials $b(\phi)$ by attaching to $b(\phi)$ the n-tuple $\tau(b(\phi)) = \tau(\phi) := (\partial_1(\phi)\dots\partial_n(\phi))$, called the type of $b(\phi)$. The set Λ of the types arising from $\phi \in \Phi_n$ is ordered lexicographically, that is to say, if $\lambda, \mu \in \Lambda$, then $\lambda < \mu$ if $\lambda_i < \mu_i$ for the least i such that $\lambda_i \neq \mu_i$. We now can state the main

DEFINITION 2.1. For every $\lambda \in \Lambda$ we define the subspace $\mathrm{gr}_n^\lambda(KF)$ of $\mathrm{gr}_n(KF)$ by the rule: $\mathrm{gr}_n^\lambda(KF) = \langle b(\phi) | \phi \in \Phi_n \text{ and } \tau(\phi) \leq \lambda\rangle$.

THEOREM 2.2. *$\{\mathrm{gr}_n^\lambda(KF)/\lambda \in \Lambda\}$ forms a finite filtration of $\mathrm{gr}_n(KF)$ by $E(Z_p)$-modules. If λ^+ is the largest type smaller than λ, then $\mathrm{gr}_n^\lambda(KF)/\mathrm{gr}_n^{\lambda^+}(KF) \simeq \bigotimes_{i=1}^n S_p^{\lambda_i}(K \otimes D_p^i(F))$.*

REMARK. Those i for which $D_p^i(F) = 0$ do not induce functions $\partial_i(\phi)$. The corresponding factors $S_p^{\lambda_i}(K \otimes D_p^i)$ must be omitted in the above formula.

PROOF. By definition of $\mathrm{gr}_n^\lambda(KF)\mu \leq \lambda$ implies $\mathrm{gr}_n^\mu(KF) \subseteq \mathrm{gr}_n^\lambda(KF)$. Further we show that $\mathrm{gr}_n^\lambda(KF)$ is an $E(Z_p)$-submodule of $\mathrm{gr}_n(KF)$. Let ε be an element of $E(Z_p)$. If $b_i \in D_p^m$, then $b_i\varepsilon = \sum n_{ij}b_j$ where the b_j belong to D_p^m. Hence for every $\phi \in \Phi_n$ such that $\tau(\phi) \leq \lambda$ one has $(b(\phi))\varepsilon = \prod(\sum n_{ij}b_j)^{\phi(i)} = \sum \nu_\alpha b_{j_1}\dots b_{j_m}$ where $\alpha = (j_1, \dots, j_m)$, $m = \sum \phi(i)$ and $\nu_\alpha \in Z_p$. If α is a non-decreasing sequence and $\operatorname{char} K = 0$, or $\operatorname{char} K = p > 0$ and for every natural μ the number of j_i equal to u is less than p, then $b_\alpha := b_{j_1}\dots b_{j_m}$ is a standard monomial of type $\tau(\phi)$. Otherwise b_α is a non-standard product and we claim that in the expansion $b_\alpha = \sum c_\psi b_{(\psi)}$ of b_α through the standard monomials $b(\psi)$ it holds that $\tau(\psi) \leq \tau(\phi)$. To this end we employ the usual 'straightening' argument. The typical step consists of shifting a factor of b_α step by step to the left. Namely, if j_x is the smallest subscript preceded by a larger subscript in α, then b_{j_x} is shifted past $b_{j_{x-1}}$ using the formula

$$(*) \qquad b_{j_{x-1}}b_{j_x} = b_{j_x}b_{j_{x-1}} + [b_{j_{x-1}}, b_{j_x}].$$

Since $D_p(F)$ is a graded Lie ring it follows that $[b_{j_{x-1}}, b_{j_x}] = \sum n_t b_t$ with $\mathrm{wt}(b_t) \geq \mathrm{wt}(b_{j_{x-1}} + \mathrm{wt}(b_{j_x})$. Therefore $b_t > b_{j_x}$ and the application of $(*)$ to b_α replaces b_α by

$$(**) \qquad b_{j_1}\dots b_{j_x}b_{j_{x-1}}\dots b_{j_m} + \sum n_t b_{\beta_t}.$$

The first monomial there has $\partial_i(\phi)$ factors of weight i for every i, whereas each product b_{β}, has $\leq \partial_i(\phi) - 1$ factors of weight $i = \mathrm{wt}(b_{j_x})$. It is also evident that every product in $(**)$ either has a lesser number of inversions or lesser number of factors. Thus by repeated applications of $(*)$ we express b_α through the standard monomials. Moreover every product appearing in the process either has $\partial_s(\phi)$ factors of weight s for every s or there exists i such that the number of factors of weight i is less than $\partial_i(\phi)$. In the latter case that number will remain less than $\partial_i(\phi)$, because only b_j's with $\mathrm{wt}(b_j) \geq i$ are getting shifted. The assertion follows.

We take up now the quotient $\bar{\mathrm{gr}}_n^\lambda(KF) := \mathrm{gr}_n^\lambda(KF)/\mathrm{gr}_n^{\lambda^+}(KF)$. Let $T = \bigotimes_{i=1}^n (K \otimes D_p^i)^{\otimes\lambda_i}$, where $V^{\otimes\alpha}$ denotes $V \otimes \dots \otimes V$ (a times).

Put $|\lambda| = \sum \lambda_i$. T has a natural basis formed by the products $b_{j_1} \otimes \cdots \otimes b_{j_{|\lambda|}}, b_{j_i} \in B$. Every such product we denote by $b_\alpha^\otimes$ with $\alpha = (j_1, \ldots, j_{|\lambda|})$. Clearly there is a linear map $\Theta\colon T \to \bar{\mathrm{gr}}_n^\lambda(KF)$ that takes $b_\alpha^\otimes$ on $b_\alpha + \mathrm{gr}_n^{\lambda^+}(KF)$. The first part of the proof shows that b_α belongs to $\mathrm{gr}_n^\lambda(KF)$ so that Θ is well defined. Moreover, another consequence of it is that the factors b_j of b_α commute modulo $\mathrm{gr}_n^{\lambda^+}(KF)$. In addition, if $p > 0$ and some b_j occurs in $b_\alpha^\otimes$ more than $(p-1)$ times, then $\Theta(b_\alpha^\otimes)$ vanishes. These remarks make it clear that Θ factors through $\bigotimes_{i=1}^n S_p^{\lambda_i}(K \otimes D_p^i)$, inducing the map $\overline{\Theta}\colon \bigotimes_{i=1}^n S_p^{\lambda_i}(K \otimes D_i^p) \to \bar{\mathrm{gr}}_n^\lambda(KF)$. Moreover a straight verification shows that both Θ and the canonical projection $T \to \bigotimes_{i=1}^n S_p^{\lambda_i}(K \otimes D_p^i)$ are $E(Z_p)$-homomorphisms and hence so is $\overline{\Theta}$. On the other hand there is an obvious $1-1$ correspondence between a basis for $\bigotimes_{i=1}^n S_p^{\lambda_i}(K \otimes D_p^i)$ built up from the selected basis in $K \otimes D_p^i$ and the cosets $b(\phi) + \mathrm{gr}_n^{\lambda^+}(KF)$, $\tau(\phi) = \lambda$. Thus $\overline{\Theta}$ is in fact an isomorphism. ◁

§3. gr(KF) for a Free Nilpotent Group of Class 2

3.1. Throughout the rest of the paper $F = F(X)$ stands for the free nilpotent group of class 2 on finite or countable infinite set $X = \{x_1, x_2, \ldots\}$ and the ground field K has characteristics 0. Our first task will be to specify the constructs of §2 in this particular case. The dimension series in F over K is

$$F > F' > 1$$

which factors F/F' and F' are free abelian groups on $\{x_i F'\}$ and $\{(x_i, x_j) | i < j\}$, respectively. Let $\Lambda^2(A)$ denote the exterior square of an R-module A, where R is either Z or K (**[GRE]**). A general decomposable element in $\Lambda^2(A)$ is denoted by $a \wedge b$, $a, b \in A$. Let $A = F/F'$. It is fairly obvious that the mapping $F/F' \otimes F/F'$, $uF' \otimes vF' \mapsto (u, v)$ is well defined and induces the natural isomorphism $\Lambda^2(F/F') \simeq F'$, $uF' \wedge vF' \mapsto (u, v)$. Moreover this isomorphism is even an $E(F/F')$-isomorphism, where $E(F/F')$ stands for the semigroup of all endomorphisms of F/F'. Tensoring with K we pass over to K-spaces $B = K \otimes F/F'$ and $K \otimes F'$. Let $E(B)$ be the semigroup of K endomorphisms of B and $G(B)$ be the group of the invertible elements of $E(B)$. Lifting the $E(F/F')$ action in $K \otimes F'$ and $K \otimes \Lambda^2(F/F')$ to an $E(B)$-action we arrive at $E(B)$-isomorphism $K \otimes F' \simeq K \otimes \Lambda^2(F/F')$. Identifying $K \otimes \Lambda^2(F/F')$ with $\Lambda^2(B)$ via $E(B)$-isomorphism $k \otimes (a \wedge b) \mapsto (k \otimes a) \wedge b$ we obtain $E(B)$-isomorphism $K \otimes F' \simeq \Lambda^2(B)$. Consequently we conclude that $K \otimes D_0(F) = B \oplus \Lambda^2(B)$ and hence $\mathrm{gr}(KF) \simeq U(B \oplus \Lambda^2 B)$. Let $\{e_i | i = 1, 2, \ldots\} = \mathscr{E}$ bc a basis for B. Notice that the cardinality of $\mathscr{E}$ equals to that of X, denoted by $|X|$. Let $|X| = n < \infty$. Then a canonical monomial of weight d has the form

$$e_{\alpha,\beta} = e_1^{\alpha_1} \ldots e_n^{\alpha_n} (e_1 \wedge e_2)^{\beta_{1,2}} \ldots (e_{r-1} \wedge e_r)^{\beta_{n-1,n}}$$

with $\alpha_i, \beta_{ij} \in N$ and $\sum_{i=1}^n \alpha_i + 2\sum_{1 \le i < j \le n} \beta_{ij} = d$. By the results of §1 $\mathrm{gr}_d(KF)$ is K-spanned by all $e_{\alpha,\beta}$ of weight d. The type of $e_{\alpha,\beta}$ is a 2-tuple

$\tau = (l, m)$, where $l = \sum \alpha_i$ and $m = \sum \beta_{ij}$. Clearly $l + 2m = d$ and to every $0 \leq m \leq [d/2]$ there corresponds a unique type $\tau_m = (d - 2m, m)$. Let $\tau = (l, m)$. Like in §2 we define $\mathrm{gr}_d^\tau(KF) = \mathrm{gr}_d^m(KF)$ by the formula: $\mathrm{gr}_d^m(KF) = \langle e_{\alpha,\beta} | \sum \beta_{ij} \geq m \rangle$. Then we deduce from Theorem 2.2.2 that the filtration

$$\mathrm{gr}_d(KF) = \mathrm{gr}_d^0(KF) > \cdots > \mathrm{gr}_3^{[d/2]}(KF) > 0$$

has quotients that are $E(B)$-isomorphic to $S^l(B) \otimes S^m(\Lambda^2 B)$, $m = 0, 1, \ldots$, $[d/2]$, where $S^k(A)$ denotes the kth symmetric power of A. Since K is an infinite field and $\mathrm{gr}_d(KF)$ affords polynomial representation of $G(B)$, every $G(B)$-submodule of $\mathrm{gr}_d(KF)$ is an $E(B)$-submodule as well. This remark enables us to replace $E(B)$ by $G(B)$ in what follows. Moreover, since $\mathrm{char}\, K = 0\, \mathrm{gr}_d(KF)$ is a semisimple $G(B)$-module, whence

$$\mathrm{gr}_d(KF) \simeq \bigoplus_{l+2m=d} S^l(B) \otimes S^m(\Lambda^2 B). \tag{1}$$

3.2. In this section we derive the decomposition of $\mathrm{gr}_d(KF)$ into the direct sum of irreducible $GL(B)$-modules. The result is thus: $\mathrm{gr}_d(KF)$ is the direct sum of all irreducible polynomial representations of $GL(B)$ of degree d, each occuring with multiplicity 1. In more detail let $\dim_K B = n$ and write G_n in place of $GL(B)$. In matters related to polynomial representations of G_n our notation is standard and tends to follow the notation of ([**J-K**]). In particular $\lambda = (\lambda_1, \ldots, \lambda_m)$ denotes a partition of the number $|\lambda| = \lambda_1 + \ldots \lambda_m$, notation $\lambda \vdash |\lambda|$. The number of nonzero parts of λ is called the length of λ, denoted $l(\lambda)$. Let $d = |\lambda|$. To every partition of d of length $\leq n$ there corresponds a unique up to isomorphism irreducible module which affords polynomial representation of degree d. We denote this by W^λ and call it the Weyl module corresponding to λ. This W^λ can be realized in many ways inside $B^{\otimes d}$. We select that particular model W^λ whose image under the natural epimorphism $-: B^{\otimes d} \to \mathrm{gr}_d(KF)$, i.e., $\overline{W}^\lambda$ is nonzero for all $\lambda \vdash d$ with $l(\lambda) \leq n$. Namely, let $\lambda = (\lambda_1, \ldots, \lambda_m)$ have $m \leq n$ nonzero parts, $d = |\lambda|$ and let S_d denote the symmetric group on the set $\underline{d} = \{1, 2, \ldots, d\}$. We recall that S_d acts on $B^{\otimes d}$ in the following way: $\pi(x_1 \otimes \cdots \otimes x_d) = x_{1\pi} \otimes \cdots \otimes x_{d\pi}$. Using this action we imbed the mth exterior power $\Lambda^m(B)$ of B in $B^{\otimes d}$ by sending $x_1 \wedge \cdots \wedge x_m$ to $\alpha_m(x_1 \otimes \cdots \otimes x_m)$, where $\alpha_m = \sum (-1)^\pi \pi$ and $(-1)^\pi = \pm 1$ if π is an even/odd permutation. Compounding these imbeddings we can regard $\Lambda^{a_1}(B) \otimes \cdots \otimes \Lambda^{a_h}(B)$ as a part of $B^{\otimes d}$ for any sequence $\alpha = (a_1, \ldots, a_h)$ satisfying $|\alpha| = a_1 + \cdots + a_h = d$. In particular fixing a basis $\{e_1, \ldots, e_n\}$ of B we denote by I_m the element $e_1 \wedge \cdots \wedge e_m$ and further we set $I_m^s = I_m \otimes \cdots \otimes I_m$ (s-times). Then for every $\lambda = (\lambda_1, \ldots, \lambda_m) \vdash d$, $l(\lambda) < n$, the element $I_\lambda := I_m^{\lambda_m} \otimes I_{m-1}^{\lambda_{m-1} - \lambda_m} \otimes \cdots \otimes I_1^{\lambda_1 - \lambda_2}$ is defined in $B^{\otimes d}$. It is well-known ([**J-K**, 8.3.4]) that I_λ generates the irreducible G_n-module corresponding to λ. This G_n-module is denoted from now on by W^λ. Notice that our definition of W^λ coincides with that given in ([**J-K**, 8.3.1]). The precise form of decomposition of $\mathrm{gr}_d(KF)$ is stated in

THEOREM 1. *In the foregoing notations*

$$\mathrm{gr}_d(KF) = \bigoplus\{\overline{W}^\lambda | \lambda \vdash d \text{ and } l(\lambda) \leq n\}.$$

As for the proof of this theorem we note that it appeared essentially under different guises in several places. For on the one hand formula (1) of §3.1 shows that if the decomposition of $S^m(\Lambda^2 B)$ into the direct sum of Weyl modules is known then an application of the Littlewood-Richardson rule will solve the problem. That decomposition is, of course, well known due to Thrall ([**T**, p. 380]). Thrall's proof is given in terms of Schur functions. It can be viewed as the decomposition into irreducibles of $S(\Lambda^2 B)$. A more general case, namely that of $S(B \oplus \Lambda^2 B)$ can be derived from one identity obtained by Littlewood ([**LI**, (11.9; 6)]; cf. also [**S**, p. 177]). This identity expands the character of G_n-module $S(B \oplus \Lambda^2 B)$ along the basis of Schur functions and is combinatorial in nature. As $\mathrm{gr}(KF)$ is isomorphic to $S(B \oplus \Lambda^2 B)$ as G_n-module the major part of Theorem 1 follows directly from these results.

On the other hand let us fix a basis $\mathscr{E} = \{e_1, \dots, e_n\}$ for B and let T_n, U_n and D_n be the lower triangular, the lower unitriangular and the diagonal subgroups of G_n regarded as the group of $n \times n$ invertible matrices relative to $\mathscr{E}$. By the fundamental fact of the rational representation theory of full linear groups (e.g., [**GR**, 3.5]) the multiplicities of Weyl modules W^λ occuring in a G_n-module V coincide with the dimensions of subspaces V_λ of relative invariants of T_n of weight λ. By definition V_λ is the set of $v \in V$ such that $v\tau = \chi_\lambda(\tau)v$ for all $\tau \in T_n$, where χ_λ is the linear character of T_n associated with λ. The computation of relative T_n-invariants of $S(\Lambda^2 B)$ and thus a new proof of Thrall's theorem can be found in ([**A-F**, C or 2.4]). However their argument relies on some results from ([**C-P**]) that are much harder than the problem at hand. In all the above mentioned proofs of Theorem 1 or $S(\Lambda^2 B)$-portion of it are rather involved. Therefore we propose below a derivation of relative T_n-invariants of $S(\Lambda^2 B)$ based on some rudimentary facts from linear algebra. We divide up the proof into a number of steps.

1. Let C stand for $\Lambda^2 B$ and let $C^* = \mathrm{Hom}_K(C, K)$ be the dual of C. C^* will be regarded as the space of the skew-symmetric forms on B. Let $P_m(C^*)$ denote the space of homogeneous polynomial functions on C^* of degree m. The natural isomorphism $C \simeq P_1(C^*)$, $v \mapsto v_*$, $v \in C$, where $v_*\colon C^* \to K$ is given by $v_*(f) = f(v)$, $f \in C^*$, extends to isomorphism $*\colon S^m(C) \simeq P_m(C^*)$, $v_1 \dots v_m \mapsto (v_1 \dots v_m)_* := (v_1)_* \dots (v_m)_*$, $v_i \in C$ for all i. The action of G_n in C on the right induces the dual action of G_n in C^* on the left: $(\varepsilon f)(c) = f(c\varepsilon)$, $f \in C^*$, $c \in C$, and further induces the action of G_n in $P_m(C^*)$, $(p\varepsilon)(f) = p(\varepsilon f)$, $p \in P_m(C^*)$, $\varepsilon \in G_n$ and f runs over C^*. Plainly $(s_*)\varepsilon = (s\varepsilon)_*$ for every $s \in S^m(C)$ showing that '$*$' is a G_n-isomorphism.

2. Let $\mathscr{E} = \{e_1, \dots, e_n\}$ be a basis for B and let $0 < B_1 < B_2 < \cdots < B_n = B$ be the complete flag in B conformal with $\mathscr{E}$, i.e., $B_k = \langle e_1, \dots, e_k \rangle$, $k \in \underline{n}$. If $f \in C^*$ we denote by f_k the restriction of f on B_k. We define the principal set $O \subset C^*$ by the rule: $f \in O$ if and only if f_k is nondegenerate

for all even $k \leq n$. Let $D(f)$ denote the discriminant of f. Then $f \in O$ if $D(f_k) \neq 0$ for all even k, which shows that O is Zariski open in C^* regarded as an $\binom{n}{2}$-dimensional K-space. Since every polynomial function is completely determined by its values on any open subset of C^*, from now on we assume that every f lies in O. We denote by $\bar{k}$ the matrix $\left(\begin{smallmatrix} 0 & k \\ -k & 0 \end{smallmatrix}\right)$, $0 \neq k \in K$, and let diag $(\bar{k}_1, \ldots \bar{k}_r)$ stand for the form associated with the matrix

$$\begin{pmatrix} \bar{k}_1 & & \\ & \ddots & \\ & & \bar{k}_r \end{pmatrix},$$

$r = n/2$ if n is even and

$$\begin{pmatrix} \bar{k}_1 & & & \\ & \ddots & & \\ & & \bar{k}_r & \\ & & & 0 \end{pmatrix},$$

$r = [n/2]$ if n is odd. The following well known fact plays the crucial role.

LEMMA 2. *$f \in O$ if and only if there exists a $\tau \in U_n$ such that $\tau f =$ diag$(\bar{k}_1, \ldots, \bar{k}_r)$. Moreover the k_i are uniquely determined by f and will be called the canonical coordinates of f.*

PROOF. In one direction, let $g = \tau f$, $\tau \in U_n$. Since B_k is invariant under $\tau g_k = \tau_k f_k$ for every k, where τ_k is the restriction of τ on B_k. Hence $D(g_k) = D(f_k)$, whence if $g \in O$, then so does f. Conversely, let $f \in O$. Write xy in place of $f(x,y)$ for brevity. Consider two cases: n is even and n is odd. (i) $n = 2r$. We define two new vectors e'_n and e'_{n-1} by $e'_n = e_n + t_{n-2}e_{n-2} + \cdots + t_1 e_1$ and $e'_{n-1} = e_{n-1} + u_{n-2}e_{n-2} + \cdots + u_1 e_1$. We determine the coefficients t_i and u_j from $n-2$ conditions $e'_n e_i = e'_{n-1} e_i = 0$, $i = 1, 2, \ldots, n-2$. Indeed, let $f_{ij} = f(e_i, e_j)$, $1 \leq i, j \leq u$, and let $M(f)$ be the matrix of f, i.e., $M(f) = (f_{ij})$. Then the system of equations $e'_n e_i = 0$, $i = 1, 2, \ldots, n-2$, has the principal $(n-2) \times (n-2)$ minor of $M(f)$ as its matrix. The latter is invertible and the claim follows. An obvious induction completes the proof. (ii) $n = 2r+1$. We note that f_{n-1} satisfies the conditions of (i) and hence τf_{n-1} has the canonical form for some $\tau \in U_n$. Utilizing the argument used in (i) we find $n-1$ elements t_i in K such that $e'_n = e_n + t_{n-1}e_{n-1} + \cdots + t_1 e_1$ is orthogonal to all e_i with $i \leq n-1$. This completes the proof.

As for the uniqueness statement it comes down to the following. Let $g^{(i)} = \text{diag}(\bar{k}_1^{(i)}, \ldots, \bar{k}_r^{(i)})$, $i = 1, 2$, and suppose that $g^{(2)} = \tau g^{(1)}$ for some $\tau \in U_n$. Then $k_i^{(1)} = k_i^{(2)}$ for all i. For by definition $k_m^{(i)} = g^{(i)}(e_{2m-1}, e_{2m})$, $1 \leq m \leq r$. Since $g^{(2)}(e_j, e_{2m}) = 0$ for all $j < 2m-1$ we obtain that $g^{(1)}(e_j\tau, e_{2m}\tau) = 0$ for $j < 2m-1$. As $e_1\tau = e_1$ we see that $g^{(1)}(e_1, e_{2m}\tau) = 0$ and an easy induction on j shows that $g^{(1)}(e_j, e_{2m}\tau) = 0$ for $j < 2m-1$. Hence $e_{2m}\tau = e_{2m} + \alpha e_{2m-1}$ for some $\alpha \in K$, whence $k_m^{(2)} = g^{(2)}(e_{2m-1}, e_{2m}) = g^{(1)}(e_{2m-1}\tau, e_{2m}\tau) = g^{(1)}(e_{2m-1}, e_{2m}) = k_m^{(1)}$. ◁

3. The canonical coordinates give rise to the functions $k_i\colon O \to K$, $f \mapsto k_i(f)$, and by the preceding lemma these are rational functions of the coordinates f_{ij}, $i < j$, of $f \in O$. The set O is clearly invariant with respect to the action of D_n in C^*. Hence one has the induced action of D_n on the set of functions on $O, \varphi \mapsto \varphi\delta$, where $\varphi\delta\colon f \mapsto \varphi(\delta f)$, $f \in O$ and $\delta \in D_n$. It is well-known and easy to see that every rational character on D_n has the form $\chi(\delta) = \delta_1^{\lambda_1} \dots \delta_n^{\lambda_n}$, $\delta = \operatorname{diag}(\delta_1, \dots, \delta_n)$, where $\lambda = (\lambda_1, \dots, \lambda_n)$ is called the weight of χ. A rational function $\varphi\colon O \to K$ is called weight function of weight λ if $K\varphi$ affords 1-dimensional representation of D_n with the character χ_λ.

LEMMA 3. *Every $k_i\colon O \to K$ is a weight function of the weight $\chi(\delta) = \delta_{2i-1}\delta_2 i$.*

PROOF. By the definition $(k_i\delta)(f) = k_i(\delta f)$, i.e., $k_i\delta$ is the ith canonical coordinate of δf. Let $\tau f = \operatorname{diag}(\bar{k}_1, \dots, \bar{k}_\tau)$. Then from $(\delta\tau\delta^{-1})(\delta f) = \delta \operatorname{diag}(\bar{k}_1 \dots \bar{k}_\tau)$ and the obvious equality $\overline{\delta \operatorname{diag}(\bar{k}_1, \dots, \bar{k}_\tau)} = \operatorname{diag}(\overline{\delta_1\delta_2 k_1}, \dots, \overline{\delta_{2\tau-1}\delta_{2\tau}k_\tau})$ we conclude that $k_i\delta = \delta_{2i-1}\delta_{2i}k_i$. ◁

Let $n = 2r$ be even. We now define the function $Pf(f)$, *pfaffian of f*. Firstly, if $f \in O$, then we put $Pf(f) = k_1 \dots k_r$, the k_i are the canonical coordinates of f. Then we observe that as $Pf(f)$ is a rational function of f_{ij} the equality $Pf(f)^2 = D(f)$ implies that $Pf(f)$ is a polynomial, i.e., $Pf(f) = p(f_{ij})$ for some $p \in K[f_{ij}]$. Hence we set $Pf(f) = p(f_{ij})$ for every $f = (f_{ij}) \in C^*$. It follows immediately that $Pf(f)$ is T_n-invariant of weight $(1, \dots, 1)$ (n units). For τf and f have the same canonical coordinates for every $\tau \in U_n$ and $f \in O$, so that $k_i\tau = k_i$, whence $(Pf)\tau = Pf$. The second assertion follows from the fact that the weight of a product of two weight functions is the sum of their weights. Another observation is that $Pf(f) = \sum \alpha_{(i)(j)} f_{i_1 j_1} \dots f_{i_r j_r}$ with $(i_1, j_1, \dots, i_r, j_r)$ being a permutation of $(1, 2, \dots, n)$. Indeed, every polynomial of f_{ij} of weight $(1, \dots, 1)$ (n units) has to have such form. With some extra effort it can be shown that $Pf(f)$ is an irreducible polynomial in $K[f_{ij}]$. In particular since Pf is a homogeneous of degree r polynomial it belongs to $P_r(C^*)$. Let P_r denote the element of $S^r(\Lambda^2 B)$ which corresponds to Pf under the '$*$'. We now can complete the computation of T_n-invariants in $S(\Lambda^2 B)$.

PROPOSITION 4. *Let $n = \dim B$, let m be an arbitrary natural number and let $\lambda = (\lambda_1, \dots, \lambda_r)$ be a partition of m of length $\leq r = [n/2]$. For every λ there exists one within a scalar factor T_n-invariant of weight $\tilde{\lambda} = (\lambda_1, \lambda_1, \dots, \lambda_r, \lambda_r)$ in $S^m(\wedge^2 B)$. That has the form $I_{\tilde{\lambda}} = P_\tau^{\lambda r} P_{\tau-1}^{\lambda_{r-1}} \dots P_1^{\lambda_1 - \lambda_2}$.*

PROOF. Let I_μ be a T_n-invariant of weight μ in $S^m(\Lambda^2 B)$. Let $(I_\mu)_*$ be the image of I_μ. Then $(I_\mu)_*$ is a T_n-invariant of weight μ in $P_m(C^*)$ and in particular $(I_\mu)_*(f) = (I_\mu)_*(\tau f)$ for every $\tau \in U_n$. Assuming $f \in O$ we select τ such that $\tau f = \operatorname{diag}(\bar{k}_1, \dots, \bar{k}_\tau)$. Then we obtain $(I_\mu)_*(f) = \sum \alpha_\lambda k_1^{\lambda_1} \dots k_r^{\lambda_r}$ with $\sum \lambda_i = m$, because $(I_\mu)_*$ is a homogeneous polynomial of degree m

in the coordinates of f. By Lemma 3 every $k_1^{\lambda_1}\dots k_r^{\lambda_r}$ is T_n-invariant of weight $(\lambda_1,\lambda_1,\dots,\lambda_r,\lambda_r) =: \tilde{\lambda}$. Hence the distinct functions $k_1^{\lambda_1}\dots k_r^{\lambda_r}$ in the above sum have distinct weights. As $(I_\mu)_*$ has weight μ it follows that $(I_\mu)_* = \alpha k_1^{\lambda_1}\dots k_r^{\lambda_r}$ for the unique $\lambda = (\lambda_1,\dots,\lambda_r)$ such that $\tilde{\lambda} = \mu$. Moreover if $f \in O$ and f_k denotes the restriction of f to B_k (see 1.), then $k_i(f) = Pf(f_{2i})/Pf(f_{2(i-1)})$. Hence $(I_{\tilde{\lambda}})_*(f) = Pf(f_2)^{\lambda_1-\lambda_2}Pf(f_4)^{\lambda_2-\lambda_3}\dots Pf(f_{2r})^{\lambda_r}$. Now $(I_\mu)_*$ is a polynomial in $f_{i,j}$. As $Pf(f_{2i})$ is irreducible polynomial in $f_{i,j}$ it becomes clear that all differences $\lambda_i - \lambda_{i+1}$ are ≥ 0. Hence $\lambda \vdash m$ and returning to $S^m(\Lambda^2 B)$ we obtain the result. ◃

4. Let $\overline{W}^{\tilde{\lambda}}$ be the G_n-submodule of $S^m(\Lambda^2 B)$ generated by the invariant $I_{\tilde{\lambda}} \cdot \overline{W}^{\tilde{\lambda}}$ is the Weyl module associated with $\tilde{\lambda}$ and we arrive at the Thrall's result: $S^m(\Lambda^2 B) = \bigoplus\{\overline{W}^{\tilde{\lambda}} | \lambda \vdash m$ and $l(\lambda) \leq [n/2]$. Turning to $\mathrm{gr}_d(KF)$ we see by (1) that $\mathrm{gr}_d(KF) \cong \bigoplus\{S^l(B) \otimes \overline{W}^{\tilde{\lambda}} | l + 2m = d\}$. Applying the Littlewood-Richardson Rule (**[J1**, §16]) we obtain that $S^l(B) \otimes \overline{W}^{\tilde{\lambda}} \simeq \bigoplus W^\mu$, where the sum on the right is extended over the set of partitions of d such that: (i) $\mu = (\lambda_1+\delta_1,\dots,\lambda_r+\delta_r,\lambda_r,\delta_{r+1})$; (ii) $0 \leq \delta_i \leq \lambda_{i-1}-\lambda_i$ for all $1 < i \leq r$; (iii) $0 \leq \delta_{r+1} \leq \lambda_r$ if $2r < n$ and $\delta_{r+1} = 0$, otherwise; and (iv) $\sum_{i=1}^{r+1} \delta_i = l$. On the other hand let μ be a partition of d with $l(\mu) \leq n$. We let m be $\sum_{i=1}^{r} \mu_{2i}$, where $r = [n/2]$ and further we denote by δ_i the difference $\mu_{2i-1}-\mu_{2i}$, $1 \leq i \leq r$, and we define $\delta_{r+1} = \mu_n$ if n is odd and zero, otherwise. Setting $l = \sum_{i=1}^{r+1} \delta_i$ and $\mu_e = (\mu_2,\mu_4,\dots,\mu_{2r}) \vdash m$ we conclude firstly that $l + 2m = d$ and secondly that W^μ occurs in $S^l(B) \otimes \overline{W}^{\tilde{\mu}_e}$. Thus $\mathrm{gr}_d(KF) \simeq \bigotimes\{W^\mu | \mu \vdash d$ and $l(\mu) \leq n\}$ which completes the proof of Theorem 1. Explicit imbeddings of W^μ into $\mathrm{gr}_d(KF)$ will be described in the next chapter.

3.3. We intend to apply the preceding result to the study of the axiomatic rank, called just rank below, of ideals I^d. We refer to the introduction for all undefined notions. There again $I = I(KF)$ with F being a free nilpotent group of class 2 on the countable set $X = \{x_i | i \in N\}$. Some more notations are in order. For a subset Z of KF we denote by $v(Z)$ the verbal ideal generated by Z. For every $n \in N$ let F_n denote the subgroup of F generated by $x_1,\dots,x_n$. We put I_n for $I(KF_n)$. More generally if V is a verbal ideal in KF we put $V_n = V \cap VF_n$. Another expression for V_n involves the projections $\pi_n\colon F \to F_n$, $n = 1,2,\dots$, defined by $x_i\pi_n = x_i$ for all $i \in \underline{n}$ and $x_i\pi = 1$, otherwise. It is easy to see (cf. [**H.N.**, p. 7]) that for every verbal ideal V of KF $V\pi_n = V_n$. In particular, since $I\pi_n = I_n$ we obtain $I^d\pi_n = (I\pi_n)^d = I_n^d$, so that $(I^d)_n = I_n^d$. We aim at the proof of the

THEOREM 1. *The chain of the verbal ideals* $I^d = v(I_d^d) > v(I_{d-1}^d) > \cdots > v(I_1^d) > 0$ *is strictly decreasing.*

REMARK. Since I^d is a K-span of $(f_1 - 1)\dots(f_d - 1)$, the f_i run over F, it is generated by $(x_1 - 1)\dots(x_d - 1)$ as the verbal ideal, whence the first equality in the above chain. Further we recall that by the definition of rank the rank of V is the smallest r such that $V = v(V_r)$. As $v(I_n^d)\pi_{n-1} = v(I_n^d)\pi_n\pi_{n-1} = I_n^d\pi_{n-1} = I_{n-1}^d$ the theorem says in effect that the rank of $v(I_n^d)$ is n for every $n = 1, 2, \ldots, d$.

PROOF OF THE THEOREM. We shall divide the proof into several steps.

1. The equality $v(I_n^d) = v(I_{n-1}^d)$ can be tested in KF_n. We mean the following. Let v' be the symbol of the verbal closure in KF_n. Then for every $Z \subset KF_n$ $v(Z)\pi_n = v'(Z)$. Indeed, every $u \in v(Z)$ has the form $u = \sum \rho_i(z_i\gamma_i)$, where $\rho_i \in KF$ and the γ_i belong to $\mathrm{End}(F)$. As $z_i \in KF_n$, clearly $z_i\pi_n = z_i$. Hence $u\pi_n = \sum(\rho_i\pi_n)(z_i\pi_n\gamma_i\pi_n)$. Now $\pi_n\gamma_i\pi_n \in \mathrm{End}(F_n)$ for every γ_i and the assertion follows. In particular, $v(I_n^d) = v(I_{n-1}^d)$ implies that $v(I_n^d)\pi_n = I_n^d = v'(I_{n-1}^d)$ and conversely if $I_n^d = v'(I_{n-1}^d)$, then clearly $v(I_n^d) = v(I_{n-1}^d)$.

2. The equality $I_n^d = v'(I_{n-1}^d)$ can be tested in $\mathrm{gr}_d(KF_n)$. The passage to $\mathrm{gr}_d(KF_n)$ is based on the following simple remarks. Let $\Theta\colon I_n^d \to I_n^d/I_n^{d+1} = \mathrm{gr}_d(KF_n)$ be the natural epimorphism. If $Z \subset KF_n$ and $V = v'(Z)$, then firstly $\Theta(V) = \langle\Theta(Z)\rangle_{G_n}$, where $\langle Y\rangle_{G_n}$ denotes the G_n-submodule generated by $Y \subset \mathrm{gr}_d(KF_n)$. Secondly, taking $Z = I_{n-1}^d$ and noticing that $I_{n-1}^d \cap I_n^{d+1} = I_{n-1}^{d+1}$ we obtain $\Theta(I_{n-1}^d) = I_{n-1}^d + I_n^{d+1}/I_n^{d+1} \simeq I_{n-1}^d/I_{n-1}^{d+1} \simeq \mathrm{gr}_d(KF_{n-1})$. Hence we must examine the equality $\langle\mathrm{gr}_d(KF_{n-1})\rangle_{G_n} = \mathrm{gr}_d(KF_n)$ in which $\mathrm{gr}_d(KF_{n-1})$ is identified with the subspace of $\mathrm{gr}_d(KF_n)$ of all elements expressable through the first $x_1, \ldots, x_{n-1}$ generators of F_n. We shall prove below that the latter equality is impossible. But before tackling that task we notice that in fact, conversely, the equality $\langle\mathrm{gr}_d(KF_{n-1}\rangle_{G_n} = \mathrm{gr}_d(KF_n)$ implies $I_n^d = v'(I_{n-1}^d)$, so that both these assertions are equivalent. Indeed, let $V = v'(I_{n-1}^d)$. As we saw above $\langle\mathrm{gr}_d(KF_{n-1}\rangle_{G_n} = \Theta(V)$, whence $\langle\mathrm{gr}_d(KF_{n-1})\rangle_{G_n} = \mathrm{gr}_d(KF_n)$ if and only if $V + I_n^{d+1} = I_n^d$. Now by a result of Gruenberg ([**G2**, Thm. 1.8]) V contains some power of I_n, say $V \supset I_n^c$. Let M be I_n^d/I_n^e and $\overline{V}$ be the image of V in M. Then $M/\overline{V} = I_n(M/\overline{V})$. By Nakayma's lemma $M/\overline{V} = 0$ as desired.

3. Here we intend to show that $\langle\mathrm{gr}_d(KF_{n-1})\rangle_{G_n} \neq \mathrm{gr}_d(KF_n)$ for every $n = 2, \ldots, d$. As we know $\mathrm{gr}_d(KF_{n-1}) = \bigotimes\{W_{n-1}^\lambda | \lambda \vdash d \text{ and } l(\lambda) \leq n-1\}$ as a G_{n-1}-module, and hence $\langle\mathrm{gr}_d(KF_{n-1})\rangle_{G_n} = \bigoplus\langle W_{n-1}^\lambda\rangle_{G_n}$. Pick a W_{n-1}^λ and let $J = \langle W_{n-1}^\lambda\rangle_{G_n}$. We claim that $J = W_n^\lambda$, i.e., J is the λth Weyl module for G_n. It will suffice to show that J is irreducible, since then J contains only one within a scalar multiple T_n-invariant, namely, that which generates W_{n-1}^λ. Suppose that $J = J_1 \oplus J_2$ with J_i being a G_n-submodule of J. Then $J\pi_{n-1} = W_{n-1}^\lambda = J_1\pi_{n-1} \oplus J_2\pi_{n-1}$ and since π_{n-1} commutes with every $\phi \in G_{n-1}$ we conclude that $J_i\pi_{n-1}$ are G_{n-1}-submodules of W_{n-1}^λ. Hence one of them, say $J_1\pi_{n-1}$, equals to W_{n-1}^λ. But then $J_1 \supset \langle J_1\pi_{n-1}\rangle_{G_n} = \langle W_{n-1}^\lambda\rangle_{G_n} = J$ and we are done. Finally, we know that $\mathrm{gr}_d(KF_n)$ contains Weyl modules W^μ with $l(\mu) = n$ among its irreducible constituents. Such μ does exist, e.g., $(d-(n-1), 1, \ldots, 1)$ ($n-1$ units). None of them occur in $\langle\mathrm{gr}_d(KF_{n-1})\rangle_{G_n}$ thereby proving our assertion. $\triangleleft$

§4. The Maximal Condition for $GL(B)$-ideals of $\text{gr}(KF)$

4.1. We continue in the notation of §3. Thus F is a free nilpotent group of class 2 on a denumerable set $X = \{x_i | i \in \mathbf{N}\}$ of free generators and $B = K \otimes F/F'$ is a K-space with a basis $\mathscr{E} = \{e_i | i \in \mathbf{N}\}$. Let V be an ideal of $\text{gr}(KF)$ invariant under the action of $G := GL(B)$. We aim at proving that V can be generated as G-ideal by a finite number of elements. The proof proceeds along the following lines. Firstly, every G-ideal is clearly graded, i.e., $V = \bigoplus V_d$, where $V_d = V \cap \text{gr}_d(KF)$. Secondly, KF_d can be regarded as the subalgebra of KF generated by $x_i^{\pm 1}$, $i = 1, 2, \dots, d$, which leads to the identification of $\text{gr}_d(KF_d)$ with the subspace of $\text{gr}_d(KF)$ of elements expressible through $e_1, \dots, e_d$. Let $B_d = \langle e_1, \dots, e_d \rangle$, $G_d = GL(B_d)$ and let $\pi_d \colon KF \to KF_d$ be defined by $x_i \pi_d = x_i$, if $i \leq d$ and 0, otherwise. It is easy to see that π_d induces an isomorphism between the lattice of G-submodules of $\text{gr}_d(KF)$ and G_d-submodules of $\text{gr}_d(KF_d)$. Replacing V_d by $\pi_d(V_d)$ we can associate with every component V_d of V the set $P_d(V)$ of partitions of d. Namely, $P_d(V) = \{\lambda | \lambda \vdash d \text{ and } W^\lambda \subset \pi_d(V_d)\}$. Let Λ be the set of all partitions of the natural numbers. Λ is clearly a graded set, $\Lambda = \bigcup_{d \in \mathbf{N}} \Lambda_d$, where Λ_d is the set of all partitions of d. The association $V \mapsto P(V) := \bigcup_{d \in \mathbf{N}} P_d(V)$ sets up a $1-1$ correspondence between G-ideals and certain subsets of Λ. Let us put an order '$\subseteq$' on Λ by saying $\lambda \subseteq \mu$ if and only if $\lambda_i \leq \mu_i$ regarding partitions as infinite sequences by appending zeros. The main point of the proof consists of showing that a subset of Λ has the form $P(V)$ if and only if it is *closed* with respect to '$\subseteq$' in the sense of Higman ([**HIG**]). It follows readily then that $(\Lambda, \subseteq)$ is a partially well-ordered set, which implies the desired result.

Finally, we remark briefly on the main technical tools used in this chapter. Let B be a finite-dimensional K-space. Let W^λ be the λth Weyl module for $G(B)$. For the $G(B)$-module $W^\lambda \otimes B$ the decomposition into the irreducible $G(B)$-modules $W^\lambda \otimes B \simeq \bigoplus W^\mu$ is a well-known simplest case of the Littlewood-Richardson rule ([**J1**, §16]). This rule, however, says nothing about the actual imbeddings $i_\mu \colon W^\mu \to W^\lambda \otimes B$. These i_μ were computed by P. Oliver ([**OL**]) using J. Towber's model of $G(B)$-irreducible modules ([**T**]). The formulas for i_μ are given in the form of a linear combination of a family of $G(B)$-homomorphisms, which is in general linearly dependent. Here we indicate the expression for i_μ in terms of a linearly independent system of $G(B)$-homomorphisms. Our work is based on various results of G. James, notably the James intersection theorem ([**J1**, §17]) and a model of Weyl modules constructed in ([**J-K**, ch. 8]).

4.2. Let $M = \text{gr}_d(KF)$ and let D be the diagonal subgroup of G. If λ is a sequence of natural numbers we define $\text{supp}(\lambda) = \{i \in \mathbf{N} | \lambda_i \neq 0\}$. Let $\mathbf{N}^{(\infty)}$ stand for the set of sequences with a finite support. For every $\lambda \in \mathbf{N}^{(\infty)}$ we define $\chi_\lambda \colon D \to K$ by $\chi_\lambda(\delta) = \prod \delta_i^{\lambda_i}$, where $\delta = \text{diag}(\delta_1, \dots, \delta_n, \dots)$. An element u of M is χ_λ-homogeneous if $u\delta = \chi_\lambda(\delta) u$ for every $\delta \in D$.

It is immediate that every G-submodule of M is a direct sum of its χ_λ-homogeneous subspaces, λ runs over $\mathbf{N}^{(\infty)}$. Let M_d, $d \in \mathbf{N}$, be the subset of M spanned by the elements whose support is $\subseteq d$. It is easy to see that M_d is a G_d-module where G_d is canonically imbedded in G and the action of G_d on M_d is by restriction. Let $\Lambda_G(M)$ and $\Lambda_{G_d}(M_d)$ denote the lattices of G- and G_d-modules, respectively.

PROPOSITION 1. *$\Lambda_G(M) \simeq \Lambda_{G_d}(M_d)$ under the mapping $p_d\colon V \mapsto V \cap M_d$.*

PROOF. Let u be a χ_λ-homogeneous element of V. We claim that the cardinality of supp(λ) is $\leq d$. For this is true with respect to $B^{\otimes d}$ since χ_λ-homogeneous elements of the latter have the form $\sum k_{(i)} e_{i_1} \dots e_{i_d}$ where for every j, $\lambda_j = \#\{k | i_k = j\}$. As M is a G-epimorphic image of $B^{\otimes d}$ and χ_λ-homogeneous elements of M are images of the same elements of $B^{\otimes d}$ the assertion follows. Hence there exists a permutation $\sigma \in G$ such that $(\mathrm{supp}(\lambda))\sigma \subseteq \underline{d}$. Now from $(u\sigma)\delta = (u\sigma\delta\sigma^{-1})\sigma = \chi_\lambda(\sigma\delta\sigma^{-1})u\sigma$ and since $\chi_\lambda(\sigma\delta\sigma^{-1}) = \delta_1^{\lambda_{1\sigma^{-1}}} \dots \delta_n^{\lambda_{n\sigma^{-1}}} \dots$ we conclude that $\chi_\lambda(\sigma\delta\sigma^{-1}) = \chi_\mu(\delta)$ where μ is defined by $\mu(i) = \lambda(i\sigma^{-1})$. Clearly supp$(\mu) \subseteq \underline{d}$ and therefore $u\sigma \in V \cap M_d$ whence $u \in \langle V \cap M_d \rangle_G$. Thus $V = \langle V \cap M_d \rangle_G$ proving that p_d is a monomorphism.

The rest of the proposition is a consequence of the following remark. Every G- (or G_d-polynomial) representation is uniquely extendable to an End(B)- (or End(B_d)-) representation. Let $\pi_d\colon B \to B_d$ be the mapping introduced in 4.1. By the above remark π_d acts in every G-submodule of V. Moreover $p_d(V) = V\pi_d$ and hence $p_d(V + U) = V\pi_d + U\pi_d = p_d(V) + p_d(U)$ which shows that p_d is a lattice homomorphism. Finally, applying the argument used in step 1 of Theorem 3.3.1 we see that $p_d(\langle U \rangle_G) = U$ for every G_d-submodule U of M_d whence p_d is an epimorphism. ◁

4.3. In this section we bring together some facts on various 'dual' S_d- and G_d-modules scattered over the literature. Let λ be a partition of d, let t be a λ-tableau and let R_t and C_t be the groups of row and column permutations of t, respectively. We associate with λ and t several objects. First, we let $\sigma_t = \sum_{p \in R_t} p$, $\alpha_t = \sum_{p \in C_t} (-1)^p p$ and $\varepsilon_t = \alpha_t \sigma_t$. Secondly, we let Φ^λ stand for the isomorphism class of $\sigma_t K S_d$, Ψ^λ for the isomorphism class of $\alpha_t K S_d$, S^λ for the isomorphism class of $\varepsilon_t K S_d$ and $\hat{S}^\lambda$ for the isomorphism class of $\sigma_t \alpha_t K S_d$. It is evident that all these S_d-modules depend on partition λ, not a particular λ-tableau t. Let A be the 1-dimensional alternative representation of S_d. By a result of Peel ([**PE**], see also [**J1**, 8.15], [**J2**, T.4]) we have the following connections between the above modules.

PROPOSITION 1. *Let K be a field of arbitrary characteristic. Then $\Psi^\lambda \simeq \Phi^{\lambda'}$ and $S^\lambda \simeq S^{\lambda'} \otimes A$. Moreover if* char $K = 0$, *then $\hat{S}^\lambda \simeq S^\lambda$ for every λ and hence $S^\lambda \simeq \hat{S}^{\lambda'} \otimes A$.*

PROOF. The first two claims follow directly from ([**PE**, L.2]). We outline a slightly different argument. Namely, let τ run over the set of representatives of C_t cosets and let a be a generator of A. Then it is immediate that both $\alpha_t KS_d$ and $(\sigma_{t'} \otimes a)KS_d$ are K-freely spanned by $\alpha_t \tau$ and $(\sigma_{t'} \otimes a)\tau$, respectively. Let Θ: $\alpha_t \tau \rightleftharpoons (\sigma_{t'} \otimes a)\tau$ be a K-isomorphism between $\Psi^\lambda \simeq \alpha_t KS_d$ and $\Phi^{\lambda'} \otimes A \simeq (\sigma_{t'} \otimes a)KS_d$. A direct verification shows that Θ is an S_d-isomorphism. Furthermore $\Theta(\varepsilon_t) = (\sigma_{t'} \otimes a)\sigma_t = (\sigma_{t'}\alpha_{t'} \otimes a)$, which shows that $S^\lambda \simeq S^{\lambda'} \otimes A$. Finally let $\hat{}: g \to g^{-1}$ be an anti-isomorphism of KS_d. Assuming that $\operatorname{char} K = 0$ we can norm ε_t so that $f^{-1}\varepsilon_t$, $f \in K$, becomes a primitive idempotent of KS_d. Then $f^{-1}\hat{\varepsilon}_t = f^{-1}\sigma_t\alpha_t$ is also a primitive idempotent. Since "$\hat{}$" preserves the conjugary classes of S_d, it also preserves the simple components of KS_d. Hence $f^{-1}\varepsilon_t$ and $f^{-1}\hat{\varepsilon}_t$ lie in the same simple component of KS_d and we are done. ◁

At this point we invoke our discussion in §3.2. Thus we let B_d be a d-dimensional K-space and with every $\lambda \vdash d$ we associate G_d-module W^λ. In fact W^λ is isomorphic to $\varepsilon_t B_d^{\otimes d}$ for every λ-tableau t. Further let μ be a sequence of positive numbers of norm d and let $\Lambda^\mu(B_d) = \Lambda^{\mu'_1}(B_d) \otimes \cdots \otimes \Lambda^{\mu'_k}(B_d)$, where μ' is the sequence conjugate to μ. We recall a well-known fact ([**J1**, 26.6]) that the multiplicity of $\hat{\varepsilon}_t B^{\otimes d}$ in $S^\mu(B_d) = S^{\mu_1}(B_d) \otimes \cdots \otimes S^{\mu_m}(B_d)$ is given by Kostka number $K_{\lambda,\mu}$ which is the number of λ-standard tableaux of weight μ. Moreover the latter coincide with the multiplicity of $\hat{S}^\lambda$ in Φ^μ. Here we are concerned with the multiplicity of W^λ in $\Lambda^\mu(B_d)$. The corresponding combinatorial objects and their number are defined as follows. Let us call a λ-tableau co-standard if the numbers are strictly increasing along every row and they are weakly increasing down every column. Let $K'_{\lambda,\mu}$ stand for the number of λ-co-standard tableaux of weight μ'. The next statement seems to be known.

PROPOSITION 2. *The dimension of* $\operatorname{Hom}_{G_d}(W^\lambda, \Lambda^\mu)$ *is* $K'_{\lambda,\mu}$.

PROOF. It is well known ([**GR**, ch. 6]) that the category of homogeneous G_d-modules of degree d is equivalent to the category of S_d-modules under the mapping $V \mapsto V^\omega$ where V^ω is the ω-weight subspace of V for $\omega = (1, \dots, 1)$ (d units). In particular $\operatorname{Hom}_{G_d}(U, V) \simeq \operatorname{Hom}_{S_d}(U^\omega, V^\omega)$ for any two U and V in our category. Furthermore it is shown in ([**GR**, 3.2]) that to every weight β for d there corresponds an idempotent e_β in the image of KG_d in $\operatorname{End}_K(B^{\otimes d})$ such that $Ve_\beta = V^\beta$, the β-weight space of V. In particular $\Lambda^\mu(B_d)^\omega = \Lambda^\mu(B_d)e_\omega \simeq \alpha_t(B^{\otimes d}e_\omega)$. Now $B^{\otimes d}e_\omega$ is the subspace of multilinear forms in $e_1, \dots, e_d$. Hence $B^{\otimes d}e_\omega \simeq KS_d$ and therefore $\Lambda^\mu(B_d)^\omega \simeq \alpha_t KS_d \simeq \Psi^\mu$.

Likewise $(W^\lambda)^\omega = W^\lambda e_\omega \simeq \alpha_t\sigma_t(B_d^{\otimes d}e_\omega) \simeq \alpha_t\sigma_t KS_d \simeq S^\lambda$, where t is a λ-tableau. It follows that $\operatorname{Hom}_{G_d}(W^\lambda, \Lambda^\mu) \simeq \operatorname{Hom}_{S_d}(S^\lambda, \Psi^\mu)$. By the preceding proposition $S^\lambda \simeq \hat{S}^{\lambda'} \otimes A \simeq S^{\lambda'} \otimes A$ and $\Psi^\mu \simeq \Phi^{\mu'} \otimes A$. Hence $\operatorname{Hom}_{S_d}(S^\lambda, \Psi^\mu) \simeq \operatorname{Hom}_{S_d}(S^{\lambda'}, \Phi^{\mu'})$. Thus $\dim_K \operatorname{Hom}_{G_d}(W^\lambda, \Lambda^\mu) = K_{\lambda',\mu'}$. Finally every standard λ'-tableau arises from a unique co-standard λ-tableau by reflecting it through the main diagonal. Hence $K'_{\lambda,\mu} = K_{\lambda',\mu'}$. ◁

4.4. Some G_d-homomorphisms between Λ^α and Λ^β.

Let $\Lambda B_d = \bigoplus_{d=0}^{d} \Lambda^p(B_d)$ be the exterior algebra on B_d and $i\colon \Lambda B_d \to \Lambda B_d$, $i(u) = (-1)^p u$, if $u \in \Lambda^p(B_d)$ is the canonical involution on ΛB_d. Let ϕ be a linear map from B_d into K. There is a standard procedure ([**GRE**]) that associates with ϕ an antiderivation ∂_ϕ of ΛB_d. Namely, for every p, $0 \le p \le d$, $\partial_\phi\colon \Lambda^p(B_d) \to \Lambda^p(B_d)$ is defined on the decomposable elements by $\partial_\phi(x_1 \wedge \cdots \wedge x_p) = \sum_{i=1}^{p}(-1)^{i-1}\phi(x_i)x_1 \wedge \cdots \wedge \hat{x}_i \wedge \cdots \wedge x_p$ where as usual a '^' over a symbol indicates that it has been deleted. A routine verification shows that $\partial_\phi(u \wedge v) = \partial_\phi(u) \wedge v + (-1)^p u \wedge \partial_\phi(v)$ for every $u \in \Lambda^p(B_d)$ and $v \in \Lambda^q(B_d)$. A particular set of antiderivations will be useful later. Namely, let e_i^*, $i = 1, 2, \ldots, d$, be the mapping dual to e_i, i.e., $e_i^*(e_j) = \delta_{ij}$. Then we put $\partial_i = \partial_{e_i^*}$. A different kind of mappings of ΛB_d is provided by the multiplication in this algebra. Namely, every $x \in B_d$ gives rise to a mapping $x\colon u \mapsto x \wedge u$, $u \in \Lambda B_d$. We now describe a family of G_d-mappings connecting various Λ^α. Let α be a sequence of natural numbers of norm d with $\alpha' = (a_1, \ldots, a_h, \ldots)$. Pick a pair i, j of natural numbers and let β be that sequence for which $\beta' = (a_1, \ldots, a_i + 1, \ldots, a_j - 1, \ldots, a_h, \ldots)$. Then a mapping $\omega_{ij}\colon \Lambda^a \to \Lambda^\beta$ is defined by the formula

$$\omega_{ij}(x_1 \otimes \cdots \otimes x_h \otimes \ldots) = \sum_{p \in \underline{d}} x_1 \otimes \cdots \otimes x_{i-1} \otimes e_p \wedge x_i \otimes \cdots \otimes \partial_p(x_j) \otimes \ldots .$$

LEMMA 1. *ω_{ij} is a G_d-homomorphism.*

PROOF. $GL(B_d)$ is generated by its diagonal subgroup D_d and the set of transvections. ω_{ij} is clearly a D_d-homomorphism so it is left to verify that $\omega_{ij}(x\tau) = (\omega_{ij}(x))\tau$ for every $x \in \Lambda^\alpha$ and every transvection τ. Of course, it will suffice to limit x to a basis for Λ^α. Furthermore, as the action of ω_{ij} affects only two factors of Λ^α we can assume that α is a two part sequence, $\alpha = (a, b)$. Put $\omega = \omega_{ij}$ and let $(i) = (i_1, \ldots, i_a)$ and $(j) = (j_1, \ldots, j_b)$ be two increasing sequences of natural numbers between 1 and d. It is clear that the elements $e_{(i)} \otimes e_{(j)}$, $e_{(i)} = e_{i_1} \wedge \cdots \wedge e_{i_a}$ and $e_{(j)} = e_{j_1} \wedge \cdots \wedge e_{j_b}$ for all possible (i) and (j) span $\Lambda^{(a,b)}$. Pick a τ, say $\tau\colon e_p \mapsto e_p + te_q$, $e_i \mapsto e_i$, $i \neq p$. Suppose that $p = j_s$. Noticing that $\omega(e_{(i)} \otimes e_{(j)}) = \sum_{k=1}^{b}(-1)^{k-1} e_{j_k} \wedge e_{(i)} \otimes e_{j_1} \wedge \cdots \wedge \hat{e}_{j_k} \wedge \cdots \wedge e_{j_b}$ we obtain $(\omega(e_{(i)} \otimes e_{(j)}))\tau = \sum_{k \neq s}(-1)^{k-1} e_{j_k} \wedge e_{(i)}\tau \otimes e_{j_1} \wedge \cdots \wedge \hat{e}_{j_k} \wedge \cdots \wedge (e_p + te_q) \wedge \cdots \wedge e_{j_b} + (-1)^{s-1} e_p \wedge e_{(i)}\tau \otimes e_{j_1} \wedge \cdots \wedge \hat{e}_{j_s} \wedge \cdots \wedge e_{j_b}$. On the other hand $e_{(i)}\tau \otimes e_{(j)}\tau = e_{(i)}\tau \otimes e_j + te_{(i)}\tau \otimes e_{j_1} \wedge \cdots \wedge e_q \wedge \cdots \wedge e_{j_b}$ from which by a direct computation one obtains $\omega(e_{(i)}\tau \otimes e_{(j)}\tau) = (\omega(e_{(i)} \otimes e_{(j)}))\tau$. The case p does not occur in (j), is simpler and can be treated similarly. ◁

Finally we remark that these ω_{ij} are wedge analogs of James's Ψ-maps ([**J1**, §17 and 26], cf. [**K**, 4.1]). This analogy plays an important rôle throughout this chapter.

4.5. A basis for one particular Hom.

Let λ be a partition of its norm $|\lambda| = \sum \lambda_i$. If μ is another partition we say that μ contains λ if $\mu_i \ge \lambda_i$ for all i, notation $\mu \supset \lambda$. Clearly $\mu \supset \lambda$ implies that the Young diagram $[\lambda]$ of λ is contained as a subset in the diagram $[\mu]$.

In this case the skew-diagram $[\mu - \lambda]$ ensues by deleting the $[\lambda]$-nodes from $[\mu]$. The norm of $[\mu - \lambda]$ is set to be $\sum_i(\mu_i - \lambda_i)$.

From now on λ denotes a partition of $d - 1$ and W^λ stands for the corresponding Weyl module for G_d. We let $E(\lambda)$ be the set of all $\mu \supset \lambda$ with $|\mu - \lambda| = 1$. We pick two special elements in $E(\lambda)$ as follows. If $\lambda = (\lambda_1, \ldots, \lambda_m)$, then we put $\lambda_r = (\lambda_1 + 1, \lambda_2, \ldots, \lambda_m)$ and $\lambda_c = (\lambda_1, \ldots, \lambda_m, 1)$. Of course, the diagram of λ_r is obtained from $[\lambda]$ by adding one node to the first row and that of λ_c by adding one node to the first column of $[\lambda]$. Let $\lambda' = (\lambda'_1, \ldots, \lambda'_h)$. Below we find it more convenient to use the exponential notation for λ'. Let $a_1 = \lambda'_1, \ldots, a_q = \lambda'_{i_q}$ be all different parts of λ' and $k_1, \ldots, k_q$ be their multiplicities. Then we write $\lambda' = (a_1^{k_1}, \ldots, a_q^{k_q})$. In this notation $\lambda'_r = (a_1^{k_1}, \ldots, a_q^{k_q}, 1)$ and $\lambda'_c = (a_1 + 1, a_1^{k_1 - 1}, \ldots, a_q^{k_q})$. Next we introduce several notions. Let μ be a partition. The maximal set of equal columns of $[\mu]$ is called a block of $[\mu]$. A node on the rim of $[\mu]$ is removable if the removal of this node leaves a proper diagram. Likewise a node on the rim of $[\mu]$ is extendable if the addition of a new node to its column produces a proper diagram. Plainly, a node is removable (extendable) if it is the end node of the ith column for some i such that $\mu'_i > \mu'_{i+1}$ ($\mu'_{i-1} > \mu'_i$ or $i = 1$). Every block has a unique removable (extendable) node. Let $\mu' = (a_1^{k_1}, \ldots, a_q^{k_q})$. Given a natural number p we say that p belongs to the ith block of $[\mu]$ if the pth column belongs to this block, i.e., $\mu'_p = a_i$. The p-removable node of $[\mu]$ is the removable node of the block to which p belongs. We now pick λ_c and define $S(\lambda_c, \lambda_r)$ to be the set of all sequences $s = (1, s_1, \ldots, s_m, h+1)$ satisfying the conditions

(i) $m \le q$

(ii) if $i < j$, then $s_i < s_j$

(iii) if $i < j$, then $(\lambda'_c)_{s_i} > (\lambda'_c)_{s_j}$

Let $St^0(\lambda_c, \lambda_r)$ stand for the set of all λ_c-co-standard tableaux of weight λ'_r. The main result is

PROPOSITION 1. *There exists a* $1 - 1$ *correspondence* $St^0(\lambda_c, \lambda_r) \rightleftharpoons S(\lambda_c, \lambda_r)$.

PROOF. In one direction for every $T \in St^0(\lambda_c, \lambda_r)$ we define $s_T \in S(\lambda_c, \lambda_r)$ as follows. $s_T = (s_0, s_1, \ldots, s_m, h + 1)$, where $s_0 = 1$ and s_i occupies the s_{i-1}-removable node for $i = 1, 2, \ldots, m$, and $h+1$ occupies the s_m-removable node of T. Conversely, given $s = (s_0 = 1, s_1, \ldots, s_m, h + 1) \in S(\lambda_c, \lambda_r)$ we define T_s to be that λ_c-tableau for which:

(a) If a block of $[\lambda_c]$ does not contain an s_i column for some $i = 0, 1, \ldots, m$, then the jth column of this block is filled with $(\lambda'_r)_j$ j's.

(b) If a block of $u, u + 1, \ldots, v$ columns of $[\lambda_c]$ contains the s_ith column, i.e., $s_i \in \{u, u+1, \ldots, v\}$, then for every j, $u \le j < s_i$, the jth column is filled with $(\lambda'_r)_j$ j's. If $s_i \le j < v$, then the jth column is filled with $(\lambda'_r)_j - 1$ j's and $j + 1$. If $j = v$, then the vth column is filled with $(\lambda'_r)_v - 1$ v's and s_{i+1}. In case $i = m$ we regard s_{m+1} as $h + 1$.

The key observation that gives rise to the above correspondences is that in a co-standard tableau T a number s can occupy a node on a t column if $s \geq t$. Now let $T \in St^0(\lambda_c, \lambda_r)$. By the preceding remark $(\lambda'_r)_1$ 1's must fill out all but the last node of the first column of T. Let s_1 be the number in the last node of the first column. Let the $(\lambda'_c)_{s_1}$-block contain $u, u+1, \ldots, v$ columns. We claim that for all $1 < j < s_1$ the nodes of a jth column must be filled with a j. Further, if $s_1 \leq j < v$ then the nodes of the jth column must be filled out with $(\lambda'_r)_j - 1$ j's and one $j+1$ and if $j = v$ then the nodes of the vth column are filled with $(\lambda'_r) - 1$ v's and one x such that $x > v$. Indeed, since all 1's occupy the 1st column, we can assume by induction on $j < s_1$ that for $j-1$ our claim is true. Then by our opening remark j cannot occupy a node on a p column for any $p > j$, whence it has to stand on the jth column. Since we have $(\lambda'_r)_j$ j's and $(\lambda'_c)_j$ nodes and $(\lambda'_r)_j = (\lambda'_c)_j$ we are done. Next consider case $s_1 \leq j < v$. By the preceding argument and using induction on j we conclude that $(\lambda'_r)_j - 1$ j's must stand in the jth column. If the last node of this column were taken by a $u > j+1$, then, as this node is not removable, $j+1$ cannot occupy the node right next to it, which forces us to put one $j+1$ into a latter column, a contradiction. Hence $u = j+1$ and the second part of the claim follows. Finally, if $j = v$, then the last node of the vth column is removable and therefore can be occupied by any $x > v$. Repeating our argument with x in place of s_1 we see that a selection of x, say $x = s_2$, will determine T up to the removable node of the $(\lambda'_c)_{s_2}$-block. It becomes clear that $s_T = (1, s_1, \ldots, s_m, h+1)$ determines T so that different tableaux have different sequences. The rest of the proposition is clear. ◁

COROLLARY 1. *Let* $l_1 = k_1 - 1$ *and* $l_i = k_i$, $i = 2, \ldots, q$. *Let* $\sigma_r(l_1, \ldots, l_q)$ *denote* $\sum\{l_{i_1} \ldots l_{i_r} | \{i_1, \ldots, i_r\} \subset \{1, \ldots, q\}\}$ *and* $\sigma_0(l_1, \ldots, l_q) = 1$. *Then* $|S(\lambda_c, \lambda_r)| = \sum_{r=0}^{q} \sigma_r(l_1, \ldots, l_q)$.

PROOF. The subset $S(\lambda_c, \lambda_r)$ of the sequences $s = (1, s_1, \ldots, s_r, h+1)$ has the cardinality of the set of r-tuples $(s_1, \ldots, s_r)$. These s_i must belong to different blocks but are arbitrary otherwise. Plainly the number of such r-tuples is $\sigma_r(l_1, \ldots, l_q)$. ◁

With each $s \in S(\lambda_c, \lambda_r)$ we shall associate a homomorphism $\omega_s \colon W^{\lambda_c} \to \Lambda^{\lambda_r}$ by defining $\omega_s = \omega_{h+1, s_m} \ldots \omega_{s_1, 1}$. The latter notation will be abbreviated to $\omega_{h+1, s_m, \ldots, s_1, 1}$. More generally the symbol $\omega_{a,b,\ldots,c,d}$ will denote the product $\omega_{a,b} \ldots \omega_{c,d}$.

THEOREM 1. *The set* $\{\omega_s | s \in S(\lambda_c, \lambda_r)\}$ *forms a basis for* $\mathrm{Hom}_{G_d}(W^{\lambda_c}, \Lambda^{\lambda_r})$.

PROOF. By the preceding proposition and Proposition 4.3.2 we know that our set has the right number of elements. It remains to show that it is linearly independent.

First, we notice that for every $\nu = (\nu_1, \ldots, \nu_m)$ the set of all decomposable vectors $x_1 \otimes \cdots \otimes x_m$, where $x_i = e_{j_1} \wedge \cdots \wedge e_{j_{\nu_i}}$, $j_1 < \cdots < j_{\nu_i}$, forms a basis for Λ^ν. Secondly, we recall that for every partition μ there corresponds an element I_μ of Λ^μ. For $\mu = \lambda_c$ with $\lambda_c' = (m_1, \dots, m_h) I_{\lambda_c} = I_{m_1} \otimes \cdots \otimes I_{m_h}$ where for every natural m, $I_m = e_1 \wedge \cdots \wedge e_m$. Let $I_m^{(i)}$ denote $e_1 \wedge \cdots \wedge \hat{e}_i \wedge \cdots \wedge e_m$. It follows from the definition of ω_s that $\omega_s(I_{\lambda_c})$ is a signed sum of the terms $T_{(i)} = T_{(i_1,\dots,i_{k+1})} = I_{m_1}^{(i_1)} \otimes \cdots \otimes (e_{i_1} \wedge I_{m_{s_1}})^{(i_2)} \otimes \cdots \otimes (e_{i_k} \wedge I_{m_{s_k}})^{(i_{k+1})} \otimes \cdots \otimes e_{i_{k+1}}$ in which $(i_1, \dots, i_{k+1})$ runs over the set of $(k+1)$-tuples such that $i_j > m_{s_j}$ for $j = 1, 2, \dots k$. Next we select an element s^0 in every given subset Z of $S(\lambda_c, \lambda_r)$ as follows. s_1^0 has the minimal value over all s_1 as s runs over Z, or $s^0 = (1, h+1)$ if the latter lies in Z. If $s_1^0, \dots, s_{k-1}^0$ are selected then s_k^0 has the minimal value over all s_k as s runs over the subset of Z of sequences $s = (1, s_1^0, \dots, s_{k-1}^0, *, \dots, *, h+1)$ or $s^0 = (1, s_1^0, \dots, s_{k-1}^0, h+1)$ if the latter lies in Z. $s^0 = (1, s_1^0, \dots, s_k^0, h+1)$ has the following property: For every $s = (1, s_1, \dots, s_r, h+1)$ either $s_p^0 \notin \{s_1, \dots, s_r\}$ for some p or $s_1 = s_1^0, \dots, s_k = s_k^0$. For if $\{s_1^0, \dots, s_k^0\} \subset \{s_1, \dots, s_r\}$, then $s_i \geq s_i^0$ for $i = 1, 2, \dots, k$ by the way s_i^0 is chosen. Since $s_i < s_j$ for $i < j$ the assertion follows. Further we select a $(k+1)$-tuple $(j_1, \dots, j_{k+1})$ corresponding to s^0. Namely, we define j_p to be a natural number satisfying $m_{s_p^0} < j_p \leq m_{s_{p-1}^0}$ for $p = 1, 2, \dots, k$, regarding $s_0^0 = 1$ and we put $j_{k+1} = 1$. We claim that $T_{(j)}$ has no like terms in $\omega_s(I_{\lambda_c})$ for every $s, s \neq s^0$. Indeed, if $s = (1, s_1, \dots, s_r, h+1)$ and $s_p^0 \notin \{s_1, \dots, s_r\}$ then every term in $\omega_s(I_{\lambda_c})$ has a factor $I_{m_{s_p^0}}$ on the s_p^0 place while $T_{(j)}$ has a factor $(e_{j_p} \wedge I_{m_{s_p^0}})^{(j_{p+1})}$ on this place. As $m_{s_p^0} > m_{s_{p+1}^0}$ we see that $j_{p+1} < j_p$ whence $(e_{j_p} \wedge I_{m_{s_p^0}})^{(j_{p+1})} \neq I_{m_{s_p^0}}$. Furthermore, if $s_p = s_p^0$ for $p = 1, 2, \dots, k$ and were $T_{(i)} = T_{(j)}$ for some term $T_{(i)}$ of $\omega_s(I_{\lambda_c})$ it would follow at once that $i_1 = j_1, \dots, i_{k+1} = j_{k+1}$. However, since $j_{k+1} = 1$, but $i_{k+1} > m_{s_{k+1}} \geq 1$ that conclusion leads to a contradiction. The assertion follows.

Finally, we remark that $\sum a_s \omega_s$, $a_s \in K$, is zero if and only if $\sum a_s \omega_s(I_{\lambda_c})$ vanishes in Λ^{λ_r}. Taking $Z = \{s | a_s \neq 0\}$, the preceding paragraphs show that $\sum a_s \omega_s = 0$ implies $a_s = 0$ for all s. ◁

4.6. The collecting process.

In the next section we find it necessary to reduce some products of homomorphisms ω_{ij} to certain normal form. This normal form will be defined here. We are also going to define the basic identities between mappings ω_{ij} by which we will be able to shift one mapping past another in a product of these. As for the definition of a normal product we say that $P = \omega_{i_1 j_1} \omega_{i_2 j_2} \dots \omega_{i_p j_p}$ is λ_c-normal or just normal if (i) i_k, j_k belong to $\underline{h+1}$, where $h = l(\lambda_c')$; (ii) $i_k > j_k$ for all k and (iii) $(\lambda_c')_{j_k} > (\lambda_c')_{i_k}$ for all k. We define two normal products to be equal if the factors and their order of appearance are the same.

LEMMA 1. *Let α be a sequence such that $\alpha' = (a_1, \dots, a_h)$. The following identities for the restrictions of ω_{ij} on $\Lambda^\alpha(B)$ hold*

(1) $\omega_{ij}\omega_{il} = -\omega_{il}\omega_{ij}$

(2) $\omega_{ij}\omega_{kj} = -\omega_{kj}\omega_{ij}$

(3) $\omega_{ij}\omega_{ki} = \omega_{kj} - \omega_{ki}\omega_{ij}$

(4) $\omega_{ij}\omega_{jl} = \omega_{il} - \omega_{jl}\omega_{ij}$

(5) $\omega_{ij}\omega_{ji} = (a_i - a_j)id_{\Lambda^a} + \omega_{ji}\omega_{ij}$
and $\omega_{ij}\omega_{kl} = \omega_{kl}\omega_{ij}$ *in all other cases.*

PROOF. The proof in all five cases follows the same pattern which we illustrate through one of them, say, the fifth. Let $x \in \Lambda^{a_i}(B)$ and $y \in \Lambda^{a_j}(B)$. Then $\omega_{ij}\omega_{ji}(x \otimes y) = \sum_{q,p\in \underline{d}}(e_q \wedge \partial_p(x)) \otimes \partial_q(e_p \wedge y)$ while $\omega_{ji}\omega_{ij}(x \otimes y) = \sum_{q,p\in \underline{d}} \partial_p(e_q \wedge x) \otimes e_p \wedge \partial_q(y)$. Now we recall that ∂_p is an antiderivation with respect to the canonical involution i. Hence $\partial_q(e_p \wedge y) = \delta_{qp}y - e_p \wedge \partial_q(y)$ from which it follows at once that $\omega_{ij}\omega_{ji}(x \otimes y) = \sum e_p \wedge \partial_p(x) \otimes y - \sum_{p,q\in \underline{d}} e_q \wedge \partial_p(x) \otimes e_p \wedge \partial_q(y)$ and $\omega_{ji}\omega_{ij}(x \otimes y) = \sum x \otimes e_p \wedge \partial_p(y) - \sum_{p,q\in \underline{d}} e_q \wedge \partial_p(x) \otimes e_p \wedge \partial_q(y)$. Moreover, setting $x = e_{i_1} \wedge \cdots \wedge e_{i_s} \wedge \cdots \wedge e_{i_{a_i}}$ to be a basis element of $\Lambda^{a_i}(B)$ we easily find out that $e_p \wedge \partial_p(x) = x$, if $p = i_s$ for some s or 0, otherwise. Hence, as p runs over $\underline{d}$ we obtain $\sum_{p\in \underline{d}} e_p \wedge \partial_p(x) \otimes y = a_i(x \otimes y)$ and likewise $\sum_{p\in \underline{d}} x \otimes e_p \wedge \partial_p(y) = a_j(x \otimes y)$. Hence $\omega_{ij}\omega_{ji} - \omega_{ji}\omega_{ij} = (a_i - a_j)id_{\Lambda^a}$. ◁

4.7. The Formula.

In the foregoing notation let λ be a partition of $d - 1$ and W^λ be the corresponding irreducible G_d-submodule. The Littlewood-Richardson rule or the Branching rule states that $W^\lambda \otimes B_d \simeq \bigoplus_{\mu\in E(\lambda)} W^\mu$, where $E(\lambda)$ is defined in §4.5. Hence for every $\mu \in E(\lambda)$ there exists a unique within a scalar multiple homomorphism $i_\mu\colon W^\mu \to W^\lambda \otimes B_d$. We aim at finding a formula for i_μ.

We recall that by definition W^λ is G_d-generated by certain element $I_\lambda \in \Lambda^\lambda(B_d)$ whence $W^\lambda \subset \Lambda^\lambda(B_d)$, from which we see that $W^\lambda \otimes B_d$ is a subset of $\Lambda^\lambda(B_d) \otimes B_d = \Lambda^{\lambda_r}(B_d)$ by the definition of λ_r. Further we know by Theorem 4.5.1 that every homomorphism from W^μ into $\Lambda^{\lambda_r}(B_d)$, in particular i_μ, has the form $i_\mu = \sum a_s\omega_s$ where $a_s \in K$ and s runs over the set of sequences $S(\mu, \lambda_r)$ associated with the co-standard μ-tableaux of weight λ_r' like in §4.5. Since $i_\mu(W^\mu)$ is G_d-generated by $i_\mu(I_\mu)$ the problem reduces to that of finding the conditions for $i_\mu(I_\mu)$ to lie in $W^\lambda \otimes B_d$. These conditions can be derived by translating James's results on certain filtrations in Φ^λ into Ψ^λ and then utilizing the connections between G_d- and S_d-modules along the lines of the proof of Proposition 4.3.2. In more detail, we recall that by ([**J1**, 17.14]) $S^\lambda \uparrow S_d$ can be identified with the Ψ^{λ,λ_r} term of the James filtration for Ψ^{λ_r}. Hence if $l(\lambda') = h$, then using ([**J1**, 17.13]) and the fact that $\operatorname{char} K = 0$ we arrive at $\Psi^{\lambda,\lambda_r} = \{x \in \Psi^{\lambda_r} | \omega_{i,i+1}(x) = 0$ for $i = 1, 2, \dots, h-1\}$ or $\Psi^{\lambda,\lambda_r} = \Psi^{\lambda_r} \cap (\bigcap_{i=1}^{h-1} \ker \omega_{i,i+1})$. On the other hand, taking the ω-weight spaces of $\Lambda^{\lambda_r}(B_d)$ and $W^\lambda \otimes B_d$ we obtain $\Lambda^{\lambda_r}(B_d)^\omega = \Psi^{\lambda_r}$ and $(W^\lambda \otimes B_d)^\omega = \bigoplus_{\alpha+\beta=\omega}(W^\lambda)^\alpha \otimes B_d^\beta$. Now if $\alpha = \omega_1 = (1, \dots, 1)(d - 1$ units) then $(W^\lambda)^{\omega_1} = S^\lambda$ and in case $\alpha = (1, \dots, 0, \dots, 1)$ (where 0 is in the ith position), ($d-1$ units) the transposition (id) takes α to ω_1 and therefore takes $(W^\lambda)^\alpha$ onto $(W^\lambda)^{\omega_1}$. It becomes evident that $(W^\lambda \otimes B_d)^\omega \simeq \bigoplus_{i=1}^{d-1}(W^\lambda)^{\omega_1}(id)$ and the latter is by definition $S^\lambda \uparrow S_d$. Lastly, let $N = \bigcap_{i=1}^{h-1} \ker \omega_{i,i+1}$. Since all $\omega_{i,i+1}$ are G_d-homomorphisms and $W^\lambda \otimes B_d$ is generated by its

ω-weight subspace, it is clear that $W^\lambda \otimes B_d \subset N$. Were that inclusion proper we would have $N^\omega \underset{\neq}{\supset} (W^\lambda \otimes B_d)^\omega = \Psi^{\lambda,\lambda_r}$, a contradiction. Thus $W^\lambda \otimes B_d = \bigcap_{i=1}^{h-1} \ker \omega_{i,i+1}$ and in particular we deduce that $i_\mu(W^\mu)$ lies in $W^\lambda \otimes B_d$ if and only if $\omega_{i,i+1} i_\mu = 0$ for all $i = 1, 2, \dots, h-1$.

REMARK. Our condition on λ is more restrictive than it is really necessary. In fact we can state a result similar to the above in the setting of group $G_n = GL(B_n)$, a partition $\lambda \vdash d$ with $l(\lambda) \leq n$, its associated Weyl module W_n^λ and $n < d$. To see that $W_n^\lambda \otimes B_n$ can still be characterized as $\bigcap_{i=1}^{h-1} \ker \omega_{i,i+1}$ we notice that the natural imbedding of $\Lambda^{\lambda_r}(B_n)$ into $\Lambda^{\lambda_r}(B_d)$ and that of $W_n^\lambda \otimes B_n$ into $W^\lambda \otimes B_d$ identifies these modules with the subspaces of $\Lambda^{\lambda_r}(B_d)$ and $W^\lambda \otimes B_d$ of elements x such that $x\pi_n = x$. Consequently, the formula for i_μ, which we will derive below, remains valid in this setting.

One more remark is in order. Suppose that $\mu \in E(\lambda)$ and $\mu \neq \lambda_c$. Then μ is obtained by adding one node to, say, the ith column of λ. Let $\check{\mu}$ be the partition conjugate to $(\lambda_i' + 1, \dots, \lambda_h')$. One can easily see by an argument verbatim to that of §4.5 that μ-co-standard tableaux of weight λ_r depend only on $[\check{\mu}]$ and so do the corresponding sequences s and mappings ω_s. Moreover if $i_{\check{\mu}}$ is a mapping that sends $W^{\check{\mu}}$ into $W^{(\lambda_i,\dots,\lambda_n,1)} \otimes B_d$, then since ω_s are defined on every $\Lambda^\alpha(B_d)$, we can regard $i_{\check{\mu}}$ as a mapping from W^μ into $\Lambda^{\lambda_r}(B_d)$. As such it satisfies the conditions $\omega_{j,j+1} i_{\check{\mu}} = 0$ for $j = 1, \dots, h-1$. For $\omega_{j,j+1} i_{\check{\mu}} = 0$ for $j \geq i$ by our hypothesis and the rest of the conditions follows from the fact that $\omega_{j,j+1}\omega_s(I_\mu) = 0$ for every $j < i$ and $s = (i, s_1, \dots, s_k, h+1)$. Since all sequences for $St^0(\mu, \lambda_r)$ are of that form we obtain $i_{\check{\mu}}$ as an imbedding of W^μ into $W^\lambda \otimes B_d$. Therefore we assume $\mu = \lambda_c$ in what follows.

We revert to our preceding notation. We recall that h is the number of non-zero parts of λ_c'. We put an equivalence relation '$\sim$' on $\underline{h}$ by defining $i \sim j$ if and only if the corresponding columns of $[\lambda_c]$ belong to the same block, i.e., $(\lambda_c')_i = (\lambda_c')_j$. We denote by $\tilde{i}$ the equivalence class of i. Let us decompose set $\underline{h}$ into the union of equivalence classes

$$\text{(D)} \qquad \underline{h} = \{1\} \cup \{2, \dots, h_2\} \cup \dots \cup \{h_{q-1}+1, \dots, h_q\}.$$

We associate with (D) the group $G(\lambda_c)$ of all permutations of $\underline{h}$ that preserve (D). Next we fix the set C of those i for which $(\lambda_c')_{i-1} > (\lambda_c')_i$. Clearly $i \in C$ if and only if the last node of the ith column is extendable. Further we associate with every $i \in C$ a number $l_i = \lambda_1' - \lambda_i' + i - 1$. We say that $s = (1, s_1, s_2, \dots, s_k, h+1)$ is a C-sequence if $\{s_1, \dots, s_k\} \subset C$. Notice that for every sequence s there exists a unique $\pi \in G(\lambda_c)$ for which $s\pi$ is a C-sequence.

THEOREM 1. *A mapping $i_{\lambda_c} = \sum a_s \omega_s$ realizes an imbedding of W^{λ_c} into $W^\lambda \otimes B_d$ provided that the coefficients a_s are defined within a scalar multiple, by the following formulas. If $s = (1, s_1, \dots, s_k, h+1)$ is a C-sequence then*

$$\text{(F)} \qquad a_s = \prod_{i=1}^{k} l_{s_i}^{-1}$$

with the understanding that the empty product is 1. *If* s *is not a* C*-sequence, but* $s\pi$, $\pi \in G(\lambda_c)$, *is, then* $a_s = a_{s\pi}$.

The theorem follows immediately from

LEMMA 1. (i) *For every* $s \in S(\lambda_c, \lambda_r)$ *and every* $\pi \in G(\lambda_c)$ *the equality* $a_s = a_{s\pi}$ *holds.* (ii) *If* $s = (1, s_1, \dots, s_k, h+1)$ *is a* C*-sequence and* $\check{s} = (1, s_2, \dots, s_k, h+1)$ *then* $l_{s_1} a_s = a_{\check{s}}$.

PROOF. We shall use repeatedly throughout the proof that $\omega_{ij}(W^\mu) = 0$ for all pairs (i, j) for which $\mu'_j \leq \mu'_i$. Of course, that condition is equivalent to $j > i$ or $j \sim i$.

(a) Here we show that for every $s = (1, s_1, \dots, h+1)$ if $u \sim s_1$, $u \in C$, then $a_s = a_{(1,u,\dots,h+1)}$. We can assume inductively that the above equality holds for all $s_1 \sim u$ and $s_1 < p$. Taking a sequence $s = (1, s_1, \dots, s_r, h+1)$ there are two possibilities: (i) no s_j equals to $p-1, p$ and (ii) $s_j = p-1, p$ for some j.

If (i) occurs, then $\omega_{p-1,p}\omega_s = \omega_s\omega_{p-1,p} = 0$ holds by an easy application of Lemma 4.6.1. In case $s_j = p-1, p$ we notice first that for every $k \neq j$ $s_k \neq p-1, p$ because $p-1 \sim p$, but all s_k should be pairwise non-equivalent in every sequence. For a given sequence s with $s_j = p-1, p$ let s_1 and s_2 be defined by $s_1 = (1, s_1, \dots, s_{j-1}, p-1)$ and $s_2 = (p, s_{j+1}, \dots, h+1)$. By an application of Lemma 4.6.1 together with the opening remark we derive $\omega_{p-1,p}\omega_s = \pm\omega_{s_2} \cdot \omega_{s_1}$ with the '+' sign corresponding to $s_j = p-1$ and '−', otherwise. It follows from this that $\omega_{p-1,p} i_{\lambda_c} = \sum (a_s - a_{s'}) \omega_{s_2}\omega_{s_1}$, where s runs over the set of sequences with $s_j = p-1$ for some j and s' is obtained from s by changing this $p-1$ into p. Now the set $\{\omega_{s_2}\omega_{s_1}\}$ of all products corresponding to these s is linearly independent. That can be seen by repeating the argument of Theorem 4.5.1 as follows. If $Z = \{(s_1, s_2)\}$ is a set of some pairs (s_1, s_2), then we define (s_1^0, s_2^0) to be that pair for which s_1^0 is minimal among all s_1's and s_2^0 is minimal among all s_2's. The pair (s_1^0, s_2^0) plays the rôle of s^0 in the proof of 4.5.1. Thus we conclude that $a_s = a_{s'}$ for every s such that $s_j = p-1$ for some j. Putting $s = (1, p-1, \dots, h+1)$ we obtain (a). We notice also that (i) of the lemma is a consequence of (a) and (ii).

(b) Here we are going to tackle (ii). We will induct on s_1. We start with case $s_1 = 2$ which serves as the basis of the induction. Using again Lemma 4.6.1 we deduce that for the product $\omega_{1,2}\omega_s, s \in S(\lambda_c, \lambda_r)$, we have three possibilities: (i) if $s_1 \not\sim 2$, $\omega_{1,2}\omega_s = \omega_{\check{s}}$, where $\check{s} = (2, s_1, s_2, \dots)$; (ii) if $s_1 = 2$, $\omega_{1,2}\omega_s = (\lambda'_1 - \lambda'_2 + 1)\omega_{\tilde{s}}$, where $\tilde{s} = (2, s_2, \dots)$ and (iii) $\omega_{1,2}\omega_s = 0$, otherwise. Therefore $\omega_{1,2} i_{\lambda_c} = \sum (l_2 a_s - a_{s'})\omega_{\tilde{s}}$, where s runs over the set of sequences $(1, 2, s_2, \dots)$ and for every such s $s' = (1, s_2, \dots)$. As the set of mappings $\{\omega_{\tilde{s}} \colon s = (1, 2, s_2, \dots)\}$ is linearly independent it follows from $\omega_{1,2} i_{\lambda_c} = 0$ that $l_2 a_s = a_{s'}$, as desired.

(c) By virtue of (a) and (b) we can assume that part (ii) of the lemma holds true for all sequences s such that $s_1 < u$, $u \in C$, and $u > 2$. We take up the product $\omega_{u-1,u}\omega_s$ and as before we find that it reduces to a few normal forms depending on s.

(i) If $s = (1, \ldots, s_{j-1}, u-1, s_{j+1}, \ldots)$ with $s_{j+1} \neq u$ and letting s_2 and s_1 be the sequences $(u, s_{j+1}, \ldots, h+1)$ and $(1, \ldots, s_{j-1}, u-1)$, respectively, we claim that $\omega_{u-1,u}\omega_s = \omega_{s_2} \cdot \omega_{s_1}$. For by definition $\omega_s = \omega_{h+1,s_r} \cdots \omega_{s_1,1}$ and all factors in that product before $\omega_{s_{j+1},u-1}$ commute with $\omega_{u-1,u}$ by Lemma 4.6.1. Applying (3) of this lemma we replace $\omega_{u-1,u}\omega_{s_{j+1},u-1}$ by $\omega_{s_{j+1},u} - \omega_{s_{j+1},u-1}\omega_{u-1,u}$. Hence $\omega_{u-1,u}\omega_s = \omega_{s_2}\omega_{s_1} - \omega_{h+1,s_r} \cdots \omega_{s_{j+1},u-1} \cdot \omega_{u-1,u} \cdot \nu$, where ν is the product of all factors of ω_s after $\omega_{s_{j+1},u-1}$. Using Lemma 4.6.1 again it is immediate that $\omega_{u-1,u}\nu = -\nu\omega_{u-1,u}$. However $\omega_{u-1,u}(W^{\lambda_c}) = 0$ so that the second term above drops out, whence $\omega_{u-1,u}\omega_s = \omega_{s_2} \cdot \omega_{s_1}$, as claimed. We note also that s_{j+1} can be assumed non-equivalent to u. For if it is, then since $\omega_{s_{j+1},u}$ commutes with ω_{s_1} and $\omega_{s_{j+1},u}(W^{\lambda_c}) = 0$ we obtain that $\omega_{u-1,u}\omega_s = 0$.

(ii) If $s = (1, \ldots, s_{j-1}, u, s_{j+1}, \ldots)$ and $s_{j-1} \not\sim u-1$ then following our argument in case of (i) one derives $\omega_{u-1,u}\omega_s = -\omega_{s_2}\omega_{s_1}$.

(iii) Let s be as in (ii), but now $s_{j-1} \sim u-1$ and $s_{j-1} \neq u-1$. Using an argument similar to that of (i) we obtain $\omega_{u-1,u}\omega_s = -\omega_{s_2} \cdot \omega_{u-1,s_{j-1}} \cdot \omega_{s_{j-1},s_{j-2}} \cdot \nu$, where ν stands for the product of all factors to the right of $\omega_{s_{j-1},s_{j-2}}$. By (3) of Lemma 4.6.1 we can replace $\omega_{u-1,s_{j-1}}\omega_{s_{j-1},s_{j-2}}$ by $\omega_{u-1,s_{j-2}} - \omega_{s_{j-1},s_{j-2}}\omega_{u-1,s_{j-1}}$ and furthermore since $\omega_{u-1,s_{j-1}}(W^{\lambda_c}) = 0$ and $\omega_{u-1,s_{j-1}}$ commutes with ν it is immediate that $\omega_{u-1,u}\omega_s = -\omega_{s_2} \cdot \omega_{s_1'}$ where $s_1' = (1, s_1, \ldots, s_{j-2}, u-1)$.

(iv) Let $s = (1, \ldots, s_{j-2}, u-1, u, s_{j+1}, \ldots)$. Then arguing as above one obtains $\omega_{u-1,u}\omega_s = -(\lambda'_{u-1} - \lambda'_u + 1)\omega_{s_2} \cdot \omega_{s_1'}$.

Summing up we let Z be the set of all sequences $s = (1, s_1, \ldots, s_{j-1}, u-1, s_{j+1}, \ldots)$ with the additional condition $s_{j+1} \not\sim u$. With the help of formulas (i)–(iv) we obtain $0 = \omega_{u-1,u} i_{\lambda_c} = \sum_{s \in Z}(a_s - a_{s'} - (\lambda'_{u-1} - \lambda'_u + 1)a_{s''} - \sum_{k \sim u-1 \,\&\, k \neq u-1} a_{s_k})\omega_{s_2}\omega_{s_1}$, where for every s in $Z s', s''$ and s_k are defined by $(1, s_1, \ldots, s_{j-1}, u, s_{j+1}, \ldots)$, $(1, s_1, \ldots, s_{j-1}, u-1, u, s_{j+1}, \ldots)$, and $(1, s_1, \ldots, s_{j-2}, k, u, s_{j+1}, \ldots)$, respectively. We remark that since $s_{j+1} \not\sim u$ all these sequences can be constructed starting from every s in Z. Because of linear independence of $\{\omega_{s_2} \cdot \omega_{s_1} | s \in Z\}$ we deduce that

$$\text{(E)} \qquad a_s - a_{s'} - (\lambda'_{u-1} - \lambda'_u + 1)a_{s''} - \sum_{k \sim u-1 \,\&\, k \neq u-1} a_{s_k} = 0.$$

Let v be the size of the equivalence class $\widetilde{u-1}$, i.e., $v = |\{k | k \sim u-1\}|$, and set $s = (1, u-1, \ldots)$. By (a) of this proof $a_{s_k} = a_{s''}$ for every k, which together with (E) implies that

$$\text{(E}'\text{)} \qquad a_s = a_{s'} + (\lambda'_{u-1} - \lambda'_u + v)a_{s''}.$$

Letting $x \in C$ be the smallest element in class $\widetilde{u-1}$ we deduce from the inductive hypothesis that $l_x a_{s''} = a_{s'}$ and $l_x a_s = a_{\check{s}}$. These equalities together with (E′) lead up to the equality $l_x^{-1} a_{\check{s}} = a_{s'}(1 + l_x^{-1}(\lambda'_{u-1} - \lambda'_u + v))$. By definition of l_x and since $x = u - v$ we obtain $l_x = \lambda'_1 - \lambda'_{u-1} + u - v - 1$, whence $l_x + (\lambda'_{u-1} - \lambda'_u + v) = l_u$. Thus $l_u a_{s'} = a_{\tilde{s}}$ which is the desired result. ◁

4.8. Descent into $\mathrm{gr}_d(KF)$.

The problem is thus. In §3 we associated with every partition λ of d an element $I_\lambda \in B^{\otimes d}$ and irreducible $G(B)$-module W^λ that is generated by I_λ. In the same §3 we showed that $\mathrm{gr}_d(KF)$ has a basis formed by the canonical monomials $e_1^{\alpha_1} \dots e_n^{\alpha_n}(e_1 \wedge e_2)^{\beta_{1,2}} \dots (e_{n-1} \wedge e_n)^{\beta_{n-1,n}}$. Since $\mathrm{gr}_d(KF)$ is an epimorphic image of $B^{\otimes d}$ under mapping $\pi\colon b_1 \otimes \dots \otimes b_d \mapsto b_1 \dots b_d$, $b_i \in B$, we can form the images of I_λ and W^λ under π. These are $\pi(I_\lambda) = \pi(I_{\lambda'_1}) \dots \pi(I_{\lambda'_h})$, if $\lambda' = (\lambda'_1, \dots, \lambda'_h)$ and $\pi(W^\lambda)$. We shall drop π and write $I_\lambda = I_{\lambda'_1} \dots I_{\lambda'_h}$ and W^λ in order to avoid cumbersome notation unless two different meaning of them can be confused. Our prime goal is to show that for every $\mu \in E(\lambda)$ and the mapping $i_\mu\colon W^\mu \to W^\lambda \otimes B$, constructed above, the image of $i_\mu(I_\mu)$ under π is non-zero. In other words, we intend to show that $W^\lambda \cdot B \subset \mathrm{gr}_{d+1}(KF)$ is the direct sum of W^μ, $\mu \in E(\lambda)$. The first step to this end is to find the expansion of I_λ along the canonical basis of $\mathrm{gr}_d(KF)$.

LEMMA 1. *Let n be a natural number. If $n = 2m$, then $\pi(I_n) = (m!)\sum_{i_k<j_k}(-1)^\sigma(e_{i_1} \wedge e_{j_1}) \dots (e_{i_m} \wedge e_{j_m})$ and if $n = 2m+1$, then $\pi(I_n) = (m!)\sum_{i_k<j_k}(-1)^\sigma(e_{i_1} \wedge e_{j_1}) \dots (e_{i_m} \wedge e_{j_m})e_r$, where $\sigma = i_1 j_1 \dots i_m j_m$ or $\sigma = i_1 j_1 \dots i_m j_m r$ runs over all permutations of $\underline{n}$ satisfying condition $i_1 < \dots < i_m$.*

PROOF. We recall that $I_n = e_1 \wedge \dots \wedge e_n$ regarded as an element of $B^{\otimes d}$ is defined by $\sum_{\sigma \in S_n}(-1)^\sigma e_{1\sigma} \otimes \dots \otimes e_{n\sigma}$. The following identities are well known and easy to check out

$$(1) \qquad I_n = \sum_{i=1}^{n}(-1)^{i-1} e_i \otimes (e_1 \wedge \dots \wedge \hat{e}_i \wedge \dots \wedge e_n)$$

$$(2) \qquad I_n = \sum (-1)^{n-i}(e_1 \wedge \dots \wedge \hat{e}_i \wedge \dots \wedge e_n)e_i$$

and

$$(3) \qquad I_n = \sum_{i<j}(-1)^{\nu_{ij}}(e_1 \wedge e_j) \otimes (e_1 \wedge \dots \wedge \hat{e}_i \wedge \dots \wedge \hat{e}_j \wedge \dots \wedge e_n)$$

where ν_{ij} is the number of inversions in permutation $ij12\dots\hat{i}\dots\hat{j}\dots n$. Using (3) we can keep decomposing wedges of length ≥ 3 until we obtain $I_n = \sum_{i_k<j_k}(-1)^\sigma(e_{i_1} \wedge e_{j_1}) \otimes \dots \otimes (e_{i_m} \wedge e_{j_m})$, if $n = 2m$ or $I_n = \sum_{i_k<j_k}(-1)^\sigma(e_{i-1} \wedge e_{j_1}) \otimes \dots \otimes (e_{i_m} \wedge e_{j_m}) \otimes e_r$, if $n = 2m+1$ and where $\sigma = i_1 j_1 \dots i_m j_m$ or $i_1 j_1 \dots i_m j_m r$. Going down into $\mathrm{gr}_d(KF)$ amounts to replacing '$\otimes$' by '$\cdot$' and assuming all $e_i \wedge e_j$ to be central. Noticing that in order to pass from $\sigma = \dots i_r j_r \dots i_s j_r \dots$ to $\sigma' = \dots i_s j_s \dots i_r j_r \dots$ one needs an even number of transposition we conclude that $m!$ permutations of the 2-wedges $e_i \wedge e_j$ in every signed product $(-1)^\sigma(e_{i_1} \wedge e_{j_1}) \dots (e_{i_m} \wedge e_{j_m})$ produce the same monomial in $\pi(I_n)$. Thus $\pi(I_n) = m! \sum_{i_k<j_k}(-1)^\sigma(e_{i_1} \wedge e_{j_1}) \dots (e_{i_m} \wedge e_{j_m})$, $1 = i_1 < i_1 < \dots < i_m$, if $n = 2m$ and similarly for $n = 2m+1$. ◁

One corollary of the lemma is worth noting. We keep in mind the explicit formula for pfaffian.

COROLLARY. *If* $f = (f_{ij})$ *is an* $2m \times 2m$ *skew-symmetric matrix then* $Pf(f) = \sum(-1)^\sigma f_{i_1 j_1} \dots f_{i_m j_m}$ *where the sum is extended over all permutations* $i_1 j_1 \dots i_m j_m$ *of* $\underline{2m}$ *such that* $i_k < j_k$ *and* $1 = i_1 < i_2 < \dots < i_m$.

PROOF. Identifying f_{ij} with $e_i \wedge e_j$ it follows from the results in §3 that $Pf(f) = k \cdot \pi(I_{2m})$ for some $k \in K$ and the coefficient of $f_{12} \dots f_{2m-12m}$ must be 1. ◁

Next we derive some additional identities in $\mathrm{gr}(KF)$. In what follows $[a, b]$ denotes $ab - ba$.

PROPOSITION 1. *Let* r *be a natural number. The following identities hold in* $\mathrm{gr}(KF)$.

(i) $[e_1 \wedge \dots \wedge e_n, e_r] = 0$ *for every* r, *if* n *is even, or* $r \in \underline{n}$.

(ii) $[e_1 \wedge \dots \wedge e_n, e_r] = \frac{1}{m+1} e_1 \wedge \dots \wedge e_n, e_r$, *if* $n = 2m + 1$ *and* $r \notin \underline{n}$.

PROOF. (i) By the above lemma $e_1 \wedge \dots \wedge e_n$ is in the center of $\mathrm{gr}(KF)$ if n is even, which proves one part of (i). Further suppose that n is odd and $r = 1$. We remark that both elements $e_1(e_1 \wedge \dots \wedge e_n)$ and $(e_1 \wedge \dots \wedge e_n)e_1$ are T_n-invariants of the same weight. Hence they must be equal within a scalar multiple. Moreover a direct computation shows that the canonical monomial $e_1^2(e_2 \wedge e_3) \dots (e_{n-1} \wedge e_n)$ occurs in the expansions of both invariants with the same coefficient $(\frac{n-1}{2})!$ and the equality $e_1(e_1 \wedge \dots \wedge e_n) = (e_1 \wedge \dots \wedge e_n)e_1$ follows. For an arbitrary $r \in \underline{n}$ applying transposition $\sigma = (1r)$ to the above equality we obtain (i).

(ii) First we assume $r = n + 1$. Using (i) it is a trivial matter to see that $[e_1 \wedge \dots \wedge e_m, e_{n+1}]$ is a T_n-invariant of weight $(1, \dots, 1)$ ($n+1$ units). Therefore it must be equal to $k(e_1 \wedge \dots \wedge e_{n+1})$ for some $k \in K$. To find k we compare the coefficients of a monomial, $(e_1 \wedge e_2) \dots (e_n \wedge e_{n+1})$ say, in the expansions of both elements through the canonical basis of $\mathrm{gr}_{n+1}(KF)$. For $e_1 \wedge \dots \wedge e_{n+1}$ this coefficients is $(m+1)!$ by Lemma 1. The same lemma applied to the other element produces equality $[e_1 \wedge \dots \wedge e_n, e_{n+1}] = \sum_{i=1}^n (-1)^{i-1} (e_i \wedge e_{n+1})(e_1 \wedge \dots \wedge \hat{e}_i \wedge \dots \wedge e_n)$. The above monomial is involved in the expansion of $(e_n \wedge e_{n+1})(e_1 \wedge \dots \wedge e_{n-1})$ only and in addition with coefficient $m!$, whence $k = \frac{1}{m+1}$. For a general $r \notin \underline{n}$ an application of $\sigma = (n + 1, r)$ to the formula for $r = n + 1$ completes the proof. ◁

We now proceed to the formulation of two main theorems of this section. Let λ be a partition of d and put $\mu = \lambda_c$ for brevity. We recall that π denotes the $G(B)$-epimorphism of $B^{\otimes d}$ onto $\mathrm{gr}_d(KF)$ introduced above.

THEOREM 1. *If* μ'_1 *is odd, then for every sequence* $s = (1, s_1, \dots, s_k, h + 1)$ *from* $S(\lambda_c, \lambda_r)$ *we have*

(i) $\pi(\omega_{(1,h+1)}(I_\mu)) = I_\mu$

(ii) $\pi(\omega_s(I_\mu)) = I_\mu$,

provided that all $\mu'_{s_1}, \dots, \mu'_{s_k}$ *are even and* $\pi(w_s(I_\mu)) = 0$, *otherwise.*

THEOREM 2. *If μ'_1 is even, then the coefficients c_s in the equalities $\pi(\omega_s(I_\mu)) = c_s I_\mu$ are nonpositive rational numbers for every $s \in S(\lambda_c, \lambda_r)$ and $c_{(1,h+1)} \neq 0$.*

COROLLARY 1. *For every λ and every $\nu \in E(\lambda) \pi(i_\nu(I_\nu)) \neq 0$.* ◃

PROOF OF THEOREM 1. (i) Let $\mu = (\mu'_1, \dots, \mu'_h)$ and put $a = \mu'_1$. By definition $I_\mu = I_{\mu'_1} \otimes \cdots \otimes I_{\mu'_h}$ and also $\omega_{(1,h+1)}(I_\mu) = \sum_{i=1}^a (-1)^{i-1} I_a^{(i)} \otimes \cdots \otimes I_{\mu'_h} \otimes e_i$, where $I_a^{(i)} := e_1 \wedge \cdots \wedge \hat{e}_i \wedge \cdots \wedge e_a$. Passing on to $\mathrm{gr}_d(KF)$ and taking into account that $I_a^{(i)}$ is central for every i we obtain $\pi(\omega_{(1,h+1)}(I_\mu)) = \sum_{i=1}^a (-1)^{i-1} I_a^{(i)} \dots I_{\mu'_h} \cdot e_i = I_{\mu'_2} \dots I_{\mu'_h} (\sum_{i=1}^a (-1)^{i-1} I_a^{(i)} e_i)$. By Lemma 1 $\sum (-1)^{i-1} I_a^{(i)} e_i = (-1)^{a+1} I_a = I_a$ while by Proposition 1 I_a commutes with every $I_{\mu'_j}$. Hence $I_{\mu'_2} \dots I_{\mu'_h} I_{\mu'_1} = I_\mu$ and (i) is done.

Before continuing with the proof we introduce the notion of the leading monomial of an $I_\mu \in \mathrm{gr}(KF)$. Namely, for every natural number a we put $m_a = (e_1 \wedge e_2) \dots (e_{a-1} \wedge e_a)$, if a is even and $m_a = (e_1 \wedge e_2) \dots (e_{a-2} \wedge e_{a-1}) e_a$, if a is odd. Given a partition μ we define m_μ as the product of all $m_{\mu'_i}$, i.e., $m_\mu = \prod_{i=1}^h m_{\mu'_i}$. Two properties of m_μ follow immediately, viz., that m_μ occurs in the expansion of I_μ through the canonical basis for $\mathrm{gr}_d(KF)$ with the coefficient $d_\mu = \prod_{i=1}^h [\frac{\mu'_i}{2}]!$, where $[\frac{a}{2}]$ denotes the largest integer in $\frac{a}{2}$, and that $e_p \wedge e_q$ is a factor of m_μ if and only if p is odd, $q = p+1$ and $p < \mu'_1$.

(ii) $\pi(\omega_s(I_\mu))$ is a T_d-invariant of weight μ in $\mathrm{gr}_d(KF)$ for every s and by virtue of that it must be equal to $c_s I_\mu$ for some $c_s \in K$. To find this c_s it will suffice to compute the coefficient of m_μ in the expansion of $\pi(\omega_s(I_\mu))$, $d_{s,\mu}$ say. Clearly, $c_s = \frac{d_{s,\mu}}{d_\mu}$, and since d_μ and $d_{s,\mu}$ are integers, c_s is a rational number. Denoting μ'_{s_i} by n_i for brevity, it follows from the definition of ω_s that $\pi(\omega_s(I_\mu))$ is a signed sum of the terms

$$(*) \qquad T_{(i)} = I_a^{(i_1)} \dots (e_{i_1} \wedge I_{n_1})^{(i_2)} \dots (e_{i_{k-1}} \wedge I_{n_k})^{(i_{k+1})} \dots I_{\mu'_h} \cdot e_{i_{k+1}}$$

taken over all $k+1$-tuples (i) such that $i_j > n_j$ for $j = 1, 2, \dots, k$. The number of significant terms can be cut down by noticing that if all n_i are even then m_μ occurs in a term $(*)$ only if $i_1 = \cdots = i_{k+1} = a_1$. For if j is the first number between 1 and k such that $i_j \neq a$, then every monomial in $(e_a \wedge I_{n_j})^{(i_j)}$ has a factor of the form $e_r \wedge e_a$ for some $r \leq n_j$. Apparently none of these factors occur in m_μ which proves our remark. Thus we must analyse only the term $T = I_a^a \cdot I_{\mu'_2} \dots I_{\mu'_h} \cdot e_a$. Plainly $T = I_a^a e_a I_{\mu'_2} \dots I_{\mu'_h} + I_a^a \cdot [I_{\mu'_2} \dots I_{\mu'_h}, e_a]$ and moreover $[I_{\mu'_2} \dots I_{\mu'_h}, e_a]$ is a sum of terms of the form $A \cdot [I_b, e_a]$, where A is a product of some $I_{\mu'_j}$ and b is an odd μ'_j. By Proposition 1 every canonical monomial involved in the expansion of $[I_b, e_a]$ has a factor $e_r \wedge e_a$ for some $r \leq b$, and therefore $I_a^a A[I_b, e_a]$ does not contain m_μ. Hence m_μ occurs only in $I_a^a e_a I_{\mu'_2} \dots I_{\mu'_h}$ and evidently with the same coefficient as that of m_μ in I_μ. Thus $\pi(\omega_s(I_\mu)) = I_\mu$ as asserted.

Next we assume that $n_j = r$ is the first odd number among n_i, $i = 1, 2, \dots, k$. Designating the subsequence $(s_j, \dots, h+1)$ by t we obtain readily from the definition of ω_s that $\pi(\omega_s(I_\mu)) = \sum \pm I_a^{(i_1)} \dots (e_{i_{j-1}} \wedge I_{n_{j-1}})^{(i_j)} \dots \pi\omega_t((e_{i_j} \wedge I_r) \dots I_{\mu'_h}))$. By the same reasons as above we have to worry only about the term $I_a^{(a)} \dots \pi\omega_t(e_a \wedge I_r) \dots I_{\mu'_h}$. Let ν' be $(r+1, \mu'_{s_{j+1}}, \dots, \mu'_h)$, that is to say that $[\nu]$ is obtained from $[\lambda]$ by removing all columns before the s_jth column and adding a node to the s_jth column of $[\lambda]$. Were $a = r+1$, then $(e_a \wedge I_r) \dots I_{\mu'_h}$ would be the standard invariant I_ν and consequently we would have $\pi\omega_t(I_\nu) = cI_\nu$ for some $c \in K$. If $a \neq r+1$, then $J := (e_{a_1} \wedge I_r) \dots I_{\mu'_h}$ is obtained from I_ν by an application of transposition $\sigma = (r+1, a)$. Since $\pi\omega_t$ is $G(B)$-mapping it becomes clear that $\pi\omega_t(J) = cJ$. Now every canonical monomial involved in J has a factor of the form $e_p \wedge e_a$, $p \leq r$, which is not a subfactor of m_μ. This implies that $I_a^{(a)} \dots \pi\omega_t(J)$ does not contain m_μ which completes the proof of (ii). ◁

Proof of Theorem 2. Given a sequence $s = (1, s_1, \dots, s_k, h+1)$ we let $t = (s_1, \dots, s_k, h+1)$. We can write again as in the proof of (ii) $\pi\omega_s(I_\mu) = \sum_{i > \mu'_i} (-1)^{i-1} I_a^{(i)} \dots \pi\omega_t(J_i)$ where $J_i = (e_i \wedge I_{\mu'_{s_1}}) I_{\mu'_{s_1+1}} \dots I_{\mu'_h}$. Using the argument given above in a similar situation we obtain $\pi\omega_t(J_i) = cJ_i$ for some $c \in K$. Next we consider two cases.

(A) μ'_{s_1} is even. Now as $\mu'_{s_1} + 1$ is odd we deduce from Theorem 1 that the constant c is 1 or 0. Let $c = 1$. Let us write $\pi\omega_s(I_\mu)$ in the form $\sum \Delta_i$, where $\Delta_i := I_a^{(i)} \dots J_i - I_a^{(i+1)} \dots J_{i+1}$ and i runs over the set of odd numbers $> \mu'_{s_1}$. It is easy to see that every monomial in the expansion of $I_a^{(i)}$ or $I_a^{(i+1)}$ except $\prod_{j \text{ odd}} (e_j \wedge e_{j+1}) e_{i+1}$ or $\prod_{j \text{ odd } \& \ j \neq i} (e_j \wedge e_{j+1}) e_i$, respectively, has a factor of the form $e_r \wedge e_{i+1}$ or $e_r \wedge e_i$, $r \neq i$, $i+1$, which does not enter m_μ. Furthermore, among all canonical monomials in the expansion of J_i only $m_{\mu'_{s_1}} \cdot e_i \dots m_{\mu'_h}$ has the right factors and also enters the expansion with a positive coefficient. Next we remark that m_a lies in the center of $\mathrm{gr}(KF)$ for every even a and $m_a = m'_a \cdot e_a$ with m'_a central, if a is odd. Hence letting $b_1, \dots, b_p$ be the odd numbers among the μ'_i with $i < s_1$ and $k_1, \dots, k_p$ be their multiplicities in μ' we conclude that m_μ can enter Δ_i via the expansion of the following element $(\prod_{j \text{ odd } \& \ j \neq i} e_j \wedge e_{j+1}) . \prod_{\mu'_i \text{ even}} m_{\mu'_i} \cdot \prod_{i=1}^p m'_{b_i} \cdot \alpha \cdot \prod_{j > s_1} m_{\mu'_j}$, where $\alpha = e_{i+1} \cdot \prod e_{b_j}^{k_j} e_i - e_i \prod e_{b_j}^{k_j} e_{i+1}$. Of course only α has to be decomposed. To this end we use the identity $[v, u_1 \dots u_p] = \sum u_1 \dots u_{j-1} \cdot [v, u_j] u_{j+1} \dots u_p$ which is valid in any ring and identity $[e_i, e_j^m] = m e_j^{m-1} [e_i, e_j]$ that holds in $\mathrm{gr}(KF)$. Thus we arrive at $\alpha = -(\prod e_{b_j}^{k_j}) \cdot (e_i \wedge e_{i+1}) + \sum c \cdot P(e_r \wedge e_q) e_r$, where $c \in K$, P is some product of powers of e_j, $r = i$, $i+1$ and $q = b_1, \dots, b_p$. A 2-wedge $e_r \wedge e_q$ enters m_μ if and only if $r = i+1$ and $q = i$. That is, if $i \notin \{b_1, \dots, b_p\}$, then every term $c \cdot P(e_r \wedge e_q) e_r$ can be discarded from α and if $i = b_v$ for some v then, since $[e_{i+1}, e_i^{k_v}] = -k_v e_i^{k_v - 1} (e_i \wedge e_{i+1})$, $\alpha \equiv -(k_v + 1) \prod e_{b_j}^{k_j} (e_i \wedge e_{i+1})$ modulo the terms which are not subfactors of

m_μ. In all, the coefficient of m_μ in Δ_i is negative and non-zero for every i and hence so is the coefficient in $\pi\omega_s(I_\mu)$. In particular, our argument goes through for $s = (1, h+1)$ resulting in $c_{(1,h+1)} < 0$.

(B) μ'_{s_1} is odd. In the foregoing notation since $\mu'_1 + 1$ is even, we can assume by induction that $\pi\omega_t(J_i) = cJ_i$ for some $c \leq 0$. Let $\nu(i)$ be $i+1$, if i is odd and $\nu(i)$ be $i-1$, if i is even. Turning to $I_a^{(i)}$ we notice that all canonical monomials involved in $I_a^{(i)}$, except $(-1)^{i-1}(\prod_{j \text{ odd } \& \ j\neq i} e_j \wedge e_{j+1})e_{\nu(i)}$ have a factor of $e_r \wedge e_{\nu(i)}$ which does not occur in m_μ. Let $T_i = (-1)^{i-1}(\prod_{j \text{ odd } \& \ j \neq i} e_j \wedge e_{j+1})e_{\nu(i)} \dots J_i$ and put $p = \mu'_{s_1}$. If $i \neq p+1$, then on the one hand $\nu(i) \neq p$ while on the other hand $e_i \wedge I_p$ is an even wedge and therefore every monomial involved in $e_i \wedge I_p$ does not have e_p as a factor. This implies that the degree of e_p in every canonical monomial occurring in T_i is at most $k_p - 1$ from which it follows that T_i does not contain m_μ. Finally, we take up T_{p+1}. Now since $e_{p+1} \wedge I_p = -I_p \wedge e_{p+1}$ we conclude that the leading monomial for J_{p+1}, viz., $m_{p+1} \dots m_{\mu'_h}$ has a negative coefficient, d say. Consequently we see that the coefficient of m_μ in $\pi\omega_s(I_\mu)$ is $c(-1)^p d \leq 0$ and the proof is completed. ◁

4.9. The Theorem.

We refer to §4.1 for notation and the preliminary discussion. In particular, F, B and $G(B)$ have the meaning introduced there.

THEOREM 1. *Every $G(B)$-ideal of* $\mathrm{gr}(KF)$ *is finitely generated.*

PROOF. We shall actually show that $\mathrm{gr}(KF)$ satisfies a.c.c. for $G(B)$-ideals by producing a lattice isomorphism between the lattice of all $G(B)$-ideals of $\mathrm{gr}(KF)$ and a lattice with a.c.c. Let $(\Lambda, \subseteq)$ be the poset formed by the set of all partitions regarded as finite weakly decreasing sequences under the containment order '$\subseteq$'. Let V be a $G = G(B)$-ideal in $\mathrm{gr}(KF)$. We associate with V the set $P(V)$ of partitions like in §4.1. By the results of §4.2 the mapping $p\colon V \to P(V)$ is a lattice isomorphism between the lattice of G-ideals of $\mathrm{gr}(KF)$ and the lattice of subsets of Λ of the form $P(V)$. Next we recall several facts about posets following ([**HIG**]). Given a poset $(Z, \leq)$, a subset X of Z is called closed if from $x \in X$ and $x \leq y$ it follows that $y \in X$. $(Z, \subseteq)$ is called partially well-ordered (p.w.o.) if every subset of Z has at least one, but no more than finite number of minimal elements. This is equivalent, by the well-known argument, to the maximal condition on the closed sets. We need the following:

LEMMA 1. *The mapping p realizes an isomorphism between the lattice of G-ideals of* $\mathrm{gr}(KF)$ *and the lattice of the closed subset of* Λ.

PROOF. Let Z be a closed subset in Λ and $\lambda \in Z$ an element of norm d. We denote by W_d^λ a copy of the λth Weyl module occurring in $\mathrm{gr}_d(KF_d)$. Let V_Z be the G-submodule of $\mathrm{gr}(KF)$ generated by all W_d^λ, as λ runs over Z, and let W^λ stand for the G-submodule generated by W_d^λ. By the results of §4.2 the mapping $Z \to V_Z$ is inverse to p. It remains to show that V_Z is an ideal of gr(KF). This comes down to showing that BW^λ and $W^\lambda B$ are contained in V_Z for every $\lambda \in Z$. Projecting into $\mathrm{gr}_{d+1}(KF_{d+1})$, by using π_{d+1}, we obtain G_{d+1}-submodules $B_{d+1}W^\lambda_{d+1}$ and $W^\lambda_{d+1}B_{d+1}$. Viewing W^λ_{d+1} as a G_{d+1}-submodule of $B^{\otimes d}_{d+1}$ generated by I_λ we see that $W^\lambda_{d+1}B_{d+1}$ is an epimorphic image of $W^\lambda_{d+1} \otimes B_{d+1}$. By the Branching Rule $W^\lambda_{d+1} \otimes B_{d+1} = \bigoplus\{W^\mu | \mu \in E(\lambda)\}$ from which it is evident that $W^\lambda_{d+1} \cdot B_{d+1} \subseteq \bigoplus\{W^\mu | \mu \in E(\lambda)\}$. Since Z is closed, $E(\lambda) \subset Z$, which implies that $W^\lambda \cdot B \subset V_Z$. Likewise, since $B_{d+1} \otimes W^\lambda_{d+1}$ is G_{d+1}-isomorphic to $W^\lambda_{d+1} \otimes B_{d+1}$, $B \cdot W^\lambda \subset V_Z$ which is the desired result.

Conversely, in order to show that $P(V)$ is closed it suffices to prove that $E(\lambda) \subset P(V)$ for every $\lambda \in P(V)$. Here we need the full strength of §4.8. Namely, by Corollary 4.8.1 the image of every W^μ_{d+1}, involved in the decomposition of $W^\lambda_{d+1} \otimes B_{d+1}$, in $\mathrm{gr}_{d+1}(KF_{d+1})$ is non-zero. Hence every $\mu \in E(\lambda)$ also lies in $P(V)$, whence $P(V)$ is closed. ◁

To complete the proof of the theorem we need only to show that $(\Lambda, \subseteq)$ is a p.w.o. Λ, as a set is a part of a larger set $\mathbf{N}^{(\infty)}$ of all finite sequences of non-negative integers. For two sequences $s = (s_1, \dots, s_r)$ and $t = (t_1, \dots, t_q)$ we define $s \leq t$ if some subsequence of t majorizes s term by term (**[HIG]**). That is, there exists an increasing function $\phi: \underline{r} \to \underline{q}$ such that $s_i \leq t_{\varphi(i)}$ for all i. By a theorem of (**[HIG]**) $(\mathbf{N}^{(\infty)}. \leq)$ is a p.w.o. We claim that the identity map $\lambda \rightleftharpoons \lambda$ is an order-isomorphism between $(\Lambda, \subseteq)$ and $(\Lambda, \leq)$. In one direction, let $\lambda \leq \mu$ and let $\phi: \underline{r} \to \underline{q}$ be the function appearing in the definition of order '$\leq$'. Since $\phi(i) \geq i$ for all i and hence $\mu_i \geq \mu_{\phi(i)}$ we conclude that $\mu_i \geq \lambda_i$ for all i or $\mu \supseteq \lambda$ and in the opposite direction verification is trivial. Thus $(\Lambda, \subseteq)$ as a subset of a p.w.o. set is itself p.w.o. ◁

References

[A-F] S. Abeasis and A. Del Fra, "Young diagrams and ideals of pfaffians," Advances in Math. **35** (1980), 158–178.

[CO] P. Cohn, *Universal Algebra, Mathematics and its Applications*, **6** Dordrecht, 1981.

[C-P] C. De Concini and C. Procesi, "A characteristic free approach to invariant theory," Advances in Math. **21** (1976), 330–354.

[D] S. Doty, "The submodule structure of certain Weyl modules for groups of type A_n," J. Algebra **95** (1985), 373–383.

[G1] K. W. Gruenberg, *Cohomological Topics in Group Theory*, Lecture Notes in Math. **143** Springer-Verlag, 1970.

[G2] K. W. Gruenberg, "The upper central series in soluble groups," Illinois J. Math. **5** (1961), 436–466.

[GR] J. A. Green, *Polynomial Representations of GL_h*, Lecture Notes in Math. **830** Springer-Verlag, 1980.

[GRE] W. H. Greub, *Multilinear Algebra*, 2nd ed., Springer-Verlag, New York, 1978.

[H] P. Hall, *Nilpotent Groups*, Canadian Math. Congress, Univ. of Alberta, 1967; republished in Queen Mary College Math. Notes, 1969.

[HART1] B. Hartley, "Topics in the theory of nilpotent groups," Group Theory: Essays for Philip Hall, 61–120, Academic Press, 1984.

[HART2] ——, "The residual nilpotence of wreath products," Proc. London Math. Soc. (3) **20** (1970), 365–392.

[HIG] G. Higman, "Ordering by divisibility in abstract algebras," Proc. London Math. Soc. **2** (1952), 326–336.

[HN] H. Neumann, *Varieties of Groups, Ergebnisse der Mathematik und ihrer Greunzgebiete,* **37** Springer-Verlag, 1967.

[J1] G. James, *The Representation Theory of the Symmetric Groups,* Lecture Notes in Math. **682** Springer-Verlag, 1978.

[J2] ——, "Some counterexamples in the theory of Specht modules," J. Algebra **46** (1977), 457–461.

[JA] N. Jacobson, "Structure of rings," Amer. Math. Soc. Colloquium **36**, Providence, R. I., 1964.

[JE] S. A. Jennings, "The structure of the group ring of a p-group over a modular field," Trans. Amer. Math. Soc. **50** (1941), 175–185.

[J-K] G. James and A. Kerber, "The representation theory of the symmetric group," Encyclopedia of Mathematics and its Applications **16** Addison-Wesley, Reading, Mass., 1981.

[K] L. Krop, "On the representations of the full matrix semigroup on homogeneous polynomials," J. Algebra **99** (1986), 370–421.

[KO] L. Kovacs, "Some representations of special linear groups," Research Report No. 42, 1986, The Australian National University.

[L] M. Lazard, "Sur les groupes nilpotents et les anneaux de Lie," Ann. École Norm. Sup. **71** (1954), 101–190.

[LA] V. Latyshev, "On a choice of basis in a T-ideal," Sibirsk. Mat. Zh., No. 5, **4** (1963), 1122–1127.

[LI] D. Littlewood, *The Theory of Group Characters,* 2nd ed., Oxford University Press, 1950.

[OL] P. Olver, "Differential hyperforms I," University of Minnesota Math. Report 82-101, 1983.

[OS] J. M. Osborn, "Varieties of algebras," Advances in Math. **8** (1972), 163–369.

[P] B. Plotkin, "Radicals and varieties in representations of groups," Latviĭsk. Mat. Ezhegodnik **10** (1972), 75–132.

[PA] I. B. S. Passi, *Group Rings and Their Augmentation Ideals,* Lecture Notes in Math. **715** Springer-Verlag, 1979.

[PE] M. H. Peel, "Duality and left ideals in a group ring," J. Algebra **71** (1981), 524–526.

[P-S] I. B. S. Passi and S. K. Sehgal, "Isomorphism of modular group algebras," Math. Z. **129** (1972), 65–73.

[Q] D. Quillen, "On the associated graded ring of a group ring," J. Algebra **10** (1968), 411–418.

[R1] Y. Razmyslov, "A certain problem of Kaplansky," Izv. Akad. Nauk. SSSR Ser. Mat. **37** (1973), 483–501.

[R2] Y. Razmyslov, "The existence of a finite basis for the identities of the matrix algebra of order 2 over a field of characteristic zero," algebra i Logika **12** (1973), 83–113.

[S] R. Stanley, "Theory and application of plane partitions: Part 1," Studies in Applied Math. **50** (1971), 167–188.

[SCH] M. Schooneveldt, *Module Structure in Certain Algebras,* Thesis, Australian National University, 1980.

[T] J. Towber, "Young symmetry, the flag manifold and representations of $GL(n)$," J. Algebra **61** (1979), 414–462.

[V] I. Volichenko, "Varieties of Lie algebras with identity $[[x_1, x_2, x_3], [x_4, x_5, x_6]] = 0$ over a field of characteristic zero," Sibirsk. Mat. Zh., No. 3, **25** (1984), 40–54.

Contemporary Mathematics
Volume **93** (1989)

On Nilpotent and Soluble Subgroups of Linear Groups Over Fields of Fractions of Enveloping Algebras and of Group Rings, I

A. I. LICHTMAN*

In memory of my dear teacher and friend, Samuil Davidovich Berman

1. Introduction. Let H be a Lie algebra over field K, $U(H)$ be its universal envelope. We will be interested in the case when the algebra H is abelian-by-finite dimensional, i.e. it contains an abelian ideal A such that the quotient algebra H/A is finite dimensional. Although $U(H)$ is not necessarily Noetherian in this case (see [**1**]) it is not difficult to prove that $U(H)$ is an Ore ring (see [**2**], Proposition 4.1, or [**3**]). The field of fractions of $U(H)$ and the linear groups over this field are the first object of our interest. Our results are new also for the case when H is a finite dimensional Lie algebra. We consider also some other classes of skew fields: the universal fields of fractions for free group rings KF or the fields of fractions of the group rings $K(F/F^{(m)})$, where $F/F^{(m)}$ is a free soluble group. *We denote these three classes of fields by D.* (The term "field" throughout this paper is used in the sense of "skew field".)

Our main results are Theorems I–III below. We will use in these theorems and throughout the paper the concept of a unipotent subgroup. We will say that a linear group U is unipotent if the elements $u - 1$ $(u \in U)$ generate a nilpotent ring. It is worth remarking that if D is any one of the fields which are considered then by combining the results of [**2**], Section 7, with Proposition 4.2 of this paper we obain that a subgroup $U \subseteq \mathrm{GL}_n(D)$ is unipotent if and only if every element $u - 1$ $(u \in U)$ is nilpotent.

We formulate now the main results of the article.

THEOREM I. *Let G be a nilpotent subgroup of* $\mathrm{GL}_n(D)$ *of nilpotency class s. Then*

*Supported in part by NSF Grant DMS 8802634.

i) *If* char $K = p > 0$ *then the commutator subgroup* G' *is locally finite, its* p'*-subgroup* V *(formed by all the elements whose order is prime to* p*) is finite, the exponent of* V *is a divisor of the number* $(n!)^{4(s-1)^2}$ *and there exists an element* $a \in \mathrm{GL}_n(D)$ *such that*

$$a^{-1}Va \subseteq \mathrm{GL}_n(K). \tag{1.1}$$

ii) *If* char $K = 0$ *then the linear envelope* $K[G']$ *is locally finite and it contains a finite dimensional subalgebra* T *such that*

$$K[G'] \cong T + J(K[G']), \tag{1.2}$$

where $J(K[G'])$ *is the nilpotent radical of* $K[G']$*. Furthermore, there exists an element* $b \in \mathrm{GL}_n(D)$ *such that*

$$b^{-1}Tb \subseteq K_{n\times n}. \tag{1.3}$$

iii) *In both the cases* i) *and* ii) G *contains a series*

$$G \supseteq A \supseteq U,$$

where U *is a unipotent normal subgroup of* G*, the quotient group* A/U *is abelian and the group* G/A *is a (locally finite)* p'*-group whose exponent divides the number* $(n!)^{2(s-1)}$.

REMARK. The representation (1.2) and the relation (1.3) are valid also in the case i). Indeed $G' \cong V \times U$ where U is the Sylow p-subgroup of G' and the radical $J(K[G'])$ is the ideal, generated by all the elements $u-1$ $(u \in U)$. Hence, if $T = K[V]$ then T satisfies (1.2), and (1.3) follows from (1.1).

We will prove in Section 7 (see Corollary of Theorem I) that the relations (1.1)–(1.3) remain true for locally nilpotent subgroups of $\mathrm{GL}_n(D)$.

Theorem I is related to the results of Zalesskii and Wehrfritz on locally nilpotent subgroups of $\mathrm{GL}_n(\Delta)$, where Δ is locally finite over its center (see **[4]**). It can be considered as their partial analog.

The length of the derived series of a soluble linear group of degree n over a commutative field does not exceed $2n$; this is Huppert's bound for Zassenhaus' Theorem (see **[5]**, 45.2). We obtain therefore the following immediate corollary of Theorem II.

COROLLARY. *Every locally nilpotent subgroup* $G \subseteq \mathrm{GL}_n(D)$ *is soluble; its solubility class does not exceed* $3n + 1$.

THEOREM II. *Assume that* char $K = 0$ *and* G *be a polycyclic-by-finite subgroup of* $\mathrm{GL}_n(D)$*. Then*

i) G *contains a normal subgroup* N *such that* $(G\colon N)|(n!)^2$*, the linear envelope* $K[N']$ *is finite dimensional and contains a subalgebra* T *such that*

$$K[N'] \cong T + J(K[N']), \tag{1.4}$$

and there exists $c \in \mathrm{GL}_n(D)$ *such that*

$$c^{-1}Tc \subseteq K_{n\times n}. \tag{1.5}$$

ii) *If $n = 1$ i.e. G is a subgroup of D^*, then it must be abelian.*

The last result, about the commutativity of polycyclic-by-finite subgroups of D^*, is new in the cases when D is the field of fractions of $U(L)$ or of $K(F/F^{(m)})$. When D is the universal field of fractions for KF a much stronger fact holds by Theorem 11.10 in Schofield's book **[6]**: two noncommuting elements of D generate a free subfield.

Theorem II is not true when $\operatorname{char} K = p > 0$: if D is the field of fractions of the Weyl algebra $A_1(K)$ then the multiplicative group contains noncommutative abelian-by-finite subgroups. Indeed one can take a subgroup, generated by the elements η and ξ subject to the relation $\eta^{-1}\xi\eta = \xi + 1$ (see [7], 0.10.5); this subgroup is abelian-by-cyclic. We prove however in this case the following result.

THEOREM III. i) *Assume that* $\operatorname{char} K = p > 0$ *and D be either the universal field of fractions for a free group ring KF or the Ore field of fractions of $K(F/F^{(m)})$, where $F/F^{(m)}$ is a free soluble group. Let G be a polycyclic-by-finite subgroup of* $\mathrm{GL}_n(D)$. *Then G contains a normal subgroup N such that $(G\colon N)|(n!)^2$, the commutator subgroup N' is finite (and hence the group N is central-by-finite and G is abelian-by-finite). The linear envelope $K[N']$ contains a separable subalgebra T such that* (1.4) *and* (1.5) *hold. Furthermore, if $n = 1$ then G is abelian.*

ii) *Let* $\operatorname{char} K = p > 0$ *and D be the field of fractions of $U(L)$. Then the statement* i) *remains true if the Hirsh number $h(G)$ of G is less than $p - 1$.*

We have the following corollary of Theorems II and III.

COROLLARY. *Let G be a polycyclic-by-finite subgroup of* $\mathrm{GL}_n(D)$ *and assume that* $\operatorname{char} K > (h(G) + 1)$. *Then the subgroup N' is an extension of a unipotent subgroup by a subgroup of* $\mathrm{GL}_n(K)$. *Hence, if G is polycyclic then the solubility class of N' does not exceed* $1 + 3n$ *and the solubility class of G does not exceed* $1 + 3n + (n!)^2$.

We conjecture that if $\operatorname{char} K = 0$ and D is the field of fractions of $U(L)$ or of the ring $K(F/F^{(m)})$ then the length of the derived series of soluble subgroups of $\mathrm{GL}_n(D)$ is bounded by a function of n. This was proven in **[8]** for the case when D is the universal field of fractions for KF. A related question is whether the derived series of a soluble subgroup in the multiplicative group of the field of fractions of $U(L)$, where L is a finite dimensional Lie algebra over a field K, $\operatorname{char} K = 0$, can have length greater than 2.

We make in our proofs an essential use of results from article **[9]**. This is first of all the result which follows from Theorem 1 and statement 12. in **[9]**: *Let D be a division algebra over K such that $D \otimes_K S$ contains no zero divisors for an arbitrary finite dimensional separable division algebra S and R be a locally finite K-subalgebra of $D_{n\times n}$. Assume that for every finite dimensional subalgebra $R_i \subseteq R$ the quotient algebra of R_i by its radical $J(R_i)$ is separable.*

Then there exists a finite dimensional separable subalgebra $T \subseteq R$ such that

$$R \cong T + J(R),$$

where $J(R)$ is the nilpotent radical of R. Furthermore, there exists an element $u \in \mathrm{GL}_n(D)$ such that

$$u^{-1}Tu \subseteq K_{n\times n}.$$

All the fields which we consider (the field of fractions of $U(L)$, the field of fractions of a free soluble ring $K(F/F^{(m)}$ and the universal field of fractions of $K(F)$) satisfy the assumption of this theorem.

Furthermore, the proofs of Theorems II and III are based on the following theorem which will be proven in Section 3.

THEOREM 3.1. *Let Δ be a field of dimension q^{2m} over its center Z, where q is a prime number, G be a polycyclic-by-finite subgroup of $\mathrm{GL}_n(\Delta)$, whose Hirsh number $h(G)$ is less than $q-1$. Then G contains a normal subgroup N such that $(G\colon N)|(n!)^2$ and the commutator subgroup N' is an extension of a unipotent normal subgroup U by a finite group V. Furthermore, if K is an arbitrary subfield of Z then the linear envelope $K[N']$ is finite dimensional and it conains a separable subalgebra $T \cong K[V]$ such that*

$$K[N'] \cong T + J(K[N']). \tag{1.6}$$

The second main step in the proofs of Theorem I–III is based on the results of Section 4. It is known (see [**1**], 11.36) that a finitely generated abelian-by-finite dimensional Lie algebra H over a field K of finite characteristic is residually finite dimensional. When $\operatorname{char} K = 0$ this is not true anymore. But we prove in Section 5 that H has some other residual properties as well as $U(H)$ and its field of fractions D. (See Theorem 4.2 and Proposition 4.3). In particular, if R is a finitely generated subring of $D_{n\times n}$ then there exist a system of ideals C_j $(j \in J)$ such that every quotient ring R/C_j is isomorphic to a subring of $(D_j)_{n\times n}$ where D_j is a field of fractions of $U(H_j)$, H_j is a finite dimensional Lie algebra over a finite field K_j; every matrix $X \in R$ is algebraic over K if and only if its images in R/C_j are algebraic over the fields K_j; the ring R is imbedded into an appropriate ultraproduct $(\prod_{j\in J} R/C_j)\mathscr{T}$ (see Propositions 4.5 and 4.6). Similar results are obtained for the case when D is the universal field of fractions for KF or the Ore field of fractions of $K(F/F^{(m)})$.

The third main step in the proofs of Theorems I–III is based on the technical results of Section 5 on ultraproducts of linear groups. We complete the proofs of Theorems I–III in Section 7.

Theorems I–III, Theorem 4.2 and the methods of this article will be applied in part II of the article to study the automorphism groups of free soluble groups and of abelian-by-finite dimensional Lie algebras. We will consider also in part II soluble subgroups of $\mathrm{GL}_n(D)$ and refine the results of the article [**8**] on the Tits alternative for subgroups of $\mathrm{GL}_n(D)$, where D is the field of fractions of $U(L)$, L is a finite dimensional Lie algebra, or the universal field of fractions for KF.

2.1 LEMMA 2.1. *Let E be a commutative finite algebraic extension of a field Z and assume that E is separable over an intermediate subfield T. Let E_1 be the maximal separable over Z subfield of E. Then* $\dim(E\colon T)$ *is a divisor of* $(E_1\colon Z)$.

PROOF. Let $\tilde{Z} \supseteq E$ be a normal algebraic extension of Z. Then the group G of all the isomorphisms of E_1 into $\tilde{Z}$ has order $(E_1\colon Z)$. On the other hand, it is easy to see that the order of G is divisible by $(E\colon T)$ and the assertion follows.

PROPOSITION 2.1. *Let M be a simple algebra of dimension m^2 over its center $Z, S = B_{k\times k}$ be a matrix subring of M, where B is a finite dimensional subalgebra of M such that $B/J(B)$ is a field of dimension r^2 over its center T. Then*

$$(rk^2)|m^2. \tag{2.1}$$

PROOF. The field $B/J(B)$ contains a maximal separable subfield E of dimension r over the field T. Let E_1 be the maximal separable over Z subfield of E and A be the inverse image of E_1 under the natural homomorphism $B \to B/J(B)$. By Wedderburn's theorem

$$A \cong E_1 + J(A)$$

and hence E_1 is isomorphically imbedded into A and therefore $(E_1)_{k\times k}$ is imbedded into M. Theorem vi.6.1 in **[10]** implies that M is a free module over $(E_1)_{k\times k}$ and hence the number $k^2(\dim(E_1\colon Z))$ is a divisor of m^2. Apply Lemma 2.1 and obtain that $k^2(\dim(E\colon T))$ is a divisor of m^2; thus (2.1) is proven. □

We will need two following facts.

LEMMA 2.2. *Let Δ be an arbitrary field, $S \subseteq D_{n\times n}$ be a ring finite dimensional over its central subfield Z. Then*

$$S \cong (S_1)_{k_1\times k_1} + (S_2)_{k_2\times k_2} + \cdots + (S_r)_{k_r\times k_r}, \tag{2.2}$$

where $S_i/J(S_i)$ is a field Δ_i $(i = 1, 2, \ldots, r)$. Furthermore, there exist natural numbers n_i such that $(S_i)_{k_i\times k_i}$, is isomorphic to a subring of $\Delta_{n_i\times n_i}$ and

$$\sum_{i=1}^{r} n_i \leq n. \tag{2.3}$$

The assertion is well known. See for instance **[8]**.

LEMMA 2.3. *Let Δ be a field of characteristic $p > 0$, G be a subgroup of $\mathrm{GL}_n(\Delta)$ such that the quotient algebra $Z_p[G]/J(Z_p[G])$ is finite dimensional over Z_p. Then G is locally finite and $Z_p[G]$ contains a semisimple subring*

$$T \cong (\Pi_1)_{k_1\times k_1} + (\Pi_2)_{k_2\times k_2} + \cdots + (\Pi_r)_{k_r\times k_r} \tag{2.4}$$

where Π_i $(i = 1, 2, \dots r)$ *are finite fields and* $\sum_{i=1}^{r} k_i \leq n$. *Furthermore,* $Z_p[G]$ *is a split extension*

$$Z_p[G] \cong T + J(Z_p[G])$$

and $(J(Z_p[G])^n = 0$. *The ring* T *is isomorphic to a subring of the matrix ring* $\Pi_{n\times n}$, *where* Π *is the composite of the fields* Π_i $(i = 1, 2, \dots r)$.

The proof is easy. See **[8]**, 2.1.

We will need in the proof of our main results some facts about nilpotent subgroups of $\mathrm{GL}_n(\Delta)$, where $\operatorname{char}\Delta = p > 0$ has Δ has dimension p^{2m} over its center (see Proposition 2.2). Recall first that if Δ is a locally finite dimensional division algebra, G is a nilpotent (or locally nilpotent) subgroup of $\mathrm{GL}_n(\Delta)$ and $u(G)$ is the maximal unipotent normal subgroup of G, $\overline{G} = G/u(G)$ and $C(\overline{G})$ is the center of $\overline{G}$, then the quotient group $\tilde{G} = \overline{G}/C(\overline{G})$ is a locally finite p'-group (see **[4]**, 3.23) and Schur's theorem implies that in this case the commutator subgroup G' is locally finite.

LEMMA 2.4. *Let* G *be a nilpotent group of class* s, A *be its abelian normal subgroup such that* G/A *has an exponent* m. *Then every given factor* $\gamma_k(G)/\gamma_{k+1}(G)$ $(k = 2, 3, \dots,)$ *has an exponent which is a divisor of* m^2.

PROOF. Apply the induction by the nilpotency class. Since by the induction hypotheses the assertion is true for the quotient group $G/\gamma_s(G)$ we obtain that it is enough to prove that the exponent of $\gamma_s(G)$ is an divisor of m^2. Take arbitrary elements $x \in \gamma_{s-1}(G)$, $g \in G$ and consider the commutator $[x, g] \in \gamma_s(G)$. Since $\gamma_{s+1}(G) = 1$ and $\gamma_s(G)$ is central we have

$$[x, g]^{m^2} = [x^m, g^m]$$

which implies, via the inclusion, $x^m \in A$, $g^m \in A$ that

$$[x, g]^{m^2} = 1 \tag{2.5}$$

But all the elements of the form $[x, g]$, $x \in \gamma_{s-1}(G)$, $g \in G$, generate $\gamma_s(G)$. We obtain therefore from (2.5) that the exponent of $\gamma_s(G)$ is a divisor of m^2 and the proof is completed.

COROLLARY. *Let the conditions of Lemma* 2.4 *hold. Then* G' *is a locally finite group whose exponent is a divisor of* $m^{2(s-1)}$.

PROPOSITION 2.2. *Let* D *be a field of characteristic* $p > 0$ *and of dimension* p^{2m} *over its center* Z, G *be a nilpotent subgroup of class* s *of* $\mathrm{GL}_n(\Delta)$. *Then the* p'-*subgroup* V *of the locally finite subgroup* G' *is finite and its exponent is a divisor of the number* $(n!)^{4(s-1)^2}$. *Furthermore, the quotient group* $\overline{G} = G/u(G)$ *is a finite extension of its center and it contains an abelian normal subgroup* $\overline{A} \supseteq C(\overline{G})$ *such that* $\overline{G}/\overline{A}$ *is a finite* p'-*group whose exponent is a divisor of* $(n!)^{2(s-1)}$. *Clearly,* $\overline{G}' \cong V$.

PROOF. Prove first the special case when the linear envelope $Z[G]$ is isomorphic to $B_{k\times k}$, where $B/J(B)$ is a field R of dimension r^2 over its center

Z_1. Thus, if $\overline{G}$ is the image of G in the ring $(B/J(B))_{k\times k} \cong R_{k\times k}$ then $Z[\overline{G}] \cong S_{k\times k}$ and hence $\overline{G}$ is an absolutely irreducible group. Since $\overline{G}$ is nilpotent it follows easily that it contains no p-elements and thus the subgroup $u(G) = G\cap(1+(J(B))_{k\times k})$ is the Sylow p-subgroup of G. Furthermore, we can consider the subgroup $\overline{G}$ as an absolutely irreducible linear subgroup of $\mathrm{GL}_{kr}(Z_1)$ and obtain from **[11]**, 8.13 that the center $C(\overline{G})$ of $\overline{G}$ has a finite index and the exponent of the group $\overline{G}/C(\overline{G})$ is a divisor of $(kr)^{s-1}$, and hence by Proposition 2.1 it is a divisor of $(p^{2m}n^2)^{s-1}$.

There exists therefore a normal subgroup $\overline{A} \supseteq \overline{C}$ such that $\overline{A}/\overline{C}$ is a p-group and $\overline{G}/\overline{A}$ is a p'-group, whose exponent divides $n^{2(s-1)}$. Since $\overline{C}$ is central in $\overline{A}$ we obtain from Schur's theorem that $\overline{A}$ is a finite p-group and the irreducibility of $\overline{G}$ implies, once again, that $\overline{A}' = 1$, i.e. $\overline{A}$ is abelian.

Corollary of Lemma 2.4 now implies that $\overline{G}'$ is a p'-group whose exponent divides $n^{4(s-1)^2}$. Furthermore, the kernel of the homomorphism $G' \to \overline{G}'$ is a p'-group $U = G' \cap u(G)$ and we conclude that the group G' is locally finite, V is its p'-subgroup, U is the Sylow p-subgroup and $\overline{G}' \cong V$. But it is known that a linear p'-group of bounded exponent is finite (see **[11]**); hence V is finite and we proved the assertion for the special case when $Z[G] \cong B_{k\times k}$.

The proof for the general case can now be completed in a routine way. Consider the decomposition (2.2) and obtain that every subdirect component G_i of G satisfies the conclusion of the assertion and, in particular, its quotient group $\overline{G}_i = G_i/U_i$ contains a normal subgroup $\overline{A}_i$ such that $\overline{G}_i/\overline{A}_i$ is a p'-group whose exponent divides $(n_i)^{2(s-1)}$. Since $\sum_i n_i \leq n$ we conclude easily that $\overline{G}$ contains a normal subgroup $\overline{A}$ such that the exponent of $\overline{G}/\overline{A}$ is a divisor of $(n!)^{2(s-1)}$ and the exponent of the group $V \cong \overline{G}'$ is a divisor of $(n!)^{4(s-1)^2}$, and the assertion follows.

COROLLARY. *Let $U = G' \cap u(G)$. Then U is the Sylow p-subgroup of G', the radical $J(Z_p[G'])$ is the nilpotent ideal, generated by all the elements $u-1$ $(u \in U)$, and $Z_p[G']$ is a split extension*

$$Z_p[G'] \cong Z_p[V] + J(Z_p[G']). \tag{2.6}$$

The subring $Z_p[V]$ is a direct sum of matrix rings over finite fields

$$Z_p[V] \cong (\Pi_1)_{n_1\times n_1} + (\Pi_2)_{n_2\times n_2} + \dots (\Pi_r)_{n_r\times n_r} \tag{2.7}$$

where

$$n_1 + n_2 + \cdots + n_r \leq n$$

and hence it is isomorphic to a subring of $\Pi_{n\times n}$, where Π is a finite field.

PROOF. The relation (2.7) follows from the fact that V is a finite p'-group. All the other statements follow easily.

PROPOSITION 2.3. *Let Δ be a field of characteristic p and of dimension p^{2m} over its center Z, G be a subgroup of the multiplicative group Δ^*. Then*

$$C(G) \cap G' = 1. \tag{2.8}$$

PROOF. Let $\Delta_1 = Z[G]$ and Z_1 be the center of Δ_1. Then

$$(\Delta\colon \Delta_1)(\Delta_1\colon Z_1)(Z_1\colon Z) = p^{2m},$$

and hence $(\Delta_1\colon Z_1) = p^{2d}$, $d \leq m$. Clearly, $C(G) \subseteq Z_1$. Consider the regular representation ρ of G in D_1. If $x \in C(G) \cap G'$ then $\rho(x) = xI_{p^d}$, and on the other hand $\det(\rho(x)) = 1$ which implies that $x^{p^{2d}} = 1$ and hence $x = 1$.

COROLLARY 1. *Let* $\operatorname{char}\Delta = p$ *and* $(\Delta\colon Z) = p^{2m}$. *Then every nilpotent subgroup of* Δ^* *is abelian.*

PROOF. Indeed, if the nilpotency class s of G is greater than 1 then $\gamma_{s-1}(G) \neq 1$ which contradicts (2.8).

3.1. LEMMA 3.1. *Let* H *be a free abelian group of finite rank* r, ϕ *be an automorphism of* H *of prime order* q. *Then* $q \leq r + 1$.

PROOF. The action of ϕ defines in the additive group H a structure of a $C[G]$-module, where G is a cyclic group of order p, C is the ring of integers. The assertion follows now from the fact that G has only one non-trivial irreducible rational representation which is obtained in the augmentation ideal $\omega(C[G])$ and has degree $q - 1$.

PROPOSITION 3.1. *Let* Δ *be a field of characteristic* $p \geq 0$ *and of dimension* q^{2m} *over its center* Z, *where* q *is a prime number (which can be equal to* p). *Let* G *be a subgroup of* $\mathrm{GL}_n(\Delta)$, H *be a normal subgroup of* G. *Let* $u(H) = H \cap (1 + J(Z[H])$ *and assume that the quotient group* $H/u(H)$ *is an abelian group, generated by* $r < q - 1$ *elements. Then* G *contains a normal subgroup* $N \supseteq H$ *such that* $(G\colon N)|(n!)^2$ *and the relative commutator subgroup* $[N, H]$ *is an extension of the unipotent normal subgroup* $u[N, H] = [N, H] \cap u(H)$ *by a finite abelian* p'-*group* V. *Furthermore the linear envelope* S *of* $[N, H]$ *is a split extension*

$$S \cong Z[V] + J(S) \tag{3.1}$$

where $Z[V]$ *is a separable subalgebra of* S, *isomorphic to an appropriate linear envelope of* V, *and* $J(S)$ *is the radical of* S.

PROOF. We have

$$Z[H] \cong R_1 + R_2 + \cdots + R_s, \tag{3.2}$$

where every ring R_i $(i = 1, 2, \ldots, s)$ is completely prime; the number s in (3.2) does not exceed n. Furthermore,

$$Z[H]/J(Z[H]) \cong K_1 + K_2 + \cdots + K_s, \tag{3.3}$$

where $K_i \cong R_i/J(R_i)$ $(i = 1, 2, \ldots, s)$ is a commutative field. The conjugation by G in $\Delta_{n\times n}$ induces a group of automorphisms in $Z[H]$ and hence in $Z[H]/J(Z[H])$. A routine argument implies an existence of a normal subgroup $H \subseteq G_1 \triangleleft G$, which stabilizes every R_i and K_i, $(G\colon G_1)|s!$ and hence $(G\colon G_1)|n!$. We see thus that for every $i = 1, 2, \ldots, s$ G_1 acts as a group of automorphisms in the field K_i over its subfield Ze_i, where e_i is the unit element of K_i. Clearly, the maximal separable subfield $T_i \subseteq K_i$ is G_1-invariant and hence G_1 induces a subgroup of the Galois group $\mathrm{Gal}(T_i\colon Ze_i)$ for every $i = 1, 2, \ldots, s$. We will prove now that every $\dim(T_i\colon Ze_i)$ is a divisor of $(n!)q^{2m}$.

Indeed, the Z-algebra

$$T = T_1 + T_2 + \cdots + T_s$$

is a separable subalgebra of $Z[H]/J(Z[H])$; it is easy to conclude from Wedderburn's theorem therefore that it is isomorphic to a subalgebra of $Z[H]$; Lemma 3.1 from **[8]** now implies that $\dim(T_i\colon Ze_i)$ is a divisor of $k_i q^{2m}$, where k_i $(i = 1, 2, \ldots, s)$ are integers such that

$$k_1 + k_2 + \cdots + k_s \leq n$$

and it is thus proven that $\dim(T_i\colon Ze_i) | ((n!)q^{2m})$ $(i = 1, 2, \ldots, s)$ and hence the orders of the groups $\mathrm{Gal}(T_i\colon Ze_j)$ $(i = 1, 2, \ldots, s)$ are divisors of $n!q^{2m}$. We obtain therefore that the group G_1 contains a normal subgroup G_2 which centralizes every T_i $(i = 1, 2, \ldots, s)$ and whose index in G_1 is a divisor of $(n!)q^{2m}$; since G_2 centralizes the maximal separable subfield T_i of K_i it acts trivially on every K_i $(i = 1, 2, \ldots, s)$ and hence on $Z[H]/(J(Z[H])$ too. It is easy to verify too that G_2 is normal in G; since $(G\colon G_1)|(n!)$ we obtain that

$$(G\colon G_2) | ((n!)^2 q^{2m}). \tag{3.4}$$

Let $\overline{H}$ be the image of H in the quotient algebra $Z[H]\ J(Z[H])$. Since G_2 acts trivially on $Z[H]/J(Z[H])$ the conjugation by G defines in a natural way the action of G/G_2 as a group of automorphisms of $\overline{H}$. Let $\mathrm{tor}(\overline{H})$ denote the torsion subgroup of $\overline{H}$. Since $\mathrm{rank}\,\overline{H} < q-1$, Lemma 3.1 implies that the group $(G\colon G_2)$ induces in the abelian group $\overline{H}/\mathrm{tor}(\overline{H})$ a group of automorphisms whose order is prime to q and (3.4) implies that it is a divisor of $(n!)^2$. Thus, G contains a normal subgroup $N \supseteq G_2 \supseteq H$ such that $(G\colon N)|(n!)^2$ and $[N, H]$ belongs to the inverse image of the subgroup $\mathrm{tor}(\overline{H})$ in H. But the subgroup $\mathrm{tor}(\overline{H})$ is finite and because $Z[\overline{H}]$ is a direct sum of commutative fields we conclude easily that $\mathrm{tor}(\overline{H})$ contains no p-elements when $\mathrm{char}\, D = p > 0$. Let $u([N, H]) = [N, H] \cap u(H)$. The quotient group $V = [N, H]/u([N, H])$ is isomorphic to a subgroup of $\mathrm{tor}(\overline{H})$ and hence is a finite abelian p'-group. Since S is the linear envelope of $[N, H]$ the quotient algebra $S/(S \cap J(Z[H]))$ is isomorphic to an appropriate linear envelope $Z[V]$. The last algebra is however a direct sum of cyclotomic fields which implies in particular that it is separable and that $S \cap J(Z[H])$ coincides with the nilpotent radical $J(S)$ of S and (3.1) follows from Wedderburn's theorem.

COROLLARY 1. *Assume that in Proposition* 3.1 $\mathrm{char}\,\Delta = p > 0$. *Then the subgroup* $[N, H]$ *is locally finite and is in fact a split extension of the Sylow* p*-subgroup* U_1 *by the group* V.

COROLLARY 2. *Let G be a subgroup of Δ^*, where Δ has dimension q^{2m} over its center, H be an abelian normal subgroup of G. Assume that H has a finite rank $r < q - 1$. Then $[G, H]$ is a cyclic group of finite order.*

PROOF. Since $n = 1$ the subgroup N in Proposition 3.1 must coincide with G. We obtain thus that $[G, H] = V$. But V now is an abelian subgroup of Δ^* and it must therefore be cyclic.

COROLLARY 3. *Let Δ be a field of characteristic $p > 0$ and of dimension p^{2m} over its center Z, H be an abelian subgroup of Δ^* of finite rank $r \leq p - 1$. Assume too that all the elements of finite order in Δ^* are central. Then the normalizer of H in Δ^* coincides with its centralizer.*

PROOF. Let $g \in \Delta^*$ be an element such that $g^{-1}Hg = H$. Corollary 2 implies that the subgroup $G = \{g, H\}$ is nilpotent. Corollary of Proposition 2.3 implies that G is abelian.

LEMMA 3.2. *Let G be a group, H be its abelian normal subgroup of finite index. Then the relative commutator subgroup $[G, H]$ has a finite index in G'.*

PROOF. Follows easily from Schur's theorem.

THEOREM 3.1. *Let Δ be a field of dimension q^{2m} over its center Z, where q is a prime number, G be a polycyclic-by-finite subgroup of $\mathrm{GL}_n(\Delta)$, whose Hirsh number $h(G)$ is less than $q - 1$. Then G contains a normal subgroup N such that $(G\colon N)|(n!)^2$ and the commutator subgroup N' is an extension of a unipotent normal subgroup U by a finite group V. Furthermore, if K is an arbitrary subfield of Z then the linear envelope $K[N']$ is finite dimensional and it contains a separable subalgebra $T \cong K[V]$ such that*

$$K[N'] \cong T + J(K[N']). \tag{3.5}$$

PROOF. The linear envelope $Z[G]$ is a finite dimensional algebra; hence the radical $J(Z[G])$ is nilpotent. Once again consider the quotient algebra $Z[\overline{G}] = Z[G]/(J(Z[G]))$. The group $\overline{G}$ is a completely reducible linear group; since it is polycyclic-by-finite it must be in fact abelian-by-finite (see **[11]**). Let $\overline{H}$ be an abelian torsion free normal subgroup of finite index in $\overline{G}$; clearly, rank $\overline{H} < q - 1$. Let H be the inverse image of H under the homomorphism $G \to \overline{G}$. Thus, $\overline{H} \cong H/u(H)$, where

$$u(H) = H \cap (1 + J(Z[G]) = H \cap (1 + J(Z[H]). \tag{3.6}$$

The second relation in (3.6) follows from the fact that $H \triangleleft G$.

Now let $N \supseteq H$ be the normal subgroup obtained in Proposition 3.1. and $\overline{N}$ be the image of N under the homomorphism $G \to \overline{G}$. Since $(\overline{G}\colon \overline{H})$ is finite so is $(\overline{N}\colon \overline{H})$ and Lemma 3.2 implies therefore that $\overline{N}'$ is a finite extension of the subgroup $[\overline{N}, \overline{H}]$. But

$$[\overline{N}, \overline{H}] \cong [N, H]/(u([N, H])) \tag{3.7}$$

is a finite abelian p'-group by Proposition 3.1. Hence, the group $\overline{N}'$ is finite. Since $\overline{N}' \cong N'/u(N')$ we obtain that N' is a finite extension of the unipotent normal subgroup $U = u(N')$.

Now let K be an arbitrary subfield of Z. The algebra $K[N']$ is a finitely generated PI-algebra; hence its radical is locally nilpotent; In fact, it must be nilpotent because it is contained in $\Delta_{n\times n}$. This implies easily that

$$(3.8) \qquad J(K[N'] \cong K[N'] \cap J(Z[N']).$$

Now denote the finite group N'/U by V and obtain from (3.8) that the algebra $K[N']/J(K[N'])$ is isomorphic to an appropriate linear envelope $K[V]$. Since $K[V]$ is finite dimensional and $J(K[N'])$ is nilpotent the algebra $K[N']$ is locally finite; but N' is finitely generated and hence $K[N']$ is finite dimensional. Furthermore, since V is finite group and $K[V]$ is semisimple it must be separable (see **[12]**, 7.10) and (3.8) follows therefore from the Wedderburn theorem.

COROLLARY. *Assume that in Proposition* 3.4 $\operatorname{char}\Delta = p > 0$. *Then the subgroup N' is finite.*

PROOF. Indeed, in this case N' is an extension of a nilpotent p-group U by finite group V.

4.1. Let H be an abelian-by-finite dimensional Lie algebra over a field K. Its universal envelope $U(H)$ has a field of fractions D (see **[2]**, **[3]**). We will develop in this section a method for construction of specializations from D into a family of fields $\Delta_i (i \in I)$ where every Δ_i is the field of fractions of the universal enveloping algebra $U(L_i)$ and L_i is a finite dimensional Lie algebra over a finite field P_i.

Assume first that $\operatorname{char} K = p > 0$. It is known that in this case every finitely generated abelian-by-finite dimensional Lie algebra is residually finite (see **[1]**). This theorem together with the results of the article **[13]** imply the following fact (see **[2]**, Corollary of Theorem 6.3).

PROPOSITION 4.1. *Let H be an abelian-by-finite dimensional Lie algebra over a field K* $\operatorname{char} K = p > 0$ *and let*

$$x_\alpha \ (\alpha = 1, 2, \ldots, k)$$

be given non-zero elements of D. Then there exists a finite dimensional Lie algebra K_i over a finite field L_i and a specialization θ_i of D into the field D_i of fractions $U(L_i)$ such that

$$x_\alpha \in T_i \ (\alpha = 1, 2, \ldots, k),$$

where T_i is the domain of θ_i.

4.2. We assume throughout this section that K is a field of characteristic zero. The following example shows that a finitely generated abelian-by-finite dimensional Lie algebra may not be residually finite in this case. Take the Lie algebra L, generated by elements x and y such that $[x,y]=z$, $[x,z]=[y,z]=0$. Let M be the left L-module, isomorphic to $U(L)/(z-1)$ and $H=M+L$ be the appropriate split extension of the abelian Lie algebra M by the Lie algebra L. Since M is cyclic H has 3 generators. We see also that the module M is in fact the module of the left regular representation of the Weyl algebra $A_1(K)$. If now M_1 is an L-submodule of M and M/M_1 is finite dimensional then we obtain in the module M/M_1 a finite dimensional representation of $U_1(K)$; it must be trivial because $U_1(K)$ is an infinite dimensional simple algebra (see [7]) and hence $M_1=M$. We see thus that the module M cannot be residually finite which implies that H is not residually finite.

In order to construct specializations in the case of a field K of characteristic zero we will obtain in this section some results about abelian-by-finite dimensional Lie algebras over K; these results will also be applied in part II of the article for study of automorphism groups of abelian-by-finite dimensional Lie algebra.

Let L be a Lie algebra of dimension n over K, $e_1,e_2,\ldots,e_n$ be its basis. Let

$$e_\alpha e_\beta=\sum_{\gamma=1}^{n}\lambda_{\alpha\beta\gamma}e_\gamma(\lambda_{\alpha\beta\gamma}\in K,\ 1\le\alpha,\beta,\gamma\le n) \tag{4.1}$$

and K_0 be an arbitrary subring (with unity) of K containing the elements $\lambda_{\alpha\beta\gamma}$ from (4.1). Then the Lie ring $L_0=K_0\langle e_1,e_2,\ldots,e_n\rangle$ over K_0 is a free K_0-module and hence has a universal envelope $U_{K_0}(L_0)$ whose basis over K_0 is given by the Poincaré-Birkhoff-Witt monomials on $e_1,e_2,\ldots,e_n$ and hence $U(L)\cong K\otimes U_{K_0}(L_0)$. If $K_0\subseteq K_1\subseteq K$ is an intermediate subring and $L_1=K_1\langle e_1,e_2,\ldots,e_n\rangle$ then

$$U_{K_1}(L_1)\cong K_1\otimes U_{K_0}(L_0).$$

THEOREM 4.1. *Let M be a $U(L)$-module with a finite system of generators $x_1,x_2,\ldots,x_m$. Then there exists a finitely generated subring $K_0\subseteq K_1\subseteq K$ and a $U_{K_1}(L_1)$ submodule M_1 of M, generated by the elements $x_1,x_2,\ldots,x_m$, such that M_1 is a free K_1-vector space and*

$$M\cong K\otimes_{K_1}M_1. \tag{4.2}$$

The proof will be based on the following assertion.

LEMMA 4.1. *Let M be a finitely generated module over $U(L)$. Then there exists a finitely generated subfield K', containing all the elements $\lambda_{\alpha\beta\gamma}$ from (4.1), a subalgebra $L'=K'\langle e_1,e_2,\ldots,e_n\rangle$ and a $U(L')$-submodule M', containing all the generators of M', such that*

$$M\cong K\otimes_{K'}(M').$$

PROOF. Let $x_1,x_2,\ldots,x_m$ be a system of generators for M and $\tilde M$ be a free $U(L)$-module with a basis $u_1,u_2,\ldots,u_m$, V be the kernel of the homomorphism $\tilde M\to M$ defined by the map $u_i\to x_i$ $(i=1,2,\ldots,m)$. Since

$U(L)$ is Noetherian the module $\tilde{M}$ is Noetherian too and hence V is finitely generated; let $v_1, v_2, \ldots, v_k$ be a system of generators for it. We find the coordinates of these elements in the basis $u_1, u_2, \ldots, u_m$ and let K' be the subfield of K, generated by these coordinates together with the elements $\lambda_{\alpha\beta\gamma}$ from (4.1), $L' = K'\langle e_1, e_2, \ldots, e_n\rangle$, M_0 be the $U_{K'}(L')$-submodule of $\tilde{M}$ generated by the systems of elements $u_1, u_2, \ldots, u_m$.

Let

$$V_0 = V \cap M_0. \tag{4.3}$$

Since $U(L')$ contains a K-basis of $U(L)$ and V_0 contains the elements v_1, $v_2, \ldots, v_k$ we obtain that V_0 contains a K-basis of V and hence

$$KV_0 \cong K \otimes V_0 \cong V.$$

We conclude now from (4.3) that the natural homomorphism $\tilde{M} \to \tilde{M}/V \cong M$ maps M_0 on $U(L')$-submodule $M' = M_0/V_0$. Since $K \otimes_{K'} M_0 = \tilde{M}$ and $K \otimes_{K'} V_0 = V$ a routine argument completes the proof.

PROOF OF THEOREM 4.1. Let K' be a finitely generated subfield of K which satisfies conclusions of Lemma 4.1. It is easy to see that it is enough to prove the assertion for the $U(L')$-module M'. In order to simplify the notations we assume that $M = M'$, i.e. that the field K is finitely generated.

Let K_0 be a finitely generated subring of K, which contains all the generators of K as well as the elements of $\lambda_{\alpha\beta\gamma}$ from (4.1), $x_1, x_2, \ldots, x_m$ be a system of generators of M and M_0 be the $U_{K_0}(L_0)$ submodule of M, generated by the elements $x_1, x_2, \ldots, x_m$. Once again, since $U_{K_0}(L_0)$ contains a K-basis of $U(L)$ we obtain that M_0 contains a K-basis of M, say y_j $(j \in J)$. It is important that K is the field of fractions of K_0; this implies easily that every K_0-linearly independent system of elements in M_0 remains K-independent and we conclude from here that the system y_j $(j \in J)$ is a maximal K_0-independent system of elements of the K_0-vector space M_0.

The maximality of the linearly independent system y_j $(j \in J)$ implies that for every element x_k $(k = 1, 2, \ldots, m)$ an element $\lambda_k \in K_0$ can be found such that $\lambda_k x_k$ is a K_0-linear combination of the elements y_j $(y \in J)$. Let $K_1 = K_0\langle\lambda_1, \lambda_2, \ldots, \lambda_k\rangle$, $L_1 = K_1\langle e_1, e_2, \ldots, e_n\rangle$, $M_1 = K_1 M_0$. Since the elements x_k $(k = 1, 2, \ldots, m)$ are generators of the $U_{K_1}(L_1)$-module M_1 and every one of these elements is a K_1-linear combination of the elements y_j $(j \in J)$ we see that every element of M_1 is a K_1-linear combination of the elements y_j $(j \in J)$. Since these elements form a K-basis of M they form a K_1-basis of M_1 and (4.2) follows. □

THEOREM 4.2. *Let H be a finitely generated abelian-by-finite dimensional Lie algebra over a field K. Then there exists a finitely generated subring $K_1 \subseteq K$ and a Lie subring H_1, which is a free K_1-module and is an extension of a K_1-free abelian Lie ring by a finite dimensional K_1-free Lie ring and*

$$H \cong K \otimes_{K_1} H_1. \tag{4.4}$$

PROOF. Let A be an abelian ideal in H such that the algebra $L = H/A$ is finite dimensional. The Lie operation in H defines in H the structure of a left H-module and we see that U is in fact an L-module. Furthermore, since H is finitely generated and H/A is finite dimensional over K we can apply the results from chapter 11.1 in **[1]** and conclude that every ideal in H is finitely generated (see **[1]**, Corollary 11.1.8) and hence A is a finitely generated L-module. Let $e_1, e_2, \dots, e_n$ be a system of elements of H which gives a basis of $L = H/A$ and let

$$[e_\alpha, e_\beta] = \sum_{j=1}^{n} \lambda_{\alpha\beta\gamma} e_\gamma + x_{\alpha\beta\gamma} \ (x_{\alpha\beta\gamma} \in A; 1 \le \alpha, \beta, \gamma \le n). \tag{4.5}$$

Now apply Theorem 3.1 to A and find a finitely generated subring $K_1 \subseteq K$ which contains all the elements $\lambda_{\alpha\beta\gamma}$ from (4.5), a free over K_1 Lie subring L_1 and a free over K_1 $U_{K_1}(L_1)$-submodule $A_1 \subseteq A$ such that

$$A \cong K \otimes_{K_1} A_1.$$

It is important that we can extend, if necessary, K_1 and assume that it contains all the coefficients of the elements $x_{\alpha\beta\gamma}$ in their representations through a given basis of A. We can therefore assume that

$$x_{\alpha\beta\gamma} \in A_1 \ (1 \le \alpha, \beta, \gamma \le n). \tag{4.6}$$

Now let $H_1 = K_1\langle A_1, e_1, e_2, \dots, e_n\rangle$. We see that

$$[e_\alpha, A_1] \subseteq A_1 \subseteq H_1; \ [e_\alpha, e_\beta] \subseteq H_1 \ (1 \le \alpha, \beta \le n) \tag{4.7}$$

which implies that H_1 is a subring of H. Clearly, $H_1/A_1 \cong L_1$. Since A_1 and L_1 are free over K_1 we conclude easily that H_1 is free over K_1. Finally, (4.4) follows now from the fact that $H = K\langle A, e_1, e_2, \dots, e_n\rangle$.

COROLLARY. *Let H be an abelian-by-finite dimensional Lie algebra over a commutative field K, D be the field of fractions of $U(H)$,*

$$x_\alpha \ (\alpha = 1, 2, \dots, k) \tag{4.8}$$

be given nonzero elements of D. Then there exists a finitely generated subring $K_1 \subseteq K$ and a finitely generated K_1-free Lie subring H_1 such that

$$x_\alpha \in D_1 \ (\alpha = 1, 2, \dots, k)$$

where D_1 is the field of fractions of $U_{K_1}(H_1)$.

PROOF. Let

$$x_\alpha = a_\alpha b_\alpha^{-1} (a_\alpha, b_\alpha \in U(H); b_\alpha \neq 0; \alpha = 1, 2, \dots, k). \tag{4.9}$$

Let H_0 a finitely generated subalgebra $H_0 \subseteq H$ such that

$$a_\alpha \in U(H_0), b_\alpha \in U(H_0) \ (\alpha = 1, 2, \dots, k).$$

The subring $K_1 \subseteq K$ and the Lie subring $H_1 \subseteq H_0$ are obtained now from Theorem 4.2.

4.2. Theorem 4.2 and its corollary reduce the study of finitely generated subrings of D to the case when D is the field of fractions of $U(H)$, where H is a finitely generated abelian-by-finite dimensional Lie algebra over a finitely generated commutative domain. The next step which is carried out in this subsection is to reduce the study of these rings to the case when H is a Lie algebra over a finite field.

We consider a more general situation when H is a soluble-by-finite dimensional Lie algebra over a finitely generated infinite commutative domain K_1. It is known that $U_{K_1}(H)$ is an Ore ring (see [**2**, 3]) and let D be its field of fractions.

PROPOSITION 4.2. *There exists an infinite system of maximal ideals $A_i \in K_1$ $(i \in I)$ such that*

i) $\bigcap_{i \in I} A_i = 0$ *and for every given non-zero elements $a_1, a_2, \ldots, a_r \in K$ a subset $I_1 \subseteq I$ can be found such that $a_\alpha \notin A_i$ $(\alpha = 1, 2, \ldots, r; i \in I_1)$ and*

$$\bigcap_{i \in I_1} A_i = 0.$$

ii) *Every ideal $A_i(U_{K_1}(H))$ defines a $\mathbb{P}$-adic valuation in the ring $U_{K_1}(H)$ and in the field D.*

iii) *Every ideal $A_i(U_{K_1}(H))$ can be localized.*

iv) *The quotient ring $(U_{K_1}(H))/A_{K_1}(H))$ is isomorphic to the universal envelope $(U_{\overline{K}_i}(\overline{H}_i))$ of a Lie algebra $\overline{H}_i = H/(A_i)H$ over the field $K_i = K_1/A_i$.*

We need first the following lemma.

LEMMA 4.2. *Let K_1 be a commutative domain, H be a Lie algebra over K_1. Assume also that $U(H)$ is an Ore ring. Let* (a) *be a principal ideal of K_1 such that*

$$\bigcap_{n=1}^{\infty} (a)^n = 0.$$

Then the ideal $(a)U_{K_1}(H)$ is localizable in $U_{K_1}(H)$.

The proof is easy and we omit it.

PROOF OF PROPOSITION 4.2. We give only a sketch of the proof since it does not differ that much from the argument in [**8**] for the case when H is finite dimensional over K_1. We can find in K_1 a family of maximal ideals which define $\mathbb{P}$-adic valuations in K_1 and satisfy i). Let A be one of these ideals. Since $U_{K_1}(L_1)$ is a free module over K_1 we conclude easily that

$$\bigcap_{n=1}^{\infty} (A(U_{K_1}(L)))^\gamma = \bigcap_{n=1}^{\infty} A^\gamma(U_{K_1}(L)) = 0. \tag{4.10}$$

It follows from Corollary 1 of Proposition 3.3 in [**2**] that the graded ring, associated to the filtration

$$(A(U_{K_1}(L)))^\gamma \ (\gamma = 1, 2, \ldots)$$

of $U_{K_1}(L)$ is a domain. This proves that the filtration (4.10) defines a $\mathbb{P}$-adic valuation on $U_{K_1}(L)$ and hence on the field of fractions of it.

In order to prove that $A(U_{K_1}(L))$ can be localized we observe first of all that the set $M = K_1 \backslash A$ is a central Ore subset and hence we can form the ring $(U_{K_1}(L)M^{-1}$ and it is enough to prove that the ideal $A(U_{K_1}(L))M^{-1} = AM^{-1}(U_{K_1}(L))$ is localizable in $(U_{K_1}(L))M^{-1}$. Since A defines a $\mathbb{P}$-adic valuation in K_1 we conclude easily that AM^{-1} is the corresponding valuation ring and hence AM^{-1} is principal ideal in K_1M^{-1}. The assertion therefore follows from Lemma 4.2.

COROLLARY 1. *Let* $M_i = (U_{K_1}(H) \backslash A_i(UK_1(H))(i \in I)$, $T_i = U_{K_1}(H)M_i^{-1}$. *Then* T_i *is a local ring,* $T_i \supseteq (U_{K_1}(H))$, *and the quotient ring* $T_i/J(T_i)$ *is isomorphic to the field* D_i *of fractions of the ring* $U_{K_1}(H_i)$.

PROOF. Follows from Proposition 4.2 via well known facts of the localization theory.

COROLLARY 2. *Let* V *be a left vector space over* R *with a basis* v_j $(j \in J)$ *and* (4.8) *be given nonzero elements of* V. *Then there exists a subset* $I_1 \subseteq I$ *such that for every* $i \in I$ *the elements* (4.8) *belong to the free* T_i-*submodule* V_i *with the basis* e_j $(j \in J)$ *and* $\bigcap_{i \in I_1} A_i = 0$.

PROOF. We can represent the elements x_α $(\alpha = 1, 2, \ldots, k)$ in the form

$$x_\alpha = b^{-1} y_\alpha \ (\alpha = 1, 2, \ldots, k) \tag{4.11}$$

where

$$y_\alpha \in \sum_{j=1}^{n} U_{K_1}(H))v_j \ (\alpha = 1, 2, \ldots, k; \ b \in U_{K_1}(H)). \tag{4.12}$$

Represent the element b as a linear combination of standard monomials with nonzero coefficients $a_1, a_2, \ldots, a_m$. Let $a = a_1, a_2 \ldots a_m$. We can find a cofinite subset $I_1 \subseteq I$ such that $a \notin A_i$ $(i \in I_1)$ and hence

$$a_\beta \notin A_i \ (i \in I_1; \ \beta = 1, 2, \ldots, m).$$

This implies that for $i \in I_1$

$$b \notin (A_i)U_{K_1}(H)(i \in I_1)$$

and hence b is invertible in the ring $T_i \supseteq U_{K_1}(H)$. The assertion follows now from (4.11) and (4.12).

Apply now Corollary 2 in the special but important case when $V = R$ and obtain the following fact.

COROLLARY 3. *Let* (4.8) *be given elements of* D. *Then there exists a subset* $I_1 \subseteq I$ *such that for every* $i \in I_1$

$$x_\alpha \in T_i \ (\alpha = 1, 2, \ldots, k). \tag{4.13}$$

and $\bigcap_{i\in I_1} A_i = 0$.

For every $i \in I_1$ consider the natural homomorphism

$$V_i \to V_i/(J(T_i))V_i.$$

Since this homomorphism is induced by the homomorphism of rings $T_i \to T/J(T_i) \cong D_i$ we will as in Proposition 4.2 and its Corollary denote the image of a subset $X \subseteq V_i$ by $\overline{X}_i$. We see that $\overline{V}_i$ is a vector space over D_i with basis $\bar{v}_j$ $(j \in J)$.

PROPOSITION 4.3. *Assume that the elements* (4.8) *of* V *are linearly independent over* K_1. *Then there exists a subset* $I_2 \subseteq I_1 \subseteq I$ *such that* $\bigcap_{i\in I_2} A_i = 0$ *and for every* $i \in I_2$ *the images* $\bar{x}_\alpha$ $(\alpha = 1, 2, \ldots, k)$ *of these elements are linearly independent in* $\overline{V}_i$ *over* K_i.

PROOF. The representation (4.11) implies immediately that the given elements are linearly independent over K_1 if and only if the elements y_α $(\alpha = 1, 2, \ldots, k)$ are linearly independent over K_1; similarly, the elements $\bar{x}_\alpha = \bar{b}^{-1}\bar{y}_\alpha$ $(\alpha = 1, 2, \ldots, k)$ are linearly independent in $\overline{V}_i$ over K_i if and only if the elements $\bar{y}_\alpha$ are linearly independent over K. We can assume that

$$(4.14) \qquad x_\alpha = \sum_{j=1}^{n} \lambda_{j\alpha} e_j (\lambda_{j\alpha} \in U_{K_1}(H);\ \alpha = 1, 2, \ldots k;\ j = 1, 2, \ldots, n),$$

that V is a free $U_{K_1}(H)$-module with basis $e_1, e_2, \ldots e_n$ and we consider now the homomorphism $V \to V/(A_i)V$.

Clearly, there exists only a finite number of standard monomials π_γ $(\gamma = 1, 2, \ldots s)$ which occur in the coordinates of the elements $\lambda_{j\alpha}$ in (4.14). We obtain therefore that the elements x_α $(\alpha = 1, 2, \ldots, k)$ are contained in a finite dimensional K_1-subspace V_0 with the basis $\pi_\gamma e_j$ $(\gamma = 1, 2, \ldots, s;\ j \in J)$. Let $\dim_{K_1}(V_0) = m$ and S be the $m \times k$ matrix formed by the coordinates of the elements x_α $(\alpha = 1, 2, \ldots, k)$. The linear independence of these elements is equivalent to the fact that $\operatorname{rank}(S) = k$. Find a cofinite subset $I_2 \subseteq I_1$ such that for every $i \in I_2$ $\operatorname{rank}(S)$ is not changed under the homomorphism: $(K_1)_{n\times n} \to (K_1/A_i)_{n\times n}$ and the assertion follows.

COROLLARY. *Let* D *be the field of fractions of* $U_{K_1}(H)$, $A_i (i \in I)$ *be the system of ideals obtained in Proposition* 4.2. *Let*

$$(4.15) \qquad X_\alpha\ (\alpha = 1, 2, \ldots, k)$$

be given matrices from $D_{n\times n}$. *Then there exists a subset* $I' \subseteq I$ *such that* $\bigcap_{i\in I'} A_i = 0$ *and for every* $i \in I'$

$$(4.16) \qquad X_\alpha \in (T_i)_{n\times n}\ (\alpha = 1, 2, \ldots, k).$$

Furthermore, the matrices (4.15) *are linearly independent over* K_1 *if and only if for every* $i \in I'$ *their images* $\overline{X}_\alpha = 1, 2, \ldots, k)$ *are linearly independent over* K_i *in the ring* $(T_i/J(T_i))_{n\times n} \cong (D_i)_{n\times n}$.

Now let X_α be a given matrix in (4.15). We apply Corollary 2 to the set of matrices

(4.17) $$1, X_\alpha, X_\alpha^2, \dots, X_\alpha^{n-1}.$$

We obtain thus a cofinite subset $I' \subseteq I$ such that for every $i \in I$ the elements (4.15) belong to T_i and are linearly dependent over K_1 if and only if their images in $(D_i)_{n\times n}$

(4.18) $$1, \overline{X}_\alpha, \dots, \overline{X}_\alpha^{n-1}$$

are linearly dependent over K_i. But the ring $K' \otimes_{K_1} D_j$ contains no zero divisors for an arbitrary extension $K' \supseteq K_i$; we can therefore apply Statement 4. from **[9]**, which implies that if the element $\overline{X}_\alpha$ in (4.18) is algebraic then its minimal polynomial has degree less or equal n and hence the linear dependence of the elements (4.18) is equivalent to the fact that $\overline{X}_\alpha$ is algebraic over K_i. Furthermore, if K is the field of fractions of K_1 then, once again by 4. in **[9]**, the linear dependence of the elements (4.17) in $D_{n\times n}$ over K is equivalent to the fact that X_α is algebraic over K. We obtained thus the following assertion.

PROPOSITION 4.4. *Let* (4.15) *be given matrices in* $D_{n\times n}$. *Then there exists a subset* $I_0 \subseteq I$ *such that* $\bigcap_{i\in I_0} A_i = 0$

$$X_\alpha \in (T_i)_{n\times n} \ (i \in I_0; \alpha = 1, 2, \dots, k)$$

and every given matrix X_α *is algebraic over* K *if and only if its image* $\overline{X}_\alpha \subseteq (D_i)_{n\times n}$ *is algebraic over* K_i *for every* $i \in I_0$.

We will need in Section 6 the analogs of Propositions 4.2–4.4 for the case when D is the field of fractions of a group ring. We formulate the appropriate result as a Corollary of Propositions 4.2–4.4 since its proof is obtained by an easy modification of the arguments which were applied in their proofs; the only difference is that instead of standard monomials one has to deal now with a group basis.

COROLLARY OF PROPOSITION 4.2–4.4. *Let* H *be a group whose group ring over any commutative field is an Ore domain,* K_1 *be a finitely generated commutative domain,* D *be the field of fractions of* K_1H. *Then the conclusions of Propositions* 4.2–4.4 *hold for the ring* $D_{n\times n}$.

4.3. Let throughout this section H be a residually finite soluble-by-finite dimensional Lie algebra over a commutative domain K. Since finitely generated abelian-by-finite dimensional Lie algebras over fields of finite characteristic are residually finite (**[1]**, 11.3) the results of this section will be applicable in this case (see Proposition 4.6). These results generalize and strengthen Proposition 4.1.

Let D be the field of fractions of $U(H)$. Let H_i $(i \in I)$ be a system of soluble ideals in H which is closed with respect to finite intersections; assume also that every quotient algebra H/H_i has a finite dimension and

$$\bigcap_{i\in I} H_i = 0.$$

Let for given $i \in I$ D_i be the field of fractions of $U(H/H_i)$. The following facts follow from Theorems 5.1 and 6.2 in **[2]**.

1) *Every homomorphism* $\phi_i\colon U(H) \to U(H/H_i)$ *is extended to a specialization* $\theta_i\colon D \to \Delta_i$ *with domain* $T_i \supseteq U(H)$.

2) *Let* (4.8) *be given elements of D. Then there exists* $i_0 \in I$ *such that*

$$x_\alpha \in T_{i_0} \ (\alpha = 1, 2, \ldots, k).$$

Now if $H_i \subseteq H_{i_0}$ then the homomorphism $\phi_i\colon U(H) \to U(H/H_i)$ can be factored through the homomorphism $\phi_{i_0}\colon U(H) \to U(H/H_{i_0})$ and hence (see [7], 7.2) $T_i \supseteq T_{i_0}$. Now the ideals $H_i \cap H_{i_0}$ form a subsystem of the system H_i $(i \in I)$; this subsystem consists from the ideals H_i $(i \in I_1)$, where $I_1 \subseteq I$, and it is closed with respect to finite intersections. We have also

$$x_\alpha \in T_i \ (i \in I_1;\ \alpha = 1, 2, \ldots, k).$$

We obtained thus the following lemma.

LEMMA 4.3. *Let* (4.8) *be a given nonzero elements of D. Then there exists a subsystem* H_i $(i \in I_1; I_1 \subseteq I)$, *closed with respect to finite intersections, and such that* $\bigcap_{i\in I_1} H_i = 0$ *and*

$$x_\alpha \in T_i \ (i \in I_1; \alpha = 1, 2, \ldots, k).$$

Now consider a left vector space V as in Corollary 2 of Proposition 4.2 and find a system of free T_i-submodules $(i \in I_1)$ such that every V_i contains the elements (4.8).

We have the following analog of Proposition 4.3.

PROPOSITION 4.3′. *Assume that the elements* (4.8) *are linearly independent in V over K. Then there exists a subsystem* H_i $(i \in I_2 \subseteq I_1 \subseteq I)$ *closed with respect to finite intersections and such that* $\bigcap_{i\in I_2} H_i = 0$ *and for every* $i \in I_2$ *the images* $\bar{x}_\alpha$ $(\alpha = 1, 2, \ldots k)$ *of these elements are linearly independent in* $\overline{V}_i$ *over K.*

PROOF. Once again, as in the proof of Proposition 4.3 we obtain (4.13) and we can assume that V is an $U(H)$-module and the homomorphism of the modules have now the form

$$V \to V/((\ker \phi_i)V) \tag{4.19}$$

where ϕ_i is the homomorphism of enveloping algebras $U(H) \to U(H/H_i)$. Once again, the elements (4.8) are contained in the finite dimensional vector subspace with basis $\pi_\gamma e_j$ $(\gamma = 1, 2, \ldots, s;\ j \in J)$. We see that only a finite number of basic elements of H, say $h_1, h_2, \ldots, h_r$ occurs in the representations of the standard monomials π_γ $(\gamma = 1, 2, \ldots, s)$. Let H_0 be the finite dimensional vector subspace of H, generated by these basic elements and M be the finite dimensional vector subspace of $U(H)$, whose basis is given by all the standard monomials in $h_1, h_2, \ldots, h_r$. Since H is finite dimensional we can find easily a subset $I_2 \subseteq I_1 \subseteq I$ such that

$$H_0 \cap H_i = 0 \ (i \in I_2) \tag{4.20}$$

and

$$\bigcap_{i \in I_2} H_i = 0.$$

The relation (4.20) implies that

$$\ker \phi_i \cap M = 0 \ (i \in I_2)$$

and hence

$$V_0 \cap (\ker \phi_i)V = 0. \tag{4.21}$$

We obtain from (4.21) that the image of V_0 under homomorphism (4.19) is a vector subspace of $V/(\ker \phi_i)V)$, which is isomorphic to V_0 and the assertion follows.

The following assertion is an analog of Proposition 4.4. It follows from Proposition 4.3′ by the same argument as Proposition 4.4 from Proposition 4.3 and we omit the proof.

PROPOSITION 4.4′. *Let* (4.15) *be given matrices in* $D_{n\times n}$. *Then there exists a subsystem* H_i $(i \in I_0)$ *closed with respect to the finite intersections and such that*

$$X_\alpha \in (T_i)_{n\times n} \ (i \in I_0; \ \alpha = 1, 2, \ldots, k)$$

and

$$\bigcap_{i \in I_0} H_i = 0$$

and every given matrix X_α *is algebraic in* $D_{n\times n}$ *over* K *if and only if for every* $i \in I_0$ *its image* $\overline{X}_\alpha$ *under the homomorphism*

$$(T_i)_{n\times n} \to (T_i/J(T_i))_{n\times n} \cong (D_i)_{n\times n}$$

is algebraic over K *in* $(D_i)_{n\times n}$.

Now consider the K-subalgebra R of $D_{n\times n}$, generated by the matrices (4.15). Thus, we have $R \subseteq T_i$ $(i \in I_0)$. Let

$$B_i = (J(T_i))_{n\times n} \cap R \ (i \in I_0). \tag{4.22}$$

We have first of all

$$R/B_i \subseteq (T_i)_{n\times n}/(J(T_i))_{n\times n} \subseteq (D_i)_{n\times n}. \tag{4.23}$$

Furthermore, if

$$Y_1, Y_2, \ldots, Y_s \tag{4.24}$$

are nonzero elements from R then there exists a subsystem H_i $(i \in \bar{I} \subseteq I_0)$, closed with respect to finite intersections and such that

$$Y_\beta \in (T_i)_{n\times n} \backslash (J(T_i))_{n\times n} \ (\beta = 1, 2, \ldots, s; \ i \in \bar{I})$$

and hence

$$Y_\beta \notin B_i \ (\beta = 1, 2, \ldots, s; i \in \bar{I}) \tag{4.25}$$

which implies easily that

$$\bigcap_{i \in I_0} B_i = 0. \tag{4.26}$$

Furthermore, if X is algebraic in R over K then for every $i \in I$ its image in R/B_i is algebraic over K. Conversely, if the image of X is algebraic in every R/B_i $(i \in I_0)$ then we apply Proposition 4.4′ to the matrix X and the system of ideals H_i $(i \in I_0)$ and obtain that X is algebraic in R over K. We proved thus the following corollary of Proposition 4.4′.

PROPOSITION 4.5. *Let R be a finitely generated subalgebra of $D_{n\times n}$. Then R contains a system of ideals B_i $(i \in I_0)$ such that* (4.23) *holds and for every given nonzero matrices* (4.24) *a subsystem H_i $(i \in \bar{I})$ closed with respect to finite intersection, can be found such that* (4.25) *and* (4.26) *hold and every given matrix $X \in R$ is algebraic over K if and only if its images are algebraic over K in all the quotient algebras R/B_i $(i \in I_0)$.*

Now let R be a finitely generated subring of $D_{n\times n}$. The corollary of Proposition 4.1 shows that there exists a finitely generated subring $K_1 \subseteq K$ and a finitely generated Lie algebra H_1 over K_1 such that $R \subseteq (D_1)_{n\times n}$ where D_1 is the field of fractions of $U_1(H)$. We fix this subring K_1 and prove now the following modification of Proposition 4.5.

PROPOSITION 4.6. *Let H be an abelian-by-finite dimensional Lie algebra over a field K, R be a finitely generated subring of $D_{\times n \times}$. Then R contains a system of ideals C_j $(j \in J)$ such that*

i) *If nonzero matrices* (4.24) *are given then there exists a subset $J_1 \subseteq J$ such that*

$$Y_\beta \notin C_j \ (\beta = 1, 2, \ldots, a; \ j \in J_1).$$

ii) *For every given $j \in J$ the quotient ring R/C_j is isomorphic to a subring of $(D_j)_{n\times n}$ where D_j is the field of fractions of $U(H_j)$, H_j is a finite dimensional Lie algebra over a finite field K_j, the image of K_1 in R/C_j.*

iii) *Every given element $X \in R$ is algebraic over K if and only if its images are algebraic over K_j in all the quotient rings R/C_j $(j \in J)$.*

PROOF. Since R and K_1 are finitely generated we can easily reduce the proof to the case when $R \supseteq K_1$. The same argument as in Proposition 4.5 shows that if now T_i $(i \in I_0)$ is a system of subrings obtained in Proposition 4.4 and the system B_i $(i \in I_0)$ is defined by (4.22) then R is a subdirect sum of rings $R_i = R/B_i$ $(i \in I_0)$ where every R_i now is a subring of $(D_i)_{n\times n}$, D_i is the field of fractions of $U_{K_1}(H_i)$, H_i is finitely generated abelian-by-finite dimensional Lie algebra over K_1, the images of the elements (4.24) are nonzero and every matrix $X \in R$ is algebraic over K_1 if and only if its images $X_i \in R_i$ are algebraic over K_1 for all $i \in I_0$. Proposition 4.5 implies that the assertion is true for every subring $R_i \subseteq (D_i)_{n\times n}$ and we obtain easily that it is true for their subdirect sum R.

COROLLARY. *Assume that in Proposition* 4.6 $\operatorname{char} K = 0$ *and let* m *be a given natural number. Then the system of ideals* C_j $(j \in J)$ *can be chosen in such a way that it satisfies the additional property:* $\operatorname{char} K_j > m$ $(j \in J)$.

PROOF. Apply Proposition 4.6 to the finitely generated subring $R_1 = \langle R, m, m^{-1}\rangle$ and obtain a system of ideals $C_j^{(1)} \subseteq R_1$ $(j \in J)$ which satisfies the conclusions i)–iii). It is easy to see that the system of ideals $C_j^{(1)} \cap R = C_j \subseteq R$ $(j \in J)$ satisfies the conclusion of the assertion.

5.1. We need now some concepts and results on ultraproducts in order to be able to apply the results of Section 5.1 in the proof of Theorems I–III.

Let A be a set of indices. A non-empty family $\mathscr{T}$ of subsets is a filter on A if

i) $\mathscr{T}$ does not contain the empty set.

ii) $\mathscr{T}$ is closed with respect to the finite intersections.

iii) If $Y \subseteq X \subseteq A$ and Y belongs to $\mathscr{T}$ then $X \in \mathscr{T}$.

The filter $\mathscr{T}$ is an ultrafilter if for every $X \in A$ either $X \in \mathscr{T}$ or $A\backslash X \in \mathscr{T}$. A family of subsets of A can be extended to an ultrafilter if it does not contain the empty subset and the intersection of a finite number of its subsets is not empty (finite intersection property).

We will need also two other properties of the ultrafilters.

iv) Let $\mathscr{T}$ be an ultrafilter on A; $A_1, A_2, \dots, A_k$ be subsets whose union belongs to $\mathscr{T}$. Then at least one of these subsets belong to $\mathscr{T}$ (see **[14]**, Corollary of Proposition 1.7.2).

v) Let B be an arbitrary subset of A which belongs to $\mathscr{T}$. Then all the subsets $Y = X \cap B (X \in \mathscr{T})$ generate an ultrafilter $\mathscr{T}_1$ on B. It is said that $\mathscr{T}_1$ is induced by $\mathscr{T}$ (see **[14]**, Proposition I.5.8).

Now let for every $\alpha \in A$ a ring R_α be given. Then an ultrafilter $\mathscr{T}$ defines an equivalency relation on $\prod_{\alpha\in A} R_\alpha$ in the following way: if $f, g \in \prod_{\alpha\in A} R_\alpha$ then $f \equiv g \pmod{\mathscr{T}}$ iff the set $\{\alpha | f_\alpha = g_\alpha\}$ belongs to $\mathscr{T}$. The appropriate quotient ring $R = (\prod_{\alpha\in A} R_\alpha)/\mathscr{T}$ is the ultraproduct of the rings R_α $(\alpha \in A)$. It is well known that if every ring R_α is a field then so is the ultraproduct $R = (\prod_{\alpha\in A} R_\alpha)/\mathscr{T}$ and that the ultraproduct of matrix rings $(R_\alpha)_{n\times n}$ is isomorphic to the ring $R_{n\times n}$. Let $U(R_\alpha)$ be the group of units of R_α. It is easy to verify that the group of units of $\prod_{\alpha\in A} R_\alpha$ is isomorphic to $\prod_{\alpha\in A} U(R_\alpha)$, that $\mathscr{T}$ induces in a natural way an ultrafilter on $\prod_{\alpha\in A} U(R_\alpha)$, which we also denote by $\mathscr{T}$, and we see that the homomorphism $\prod_{\alpha\in A} R_\alpha \to (\prod_{\alpha\in A} R_\alpha)/\mathscr{T}$

induces a group homomorphism $\prod_{\alpha\in A} U(R_\alpha) \to (\prod_{\alpha\in A} U(R_\alpha))/\mathscr{T}$. We denote by $\overline{X}$ the image of a subset $X \subseteq \prod_{\alpha\in A} R_\alpha$ (or $X \subseteq \prod_{\alpha\in A} U(R_\alpha)$) under the homomorphism of rings $\prod_{\alpha\in A} R_\alpha \to (\prod_{\alpha\in A} R_\alpha)/\mathscr{T}$ (or, correspondingly under the homomorphism of groups $\prod_{\alpha\in A} U(R_\alpha) \to (\prod_{\alpha\in A} U(R_\alpha))/\mathscr{T}$.

We need the following fact.

LEMMA 5.1. *Let every R_α $(\alpha \in A)$ be a field and let $R = (\prod_{\alpha\in A} R_\alpha)/\mathscr{T}$. For every $\alpha \in A$ pick a subring $S_\alpha \subseteq (R_\alpha)_{n\times n}$ and a subgroup $G_\alpha \in \mathrm{GL}_n(R_\alpha)$. Let $S = (\prod_{\alpha\in A} S_\alpha)/\mathscr{T}$ and $G = (\prod_{\alpha\in A} G_\alpha)/\mathscr{T}$. Denote by S the subring $(\prod_{\alpha\in A} S_\alpha)/\mathscr{T} \subseteq R_{n\times n}$ and by G the subgroup $(\prod_{\alpha\in A} G_\alpha)/\mathscr{T} \subseteq \mathrm{GL}_n(R)$. Then*

i) *If every ring S_α $(\alpha \in A)$ is nilpotent then so is S.*

ii) *If every subgroup G_α $(\alpha \in A)$ is unipotent in $\mathrm{GL}_n(R_\alpha)$ then G is unipotent in $\mathrm{GL}_n(R)$.*

iii) *If there exists a law such that $v(G_\alpha) = 1$ then the group G also satisfies the law v.*

PROOF. If a subring $S_\alpha \subseteq \mathrm{GL}_n(R_\alpha)$ is nilpotent then $(S_\alpha)^n = 0$ (see **[15]**, 4.13). This implies easily that $S^n = 0$ if every S_α $(\alpha \in A)$ is nilpotent. Furthermore if for every $\alpha \in A$ there exists a nilpotent subring S_α such that $G_\alpha \subseteq (1 + S_\alpha)$ then we conclude that $(\prod_{\alpha\in A} G_\alpha)/\mathscr{T} \subseteq (\prod_{\alpha\in A}(1 + S_\alpha))/\mathscr{T} \equiv 1+(\prod_{\alpha\in A} S_\alpha)/\mathscr{T} = 1+S$, where S is nilpotent. We proved thus the statements i) and ii). The statement iii) follows immediately.

LEMMA 5.2. *Let $G = (\prod_{\alpha\in A} G_\alpha)/\mathscr{T}$ be an ultraproduct of groups. Assume that there exists a number m such that for all $\alpha \in A$ $(G_\alpha\colon 1) \le m$. Then G is isomorphic to one of the groups G_α.*

PROOF. We point out first of all that the condition $(G_\alpha\colon 1) \le m(\alpha \in A)$ implies that there exists a finite number, say k, of classes nonisomorphic groups among the groups G_α $(\alpha \in A)$. Let $G_1, G_2, \ldots, G_k$ be representatives of these classes. Define for every $1 \le i \le k$ the subset $A_i = \{\alpha \in A | G_\alpha \cong G_i\}$. The subsets A_i $(i = 1, 2, \ldots, k)$ form a partition of A and we conclude from property iv) that at least one (and in fact only one) of these subsets, say A_1, belongs to the filter. We can replace therefore all the groups G_α $(\alpha \notin A_1)$ by 1 and we conclude easily that

$$\left(\prod_{\alpha\in A} G_\alpha\right)/\mathscr{T} \cong \left(\prod_{\alpha\in A_1} G_\alpha\right)/\mathscr{T}_1 \tag{5.1}$$

where $\mathscr{T}_1$ is the ultrafilter, induced by $\mathscr{T}$ on A_1, (see property v) and every subgroup in the right side of (5.1) is isomorphic to G_1. More precisely, we can represent every such factor G_α in a form $\pi_j(G_1)$, where π_j $(j = 1, 2, \ldots, r)$ is a one-to-one map $G_1 \to G_1$. Once again, we define for every given j the subset $A_{1j} = \{\alpha \in A_1 | G_\alpha = \pi_j(G_1)\}$ and, because A_{1j} $(j = 1, 2, \ldots, r)$ is a partition of A_1, conclude that there exists a subset A_{1j} and an ultrafilter $\mathscr{T}_2$, induced by $\mathscr{T}_1$, such that

$$\left(\prod_{\alpha\in A_1} G_\alpha\right)/\mathscr{T}_1 \cong \left(\prod_{\alpha\in A_{1j}} G_\alpha\right)/\mathscr{T}_2 \tag{5.2}$$

where $G_\alpha = \pi_j(G_1)(\alpha\in A_{1j})$ and hence $(\prod_{\alpha\in A_{1j}} G_\alpha)/\mathscr{T}_2 \cong G_1$ and we obtain from (5.2) and (5.1) that $(\prod_{\alpha\in A} G_\alpha)/\mathscr{T} \cong G_1$. This completes the proof.

6.1. Let F be a free group, $F^{(m)}$ be the mth term of its derived series. We study in this section the universal field of fractions of KF and consider also some other classes of fields, mainly the field of fractions of the group ring $K(F/F^{(m)})$ of the free soluble group $F/F^{(m)}$. It is known that free soluble groups are residually finite p-groups (see **[18]**, 26.32); this fact plays an important role in the proof of Proposition 6.4. below, which is the main result of this subsection.

We recall (see **[19]**) that a group G is discriminated by a family of groups G_j $(j\in J)$ if for every given non-unit elements $h_1, h_2, \dots, h_m$ an index $j\in J$ and a homomorphism $\phi\colon G\to G_i$ can be found such that

$$\phi(h_\alpha)\neq 1 \ (\alpha = 1, 2, \dots, n).$$

We need the following assertion which is a modification of Dunwoody's Theorem in **[20]**. Its proof is essentially the same as Dunwoody's and we omit it.

PROPOSITION 6.1. *Let F be a free group, $S\lhd F$ and S_i $(i\in J)$ be a family of normal subgroups such that the group F/S is discriminated by the family of groups F/S_i $(i\in I)$. Let $v(S)$ be a verbal subgroup of S. Then the group $F/v(S)$ can be discriminated by the family of groups $F/v(S_i)$.*

Now let $S\lhd F$, S_i be a family of normal subgroups such that $\bigcap_{i\in I} S_i = S$ and that the family S_i $(i\in I)$ is closed with respect to finite intersections which implies easily that the family of groups F/S_i $(i\in I)$ discriminates F/S. We obtain thus the following corollary.

COROLLARY 1. *The group F/S' is discriminated by the family of groups F/S'_i $(i\in I)$.*

It is known that all the groups F/S' and F/S'_i $(i\in I)$ are torsion free. We will be especially interested in the case when F/S is a residually finite p-group. We have in this case the following fact.

COROLLARY 2. *Let S_i $(i\in I)$ be a system of normal subgroups in F, closed with respect to finite intersections, and such that every F/S_i $(i\in I)$ is a finite p-group. Assume also that $\bigcap_{i\in I} S_i = S$. Then the quotient group F/S' is discriminated by the family of groups F/S_i $(i\in I)$. Every group F/S'_i is torsion free and is an extension of a free abelian group S_i/S'_i by a finite p-group which is isomorphic to F/S_i.*

Apply now Corollary 2 to the free soluble group $F/F^{(m)}$. Since $F^{(m)} = (F^{(m-1)})$ and the free soluble group $F/F^{(m-1)}$ is residually a finite p-group for every prime p we obtain easily the following fact.

COROLLARY 3. *Let S_i $(i \in I)$ be a system of normal subgroups in F, which is closed with respect to finite intersections and such that every quotient group F/S_i is a finite p-group and $\bigcap_{i \in I} S_i = F^{(m-1)}$. Then the free soluble group $F/F^{(m)}$ is discriminated by the family of groups F/S_i' $(i \in I)$.*

We need the following known fact (see **[21]**),

LEMMA 6.1. *Let F be a free group, $S \triangleleft F$ be a normal subgroup such that F/S is soluble. Then F/S' has a series of normal subgroups*

$$F/S' = H_1 \supseteq H_2 \supseteq \cdots \supseteq H_k \supseteq H_{k+1} = 1$$

with free abelian factors H_j/H_{j+1} $(j = 1, 2, \ldots, k)$.

COROLLARY 1. *Let the conditions of Lemma* 6.1 *hold. Then the group F/S' is a unique product group and for every field K the group ring $K(F/S')$ is an Ore domain.*

PROOF. Both statements are well known facts about groups which have a series with free abelian factors.

LEMMA 6.2. *Let F be a noncyclic free group, S be a normal subgroup of an infinite index, K be an arbitrary commutative field. Then the group ring $K(F/S')$ has a trivial center.*

PROOF. Denote the group F/S' by $\overline{F}$. We have to prove that $\Delta(\overline{F}) = 1$, where $\Delta(\overline{F})$ is the characteristic subgroup formed by all the elements of $\overline{F}$, which have a finite number of conjugates (see **[22]**, 4.1). Assume thus that $\Delta(\overline{F}) \neq 1$. It is known (see **[23]**) that the normal subgroup $\overline{S} = S/S'$ coincides with its centralizer in $\overline{F}$. If now $\Delta(\overline{F}) \not\subseteq \overline{S}$ then the relative commutator subgroup $[\Delta(\overline{F}), \overline{S}] \neq 1$ and is contained in $\overline{S}$. Thus, in both the cases, when $\Delta(\overline{F})$ is contained in $\overline{S}$ or not, we obtain that $\overline{S}$ contains an element $\bar{s} \neq 1$, whose centralizer in $\overline{F}$ has a finite index. Clearly, this centralizer contains $\overline{S}$ and we can find easily a subgroup $F \supseteq F_1 \supseteq S$ such that $(F : F_1)$ is finite and $\bar{s}$ is in the center of the quotient group $\overline{F}_1 = F_1/S'$. Now imbed $\overline{F}_1$ into the wreath product $W = \overline{S}\,\mathrm{wr}(F_1/S)$ (the Magnus imbedding). Since F_1/S must be infinite it is known (see **[18]**, 24.23) that in this case the center of W is trivial. But it follows easily that $\bar{s}$ must be in the center of W and in this contradiction shows that $\Delta(\overline{F}) = 1$.

Assume now that F/S is soluble-by-finite. It was proven in **[24]** that the group ring $K(F/S')$ is an Ore domain. (The recent results of **[25]** imply that this is true for an arbitrary soluble-by-finite torsion free group.) We have the following corollary of Lemma 6.2.

COROLLARY. *Let F be a noncyclic free group, F/S be an infinite soluble-by-finite group. Then the field D of fractions of $K(F/S')$ has a trivial center.*

PROOF. Follows from Lemma 6.2 via the fact (see **[22]**, 4.5) that the center of D coincides with the field of fractions of $K(F/S')$.

LEMMA 6.3. *Let H be a group, N be its normal subgroup. Assume that both the groups H and H/N have finite series of normal subgroups with free abelian factors. Let D and Δ be the Ore fields of fractions of KH and of $K(H/N)$ correspondingly. Then the homomorphism $\phi\colon H \to H/N$ can be extended to a specialization $\theta\colon D \to \Delta$.*

PROOF. Let

$$H = H_1 \supseteq H_2 \supseteq \cdots \supseteq H_k \supseteq H_{k+1} = 1 \tag{6.1}$$

be a series of normal subgroups in H with free abelian factors. Combining the induced series $H_j \cap N$ $(j = 1, 2, \ldots, k+1)$ with a series in H/N we can construct a series in H such that N coincides with one of its terms, say $N = H_{j_0}$. We assume that the series (6.1) itself has this property and that all the inclusions in this series are strict; the number k is the length of this series. Since the group H_k is free abelian and H/H_k is a unique product group we obtain from **[13]**, that the homomorphism $\phi_1\colon H \to H/H_k$ is extended to a specialization $\theta_1\colon D \to \Delta_1$, where Δ_1 is the Ore field of fractions of $K(H/H_k)$. We obtain from (6.1) an appropriate series for the group H/H_k; the length of this series is $k-1$ and we can apply an induction argument and to assume that the homomorphism $\phi_2\colon H/H_k \to (H/H_k)/(N/H_k) \cong H/N$ can be extended to a specialization: $\Delta_2 \to \Delta$. We obtain therefore from **[7]**, 7.2 that the composition of homomorphisms $\phi = \phi_1\phi_2\colon H \to H/N$ can be extended to a specialization $\theta\colon D \to \Delta$.

COROLLARY. *Let F be a free group, $N \supseteq S$ be two normal subgroups of F such that F/S (and hence F/N) is soluble. Let D and Δ be the fields of fractions of $K(F/S')$ and $K(F/N')$ correspondingly. Then the homomorphism $\phi\colon F/S' \to F/N'$ with the kernel N'/S' can be extended to a specialization $\theta\colon D \to \Delta$.*

PROOF. Follows from Lemmas 6.1 and 6.3.

PROPOSITION 6.2. *Let F be a free group, $S_i \triangleleft F$ $(i \in I)$ be a system of normal subgroups closed with respect to finite intersections and such that every F/S_i is a finite group. Let $\bigcap_{i\in I} S_i = S$, and assume that F/S is soluble; for every $i \in I$ let D_i be the field of fractions of $K(F/S_i)$ and D be the field of fractions of $K(F/S')$. Then*

i) *The dimension of D_i over its center Z_i is m_i^2 where $m_i = (F : S_i)$.*

ii) *Every homomorphism $\phi_1\colon F/S' \to F/S_i'$ can be extended to a specialization $\theta_i\colon D \to D_i$ with a domain $T_i \subseteq D$.*

iii) *For every given elements* (6.1) *in* D *a subset* $I_1 \subseteq I$ *can be found such that*

$$x_\alpha \in T_i \ (\alpha = 1, 2, \ldots, k; \ i \in I_1) \tag{6.2}$$

and

$$\bigcap_{i \in I_1} S_i' = S'.$$

PROOF. It follows from Corollary 2 of Proposition 6.2 that the dimension of $K(F/S_i')$ over the commutative subring $K(S_i/S_i')$ is m_i; this implies easily that D_i has dimension m_i over the subfield V_i, generated by $K(S_i/S_i')$. It is well known that S_i/S_i' coincides with its centralizer in F/S_i' and this implies easily that V_i coincides with its centralizer in D_i, i.e. V_i is maximal and hence $(D_i \colon Z_i) = m_i^2$.

The statement ii) follows from Corollary to Lemma 6.3. The proof of ii) does not differ from the proof of Lemma 6.3 and we omit it.

6.2. Let throughout this section F be a free group, D be the universal field of fractions of the group ring KF. We refer the reader to Cohn's book [**6**], chapter 7, for properties of D; here we recall only the following facts: D is generated by KF; if there is a homomorphism $\phi \colon KF \to R$, where the ring R is imbedded into a field Δ and generates it, then ϕ is extended to a specialization $D \to \Delta$. Furthermore, the center of D coincides with K (Lewin [**16**]).

Let for given j $\gamma_j(F)$ denote the jth term of the lower central series. It was proven in [**17**] that for every given elements

$$x_1, x_2, \ldots, x_k \tag{6.3}$$

of D there exists j_0 such that for every $j \geq j_0$ the elements (6.1) belong to the local subring T_j, the domain of the specialization $\theta_j \colon D \to \Delta_j$, where Δ_j is the field of fractions of $K(F/F_j)$. We need now another system of homomorphisms and specializations. Let for a given $j \geq j_0$ $F^{(j)}$ be the jth derived subgroup of F. Clearly, the homomorphism $\phi_j \colon F \to F/\gamma_j(F)$ can be factored through the homomorphism $\phi^{(j)} \colon F \to F^{(j)}$. But the free soluble group $F/F^{(j)}$ and the free nilpotent group $F/\gamma_j(F))$ satisfy the conditions of Lemma 6.3. Let D_j and Δ_j be the fields of fractions of $K(F/F^{(j)})$ and $K(F/\gamma_j(F))$ correspondingly; Lemma 6.3 implies therefore that the homomorphism $\phi \colon K(F/F^{(j)}) \to K(F/\gamma_j(F))$ can be extended to a specialization $\theta \colon D_j \to \Delta_j$. Let $\theta^{(j)} \colon D \to D_j$ be the specialization which extends the homomorphism $\phi^{(j)} \colon KF \to K(F/F^{(j)})$ and $T^{(j)}$ be the domain of $\theta^{(j)}$. We obtain from [**7**], chapter 7, that in this situation $T^{(j)} \supseteq T_j$ and we proved thus the following assertion.

PROPOSITION 6.3. *Let* (6.3) *be given nonzero elements in* D. *Then there exists an index* j_0 *such that for every* $j \geq j_0$ *these elements belong to the local subring* $T^{(j)}$.

PROPOSITION 6.4. *Let D be the universal field of fractions of KF and* (6.3) *be given nonzero elements of D. Let p be a given prime number. Then there exists a system of normal subgroups $S_i \triangleleft F$ $(i \in I)$ closed with respect to finite intersections and such that*

i) $(F : S_i) = p^{m_i}$ $(i \in I)$.

ii) $\bigcap_{i \in I} S_i = 1$.

iii) *For every $i \in I$ the elements* (6.3) *belong to the domain T_i of the specialization θ_i, extending the natural homomorphism $\phi_i \colon F \to F/S_i'$.*

PROOF. Apply first Proposition 6.3 and find an index j_0 such that for every $j \geq j_0$ be elements (6.3) belong to the domain $T^{(j)}$ of the specialization $\theta^{(j)} \colon D \to D_j$ extending the natural homomorphism $\phi^{(j)} \colon F \to F/F^{(j)}$. Let for a given $j \geq j_0$

$$\bar{x}_\alpha \ (\alpha = 1, 2, \dots, k)$$

be the images of the elements (6.3) in $T^{(j)}/J(T^{(j)}) \cong D_j$. Apply Proposition 6.2 to the case $S = F^{(j-1)}$, $F/S' = F/F^{(j)} = \overline{F}$ and find a system of normal subgroups $\overline{S}_{ij} \triangleleft \overline{F}$ $(i \in I_j)$ such that

$$(6.4) \qquad (\overline{F} : \overline{S}_{ij}) = p^{m_{ij}}; \bigcap_i \overline{S}_{ij} = \overline{F}^{(j-1)}; \bigcap_i \overline{S}'_{ij} = 1,$$

every ring $K(\overline{F}/\overline{S}'_{ij})$ has a field of fractions D_{ij} and

$$(6.5) \qquad \bar{x}_\alpha \in \overline{T}_{ij} \ (\alpha = 1, 2, \dots, k)$$

where $\overline{T}_{ij}$ is the domain of the specialization $\theta_{ij} \colon D_j \to D_{ij}$, extending the homomorphism $K(\overline{F}) \to K(\overline{F}/\overline{S}'_{ij})$. Now let T_{ij} be the inverse image of $\overline{T}_{ij}$ under the natural homomorphism $T^{(j)} \to T^{(j)}/J(T^{(j)}) \cong D_j$ and S_{ij} be the inverse image of the subgroup $S_{ij} \subseteq \overline{F}$ under the homomorphism of groups $F \to \overline{F} \cong F/F^{(j)}$. Then

$$(6.6) \qquad x_\alpha \in T_{ij} \ (\alpha = 1, 2, \dots, k).$$

It can be shown easily or read off from **[6]**, 7.1, that T_{ij} is the domain of the specialization $\theta_{ij} \colon D \to D_{ij}$, which extends the homomorphism $\phi_{ij} \colon F \to F/S'_{ij} \cong \overline{F}/\overline{S}'_{ij}$. Finally, take $I = \bigcup_{j \geq j_0} I_j$ and obtain from (6.4) and (6.6) (via the relation $\bigcap_{j \geq j_0} F^{(j)} = 1$) that the system of normal subgroups S_i $(i \in I)$ satisfies the conclusions i)–iii); one can also join to this system, if necessary, all the finite intersection of subgroups S_i $(i \in I)$.

6.3. We need the following known fact whose proof we include for the sake of completeness.

LEMMA 6.4. *Let R_i $(i \in I)$ be a family of rings and R be a ring such that for every given non-zero elements*

$$(6.7) \qquad x_1, x_2, \dots, x_s$$

of it a homomorphism $\phi_i\colon R \to R_i$ can be found such that

$$\phi_i(x_\alpha) \neq 0 \ (\alpha = 1, 2, \dots, s). \tag{6.8}$$

Then R can be imbedded into a suitable ultraproduct $(\prod_{i\in I} R_i)/\mathcal{T}$.

PROOF. Take an arbitrary $0 \neq r \in R$. Let

$$I_r = \{i \in I | \phi_i(r) \neq 0\}.$$

The system of subsets I_r has the finite intersection property. Indeed, if r_1 and r_2 are two non-zero elements in R then we can find a homomorphism ϕ_j such that

$$\phi_j(r_1) \neq 0, \ \phi_j(r_2) \neq 0$$

and hence $j \in (I_{r_1} \cap I_{r_2})$. The system of ideals I_r generate therefore an ultrafilter $\mathcal{T}$. For every $r \in R$ define a map ϕ of R into the direct product $(\prod_{i\in I} R_i)$

$$\phi(r) = \{\phi_i(r) | i \in I\}.$$

It is immediate that is a homomorphism. The composition of this homomorphism together with the natural homomorphism $(\prod_{i\in I} R_i) \to (\prod_{i\in I} R_i)/\mathcal{T}$ gives a homomorphism $\psi\colon R \to (\prod_{i\in I} R_i)/\mathcal{T}$. If $\psi(r) = 0$ then the set of the zero coordinates of $\phi(r)$ belongs to $\mathcal{T}$ and because $\mathcal{T}$ is an ultrafilter the set I_r of the coordinates which are not equal to zero does not belong to $\mathcal{T}$. This contradicts the definition of $\mathcal{T}$ and the assertion follows.

COROLLARY. *Let R be a domain, A_i $(i \in I)$ be a system of ideals such that*

$$\bigcap_{i\in I} A_i = 0. \tag{6.9}$$

Then R can be imbedded into a suitable ultraproduct $(\prod_{i\in I} R/A_i)/\mathcal{T}$.

PROOF. Let $R_i = R/A_i$ $(i \in I)$ and for every given $i \in I$ let ϕ_i be the natural homomorphism $R \to R/A_i$. Since R is a domain the condition (6.8) for given nonzero elements (6.7) follows easily from (6.9) and the assertion follows from the lemma.

PROPOSITION 6.5. *Let D be a central division algebra over a field K, $R \subseteq D$ be a subring, such that the subfield, generated by it, coincides with D. Assume that there exists a system of ideals $A_i \subseteq R$ $(i \in I)$ such that*

$$\bigcap_{i\in I} A_i = 0$$

and every quotient ring $R_i = R/A_i$ is imbedded into a field D_i. Let

$$X_1, X_2, \dots, X_k \tag{6.10}$$

be given matrices in $R_{n\times n}$. Then these matrices are linearly dependent over K if (and only if) for every $i \in I$ their images in $(R_i)_{n\times n}$ are linearly dependent over the image K_i of the subring $K_0 = R \cap K$.

PROOF. Let $\tilde{K}_i$ be the subfield of D_i, generated by K_i. If the images

$$X_1^{(i)}, X_2^{(i)}, \dots, X_k^{(i)} \tag{6.10$'$}$$

of the elements (6.10) are linearly dependent over K_i then one of them is a $\tilde{K}_i$-linear combination of the other elements. Lemma 6.3 implies that $R \subseteq (\prod_{i\in I}(R_i))/\mathscr{T} \subseteq (\prod_{i\in I} D_i)/\mathscr{T}$, where $\mathscr{T}$ is a suitable ultrafilter on the set of indices I and hence $R_{n\times n} \subseteq (\prod_{i\in I}(R_i)_{n\times n})/\mathscr{T} \subseteq (\prod_{i\in I}(D_i)_{n\times n})/\mathscr{T}$. Denote by I_j the subset of I which includes those indices i for which $X_j^{(i)}$ is a linear combination of the other elements (6.10′). We see that $I = I_1 \cup I_2 \cdots \cup I_k$ and hence property iv of the ultrafilters implies that one of these subsets, say I_1, is contained in $\mathscr{T}$. Once again, as in the proof of Lemma 5.3 we can replace all the rings R_i $(i \notin I_1)$ by 0 and the ultraproduct $(\prod_{i\in I} R_i)/\mathscr{T}$ is isomorphic in fact to the ultraproduct $(\prod_{i\in I_1} R_i)/\mathscr{T}_1$, where $\mathscr{T}_1$ is the ultrafilter induced by $\mathscr{T}$ on I_1. We have also for every $i \in I_1$

$$X_1^{(i)} = \sum_{j=1}^{k} c_j^{(i)} X_j^{(i)} (c_j^{(i)} \in \tilde{K}_i;\ j = 2, \ldots, k). \tag{6.11}$$

The relation (6.11) implies

$$X_1 = \sum_{j=1}^{k} c_j X_j \tag{6.11′}$$

where c_j $(j = 2, \ldots, k)$ are elements of the ultraproduct $(\prod_{i\in I_1} \tilde{K}_i)/\mathscr{T}$, whose coordinates are defined by (6.11). It is well known that the ultraproducts $\tilde{D} = (\prod_{i\in I_1} D_i)/\mathscr{T}$ and $\tilde{K} = (\prod_{i\in I_1} \tilde{K}_i)/\mathscr{T}$ are fields and K is central in D because every K_i is central in D_i. Since $R \subseteq \tilde{D}$ and R generate D we obtain that $D \subseteq \tilde{D}$. Now let K' be the subfield of $\tilde{D}$ generated by K and $\tilde{K}$. The relation (6.11′) implies that the elements (6.10) are linearly dependent over $\tilde{K}$ and hence over K'. Finally, the K-subalgebra of $\tilde{D}_{n\times n}$, generated by K' and $D_{n\times n}$ is a homomorphic image of the tensor product $K' \otimes_K D_{n\times n}$ and the linear dependence of the elements (6.10) over K' implies their dependence over K. The proof is completed.

COROLLARY. *Let under the conditions of Proposition* 6.5 $K' \otimes_K D$ *be a division algebra for every finite algebraic extension* $K' \supseteq K$ *and every* $K_i' \otimes_{K_i} D_i$ *be a division algebra for every field* K_i', *which is a finite algebraic extension of the field* $\tilde{K}_i$ *of fractions of* K_i. *Then a given matrix* $X \in R_{n\times n}$ *is algebraic over* K *if and only if for every* $i \in I$ *its image* X_i *in the ring* R_i *is algebraic over* K_i.

PROOF. Statement 4. in **[9]** implies that X is algebraic over K if and only if the matrices $1, X, X^2, \ldots, X^{n-1}$ are linearly dependent; the same is true for all the images X_i of X. The assertion follows therefore from Proposition 6.5.

THEOREM 6.1. *Let* p *be a given prime number,* D *be the universal field of fractions of a free group ring* KF *or the Ore field of fractions of* $K(F/S')$ *where* F/S *is a soluble residually finite* p*-group. Let* R *be a finitely generated subalgebra of* $D_{n\times n}$. *Then there exists in* R *a system of ideals* B_i $(i \in I)$ *such that*

i) *for every given nonzero matrices*

(6.12) $$Y_\beta \in R \ (\beta = 1, 2, \ldots, s),$$

an ideal B_i *can be found such that*

$$Y_\beta \notin B_i \ (\beta = 1, 2, \ldots, s).$$

ii) *For every* $i \in I$ *the quotient algebra* R/B_i *is a subalgebra of* $(D_1)_{n\times n}$, *where* D_i *is the field of fractions of the group ring* $K(F/S_i')$, F/S_i *is a finite* p-*group and hence* D_i *has dimension* p^{2m_i} *over its center.*

iii) *A given matrix* $X \in R$ *is algebraic over* K *if and only if for every* $i \in I$ *its image* X_i *is algebraic in the quotient algebra* R/A_i.

PROOF. The proof of i) and ii) is completely similar to the proof of Proposition 4.5; one has to apply now Proposition 6.2–6.4 instead of Propositions 4.3′ and 4.4′.

To prove iii) we point out first of all that a routine argument reduces the proof to the case when F is finitely generated and $KF \subseteq R$. Assume first that F is non-cyclic and F/S is infinite. Then the field D is central over K by Lewin's result **[16]** for the universal field of fractions of KF and by Corollary of Lemma 6.2 for the Ore field of fractions of $K(F/S')$. The assertion follows in these cases from Corollary of Proposition 6.5.

Finally, if F/S is finite then it must be a p-group, because it is a residually finite p-group and the theorem is true in this case if we take $S_i = S$ $(i \in I)$ and $B_i = 0$ $(i \in I)$. Similarly, the theorem is trivial when F is an infinite cyclic group.

We need also an analog of Proposition 4.6. Let D be as in Theorem 6.1 and R be a finitely generated subring of $D_{n\times n}$. Clearly, we can find a finitely generated subring $K_1 \subseteq K$ such that $R \subseteq (D_1)_{n\times n}$, where D_1 is the field of fractions of K_1F (or $K_1(F/F^{(m)})$ correspondingly). We fix this subring K_1 and obtain the following corollary of Theorem 6.1.

COROLLARY. *Let* D *be the universal field of fractions for* KF *or the Ore field of fractions of* $K(F/F^{(m)})$ *and* R *be a finitely generated subring of* $D_{n\times n}$. *Then there exists in* R *a system of ideals* C_j $(i \in I)$ *such that*

i) *For every given nonzero matrices* (6.12) *an ideal* C_j $(j \in J)$ *can be found such that*

$$Y_\beta \notin C_j \ (\beta = 1, 2, \ldots, s).$$

ii) *For every* $j \in J$ *the quotient ring* R/C_j *is a subring of* $(D_j)_{n\times n}$, *where* D_j *is the field of fractions of* $K_j(F/S_j')$, F/S_j *is a finite* p_j-*group,* K_j *is the image of* K_1 *in* R/C_j *and* $\operatorname{char} K_j = p_j$.

iii) *A given matrix* $X \in R$ *is algebraic over* K *if and only if for every* $j \in J$ *its image is algebraic in* R/C_j *over* K_j.

PROOF. The proof is similar to the proof of Proposition 4.6 and we give only the outline of the argument. First, consider the case when D is the field of fractions of $K(F/F^{(m)})$. Apply now an argument similar to the one which was used in Proposition 4.2 and reduce the proof to the case when field K is finite and has characteristic p. Since the group $F/F^{(m)}$ is a residually finite p-group for every p we can apply now Theorem 6.1 to the case $S = F^{(m)}$ and the proof is completed for this case as in Proposition 4.6.

Now let D be the universal field of fractions for KF. Apply Proposition 6.1 and reduce the assertion to the case of a free soluble group and the assertion follows.

7. Proof of Theorems II and III. We will prove only the existence of a normal subgroup $N \triangleleft G$ such that $(G\colon N)|(n!)^2$ and N' is finite in the case of $\operatorname{char} K = p > 0$ or, correspondingly, $K[N']$ is finite dimensional when $\operatorname{char} K = 0$. The existence of the separable subalgebra T and of an element $c \in \mathrm{GL}_n(D)$ such that

$$c^{-1}Tc \subseteq K_{n\times n}$$

will follow from the result from **[9]** which was quoted in the introduction.

Prove first Theorem III *for the case when D is the field of fractions of* $U(L)$ (Statement ii). Since the group G is finitely generated we can assume (see Corollary of Theorem 4.2) that the field K is finitely generated. If now $\operatorname{char} K = p > 0$ we apply Proposition 4.5 to the linear envelope $R = K[G]$ of G. Since every field D_i in this proposition has dimension p^{2m_i} over its center and it was assumed that $p > h(G) + 1$ we obtain that for every $i \in I$ the image G_i of G in the quotient ring $R/B_i \subseteq (D_i)_{n\times n}$ satisfies conditions of Theorem 3.1. Thus, every group G_i in this case contains a normal subgroup N_i such that $(G_i\colon N_i)|(n!)^2$ and N_i is an extension of a nilpotent p-group U_i by a finite group V_i, which is isomorphic to a subgroup of $\mathrm{GL}_n(K)$. But it is known that if K is a finitely generated commutative field then there exists a bound $f(n)$ for orders of finite subgroups of $\mathrm{GL}_n(K)$. We apply now Lemma 6.4 and Proposition 4.5 and imbed R into a suitable ultraproduct $(\prod_{i\in I} R/B_i)/\mathscr{T}$ and G into an ultraproduct $(\prod_{i\in I} G_i)/\mathscr{T}$. Lemmas 5.1 ii) and 5.2 imply that the group $(\prod_{i\in I} N_i')/\mathscr{T}$ is an extension of the unipotent group $U = (\prod_{i\in I} U_i)/\mathscr{T}$ by the group $(\prod_{i\in I} V_i)/\mathscr{T}$ which is finite of order less or equal than $f(n)$. Since $\operatorname{char} K = p > 0$, U is a p-group and hence the group $(\prod_{i\in I} N_i')/\mathscr{T}$ is locally finite.

Now let $N = G \cap (\prod_{i\in I} N_i)/\mathscr{T})$. Clearly, $(G\colon N)|(n!)^2$. We see also that N' is contained in the commutator subgroup of $(\prod_{i\in I} N_i)/\mathscr{T}$. But $(\prod_{i\in I} N_i)' \subseteq \prod_{i\in I} N_i'$ which implies easily that

$$\left(\left(\prod_{i\in I} N_i\right)/\mathscr{T}\right)' \subseteq \left(\prod_{i\in I} N_i'\right)/\mathscr{T}$$

and hence

$$N' \subseteq \left(\prod_{i\in I} N_i'\right)/\mathscr{T}. \tag{7.1}$$

We have already shown that the right side of (7.1) is locally finite. Since G is polycyclic-by-finite we obtain that N' is finite. This completes the proof of Theorem III for the case when D is the field of fractions of $U(L)$.

To prove Theorem II for this case we apply the corollary of Proposition 4.6 and find a system of ideal C_j $(j \in J)$ such that it satisfies all the conclusions of this assertion and in particular every K_j is a finite field of characteristic greater than $h(G) + 1$. We can now apply the same argbument as in the proof of Theorem III, changing only the following point. Since every group N_i' $(i \in I)$ is finite every algebra $K_i[N_i']$ is finite dimensional over K_i and Corollary of Proposition 4.6 implies that every element of $K[N']$ is algebraic over K. We obtain therefore from **[9]** that $K[N']$ is locally finite over K and, because N' is finitely generated the algebra $K[N']$ is in fact finite dimensional. This completes the proof of Theorem II for the case when D is the field of fractions of $U(L)$.

We have to consider now the case when the field D in Theorems II and III is the field of fractions of $K(F/F^{(m)})$ or the universal field of fractions for KF. Find a prime number $q > h(G) + 1$ and apply Theorem 6.1. The proof is obtained now by the same argument as above with the use of Theorem 6.1 instead of Proposition 4.6. The proof of Theorems II and III is completed.

7.2. LEMMA 7.1. *Let R be a field, G be a subgroup of* $\mathrm{GL}_n(R)$. *Assume that there exists a word v such that for every finitely generated subgroup G_j $(j \in J)$ the subgroup $v(G_j)$ is unipotent. Then the group $v(G)$ is also unipotent.*

The proof is easy and we omit it.

PROOF OF THEOREM I. Prove first the statement iii) i.e. that G contains a series of normal subgroups

$$G \supseteq A \supseteq U, \tag{7.2}$$

where the subgroup U is unipotent, the quotient group A/U is abelian and G/A has an exponent which divides $(n!)^{2(s-1)}$; in other words show that the verbal subgroup $[G^{(n!)^{2(s-1)}}, G^{(n!)^{2(s-1)}}]$ is unipotent. Lemma 7.1 reduces the proof of this fact to the case when G is finitely generated. Proposition 2.2 implies that this fact is true for nilpotent subgroups of $\mathrm{GL}_n(\Delta)$, where Δ is either the field of fractions of $U(L)$, L_i is a finite dimensional Lie algebra over a field K_i of characteristic p, or is the field of fractions of the group ring $K_i(F/S_i')$, where F/S_i is a finite p-group. Once again, as in the proofs of Theorems II and III Lemma 6.4 together with Proposition 4.6 or Corollary of Theorem 6.1 imply that G is imbedded into an appropriate ultraproduct $(\prod_{i \in I} G_i)/\mathscr{T}$, where G_i is a nilpotent subgroup of $\mathrm{GL}_n(\Delta_i)$. The existence of the series (7.2) now follows from Lemma 5.1.

The nilpotent group $\overline{G} = G/U$ is thus abelian-by-periodic and it is well known that this implies its commutator subgroup is locally finite. (This follows also from the corollary of Lemma 2.4.) We conclude therefore that the commutator subgroup G' of G is an extension of a unipotent subgroup

$U_0 = G' \cap U$ by a locally finite group; this implies first of all that in the case of $\operatorname{char} K = p > 0$ the group G' is locally finite; second, we obtain that in the case of an arbitrary characteristic the linear envelope $K[G']$ is locally finite.

If now $\operatorname{char} K = 0$ we can apply the result from **[9]** which was quoted in the introduction and to obtain (1.2) and (1.3). This completes the proof for the case when $\operatorname{char} K = 0$.

We have already proven that if $\operatorname{char} K = p > 0$ then G' is locally finite. Hence,

$$G' \cong U_0 \times V,$$

where U_0 is the Sylow p-subgroup of G' and V is the p'-subgroup of G'. The group V is locally finite and if V_1 is an arbitrary finite subgroup of it then the subalgebra $K[V_1]$ is separable (see **[12]**, 7.10). Once again, the results from **[9]** imply that there exists $a \in \mathrm{GL}_n(D)$ such that

$$a^{-1}Va \in \mathrm{GL}_n(K),$$

which proves (1.1).

Finally, prove that V is a finite p'-group whose exponent is a divisor of $(n!)^{4(s-1)^2}$. Let V_1 be an arbitrary finite subgroup of V. We can take a finitely generated subgroup $G_1 \subseteq G$ such that $V_1 \subseteq G_1'$. Once again consider the imbedding of G into the ultraproduct $(\prod_{i\in I} G_i)/\mathcal{T}$, which was used in the proof of (7.2). It is important that the images V_i of V under the homomorphisms $G \to G_i$ are contained in the p'-subgroups of G_i'; hence, by Proposition 2.1, the exponents of all the subgroups V_i are divisors of $(n!)^{4(s-1)^2}$ and Lemma 5.2 implies that the exponent of V is a divisor of $(n!)^{4(s-1)^2}$. But V is isomorphic via (1.1) to a subgroup of $\mathrm{GL}_n(K)$; since it is a p'-group of a bounded exponent it must be finite (see **[11]**), and the assertion follows.

COROLLARY. *Let G be a locally nilpotent subgroup of* $\mathrm{GL}_n(D)$, *then*

i) *If* $\operatorname{char} K = p > 0$ *then the commutator subgroup G' is locally finite and if V is the p'-subgroup of G' then there exists an element $a \in \mathrm{GL}_n(D)$ such that*

$$a^{-1}Va \subseteq \mathrm{GL}_n(K).$$

ii) *If* $\operatorname{char} K = 0$ *then the linear envelope $K[G']$ is locally finite and contains a finite dimensional subalgebra T such that*

$$K[G'] \cong T + J(K[G'])$$

where $J(K[G'])$ is the nilpotent radical of $K[G']$. Furthermore, there exists an element $b \in \mathrm{GL}_n(D)$ such that

$$b^{-1}Tb \subseteq K_{n\times n}.$$

PROOF. Theorem I implies that every finitely generated subgroup of G satisfies the conclusions of the corollary; the assertion now follows from Statement 12. in **[9]**.

Acknowledgement

I am grateful to B. A. F. Wehrfritz for useful remarks.

References

1. R. Amayo and I. Stewart, *Infinite-dimensional Lie algebras*, Noordhoff International Publishing, 1974.

2. A. I. Lichtman, *Matrix rings and linear groups over a field of fractions of enveloping algebras and group rings, I*, J. Algebra **88** (1984), 1–37.

3. ——, *Growth in enveloping algebras*, Isr. J. Math. **47** (1984), N4, 296–304.

4. M. Shirvani and B. A. F. Wehrfritz, *Skew linear groups*, LMS Lecture Notes Series 118, Cambridge University Press, 1986.

5. Yu. I. Merzljakov, *Rational groups*, Nauka, 1980. (Russian)

6. A. H. Schofield, *Representations of rings over skew fields*, LMS Lecture Notes Series 92, Cambridge University Press, 1985.

7. P. M. Cohn, *Free rings and their relations*, Academic Press, 1985.

8. A. I. Lichtman, *Free subgroups in linear groups over some skew fields*, J. Algebra **105**, N1 (1987), 1–27.

9. A. I. Lichtman and B. A. F. Wehrfritz, *Finite dimensional subalgebras in matrix rings over transcendental division algebras*, Proc. Amer. Math. Soc. (to appear).

10. N. Jacobson, *Structure of rings*, Amer. Math. Soc. Colloq. Publ., vol. 37, Amer. Math. Soc., Providence, R. I. 1956.

11. B. A. F. Wehrfritz, *Infinite linear groups*, Springer-Verlag, 1973.

12. C. W. Curtis and I. Reiner, *Methods of representation theory*, Wiley, New York, 1981.

13. A. I. Lichtman, *On matrix rings and linear groups over fields of fractions of groups rings and enveloping algebras, II*, J. Algebra **90**, N2, (1984), 516–526.

14. N. Bourbaki, *Groups et algebres de Lie*, Chaps. 1–3, Hermann, Paris, 1972.

15. N. Jacobson, *The theory of rings*, Amer. Math. Soc., Providence, R. I., 1943.

16. J. Lewin, *Fields of fractions for group algebras of free groups*, Trans. Amer. Math. Soc. **192** (1974), 339–346.

17. A. Eisenbud and A. I. Lichtman, *On embedding of group rings of residually torsion free nilpotent groups into skew fields*, Trans. Amer. Math. Soc. **199** (1987), N1, 373–386.

18. H. Neumann, *Varieties of groups*, Springer-Verlag, 1967.

19. G. Baumslag, *Wreath products and extensions*, Math. Z. **81** (1963), 286–299.

20. M. J. Dunwoody, *On verbal subgroups of free groups*, Arch. Math. **16** (1965), 153–157.

21. D. M. Smirnov, *On generalized soluble groups and their group rings*, Mat. Sb. **67** (1965), 366–383. (Russian)

22. D. S. Passman, *The algebraic structure of group rings*, Wiley, 1971.

23. I. B. S. Passi, *Annihilators of relation modules, II*, Journal Pure Appl. Algebra **6** (1975), 235–237.

24. A. I. Lichtman, *On imbedding of group rings in division rings*, Isr. J. Math. vol. 23 (1976), 151–160.

25. P. H. Kropholler, P. A. Linnel and J. A. Moody, *Applications of a new K-theorem to soluble group rings*, Proc. Amer. Math. Soc. **104** (1988), 675–684.

Contemporary Mathematics
Volume 93 (1989)

The Structure of G_0 for Certain Polycyclic Group Algebras and Related Algebras

MARTIN LORENZ AND D. S. PASSMAN

0. Introduction.

Recently there has been considerable interest in the structure of the Grothendieck group $G_0(K[\Gamma])$ of finitely generated modules over the group algebra of a polycyclic-by-finite group Γ. The motivation has been partly topological (e.g., **[Fa-Hs]**, **[Q]**) and partly algebraic, the latter especially in connection with the so-called Goldie rank conjecture (e.g., **[F]**, **[P2]**, **[R]**, **[Br-H-Lo]**). The present highlight in this area is the thesis of J. Moody (**[Mo 1]**, see also **[Mo 2]**) which solves, amongst other things, the aforementioned Goldie rank conjecture. Moody's principal result states that $G_0(K[\Gamma])$ is generated by the subgroups $\mathrm{Ind}_{K[H]}^{K[\Gamma]}(G_0(K[H]))$, where H runs over the *finite* subgroups of Γ and $\mathrm{Ind}_{K[H]}^{K[\Gamma]}$ is the usual induction map given by $(\cdot)\bigotimes_{K[H]} K[\Gamma]$ (cf. also Theorem 2.8 below). Since the groups $G_0(K[H])$ can be dealt with by classical methods, the determination of $G_0(K[\Gamma])$ has now come within reach. The goal of this article is to present an explicit description of $G_0(T)$ for certain polycyclic group algebras $T = K[\Gamma]$, as well as for other related algebras T, such as the Zalesskii-Neroslavskii example **[Z-N]**.

Most of our work is directed towards determining the structure of the cokernel of the Cartan map $c\colon K_0(T) \to G_0(T)$. This cokernel will be denoted by

$$\overline{G}_0(T)$$

in the sequel. The results about $G_0(T)$ then follow easily. Besides Moody's theorem, our principal technical tool is a refined version of the usual restriction map. This modified restriction map, denoted θ, depends on the notion of almost injective modules which form the subject of Section 1. In Section 2, the map θ is used to exhibit direct summands of $\overline{G}_0(T)$ for various classes of right Noetherian K-algebras T. Moody's theorem is then used to show that the direct summands constructed by these methods do indeed generate all of $\overline{G}_0(T)$, for the algebras T in question. The final Section 3 contains two appendices. The first describes an alternative cohomological approach to some

0271-4132/89 \$1.00 + \$.25 per page

of the work on almost injective modules in Section 1. The second appendix offers a very special result on the question of when the fixed subring R^G of the action of a finite group G on a Noetherian ring R is again Noetherian. This result is needed for our discussion of the Zalesskii-Neroslavskii example and is based on ideas taken from **[H-O]**. It can presumably be further generalized.

Notations and Conventions.

All modules will be understood to be right modules. Sometimes we explicitly indicate the acting ring as an index, e.g., V_T.

Our notation concerning G_0 and K_0 follows **[B]**. In particular, if V is a finitely generated module over a right Noetherian ring T, then [V] denotes the corresponding element of $G_0(T)$. Moreover, $\overline{G}_0(T)$ denotes the cokernel of the Cartan map $c: K_0(T) \to G_0(T)$.

The following notations will be fixed throughout.

K	will be a commutative field,
$K^\bullet, D^\bullet$	denotes the multiplicative group of K, resp. of the division ring D,
$U(T)$	is the group of units of the ring T,
$E(V)$	denote the injective hull of the module V,
Rad V, Soc V	is the radical resp., socle of the module V,
B^G, V^G, R^G	denote the fixed point set (resp., module, ring) of the action of a finite group G on a set B (resp., module V, ring R),
$\mathrm{Tr} = \mathrm{Tr}_G : V \to V^G$	will be the trace map, $\mathrm{Tr}(v) = \sum_{g \in G} vg$.
$\#, \lvert\ \rvert$	are both used to denote cardinalities,
(a, b)	denotes the greatest common divisor of a and b.

§1. Almost injective modules.

In the first part of this section, we will define almost injective modules over a quasi-Frobenius ring S and develop their basic properties. The second half is devoted to the special case where $S = K[G]$ is the group algebra of a finite group G.

Recall that a ring S is said to be quasi-Frobenius (QF) if the projective S-modules coincide with the injective S-modules. In this case, S is Artinian on both sides ([**K**, 13.1.1 and 13.6.1]).

DEFINITION. A module V over a QF-ring S will be called *almost injective* (*a.i.*) if $E(V)/V$ is finitely generated over S. Here, $E(V)$ denotes the injective hull of V.

LEMMA 1.1. *Let V be a module over a QF-ring S.*

i. *V has a decomposition $V = V_1 \oplus V_2$, where V_2 is injective and V_1 has no nonzero injective submodules. Moreover, V is a.i. if and only if V_1 is a.i.*

ii. *Assume that V has no nonzero injective submodules. Then $V \subseteq \operatorname{Rad} E(V)$, and*

$$V \text{ is a.i.} \Leftrightarrow E(V) \text{ is finitely generated} \\ \Leftrightarrow V \text{ is finitely generated.}$$

PROOF. (i) By Zorn's lemma, we can define V_2 to be a maximal direct sum of indecomposable injective submodules of V (cf. [**K**, 6.6.7(a)]). Since S is Noetherian, V_2 is injective and so we have $V = V_1 \oplus V_2$ for a suitable submodule V_1 of V. Clearly, V_1 has no nonzero injective submodules. Finally, $E(V) = E(V_1) \oplus V_2$ and so $E(V)/V \cong E(V_1)/V_1$.

(ii) Write $E(V) = \bigoplus_{\lambda \in \Lambda} E_\lambda$ with E_λ indecomposable injective (cf. [**K**, 6.6.4]), and let $\pi_\lambda \colon E(V) \twoheadrightarrow E_\lambda$ denote the canonical projection. Since $\operatorname{Rad} E_\lambda$ is the unique largest proper submodule of E_λ, we must have $\pi_\lambda(V) \subseteq \operatorname{Rad} E_\lambda$. Otherwise since E_λ is projective, V would contain an isomorphic copy of E_λ. Therefore

$$V \subseteq \bigoplus_{\lambda \in \Lambda} \operatorname{Rad} E_\lambda = \operatorname{Rad} E(V).$$

Moreover, if $E(V)/V$ is finitely generated then so is $E(V)/\operatorname{Rad} E(V) \cong \bigoplus_{\lambda \in \Lambda} E_\lambda/\operatorname{Rad} E_\lambda$, whence Λ must be finite. Thus $E(V)$ is finitely generated, and hence so is V. Conversely, if V is finitely generated, then $E(V) \cong E(\operatorname{Soc} V)$ is a finite direct sum of indecomposable injectives. So $E(V)$ is also finitely generated. □

DEFINITION AND REMARK. For any a.i. module V over a QF-ring S, $[E(V)/V]$ is a well-defined element of $G_0(S)$. In fact, if V is decomposed as $V = V_1 \oplus V_2$ with V_1 finitely generated and V_2 injective, as in Lemma 1.1, then $[E(V)/V] = [E(V_1)] - [V_1]$. We let $\theta(V)$ denote the image of $-[E(V)/V]$ in $\overline{G}_0(S)$. Thus

$$\theta(V) = -[E(V)/V] \bmod cK_0(S) \in \overline{G}_0(S) = G_0(S)/cK_0(S).$$

Using the above decomposition $V = V_1 \oplus V_2$, we have

$$\theta(V) = [V_1] \bmod cK_0(S) \in \overline{G}_0(S),$$

since $E(V_1)$ is a finitely generated projective module and hence $[E(V_1)] \in cK_0(S)$.

THEOREM 1.2. *Let $V' \hookrightarrow V \twoheadrightarrow V''$ be a short exact sequence of modules over a QF-ring S. If two of $\{V, V', V''\}$ are a.i. then so is the third and, in this case, the equality*

$$\theta(V) = \theta(V') + \theta(V'')$$

holds in $\overline{G}_0(S)$.

PROOF. First assume that V' and V'' are a.i.. By [**C-E**, Prop. 3.5 on p. 10], the given exact sequence can be imbedded in a commutative diagram of

S-modules

$$\begin{array}{ccccc} V' & \hookrightarrow & V & \twoheadrightarrow & V'' \\ \downarrow & & \downarrow & & \downarrow \\ E(V') & \hookrightarrow & E & \twoheadrightarrow & E(V'') \\ \downarrow & & \downarrow & & \downarrow \\ N' & \hookrightarrow & N & \twoheadrightarrow & N'' \end{array}$$

where E is injective. In particular, $V \subseteq E(V) \underset{\sim}{\subseteq} E \cong E(V) \oplus C$. By assumption, N' and N'' are finitely generated, and hence so are $N \cong E/V$, $E(V)/V$, and C. Thus V is a.i., and in $G_0(S)$ we have

$$\begin{aligned} [N'] + [N''] = [N] = [E/V] &= [E(V)/V] + [C] \\ &\equiv [E(V)/V] \bmod cK_0(S) \end{aligned}$$

since C is injective and hence projective. This proves the equality $\theta(V) = \theta(V') + \theta(V'')$.

Next assume that V' and V are a.i.. Observe that $V' \subseteq V \subseteq E(V)$ implies $V' \subseteq E(V') \underset{\sim}{\subseteq} E(V) \cong E(V') \oplus D$ and so

$$V'' \cong V/V' \underset{\sim}{\subseteq} E(V)/V' \cong E(V')/V' \oplus D =: M$$

and

$$M/V'' \cong E(V)/V.$$

Since $E(V')/V'$ is finitely generated and D is injective, it follows that $E(M)/M$ is finitely generated. Moreover, as $M/V'' \cong E(V)/V$ is finitely generated, we conclude that $E(M)/V''$ is finitely generated. Therefore, $E(V'')/V''$ is also finitely generated, being a submodule of $E(M)/V''$.

Finally, assume that V and V'' are a.i.. Then, using the preceding paragraph and its notation, we see that M/V'' is finitely generated, and hence M/V'' is a.i.. Applying the result of the first paragraph to $V'' \hookrightarrow M \twoheadrightarrow M/V''$, we deduce that M is a.i.. Inasmuch as $E(V')/V'$ is a direct summand of M, $E(V')/V'$ is also a.i.. Moreover, $E(V')/V'$ has no nonzero projective submodules, since V' is essential in $E(V')$. Thus, by Lemma 1.1(ii), $E(V')/V'$ is finitely generated, whence V' is a.i.. □

COROLLARY 1.3. *Let $S \to T$ be a homomorphism of rings, with S a* QF *ring and T right Noetherian. Assume that all finitely generated T-modules are a.i. over S. Then the map θ yields a homomorphism*

$$G_0(T) \to \overline{G}_0(S), \ [V_T] \mapsto \theta(V_S).$$

If, in addition, T_S is projective then this map factors through $\overline{G}_0(T)$.

PROOF. The map $G_0(T) \to \overline{G}_0(S)$ is well-defined by the additivity of θ. Moreover, if T_S and V_T are projective, then V_S is projective and so $\theta(V_S) = 0$. □

The above maps $G_0(T) \to \overline{G}_0(S)$, resp. $\overline{G}_0(T) \to \overline{G}_0(S)$, will also be denoted by θ.

For the remainder of this section, we will concentrate on the case where $S = K[G]$ is the group algebra of a finite group G. Of course, if $\operatorname{char} K \nmid |G|$

then all $K[G]$-modules are injective, hence a.i.. Thus the following is of interest only in the case where $|G|$ is divisible by char K.

In the next lemma and corollary, we isolate some basic facts about $K[G]$-modules. Recall that, for any $K[G]$-module V, $V^G = \{v \in V | vg = v$ for all $g \in G\}$ and $\mathrm{Tr}\colon V \to V^G$, $v \mapsto \sum_{g\in G} vg$, is the trace map. Furthermore, $V^* = \mathrm{Hom}_K(V, K)$ becomes a $K[G]$-module via $(fg)(v) = f(vg^{-1})$ $(f \in V^*, g \in G, v \in V)$. Finally, if V and W are $K[G]$-modules, then $V \otimes_K W$ is a $K[G]$-module with G acting diagonally, that is $(v \otimes w)g = vg \otimes wg$ $(v \in V, w \in W, g \in G)$.

LEMMA 1.4. *Let G be a finite group, and let V and X be modules over $K[G]$. Assume that X is irreducible.*

i. $\dim_K(X^* \otimes_K V)^G < \infty$ *if and only if* $\mathrm{Soc}\, V$ *has a finite-dimensional X-homogeneous component.*

ii. $V \supseteq E(X)$ *if and only if* $\mathrm{Tr}(X^* \otimes_K V) \neq 0$.

PROOF. (i) The map $X^* \otimes_K V \to \mathrm{Hom}_K(X, V)$ given by $f \otimes v \mapsto (x \mapsto f(x)v)$, is an isomorphism of $K[G]$-modules, where G acts on $H := \mathrm{Hom}_K(X, V)$ via $(fg)(x) = f(xg^{-1})g (f \in H, g \in G, x \in X)$. Thus $(X^* \otimes_K V)^G$ is isomorphic to H^G which in turn is just $\mathrm{Hom}_{K[G]}(X, V)$. Part (i) follows easily from this.

(ii) By [**L**, Theorem 2.3 on p. 95], $T := \mathrm{Tr}(\mathrm{Hom}_K(X, V))$ is the set of those $f \in \mathrm{Hom}_{K[G]}(X, V)$ which factor through a projective module. Thus, if $E(X) \subseteq V$ then the embeddings $X \subseteq E(X) \subseteq V$ yield a nonzero element of T. Conversely, if $0 \neq f \in T$ then f is mono, since X is irreducible, and for a suitable projective $K[G]$-module P we have a commutative diagram

$$\begin{array}{ccc} X & \overset{f}{\hookrightarrow} & V \\ & \searrow \quad \nearrow_{\varphi} & \\ & P & \end{array} .$$

Now $X \subseteq E(X) \underset{\sim}{\subseteq} P$ and so we may take $P = E(X)$. In this case, φ becomes mono, since $\mathrm{Ker}\, \varphi \cap X = \mathrm{Ker}\, f = 0$. Part (ii) now follows, because $T \cong \mathrm{Tr}(X^* \otimes_K V)$, as in (i). □

COROLLARY 1.5. *Let G be a finite group and let V be a $K[G]$-module.*

i. *V is finite-dimensional if and only if $(X \otimes_K V)^G$ is finite-dimensional for all irreducible $K[G]$-modules X.*

ii. *V contains no nonzero injective submodules if and only if* $\mathrm{Tr}(X \otimes_K V) = 0$ *holds for all irreducible $K[G]$-modules X.*

iii. *V is a.i. if and only if $(X \otimes_K V)^G/\mathrm{Tr}(X \otimes_K V)$ is finite-dimensional for all irreducible $K[G]$-modules X.*

PROOF. (i) The condition on $(X \otimes_K V)^G$ is certainly necessary. Conversely, assume it is satisfied. Then Lemma 1.4 (i) implies that $\mathrm{Soc}\, V$ is finite-dimensional, and hence so is $E(\mathrm{Soc}\, V) \cong E(V) \supseteq V$. (Note that X^* is irreducible if and only if X is irreducible.)

(ii) This is immediate from Lemma 1.4(ii). Just note that any nonzero injective $K[G]$-module E contains an irreducible submodule, say X, and hence $E(X) \subseteq E$.

(iii) Write $V = V_1 \oplus V_2$, where V_2 is injective and V_1 has no nonzero injective submodules, as in Lemma 1.1(i). Then, for any $K[G]$-module X, $X \otimes_K V_2$ is projective. Since $(K[G])^G = \mathrm{Tr}(K[G])$ it follows that $(X \otimes_K V_2)^G = \mathrm{Tr}(X \otimes_K V_2)$ and therefore

$$\begin{aligned}(X \otimes_K V)^G/\mathrm{Tr}(X \otimes_K V) &\cong (X \otimes_K V_1)^G/\mathrm{Tr}(X \otimes_K V_1)\\ &\cong (X \otimes_K V_1)^G,\end{aligned}$$

where the last isomorphism comes from part (ii). Thus, by part (i), we conclude that $(X \otimes_K V)^G/\mathrm{Tr}(X \otimes_K V)$ is finite-dimensional for all irreducible $K[G]$-modules X if and only if V_1 is finite-dimensional, and this in turn is equivalent to V being a.i., by Lemma 1.1. □

REMARKS 1.6. (a) For any $K[G]$-module V, $V^G/\mathrm{Tr}(V)$ is the Tate cohomology group $\widehat{H}^0(G, V)$. Thus part (iii) above can be viewed as a cohomological characterization of a.i. $K[G]$-modules. We will give a quick proof of part of Theorem 1.2 for $S = K[G]$ along these lines in Appendix A.

(b) Later on, we will be mostly interested in the case where G is a finite p-group and $\operatorname{char} K = p$. In this case, only $X = K$, the trivial $K[G]$-module, is relevant in Lemma 1.4 and Corollary 1.5 above. The proof of Corollary 1.5(iii) simplifies as follows. Write $V = V_1 \oplus V_2$ as in Lemma 1.1(i) so that, in particular, $V_1 \subseteq \operatorname{Rad} E(V_1)$ and $V_2^G = \mathrm{Tr}(V_2)$. Now $\operatorname{Rad} E(V_1) = E(V_1) \cdot \operatorname{Rad} K[G]$ is annihilated by the element $\hat{G} = \sum_{g \in G} g \in K[G]$. Therefore, $\mathrm{Tr}(V_1) = V_1 \cdot \hat{G} = 0$ and so we have

$$V^G/\mathrm{Tr}(V) = V_1^G/\mathrm{Tr}(V_1) \cong V_1^G \cong \operatorname{Soc} V_1.$$

The condition on V thus is satisfied if and only if $\operatorname{Soc} V_1$, and hence V_1, are finite-dimensional or, equivalently, V is a.i..

We now apply the foregoing to modules over skew group rings. Let G be a finite group acting by automorphisms on a ring R. Then the *skew group ring* of G over R, denoted RG, is a free right R-module with G as basis. Thus each element $r \in R$ can be uniquely written in the form

$$r = \sum_{g \in G} \mathrm{g}r_g \ (r_g \in R).$$

Furthermore, multiplication in RG is defined by the rule

$$(g_1 r_1) \cdot (g_2 r_2) = (g_1 g_2)(r_1^{g_2} r_2) \ (g_i \in G, r_i \in R),$$

where we use the exponential notation for the action of G on R.

THEOREM 1.7. *Let R be a K-algebra and let G be a finite group acting on R by K-algebra automorphisms. Assume that*

(1) *R is Noetherian as a right module over the fixed subring R^G, and*

(2) *$R^G/\mathrm{Tr}(R)$ is finite-dimensional over K.*
Then all finitely generated modules over the skew group ring RG are a.i. over $K[G] \subseteq RG$.

PROOF. Let V be a finitely generated RG-module. Then V is also finitely generated over R, and hence (1) implies that V is Noetherian over R^G. The module action of RG on V makes $V^G/\mathrm{Tr}(V)$ a module over $R^G/\mathrm{Tr}(R)$ which is finitely generated, by the foregoing. Hence by (2), $V^G/\mathrm{Tr}(V)$ is finite-dimensional for all finitely generated RG-modules V.

Finally, if X is any finite-dimensional $K[G]$-module, then $X \otimes_K V$ becomes a finitely generated RG-module by the RG-action

$$(x \otimes v) \cdot \sum_{g \in G} g r_g := \sum_{g \in G} xg \otimes v\, g r_g \quad (x \in X, v \in V, r_g \in R).$$

Thus the foregoing yields that $(X \otimes_K V)^G/\mathrm{Tr}(X \otimes_K V)$ is finite-dimensional and, since X was arbitrary, Corollary 1.5 (iii) implies that V is a.i. over $K[G]$. □

REMARKS 1.8. (a) R is a module over RG via $r \cdot \sum_{g \in G} g r_g = \sum_{g \in G} r^g r_g$ $(r, r_g \in R)$. In fact, $R \cong \hat{G} \cdot RG$, where $\hat{G} = \sum_{g \in G} g \in RG$. In view of Corollary 1.5(iii), assumption (2) above is certainly necessary, even for R to be a.i. over $K[G]$.

(b) Assumption (1) is satisfied, by E. Noether's theorem, if R is an affine commutative K-algebra, and also in the case (uninteresting for our purposes) where $|G|^{-1} \in R$ and R is right Noetherian, by **[Lo-P**, Theorem 3.3 (ii)]. Beyond these cases, not very much appears to be known (cf. [**M2**, Problems 6 and 8]). Proposition 3.4 in Appendix B will yield some further examples. Here we just note that the equivalence (a) ⇔ (b) in Lemma 3.3 shows that if R is right Noetherian and if all RG-submodules of R (i.e., all G-stable right ideals of R) are a.i. over $K[G]$, then R^G must be right Noetherian. Often (1) follows when R^G is right Noetherian (cf. [**M1**, Cor. 5.8 on p. 79]). So, for example, if R is semiprime right Noetherian and G is X-outer, then (1) is necessary for all finitely generated RG-modules to be a.i. over $K[G]$.

We conclude this section by discussing two examples.

Example 1.9. Suppose that G acts on the K-algebra R by permuting a K-basis B of R. For each G-orbit $C \in B/G$ put

$$\hat{C} := \sum_{b \in C} b \in R$$

and

$$|G\colon C| := |G|/|C| = |\mathfrak{C}_G(b)| \text{ for any } b \in C.$$

Then

$$R^G = \bigoplus_{C \in B/G} K \cdot \hat{C} \quad \text{and} \quad \mathrm{Tr}(R) = \bigoplus_{\substack{C \in B/G \\ |G:\, C| \text{ nonzero in } K}} K \cdot \hat{C}.$$

To see the latter, note that, for $b \in B$, $\mathrm{Tr}(b) = |\mathbb{C}_G(b)| \cdot \hat{C}_b$, where $C_b \in B/G$ is the G-orbit of b. Therefore, in the present situation, condition (2) in Theorem 1.7 is equivalent with

(2′) For almost all $b \in B$, $|\mathbb{C}_G(b)|$ is nonzero in K.

Example 1.10: *The Zalesskii-Neroslavskii example.* Let $\mathrm{char}\, K = 2$ and assume that $K^\bullet$ contains an element λ which is not a root of unity. Put $R = R_\lambda := K\{x^{\pm 1}, y^{\pm 1}\}/(xy - \lambda yx)$. Then $\sigma\colon x \mapsto x^{-1}$, $y \mapsto y^{-1}$ yields an automorphism of R which permutes the K-basis $B = \{x^i y^j | i, j \in \mathbb{Z}\}$ of R. Moreover, for $G = \langle\sigma\rangle \cong C_2$, we have $\mathbb{C}_G(x^i y^j) = \langle 1 \rangle$ unless $i = j = 0$. Thus (2′) above is satisfied. (In fact, $R^G = K \oplus \mathrm{Tr}(R)$.) Furthermore, the main result of **[H-O]** shows that (1) is also satisfied. Thus Theorem 1.7 ensures that all finitely generated modules over the skew group ring RG are a.i. over $K[G]$.

Of course, all this is still valid, but much less interesting, if $\mathrm{char}\, K \neq 2$ or λ is a root of unity in $K^\bullet$. In the latter case, R is a finite module over its center and so (1) follows from Noether's theorem.

§2. Applications to G_0.

In this section, we use the maps θ defined in Corollary 1.3 to study $\overline{G}_0(T)$ for various classes of right Noetherian K-algebras T. Besides θ, our principal tools will be decompositions of Mackey type and Moody's theorem (Theorem 2.8 below). Throughout, we assume that

$$p = \mathrm{char}\, K > 0$$

(otherwise all $\overline{G}_0(S)$ would be zero) and, for simplicity, we will concentrate on the case where $S = K[G]$ is the group algebra of a finite p-group G. Usually, we will use several such subalgebras S_i for a given K-algebra T. In this case,

$$\theta_i$$

denotes the map θ of Corollary 1.3 for $S = S_i$ (when defined).

LEMMA 2.1. *Let T be a right Noetherian K-algebra and let G_1 and G_2 be finite p-subgroups of the group of units $U(T)$. Assume that T has a K-basis B such that the following conditions are satisfied.*

(1) $G_1 \cdot B \cdot G_2 \subseteq B$, *and*

(2) *For all* $b \in B$, $l.\mathrm{Stab}_{G_1}(b) = \{g \in G_1 | gb = b\} = \langle 1 \rangle$ *and*
$r.\mathrm{Stab}_{G_2}(b) = \{g \in G_2 | bg = b\} = \langle 1 \rangle$.

Let S_i denote the subalgebra of T generated by G_i ($i = 1, 2$). Then $S_i \cong K[G_i]$, and the modules ${}_{S_1}T$ and T_{S_2} are free.

Assume further that all finitely generated T-modules are a.i. over S_2 and consider the map

$$\varphi\colon \overline{G}_0(S_1) \overset{\mathrm{Ind}}{\to} \overline{G}_0(T) \overset{\theta_2}{\to} \overline{G}_0(S_2) \cong \mathbb{Z}/|G_2|\mathbb{Z},$$

where Ind *is given by* $(.) \otimes_{S_1} T$, *as usual. Note that* G_2 *acts by right multiplication on the set* $G_1 \backslash B$ *of left* G_1*-orbits in* B. *It follows that the fixed point set* $(G_1 \backslash B)^{G_2}$ *is finite.*

If $(G_1 \backslash B)^{G_2}$ *has order prime to* p, *then* φ *is onto. Otherwise* $\operatorname{Im} \varphi$ *is contained in* $p \cdot \overline{G}_0(S_2)$, *the unique largest proper subgroup of* $\overline{G}_0(S_2)$.

PROOF. The assertions concerning the structure of the subalgebras S_i and of the modules ${}_{S_1}T$ and T_{S_2} are clear. Note also that, since ${}_{S_1}T$ is free, Ind is well-defined on $\overline{G}_0(S_1)$.

Let W be a finitely generated S_1-module. Fix a representative system $B_0 \subseteq B$ for $G_1 \backslash B / G_2$. Then

$$V := \operatorname{Res}^T_{S_2} \circ \operatorname{Ind}^T_{S_1}(W) = \bigoplus_{b \in B_0} W \otimes_{S_1} S_1 b S_2.$$

For each $b \in B_0$, put $G_2(b) := \operatorname{Stab}_{G_2}(G_1 b) = \{g \in G_2 | bg \in G_1 b\}$ and $W(b) := W \otimes_{S_1} S_1 b S_2$. Then $W(b) \supseteq W \otimes_{S_1} b$, $W \otimes_{S_1} b$ is a module over the group algebra $K[G_2(b)] \subseteq S_2$, and

$$W(b) \cong \operatorname{Ind}^{S_2}_{K[G_2(b)]}(W \otimes_{S_1} b).$$

To see the latter, just note that $S_1 b S_2 = \bigoplus_g S_1 b g$, where g runs over a transversal for $G_2(b)$ in G_2. Therefore,

$$V \cong \bigoplus_{b \in B_0} \operatorname{Ind}_{K[G_2(b)]}(W \otimes_{S_1} b).$$

This is just Mackey decomposition. Since V is a.i. over $S_2 = K[G_2]$, by assumption, almost all summands on the r.h.s. are free over S_2.

If $G_2(b) \neq G_2$, then p divides the dimension of the corresponding summand on the r.h.s.. Hence its image in $G_0(S_2) \cong \mathbb{Z}$ belongs to $p \cdot G_0(S_2)$.

If $G_2(b) = G_2$, that is if $G_1 b \in (G_1 \backslash B)^{G_2}$, then the corresponding image in $G_0(S_2) \cong \mathbb{Z}$ equals $\dim W$.

Taking $W = K$, the trivial S_1-module, in the above, we see that V has a direct summand of the form K^α with $\alpha := \#(G_1 \backslash B)^{G_2}$. Therefore, α must be finite. Summarizing, we have shown that

$$\varphi([W] \bmod cK_0(S_1)) \equiv \alpha \cdot \dim W \bmod p \cdot \overline{G}_0(S_2).$$

The assertions about $\operatorname{Im} \varphi$ follow from this. □

REMARKS 2.2. (a) The assumption that G_1 is a p-group was irrelevant in the above proof.

(b) Certainly, all finitely generated T-modules are a.i. over S_2 whenever T has a G_2-stable subalgebra R such that T is finitely generated as right R-module and conditions (1) and (2) in Theorem 1.7 are satisfied. Indeed, in this case all finitely generated T-modules are finitely generated over the subalgebra of T that is generated by R and G_2, and the latter is a homomorphic image of the skew group ring RG_2.

COROLLARY 2.3. *Let T be a right Noetherian K-algebra and let $\{G_i | i \in I\}$ be a finite collection of p-subgroups of $U(T)$. Assume that there exist K-bases B_{ij} $(i, j \in I)$ for T such that, for all i and j,*

(1′) $G_i \cdot B_{ij} \cdot G_j \subseteq B_{ij}$, and

(2′) $l.\mathrm{Stab}_{G_i}(b) = \langle 1 \rangle = r.\mathrm{Stab}_{G_j}(b)$ holds for all $b \in B_{ij}$.

Assume further that all finitely generated T-modules are a.i. over $S_i = K[G_i] \subseteq T$ for all i, and that

$$(3) \qquad (p, \#(G_i \backslash B_{ij})^{G_j}) = \begin{cases} 1 & \text{for } i = j \\ p & \text{for } i \neq j. \end{cases}$$

Then the map

$$\varphi \colon \bigoplus_{i \in I} \overline{G}_0(S_i) \overset{\oplus \mathrm{Ind}}{\to} \overline{G}_0(T) \overset{\oplus \theta_i}{\to} \bigoplus_{i \in I} \overline{G}_0(S_i)$$

is an automorphism of $\bigoplus_{i \in I} \overline{G}_0(S_i)$. In particular, $\bigoplus_{i \in I} \overline{G}_0(S_i)$ is isomorphic to a direct summand of $\overline{G}_0(T)$.

PROOF. Put $A := \bigoplus_{i \in I} \overline{G}_0(S_i) \cong \bigoplus_{i \in I} \mathbb{Z}/|G_i|\mathbb{Z}$. Then Lemma 2.1 implies that

$$\varphi(\overline{G}_0(S_i)) + p \cdot A = \overline{G}_0(S_i) + p \cdot A.$$

Therefore, $A = \operatorname{Im} \varphi + p \cdot A$ and so $A = \operatorname{Im} \varphi$, since p is nilpotent on A. The corollary follows. $\square$

Example A: *Skew group rings.* Let G be a finite p-group acting on a right Noetherian K-algebra R by permuting a K-basis B_0 of R, and let $T = RG$ denote the skew group ring. Then we can take $B = B_0 G = \{bg | b \in B_0, g \in G\}$ and $G_1 = G_2 = G$ in Lemma 2.1. Conditions (1) and (2) are clearly satisfied. Moreover,

$$G \backslash B \overset{1-1}{\leftrightarrow} B_0 \text{ and } (G \backslash B)^G \overset{\leftrightarrows}{\to} B_0^G.$$

These observations will be used in the following.

THEOREM 2.4. *Let G be a finite p-group acting on a right Noetherian K-algebra R by permuting a K-basis B_0 of R. Assume that the following conditions are satisfied:*

(1) R is Noetherian as a right module over the fixed subring R^G,

(2) For almost all $b \in B_0$, $\mathbb{C}_G(b) = \langle 1 \rangle$, and

(3) $(p, \#B_0^G) = 1$.

$\overline{G}_0(RG)$ has a direct summand which is isomorphic to $\overline{G}_0(K[G]) \cong \mathbb{Z}/|G|\mathbb{Z}$.

PROOF. By Theorem 1.7 and Example 1.9, (1) and (2) insure that all finitely generated RG-modules are a.i. over $K[G] \subseteq RG$. Thus, in view of our above remarks, condition (3) guarantees that the map

$$\varphi \colon \overline{G}_0(K[G]) \overset{\mathrm{Ind}}{\to} \overline{G}_0(RG) \overset{\theta}{\to} \overline{G}_0(K[G])$$

of Lemma 2.1 is an automorphism of $\overline{G}_0(K[G])$. $\square$

The above theorem applies readily to the Zalesskii-Neroslavskii example RG (Example 1.10) and shows that $\overline{G}_0(RG)$ has a direct summand isomorphic to $\mathbb{Z}/2\mathbb{Z}$. Indeed, conditions (1) and (2) have been checked in Example 1.10, and (3) is clear, because $\#B_0^G = 1$ holds for the basis considered there.

As a second application, we discuss the group algebra of the infinite dihedral group. We remark, however, that both examples will also be subsumed under our later work on twisted group algebras.

Example 2.5: *The infinite dihedral group.* Let $\Gamma = \langle x, y | xy = yx^{-1}, y^2 = 1\rangle$ be the infinite dihedral group. Then Γ is the free product $\Gamma = G_1 * G_2$ with $G_1 = \langle y\rangle$ and $G_2 = \langle xy\rangle$, both of order 2. Thus, for any field K, the group algebra $K[\Gamma]$ is the free product of $K[G_1]$ and $K[G_2]$ over K,

$$K[\Gamma] \cong K[G_1] \underset{K}{*} K[G_2].$$

It follows from [**W**, Proposition 4.1] that the map

$$\bigoplus_{i=1,2} G_0(K[G_i]) \xrightarrow{\bigoplus_{i=1,2} \mathrm{Ind}_{K[G_i]}^{K[\Gamma]}} G_0(K[\Gamma])$$

is onto. (This also follows from Moody's theorem.)

Now assume that $\operatorname{char} K = 2$. Then $G_0(K[G_i])$ is infinite cyclic with generator $[K]$, corresponding to the trivial $K[G_i]$-module. Therefore, putting $\alpha_i = [\mathrm{Ind}_{K[G_i]}^{K[\Gamma]}(K)] \in G_0(K[\Gamma])$, we see that α_1 and α_2 generate $G_0(K[\Gamma])$. Moreover, in $G_0(K[G_i])$, we have $2[K] = [K[G_i]]$, and so $2\alpha_i = [\mathrm{Ind}_{K[G_i]}^{K[\Gamma]}(K[G_i])] = [K[\Gamma]]$ holds in $G_0(G[\Gamma])$. Therefore, putting $\beta = \alpha_1 - \alpha_2$, we have $G_0(K[\Gamma]) = \langle \alpha_1, \beta\rangle$ and $2\beta = 0$. In fact, since the augmentation ideal $\omega\Gamma$ of $K[\Gamma]$ has the form $\omega\Gamma = (1+y)K[\Gamma] \oplus (1+xy)K[\Gamma]$, we see that $[\omega\Gamma] = \alpha_1 + \alpha_2$ holds in $G_0(K[\Gamma])$, whence $\beta = [K]$ is the element corresponding to the trivial $K[\Gamma]$-module. Also, Goldie's reduced rank function yields an explicit homomorphism of $G_0(K[\Gamma])$ onto $\mathbb{Z}$, for any polycyclic-by-finite group Γ, and so α_1 must have infinite order. Finally, $K[\Gamma]$ can also be viewed as the skew group ring $K[\Gamma] = RG_1$ with $R = K[\langle x\rangle]$. All assumptions of Theorem 1.7 and Example 1.9 are readily verified so we conclude that $\theta\colon G_0(K[\Gamma]) \to \overline{G}_0(K[\langle y\rangle]) \simeq \mathbb{Z}/2\mathbb{Z}$ exists. Furthermore since $\theta(\beta) = [K] \in \overline{G}_0(K[\langle y\rangle])$ has order 2, we conclude that β has order 2. Summarizing, we have shown that

$$G_0(K[\Gamma]) = \langle [\mathrm{Ind}_{K[\langle y\rangle]}^{K[\Gamma]}(K)], [K]\rangle \cong \mathbb{Z} \oplus \mathbb{Z}/2\mathbb{Z}.$$

For completeness, we remark that Gersten's theorem [**B**, p. 699] implies that

$$K_0(K[\Gamma]) \cong \mathbb{Z}.$$

In fact, it follows from more general results of G. Bergman [**Be**] that all projective $K[\Gamma]$-modules are free.

In the case when $\operatorname{char} K \neq 2$, $K[\Gamma]$ has finite global dimension and so $G_0(K[\Gamma]) = K_0(K[\Gamma])$. Each $G_0(K[G_i])$ is isomorphic to $\mathbb{Z}\oplus\mathbb{Z}$, with generators $[K[G_i]]$ and $[K]$. As before, it follows that $G_0(K[\Gamma])$ has generators α_1, α_2, and $[K[\Gamma]]$. Finally, Gersten's theorem implies that

$$G_0(K[\Gamma]) \cong \mathbb{Z} \oplus \mathbb{Z} \oplus \mathbb{Z}.$$

Example B: *Twisted group algebras.* Let $T = K^t[\Gamma]$ be a twisted group algebra of the group Γ over K. (Say Γ is polycyclic-by-finite to insure that T is Noetherian.) Thus T has a K-basis $\overline{\Gamma} = \{\bar{x} | x \in \Gamma\}$, and multiplication in T is determined by the rule $\bar{x} \cdot \bar{y} = t(x, y)\overline{xy}$ for $x, y \in \Gamma$, where $t: \Gamma \times \Gamma \to K^\bullet$ is a fixed 2-cocycle of Γ in $K^\bullet$.

Assume that H_1 and H_2 are finite p-subgroups of Γ such that the classes of $t|_{H_i \times H_i}$ in $H^2(H_i, K^\bullet)$ are trivial, i.e. the group extensions

$$1 \to K^\bullet \to K^\bullet \overline{H}_i \to H_i \to 1 \text{ with}$$
$$K^\bullet \overline{H}_i = \{\xi\bar{x} | \xi \in K^\bullet, x \in H_i\} \subseteq U(T)$$

split $(i = 1, 2)$. We remark that this assumption is automatically satisfied if the field K is perfect ([**P1**, Lemma 1.2.10]). Fix splitting subgroups $G_i \subseteq K^\bullet \overline{H}_i$, $G_i \cong H_i$, and put

$$B = G_1 \overline{X} G_2 \subseteq K^\bullet \overline{\Gamma},$$

where $X \subseteq \Gamma$ is a fixed representative system for the double cosets $H_1 \backslash \Gamma / H_2$. Then B is a K-basis of T which clearly satisfies conditions (1) and (2) in Lemma 2.1. Moreover, for $b \in B$, we have

$$G_1 b \in (G_1 \backslash B)^{G_2} \Leftrightarrow G_1 b G_2 = G_1 b \Leftrightarrow b G_2 \subseteq G_1 b.$$

The canonical map $K^\bullet \overline{\Gamma} \to \Gamma$, $\xi\bar{x} \mapsto x$, is mono on B. So, letting $x \in \Gamma$ denote the image of b above, the condition $bG_2 \subseteq G_1 b$ becomes $xH_2 \subseteq H_1 x$, or $H_2 \subseteq H_1^x$. Thus:

$$(G_1 \backslash B)^{G_2} \overset{1-1}{\leftrightarrow} H_1 \backslash \{x \in \Gamma | H_2 \subseteq H_1^x\}.$$

The following result uses these facts.

THEOREM 2.6. *Let $T = K^t[\Gamma]$ be a twisted group algebra with* char $K = p$ *and Γ polycyclic-by-finite. Assume that Γ has a torsion-free normal subgroup N such that Γ/N is a finite p-group.*

Let $\{H_i | i = 1, \ldots, r\}$ be a collection of non-conjugate maximal finite subgroups on Γ such that, for all i, the following assumptions are satsified:

(1) *The group extension $1 \to K^\bullet \to K^\bullet \overline{H}_i \to H_i \to 1$ splits. (This is automatic if K is perfect.)*

(2) *For all $1 \neq x \in N$, $\mathbb{C}_{H_i}(x) = \langle 1 \rangle$.*
Put $R := K^t[N] \subseteq T$ and note that R is stable under conjugation by the group $K^\bullet \overline{\Gamma} \subseteq U(T)$. In fact, Γ acts on R. Assume that, for all i,

(3) *R is Noetherian as a right module over the fixed ring R^{H_i}. (This holds, e.g., if R is commutative. See also Prop.* 3.4.)
Then the map

$$\bigoplus_{i=1}^{r} \overline{G}_0(K^t[H_i]) \overset{\oplus \mathrm{Ind}}{\longrightarrow} \overline{G}_0(T) \overset{\oplus \theta_i}{\longrightarrow} \bigoplus_{i=1}^{r} \overline{G}_0(K^t[H_i])$$

is an automorphism of $\bigoplus_{i=1}^r \overline{G}_0(K^t[H_i]) \cong \bigoplus_{i=1}^r \mathbb{Z}/|H_i|\mathbb{Z}$. *In particular,* $\overline{G}_0(T)$ *has a direct summand isomorphic to* $\bigoplus_{i=1}^r \mathbb{Z}/|H_i|\mathbb{Z}$.

PROOF. The H_i's are finite p-groups with $N \cap H_i = \langle 1 \rangle$. Moreover, by (1), we can choose splitting subgroups $G_i \subseteq K^\bullet \overline{H}_i$ and construct K-bases B_{ij} of T as in the remarks preceding the theorem. Then conditions (1′) and (2′) in Corollary 2.3 are satisfied . For (3), we have to determine $\#(G_i \backslash B_{ij})^{G_j} = \#H_i \backslash \{t \in \Gamma | H_j \subseteq H_i^t\}$. But, since the H_i's are maximal and non-conjugate, we have

$$\{t \in \Gamma | H_j \subseteq H_i^t\} = \begin{cases} \varnothing & \text{if } i \neq j \\ \mathbb{N}_\Gamma(H_i) & \text{if } i = j. \end{cases}$$

Since $\mathbb{N}_\Gamma(H_i) \cap N = \mathbb{C}_N(H_i) = \langle 1 \rangle$, by (2), $\mathbb{N}_\Gamma(H_i)$ is finite. Thus we must have $H_i = \mathbb{N}_\Gamma(H_i)$ and so

$$\#(G_i \backslash G_{ij})^{G_j} = \begin{cases} 0 & \text{for } i \neq j \\ 1 & \text{for } i = j. \end{cases}$$

Hence condition (3) in Corollary 2.3 is also satisfied. It remains to check that all finitely generated T-modules are a.i. over $S_i := K[G_i] = K^t[H_i] \subseteq T$. In view of Remark 2.2(b), it suffices to verify conditions (1) and (2) in Theorem 1.7 for the skew group ring $RG_i \subseteq T$. But (1) corresponds to assumption (3) above, because the actions of H_i and G_i on R coincide. For (2), note that assumption (2) above implies that, for each $1 \neq x \in N$, the H_i-orbit $\bar{x}^{H_i} = \{\bar{x}^h | h \in H_i\}$ of $\bar{x}$ has size $|H_i|$ and consists of linearly independent elements of R. Thus H_i permutes the K-basis $B_i = \{\bar{1}\} \cup \bigcup_x \bar{x}^{H_i}$ of R, where x runs over a representative set of the H_i-conjugacy classes in $N \backslash \{1\}$, and for all $b \in B_i \backslash \{1\}$ we have $\mathbb{C}_{H_i}(b) = \langle 1 \rangle$. Thus (2′) in Example 1.9 is satisfied, and hence Theorem 1.7 applies. Corollary 2.3 now yields the desired conclusion. □

COROLLARY 2.7. *Let* $T = K^t[\Gamma]$ *and* $\{H_i | i = 1, \dots, r\}$ *be as in Theorem* 2.6. *Assume, in addition to* (1), (2) *and* (3) *in Theorem* 2.6, *that*

$$\bigoplus_{i=1}^r G_0(K^t[H_i]) \overset{\oplus \mathrm{Ind}}{\to} G_0(T)$$

is onto. Then

$$\overline{G}_0(T) \cong \bigoplus_{i=1}^r \mathbb{Z}/|H_i|\mathbb{Z}.$$

Moreover, if $|H_1| \geq |H_i|$ *for all* i, *then*

$$G_0(T) \cong \mathbb{Z} \oplus \bigoplus_{i=2}^r \mathbb{Z}/|H_i|\mathbb{Z}.$$

PROOF. The structure of $\overline{G}_0(T)$ follows immediately from our assumption on $G_0(T)$ and Theorem 2.6, because all $G_0(K^t[H_i])$ are cyclic. Indeed, a generator of $G_0(K^t[H_i])$ is $[K_{K[G_i]}]$, the element corresponding to the trivial

$K[G_i]$-module, where $K[G_i] = K^t[H_i] \subseteq T$ is as in the proof of Theorem 2.6. Thus, by our assumption on $G_0(T)$, the elements $\alpha_i := [\mathrm{Ind}_{K[G_i]}^T(K)]$ generate $G_0(T)$. Moreover, $|H_i| \cdot \alpha_i = [T]$ holds in $G_0(T)$. By Theorem 2.6, the image $\bar{\alpha}_i$ of α_i in $\overline{G}_0(T)$ has order precisely $|H_i|$, and $\overline{G}_0(T) = \bigoplus_{i=1}^r \langle \bar{\alpha}_i \rangle$. Putting $\beta_i = \alpha_i - |H_1| \cdot |H_i|^{-1} \cdot \alpha_1$ $(i > 1)$ we see that β_i has order precisely $|H_i|$ in $G_0(T)$, and the sum $\sum_{i=2}^r \langle \beta_i \rangle$ is direct. Finally, as $G_0(T)$ is generated by $\alpha_1, \beta_2, \ldots, \beta_r$. Goldie's reduced rank must be nonzero for α_1. (This can of course be checked directly.) So

$$G_0(T) = \langle \alpha_1 \rangle \oplus \langle \beta_2 \rangle \oplus \cdots \oplus \langle \beta_r \rangle \cong \mathbb{Z} \oplus \bigoplus_{i=2}^r \mathbb{Z}/|H_i|\mathbb{Z},$$

as we have claimed. $\square$

The hypothesis in Corollary 2.7 above, that $G_0(T)$ is generated by the images of all $G_0(K^t[H_i])$ under induction, can often be verified by invoking the following result, due to J. Moody.

THEOREM 2.8. *(Moody* **[Mo2]***). Let $R^*\Gamma$ be a crossed product with R right Noetherian and with Γ a polycyclic-by-finite group. If $\{H_i | i \in I\}$ denotes a representative set for the conjugacy classes of maximal finite subgroups in Γ, then the map*

$$\bigoplus_{i \in I} G_0(R^* H_i) \stackrel{\oplus \mathrm{Ind}}{\rightarrow} G_0(R^* \Gamma)$$

is onto.

Example 2.9: *The Zalesskii-Neroslavskii example (continued).* We retain the notations $R = R_\lambda$, $G = \langle \sigma \rangle$ etc. from Example 1.10, with $\operatorname{char} K = 2$. Then $T := RG$ can also be viewed as the twisted group algebra

$$\begin{aligned} T = K^t[\Gamma] \text{ with } \Gamma &= \langle a, b, c | a^c = a^{-1}, b^c = b^{-1}, ab = ba, c^2 = 1 \rangle \\ &\cong (\mathbb{Z} \oplus \mathbb{Z}) \rtimes C_2, \end{aligned}$$

and $\overline{a^i b^j c^l} = x^i y^j \sigma^l$. Put $A := \langle a, b \rangle \cong \mathbb{Z} \oplus \mathbb{Z}$. Then $H^1(\langle c \rangle, A) = \operatorname{ann}_A(c+1)/A^{c-1} = A/2A \cong (\mathbb{Z}/2\mathbb{Z})^2$, and so Γ has 4 conjugacy classes of maximal finite subgroups, all of order 2:

$$H_1 = G = \langle c \rangle, \ H_2 = \langle ac \rangle, \ H_3 = \langle bc \rangle, \ H_4 = \langle abc \rangle.$$

For $i = 1, 2, 3$, the extension $1 \to K^\bullet \to K^\bullet \overline{H}_i \to H_i \to 1$ splits, because σ, $x\sigma$, and $y\sigma$ have order 2 in $K^\bullet \overline{H}_i \subseteq T$. But for X_4, we have $(xy\sigma)^2 = \lambda$ and so the subalgebra $K^t[H_4]$ of T has the form

$$K^t[H_4] = \begin{cases} \text{a commutative subfield of } T \text{ if } \lambda \notin (K^\bullet)^2 \\ K[\mu x y \sigma] \cong K[C_2] \text{ if } \lambda^{-1} = \mu^2, \ \mu \in K^\bullet. \end{cases}$$

Putting $G_1 = \langle \sigma \rangle$, $G_2 = \langle x\sigma \rangle$, $G_3 = \langle y\sigma \rangle$, and $G_4 = \langle \mu x y \sigma \rangle$, if defined, we are in the situation of Theorem 2.6, and (1) and (2) are satisfied. For (3), we quote Proposition 3.4 in Appendix B. By Moody's theorem, $G_0(T)$ is generated by the images of $G_0(K^t[H_i])$ $(i = 1, \ldots, 4)$. If $\lambda \notin (K^\bullet)^2$, then

$G_0(K^t[H_4]) = \langle [K^t[H_4]] \rangle$ maps to $\langle [T] \rangle$ and can be ignored. So, in this case, the images of $G_0(K^t[H_i])$ $(i = 1, 2, 3)$ already generate $G_0(T)$. Corollary 2.7 now yields

$$\overline{G}_0(T) \cong \begin{cases} (\mathbb{Z}/2\mathbb{Z})^3 & \text{if } \lambda \notin (K^\bullet)^2 \\ (\mathbb{Z}/2\mathbb{Z})^4 & \text{if } \lambda \in (K^\bullet)^2 \end{cases}$$

and

$$G_0(T) \cong \begin{cases} \mathbb{Z} \oplus (\mathbb{Z}/2\mathbb{Z})^2 & \text{if } \lambda \notin (K^\bullet)^2 \\ \mathbb{Z} \oplus (\mathbb{Z}/2\mathbb{Z})^3 & \text{if } \lambda \in (K^\bullet)^2. \end{cases}$$

We finish with just one further explicit example, although clearly many others could be constructed that fit naturally in the framework of Theorem 2.6 and Corollary 2.7.

Example 2.10. Let $\operatorname{char} K = 2$, and let $\Gamma = A \rtimes \langle x \rangle$ with $A = \mathbb{Z} \oplus \mathbb{Z}$ and $x = \begin{bmatrix} 0 & 1 \\ -1 & 0 \end{bmatrix} \in \mathrm{GL}_2(\mathbb{Z})$. Then x has order 4, and $x^2 = -\mathrm{Id}$. Thus, letting $\{H_i | i \in I\}$ denote a full representative set of the conjugacy classes of maximal finite subgroups of Γ, condition (2) in Theorem 2.6 is surely satisfied, with $N = A$. We will consider the ordinary group algebra $T = K[\Gamma]$. Thus the splitting condition (1) is trivial and (3), with $R = K[A]$, follows from Noether's theorem. Also, by Theorem 2.8, $G_0(T)$ is generated by the images of all $G_0(K[H_i])$ under induction. Thus, in order to apply Corollary 2.7, we have to determine the orders of the H_i's.

First, $H^1(\langle x \rangle, A) = \operatorname{ann}_A(1 + x + x^2 + x^3)/A^{x-1} = A/A^{x-1}$ has order 2, because $1 + x + x^2 + x^3 = (1 + x)(1 + x^2) = 0$ and $\det(x - 1) = 2$. Thus Γ has two conjugacy classes of subgroups of order 4, both clearly maximal. Similarly, since $H^1(\langle x^2 \rangle, A) = \operatorname{ann}_A(1+x^2)/A^{x^2-1} = A/2A$ has order 4, there are 4 conjugacy classes of subgroups of order 2 in $A \rtimes \langle x^2 \rangle$. Since every subgroup of order 4 in Γ contains a unique subgroup of order 2 and since the centralizer of every subgroup of order 2 is finite, it follows that two of these classes of subgroups of order 2 consist of maximal finite subgroups of Γ. Moreover, they become conjugate in Γ. Therefore, Corollary 2.7 yields

$$\overline{G}_0(K[\Gamma]) \cong (\mathbb{Z}/4\mathbb{Z})^2 \oplus (\mathbb{Z}/2\mathbb{Z})^1$$

and

$$G_0(K[\Gamma]) \cong \mathbb{Z} \oplus \mathbb{Z}/4\mathbb{Z} \oplus (\mathbb{Z}/2\mathbb{Z})^1.$$

§3. Appendices.

(A) *A cohomological characterization of almost injective modules over* $K[G]$.

Throughout, let G denote a finite group, and let $\widehat{H}^n(G, \cdot)$, $n \in \mathbb{Z}$, denote the Tate cohomology groups of G ([**C-E**, p. 236]). Then we have the following extension of Corollary 1.5 (iii).

LEMMA 3.1. *The following are equivalent for any* $K[G]$*-module* V:

(a) V *is a.i.,*

(b) *For all finite-dimensional* $K[G]$*-modules* X*, the Tate cohomology groups* $\widehat{H}^n(G, X \otimes_K V)$*,* $n \in \mathbb{Z}$*, are finite dimensional over* K*,*

(c) *For all irreducible $K[G]$-modules X, $\widehat{H}^0(G, X \otimes_K V)$ is finite-dimensional over K.*

PROOF. The equivalence (a)⇔(c) is Corollary 1.5 (iii). As (b)⇒(c) is clear, it suffices to prove (a)⇒(b). For this, write $V = V_1 \oplus V_2$ with V_1 finite dimensional and V_2 projective, as in Lemma 1.1. Then $X \otimes_K V = W_1 \oplus W_2$ with $W_1 = X \otimes_K V_1$ finite dimensional and $W_2 = X \otimes_K V_2$ projective. Therefore, $\widehat{H}^n(G, W_2) = 0$ for all n, and so $\widehat{H}^n(G, X \otimes_K V) \cong \widehat{H}^n(G, W_1)$ is finite dimensional. □

COROLLARY 3.2 (cf. Theorem 1.2). *Let $V' \hookrightarrow V \twoheadrightarrow V''$ be a short exact sequence of $K[G]$-modules. If two of $\{V, V', V''\}$ are a.i., then so is the third.*

PROOF. For each $K[G]$-module X, the sequence $X \otimes_K V' \hookrightarrow X \otimes_K V \twoheadrightarrow X \otimes_K V''$ is exact. The long exact cohomology sequence ([**C-E**, p. 236])

$$\cdots \to \widehat{H}^{n-1}(G, X \otimes V'') \to \widehat{H}^n(G, X \otimes V') \to \widehat{H}^n(G, X \otimes V) \to \widehat{H}^n(G, X \otimes V'') \to \dots$$

together with the equivalence (a)⇔(b) in Lemma 3.1 now proves the corollary. □

(B) *Noetherian fixed rings.*

Let G denote a finite group acting on a ring R, and let $T = RG$ denote the corresponding skew group ring. Recall that, for any T-module V, Tr: $V \to V^G$ is the trace map, $\operatorname{Tr}(v) = \sum_{g \in G} vg$.

LEMMA 3.3. *Let V be a Noetherian T-module. Then the following are equivalent*:

(a) *V^G is Noetherian over R^G,*

(b) *For all T-submodules $U \subseteq V$, $U^G/\operatorname{Tr}(U)$ is Noetherian over $R^G/\operatorname{Tr}(R)$,*

(c) *For all R^G-submodules $X \subseteq V^G$, $(XR)^G/X$ is Noetherian over $R^G/\operatorname{Tr}(R)$.*

PROOF. (a)⇒(b) is clear. Note that $U^G \cdot \operatorname{Tr}(R) \subseteq \operatorname{Tr}(U)$ so that $U^G/\operatorname{Tr}(U)$ is indeed a module over $R^G/\operatorname{Tr}(R)$.

(b)⇒(c). Let $X \subseteq V^G$ be an R^G-submodule and put $U = XR \subseteq V$. Then U is a T-submodule of V with $X \subseteq U^G$, and

$$\operatorname{Tr}(U) = X \cdot \operatorname{Tr}(R) \subseteq X.$$

By (b), $U^G/\operatorname{Tr}(U)$ is Noetherian over $R^G/\operatorname{Tr}(R)$, and hence so is U^G/X.

(c)⇒(a). Let $X = X_0 \subseteq X_1 \subseteq \dots$ be a chain of R^G-submodules of V^G. Then each X_iR is a T-submodule of V and, since V is Noetherian, we may assume that $X_iR = XR$ holds for all i. Hence $X \subseteq X_i \subseteq (XR)^G$ and so the given chain must be finite, by (c). □

We will now consider crossed products of the form $R = D^*\Gamma$ with $\Gamma \cong \mathbb{Z}^r$ for some r and D a division ring. Note that R can be viewed as an iterated skew Laurent extension, and hence is a Noetherian domain. By definition, R

has (left) D-basis $\overline{\Gamma} = \{\bar{x} | x \in \Gamma\}$ and so each element $r \in R$ can be uniquely expressed as

$$r = \sum_{x \in \Gamma} d_x \bar{x}, \text{ with } d_x \in D \text{ and}$$
$$\mathrm{Supp}(r) = \{x \in \Gamma | d_x \neq 0\} \text{ a finite subset of } \Gamma.$$

The group of units $U(R)$ equals $U(R) = D^\bullet \overline{\Gamma}$ and $D^\bullet = D^\bullet \overline{1}$ is normal in $U(R)$ with $U(R)/D^\bullet \cong \Gamma$. In particular, any group of automorphisms of R which normalizes D will also act on Γ.

The following proposition extends the main result of **[H-O]**. Our proof essentially follows the arguments in **[H-O]**.

PROPOSITION 3.4. *Let $R = D^*\Gamma$ be a crossed product with $\Gamma \cong \mathbb{Z}^r$ for some r and with D a division ring. Let $G = \langle \sigma | \sigma^2 = 1 \rangle$ act on R so that D is centralized and the action of G on $\Gamma = U(R)/D^\bullet$ is inversion, $x \mapsto x^{-1}$. Then R is right and left Noetherian as a module over the fixed subring R^G.*

(Note that the possibility $\operatorname{char} D = 2$ is not excluded. In fact, for $\operatorname{char} D \neq 2$, the result is clear, e.g. by **[Lo-P**, Theorem 3.3 ii].)

PROOF. Put $U = R^G$. We will consider ${}_U R$.

Step 1. An ordering of Γ. (Here we think of $\Gamma = \mathbb{Z}^r$ as being written additively.)

For each $s \geq 0$, let $\Gamma_s = \{x = (x_1, \dots, x_r) \in \Gamma | |x| := \sum |x_i| = s\}$ denote the r-dimensional cube of diameter $2s$, centered at the origin and with corners on the axes. Then $\Gamma = \dot{\bigcup}_{s \geq 0} \Gamma_s$, and we can linearly order the *set* Γ as follows:

For $x, y \in \Gamma$, say $x \in \Gamma_s$ and $y \in \Gamma_t$, define $x \leq y :\Leftrightarrow s < t$ or $s = t$ and $x \leq_l y$, where $\leq_l$ denotes the lexicographical order on Γ. Clearly, Γ contains no infinite decreasing sequence with respect to this ordering. For each multi-sign $\varepsilon = (\varepsilon_1, \dots, \varepsilon_r) \in \{\pm\}^r$, put

$$Q_\varepsilon = \{(x_1, \dots, x_r) \in \Gamma | x_i \geq 0 \text{ if } \varepsilon_i = +, x_i < 0 \text{ if } \varepsilon_i = -\}.$$

Furthermore, let $e(1), \dots, e(r)$ be the canonical $\mathbb{Z}$-basis of Γ.

Claim. Let $\emptyset \neq M \subseteq \Gamma$ be a finite subset of Γ. If $m = \max M \in Q_\varepsilon$, $\varepsilon = (\varepsilon_1, \dots, \varepsilon_r)$, then

$$m + \varepsilon_i e(i) = \max(M + \{e(i), -e(i)\}).$$

PROOF. Clearly, $m + \varepsilon_i e(i) \leq \max(M + \{e(i), -e(i)\})$. Moreover, for all $x \in \Gamma$, $|x \pm e(i)| \in \{|x| + 1, |x| - 1\}$. In particular, if $x \in M$, then $|x \pm e(i)| \leq |m| + 1 = |m + \varepsilon_i e(i)|$. Here, the latter equality uses the fact that $m \in Q_\varepsilon$. Also, for $x \in M$, $|x \pm e(i)| = |m| + 1$ implies $|x| = |m|$. But then we must have $x \leq_l m$. Thus, if $\varepsilon_i = +$ then $x \pm e(i) \leq_l m + \varepsilon_i e(i)$. So we may assume that $\varepsilon_i = -$. If $x_j < m_j$ for some $j < i$ then, clearly, $x \pm e(i) <_l m + \varepsilon_i e(i)$. On the other hand, if $x_j \geq m_j$ for all $j < i$ then $x_j = m_j$ for all $j < i$, and $x_i \leq m_i < 0$. Hence $|x + e(i)| = |x| - 1 < |m| + 1$, and $x - e(i) \leq_l m + \varepsilon_i e(i)$. Thus, for all $x \in M$, we have $x \pm e(i) \leq m + \varepsilon_i e(i)$, as we have claimed.

Step 2. Leading terms.

For each nonzero element $r \in R$ put

$$\lambda(r) := \max \operatorname{Supp}(r) \in \Gamma,$$

using the ordering of Γ introduced in Step 1. For each nonzero U-submodule I of R put

$$\lambda I := \lambda(I \backslash \{0\}) \subseteq \Gamma.$$

Claim. If $I \subseteq J$ are nonzero U-submodules of R, then $\lambda I \subseteq \lambda J$. If $I \subsetneqq J$, then $\lambda I \subsetneqq \lambda J$.

PROOF. The first assertion is clear. Suppose that $I \subsetneqq J$ but $\lambda I = \lambda J$. Pick $r \in J \backslash I$ with $\lambda(r)$ as small as possible. By assumption, $\lambda(r) = \lambda(t) =: x$ for some $t \in I$. But then, for some $d \in D^{\bullet}$, $x \notin \operatorname{Supp}(r - dt)$ and so $\lambda(r - dt) < \lambda(r)$. Since $r - dt \in J \backslash I$, this contradicts the minimality of r.

For each basis vector $e(i) \in \Gamma$, $i = 1, \ldots, r$, put

$$b_i = \overline{e(i)} + \overline{e(i)}^{\sigma} \in U.$$

Note that $\overline{e(i)}^{\sigma} = \overline{d_i e(i)^{-1}}$ for some $d_i \in D^{\bullet}$. (Here we are temporarily reverting to the multiplicative notation of Γ.) Thus $\operatorname{Supp}(b_i r) \subseteq \operatorname{Supp}(r) \pm \{e(i)\}$ holds for any nonzero $r \in R$, and the claim in Step 1 implies easily that

$$\lambda(r) \in Q_{\varepsilon}, \varepsilon = (\varepsilon_1, \ldots, \varepsilon_r) \Rightarrow \lambda(b_i r) = \lambda(r) + \varepsilon_i e(i) \in Q_{\varepsilon}.$$

Therefore, if $\lambda(r) \in Q_{\varepsilon}$, as above, then

$$\lambda U r \supseteq \lambda(r) + \tilde{Q}_{\varepsilon}, \text{ where } \tilde{Q}_{\varepsilon} = (\varepsilon_1 \mathbb{N}_0, \ldots, \varepsilon_r \mathbb{N}_0) \subseteq \Gamma$$
$$(\mathbb{N}_0 = \{0, 1, 2, \ldots,\}).$$

Step 3. Conclusion.

Suppose that $0 \subsetneqq I_1 \subsetneqq I_2 \subsetneqq \ldots$ is an infinite strictly ascending chain of U-submodules of R. Then $\lambda I_1 \subsetneqq \lambda I_2 \subsetneqq \ldots$ and so we can select elements $a(i) \in \lambda I_i \backslash \lambda I_{i-1}$. By considering a suitable subchain, if necessary, we may assume that all $a(i)$ belong to the same Q_{ε}, $\varepsilon = (\varepsilon_1, \ldots, \varepsilon_r)$. By Step 2, we have $a(i) + \tilde{Q}_{\varepsilon} \subseteq \lambda I_i$, whence $a(t) \notin a(i) + \tilde{Q}_{\varepsilon}$ for $t > i$.

Write $a(i) = (\varepsilon_1 a_{1,i}, \ldots, \varepsilon_r a_{r,i})$ with $a_{i,j} \in \mathbb{N}_0$ and put

$$a(i)^* := (a_{1,i}, \ldots, a_{r,i}) \in \mathbb{N}_0^r.$$

Consider the (partial) product ordering of $\mathbb{N}_0^r$ given by the natural ordering of $\mathbb{N}_0$: $(x_1, \ldots, x_r)$ precedes $(y_1, \ldots, y_r)$ if and only if $x_i \leq y_i$ holds for all i. Then the set $\{a(i)^* | \text{all } i\} \subseteq \mathbb{N}_0^r$ has finitely many minimal elements, say $a(i_1)^*, \ldots, a(i_p)^*$, and each $a(i)^*$ majorizes at least one of these (cf. [**D**, Lemma 2.6.2]).

Finally let $t > \max_{1 \leq k \leq p} i_k$. Since $a(t) \notin a(i_k) + \tilde{Q}_{\varepsilon}$, we must have $a_{j_k,t} < a_{j_k,i_k}$ for some $j_k \in \{1, \ldots, r\}$. Consequently $a(t)^*$ majorizes none of $a(i_1)^*, \ldots, a(i_p)^*$, contradiction. This completes the proof of the proposition. □

Acknowledgement. The first author's research was supported by the Deutsche Forschungsgemeinschaft/Heisenberg Program (Lo 261/2-2). He would also like to thank the University of Wisconsin at Madison and the University of Granada for their hospitality and support while working on this article.

The second author's research was supported in part by NSF Grant No. DMS 8521704.

Both authors would like to thank L. Levy for suggesting that the material at the beginning of Section 1 be formulated in the more general context of quasi-Frobenius rings.

REFERENCES

[B] H. Bass, *Algebraic K-theory*, Benjamin, New York, 1968.

[Be] G. M. Bergman, *Modules over coproducts of rings*, Trans. Amer. Math. Soc. **200** (1974), 1–32.

[Br-H-Lo] K. A. Brown, J. Howie, and M. Lorenz, *Induced resolutions and Grothendieck groups of polycyclic-by-finite groups*, J. Pure and Appl. Algebra **53** (1988), 1–14.

[C-E] H. Cartan and S. Eilenberg, *Homological Algebra*, Princeton Univ. Press, Princeton, 1956.

[D] J. Dixmier, *Enveloping algebras*, North-Holland, Amsterdam, 1977.

[F] D. R. Farkas, *Group rings: an annotated questionnaire*, Commun. in Algebra **8** (1980), 585–602.

[Fa-Hs] F. T. Farrell and W. C. Hsiang, *The Whitehead group of poly (finite-or-cyclic) groups*, J. London Math. Soc. (2) **24** (1981), 308–324.

[H-O] T. J. Hodges and J. Osterburg, *A rank two indecomposable projective module over a Noetherian domain of Krull dimension one*, Bull. London Math. Soc. **19** (1987), 139–144.

[K] F. Kasch, *Modules and rings*, London Math. Soc. Monograph No. 17, Academic Press, London, 1982.

[L] P. Landrock, *Finite group algebras and their modules*, London Math. Soc. Lect. Notes Series, Vol. 98, Cambridge Univ. Press, Cambridge, 1986.

[Lo-P] M. Lorenz and D. S. Passman, *Observations on crossed products and fixed rings*, Commun. in Algebra **8** (1980), 743–779.

[M1] S. Montgomery, *Fixed rings of finite automorphism groups of associative rings*, Lecture Notes in Math., Vol. 818, Springer, Berlin, 1980.

[M2] ——, *Group actions on rings: some classical problems, in: Methods in ring theory* (F. van Oystaeyen, ed.), Reidel, Dordrecht, 1984.

[Mo 1] J. A. Moody, Ph. D. thesis, Columbia University, 1986.

[Mo 2] ——, *A Brauer induction theorem for G_0 of certain infinite groups*, J. Algebra (to appear).

[P1] D. S. Passman, *The algebraic structure of group algebras*, John Wiley & Sons, New York, 1977.

[P2] ——, *On the Goldie rank of group algebras*, Commun. in Algebra **13** (1985), 1305–1312.

[Q] F. Quinn, *Algebraic K-theory of poly-(finite or cyclic) groups*, Bull. Amer. Math. Soc. **12** (1985), 221–226.

[R] S. Rosset, *The Goldie rank of virtually polycyclic groups*, in: Lect. Notes in Math., Vol. 844, pp. 35–45, Springer, Berlin, 1981.

[**W**] F. Waldhausen, *Algebraic K-theory of generalized free products, Part 1*, Annals Of Math. **108** (1978), 135–204.

[**Z-N**] A. E. Zalesskii and O. M. Neroslavskii, *There exists a simple Noetherian ring with divisors of zero but without idempotents*, Commun. in Algebra **5** (1977), 231–245. (Russian)

DEPARTMENT OF MATHEMATICAL SCIENCES
NORTHERN ILLINOIS UNIVERSITY
DEKALB, ILLINOIS 60115

AND

DEPARTMENT OF MATHEMATICS
UNIVERSITY OF WISCONSIN-MADISON
MADISON, WISCONSIN 53706

Contemporary Mathematics
Volume **93** (1989)

The Direct Factor Problem for Modular Abelian Group Algebras

WARREN MAY

ABSTRACT. Let F be a field of characteristic p. Some results are proved concerning whether an abelian group G is a direct factor of the unit group of FG.

1. Introduction.

Let F be a commutative ring with identity. All groups considered will be abelian. If G is an abelian group, let $V(G)$ denote the group of normalized units (i.e., of augmentation 1) in the group algebra FG. We say that the group algebra FG splits if there exists an F-isomorphism $FG \approx FG'$, where G' splits as a mixed abelian group. In other words, FG has a group basis which splits. S. D. Berman and T. Mollov investigated this phenomenon in [**2**, **3**], and obtained theorems in which countability plays a significant role. Since the question of whether FG splits is a special case of the isomorphism question, we would expect to be able to draw strong conclusions under circumstances which favor strong isomorphism results. In particular, suppose that F is a field of prime characteristic p, and that the torsion subgroup tG of G contains only p-torsion. Then Berman and Mollov proved the following theorem (slightly modified).

THEOREM. ([**3**]) *Let F be a field of characteristic p and G an abelian group such that tG is a p-group and G/tG is countable. Then FG splits if and only if G splits.*

One might hope that the hypothesis of countability can be deleted, thus leading to a splitting theorem for characteristic p. What we wish mainly to discuss in this paper is not the splitting question, but another problem which is relevant under the same hypotheses. The direct factor problem asks whether G is a direct factor of $V(G)$. Various aspects of this question are discussed in [**4**, **6**, **7**]. A connection between the splitting problem and the direct factor problem is the following.

1980 *Mathematics Subject Classification* (1985 *Revision*). 20K21; Secondary 20C07.

PROPOSITION A. *Let F be a field of characteristic p, and suppose that tG is a p-group. If G is a direct factor of $V(G)$, then FG splits if and only if G splits.*

We will prove two theorems on the direct factor problem, one of which will yield a result on splitting. We remark that there is no significant advantage in discussing coefficient rings in which p is not invertible, thus we will stay with fields of characteristic p.

THEOREM B. *Let F be a field of characteristic p.*

(1) *If tG is the direct sum of cyclic p-groups, then G is a direct factor of $V(G)$ with complement a direct sum of cyclic p-groups.*

(2) *If G and F are both countable, and if tG is p-torsion, then G is a direct factor of $V(G)$ with complement a simply presented p-group.*

By Proposition A, statement (1) yields the

COROLLARY. *If tG is a direct sum of cyclic p-groups, then FG splits if and only if G splits.*

We note that (2) is not strong enough to imply the theorem of Berman and Mollov; we have not seen a way to assume that only G/tG is countable. Nevertheless, (2) improves Theorem 5.2 in **[1]** by considering mixed groups.

Now we restrict our attention to p-groups. It is not hard to show that if a mixed group G with only p-torsion is a direct factor of $V(G)$, then tG is a direct factor of $V(tG)$. Thus the direct factor problem for p-groups is an important special case. It is proved in **[8]** that if T is a simply presented p-group, then T is a direct factor of $V(T)$. In many investigations of abelian group algebras, it seems to be difficult to deal with torsion groups that are not simply presented. For our second theorem, we shall exploit the fact (see Lemma 5) that there exist more nice subgroups between T and $V(T)$ than one might expect. This will enable us to handle p-groups of cardinal $\aleph_1$ and length not exceeding ω_1, a class which includes many groups which are not simply presented.

THEOREM C. *Let F be a field of characteristic p and T a p-group of cardinal not exceeding $\aleph_1$, and p-length not exceeding ω_1. Then T is a direct factor of $V(T)$. If F is either perfect or countable, then the complement is simply presented.*

One expects a complement for T in $V(T)$ to be of the same complexity as T. A surprise in this case (under the further assumption on F) is that the complement is simply presented even if T is not. The general question of whether G must be a direct factor of $V(G)$ whenever F is a field of characteristic p and tG is a p-group seems unapproachable at this time. Two obvious cases for further examination are when tG is simply presented, and when G is p-torsion.

The next section consists of lemmas on p-heights in group algebras. We prove the proposition and theorems in the final section.

2. Lemmas on p-Height.

Let F be the field of characteristic p throughout this section. If G is an abelian group and $\alpha \in FG$, then $\operatorname{aug}\alpha$ denotes the augmentation of α, and $\operatorname{Supp}\alpha$ the support of α. The ideal of FG consisting of all α with $\operatorname{aug}\alpha = 0$ is denoted $I(G)$.

If $H \leq G$, then we define $K(H)$ to be the kernel of the map $V(G) \to V(G/H)$ induced by the natural map $G \to G/H$. Let $\{g_i\colon i < \nu\}$ be a complete family of coset representatives of H in G. Then every element of $K(H)$ has a unique representation $1 + \sum_{i<\nu} g_i\beta_i$ with $\beta_i \in I(H)$. If H is a p-group, then every element of this form lies in $K(H)$ since such an element is clearly p-torsion. In this case, $V(H)$ is a p-group and consists of all elements of FH of augmentation 1.

To define p-height, let A be either a commutative ring of characteristic p, or an abelian group. We define A^σ for every ordinal σ by induction: $A^0 = A$; $A^{\sigma+1} = (A^\sigma)^p$; and if σ is a limit ordinal, then $A^\sigma = \bigcap_{\tau<\sigma} A^\tau$. If λ is the smallest ordinal such that $A^\lambda = A^{\lambda+1}$, we define $A^\infty = A^\lambda$. Let $\alpha \in A$. We define the *p-height* $h(\alpha)$ to be ∞ if $\alpha \in A^\infty$, or the largest ordinal σ such that $\alpha \in A^\sigma$ otherwise. Since the prime p is fixed, we will simply refer to heights. It is easily shown by induction that $(FG)^\sigma = F^\sigma G^\sigma$ and $V(G)^\sigma = V(G) \cap F^\sigma G^\sigma$. If $g_i \in G$ $(1 \leq i \leq n)$, are distinct, then $h(\sum a_i g_i) = \min\{h(a_i g_i)\} = \min\{h(a_i), h(g_i)\colon a_i \neq 0\}$. If F is perfect, then $F^\sigma = F$ for every σ, thus $(FG)^\sigma = FG^\sigma$ and $h(\sum a_i g_i) = \min\{h(g_i)\colon a_i \neq 0\}$.

A subgroup H is called isotype in G if $H \cap G^\sigma = H^\sigma$ for every ordinal σ.

LEMMA 1. *Let H be isotype in G, and let $g \in G$, $\gamma \in I(H)$ such that $g\gamma \in (FG)^{\sigma+1}$. Then there exist $g' \in G$ and $\gamma' \in I(H)$ such that $g\gamma = (g'\gamma')^p$, and $g'\gamma' \in (FG)^\sigma$.*

PROOF. We may write $\gamma = \sum_{i=1}^n a_i h_i$, where $a_i \in F^{\sigma+1}$ and $gh_i \in G^{\sigma+1}$. Thus $h_1^{-1}h_j \in G^{\sigma+1} \cap H = H^{\sigma+1}$ for every j, consequently $h_1^{-1}h_j = h_j'^p$ for some $h_j' \in H^\sigma$. But $gh_1 \in G^{\sigma+1}$ implies $gh_1 = g'^p$ for some $g' \in G^\sigma$, therefore $gh_j = (g'h_j')^p$ and $g'h_j' \in G^\sigma$. Also, $a_i = a_i'^p$, $a_i' \in F^\sigma$ for every i. Put $\gamma' = \sum a_i'h_i'$. Then $g'\gamma' \in (FG)^\sigma$ and $(g'\gamma')^p = g\gamma$. $\square$

LEMMA 2. *Let H be isotype in G. Then H, $V(H)$, $K(H)$ and $GV(H)$ are all isotype in $V(G)$.*

PROOF. $H \cap V(G)^\sigma = H \cap G^\sigma = H^\sigma$, thus H is isotype in $V(G)$. To show $V(H)$ isotype, we use the fact that $V(H) \cap FG^\sigma = V(H \cap G^\sigma)$, thus $V(H) \cap V(G)^\sigma = V(H \cap G^\sigma) \cap F^\sigma G^\sigma = V(H^\sigma) \cap F^\sigma G^\sigma = V(H)^\sigma$.

Next, we show $K(H) \cap V(G)^\sigma = K(H)^\sigma$ by induction on σ. We need only consider the step from σ to $\sigma+1$, therefore let $\alpha \in K(H) \cap V(G)^{\sigma+1}$. Then $\alpha = 1 + \sum g_i\gamma_i$, where the g_i represent distinct cosets of H in G, $\gamma_i \in I(H)$, and $g_i\gamma_i \in (FG)^{\sigma+1}$. By Lemma 1, $g_i\gamma_i = (g_i'\gamma_i')^p$, where $\gamma_i' \in I(H)$, $g_i'\gamma_i' \in (FG)^\sigma$. Put $\beta = 1 + \sum g_i'\gamma_i'$. Then $\beta^p = \alpha$ and $\beta \in V(G)^\sigma \cap K(H) = K(H)^\sigma$ by induction. Therefore $\alpha \in K(H)^{\sigma+1}$.

Finally, we show $GV(H) \cap V(G)^\sigma = (GV(H))^\sigma$ by induction. Suppose $g\beta \in V(G)^{\sigma+1}$, where $g \in G$ and $\beta \in V(H)$. By altering the factor g, we may assume that $\beta = ae + \gamma$, where $a \in F$, $a \neq 0$, and $e \notin \sup \operatorname{Supp} \gamma$. Therefore $g\beta = ag + \gamma g$, and hence $g \in G^{\sigma+1}$. Since $V(H)$ is isotype in $V(G)$, we have $\beta \in V(G)^{\sigma+1} \cap V(H) = V(H)^{\sigma+1}$. Thus $g\beta \in (GV(H))^{\sigma+1}$. □

An element of a group is called proper with respect to a subgroup if it has maximum height among the elements of its coset. The subgroup is called nice if every coset contains a proper element.

LEMMA 3. *Let $S \leq G$, $N \leq GV(S)$, and $\alpha \in V(G)$. Assume that $\alpha = ae + \beta$, where $a \in F$, $a \neq 0$, and* $(\operatorname{Supp} \beta) \cap S = \phi$. *Then α is proper with respect to N.*

PROOF. Let $g \in G$, $\gamma \in V(S)$. It suffices to show that $h(g\gamma\alpha) \leq h(\alpha)$. We have $g\gamma\alpha = ag\gamma + g\gamma\beta$, where the supports of the two summands are disjoint. Thus $h(g\gamma\alpha) \leq h(ag\gamma)$. Put $\gamma = \sum b_i s_i$ $(b_i \in F, s_i \in S)$. Then $ag\gamma = \sum ab_i g s_i$. Suppose it were the case that $h(ag\gamma) > h(g\gamma)$. Then $h(ab_i) > h(g\gamma)$ for every i. Since $\operatorname{aug} \gamma = 1$, $a = \sum ab_i$, thus $h(a) > h(g\gamma)$. But then $h(ag\gamma) = h(g\gamma)$, a contradiction. Thus $h(ag\gamma) \leq h(g\gamma)$ and $h(g\gamma\alpha) \leq h(ag\gamma) \leq h(g\gamma)$. Therefore $h(\alpha) \geq h(g\gamma\alpha)$. □

A subgroup is called balanced if it is both isotype and nice.

LEMMA 4. *If H is isotype in G, then H is balanced in $K(H)$.*

PROOF. By Lemma 2, H is isotype in $V(G)$, hence in $K(H)$. But Lemma 2 also implies that $K(H)$ is isotype in $V(G)$, hence to show that H is nice in $K(H)$, it suffices to show that if $\alpha \in K(H)$, then αH contains an element of maximum height in $V(G)$. Note that $\alpha \in K(H)$ implies there exists $h \in H \cap (\operatorname{Supp} \alpha)$. Replacing α by αh^{-1}, we may assume that $\alpha = ae + \beta$, where $a \in F$, $a \neq 0$, and $e \notin \operatorname{Supp} \beta$. Now apply Lemma 3 with $S = 1$, $N = H$. □

The next lemma is the key to Theorem C since there is no requirement that S be nice in G.

LEMMA 5. *Let $S \leq G$, S p-torsion.*

(1) *$GV(S)$ is nice in $V(G)$.*

(2) *If F is perfect, $A \leq G$ is finite p-torsioin, and $GV(S) \leq N \leq GV(S)V(A)$, then N is nice in $V(G)$.*

PROOF. Suppose we are given $g_1, \dots, g_n \in G$ representing distinct cosets of S, and a finite subgroup $S_0 \leq S$. Put $\Gamma = \{h(g_i^{-1} g_j s) \colon i \neq j, s \in S_0\}$. Assume that $\alpha' \in V(G)$, and that $\alpha' = g_1\beta_1 + \cdots + g_n\beta_n$ for some $\beta_i \in FS_0$. Since $\operatorname{aug} \alpha' = 1$, we may assume that $\operatorname{aug} \beta_1 = a \neq 0$. Then $a(g_1\beta_1)^{-1} \in GV(S)$ and $a(g_1\beta_1)^{-1}\alpha' = ae + \gamma$, where $\operatorname{Supp} \gamma \subseteq \bigcup_{i=2}^n g_1^{-1} g_i S_0$. Then Lemma 3 implies that $ae + \gamma$ is proper with respect to $GV(S)$. Observe that if F is perfect, then $h(ae + \gamma) \in \Gamma$.

Now (1) follows immediately, since given $\alpha' \in V(G)$, we may find suitable $g_1, \dots, g_n$ and S_0, hence $ae + \gamma$ is a proper element in $\alpha' GV(S)$. To show

(2), fix $\alpha \in V(G)$. Then one can find suitable $g_1, \dots, g_n$ and S_0 such that $\alpha V(A) \subseteq \{g_1\beta_1 + \cdots + g_n\beta_n : \beta_i \in FS_0\}$. Let $\alpha' \in \alpha V(A)$. Then the coset $\alpha' GV(S)$ contains a proper element whose height lies in the finite set Γ. It follows that an element of maximal height will exist in αN, thus N is nice in $V(G)$. □

LEMMA 6. *Let T be a countable p-group and $S \leq T$. If F is either countable or perfect, then $V(T)/TV(S)$ is a simply presented p-group.*

PROOF. If F is countable, then so is $V(T)$, hence $V(T)/TV(S)$ is simply presented. Therefore, assume F is perfect. Choose finite groups $1 = A_0 \leq A_1 \leq A_2 \leq \dots$ with $\bigcup_{i<\omega} A_i = T$. Then $TV(S) \leq TV(S)V(A_1) \leq \dots$ with $\bigcup_{i<\omega} TV(S)V(A_i) = V(T)$. By Lemma 5, we may interpolate a nice composition series between $TV(S)V(A_i)$ and $TV(S)V(A_{i+1})$ for every i, hence between $TV(S)$ and $V(T)$. □

LEMMA 7. *If tG is a direct sum of cyclic p-groups, then so is $K(tG)/tG$.*

PROOF. Let $T = tG$. First assume that F is perfect. Since T is a direct sum of cyclics, there exist $T_0 \leq T_1 \leq T_2 \leq \dots$ with $T = \bigcup_{k<\omega} T_k$, and such that $h(t) \leq k$ for every $t \in T_k$, $t \neq 0$. Clearly $K(T_0) \leq K(T_1) \leq \dots$ with $K(T) = \bigcup_{k<\omega} K(T_k)$. Thus $K(T)/T = \bigcup_{k<\omega}(K(T_k)T/T)$. Let $\alpha \in K(T_k)$, $\alpha \notin T$. By Kulikov's criterion ([**5**, Theorem 17.1]), it suffices to show that αT has height $\leq k$ in $K(T)/T$. Since T is isotype in G, Lemma 4 implies that T is nice in $K(T)$, hence it suffices to show that αT contains a proper element with respect to T of height $\leq k$. Since $\alpha \in K(T_k)$, there exists $t \in T_k \cap \operatorname{Supp} \alpha$. Replacing α by αt^{-1}, we may assume that $e \in \operatorname{Supp} \alpha$, thus α is proper with respect to T by Lemma 3. We may write $\alpha = 1 + \sum g_i \beta_i$, where the g_i represent distinct cosets of T_k, and $\beta_i \in I(T_k)$ for every i. But $\alpha \notin T$, therefore $\beta_i \neq 0$ for some i. Since $\operatorname{aug} \beta_i = 0$, there exist t, $t' \in \operatorname{Supp} \beta_i$, with $t\ t' \in T_k$, $t \neq t'$. If $h(g_i t) > k$, then $h(g_i t') \leq k$ since $h(t^{-1}t') \leq k$. In any case, it follows that $h(\alpha) \leq k$, as desired.

Now assume that F is not perfect. Then $F \subseteq \overline{F}$ for some perfect field $\overline{F}$. We have an induced inclusion $(K_F(T)/T) \leq (K_{\overline{F}}(T)/T)$. But $K_{\overline{F}}(T)/T$ is a direct sum of cyclic p-groups by the above, hence so is $K_F(T)/T$. □

3. Proofs of A, B and C.

LEMMA 8. *Let F be any field. Then $V(G) = GK(tG)$ and $G \cap K(tG) = tG$.*

PROOF. The natural map $G \to G/tG$ induces $V(G) \to V(G/tG)$ with kernel $K(tG)$. It is well-known that $V(G/tG) = G/tG$ since G/tG is torsion-free, hence $V(G) = GK(tG)$. Clearly $G \cap K(tG) = tG$. □

PROOF OF A. We may assume that $FG \approx FH$, where H splits. Thus $H = tH \times C$ and we have $V(H) = HK(tH) = C \times K(tH)$. There are no nontrivial idempotents in FG since tG is a p-group, hence tH must also be a p-group. Thus $K(tH)$ is the torsion subgroup of $V(H)$, and we see that $V(H)$ splits as a mixed abelian group. Since $V(G) \approx V(H)$, we conclude that there is a projection $V(G) \to K(tG)$. By hypothesis, there is a projection $V(G) \to G$. The composition $V(G) \to G$ has image in tG since $K(tG)$ is torsion. Thus it restricts to a projection $G \to tG$. $\square$

In the proofs of Theorems B and C, we will use the following fact (see **[5]**, Theorem 81.9 and Exercise 81.11). If a p-group modulo a balanced subgroup is simply presented, then the subgroup is a direct factor.

PROOF OF B. By Lemma 4, tG is balanced in $K(tG)$. In the case of (1), Lemma 7 implies by the above fact that $K(tG) = tG \times C$ with C a direct sum of cyclic p-groups. In case (2), $K(tG)/tG$ is countable, hence simply presented. Thus $K(tG) = tG \times C$ with C simply presented. In either case, we are done since Lemma 8 implies $V(G) = GK(tG) = G \times C$. $\square$

PROOF OF C. By the cardinal restriction on T, we may write $T = \bigcup_{i<\omega_1} T_i$, where $\{T_i : i < \omega_1\}$ is a smooth chain of countable subgroups of T with $T_0 = 1$. Thus $T_j \leq T_i$ if $j < i$, and $T_i = \bigcup_j T_j$ if i is a limit ordinal. By the length restriction and [**5**, Exercise 80.5], we may assume that all T_i are isotype in T. Lemmas 2 and 5 imply that $T_{i+1}V(T_i)$ is balanced in $V(T_{i+1})$ for $i < \omega_1$. First suppose that F is perfect or countable. By Lemma 6 and the fact above, $V(T_{i+1}) = T_{i+1}V(T_i) \times C_i$, with C_i simply presented. Since $V(T_i) \cap T_{i+1} = T_i$, it is easily seen by induction that $V(T_i) = T_i \times (\coprod_{j<i} C_j)$, hence $V(T) = T \times (\coprod_{j<\omega_1} C_j)$. Therefore T is a direct factor of $V(T)$ with simply presented complement.

Now suppose that F is arbitrary. We may assume that $F \subseteq \overline{F}$, where $\overline{F}$ is a perfect field. Then T is a direct factor of $V_{\overline{F}}(T)$, hence of $V_F(T)$ since $V_F(T) \leq V_{\overline{F}}(T)$. $\square$

REFERENCES

1. D. Beers, F. Richman and E. Walker, *Group algebras of abelian groups*, Rend. Sem. Mat. Univ. Padova **69** (1983), 41–46.

2. S. D. Berman and T. Mollov, *Group rings of mixed abelian groups* (Russian), C. R. Acad. Bulgare Sci. **28** (1975), 879–881.

3. ——, *Splitting of bases and isomorphisms of group rings of abelian groups* (Russian), Pliska **2** (1981), 6–22.

4. R. K. Dennis, *Units in group rings*, J. Algebra **43** (1976), 655–664.

5. L. Fuchs, "Infinite Abelian Groups," Vols. I and II, Academic Press, New York, 1970 and 1973.

6. G. Karpilovsky, "Commutative Group Algebras," Marcel Dekker, New York, 1983.

7. W. May, *Unit groups and isomorphism theorems for commutative group algebras*, in "*Group and Semigroup Rings*," North-Holland Mathematics Studies No. 126, North-Holland, Amsterdam, 1986, 163–178.

8. ——, *Modular group algebras of simply presented abelian groups*, Proc. Amer. Math. Soc. **104** (1988), 403–409.

DEPARTMENT OF MATHEMATICS
UNIVERSITY OF ARIZONA
TUCSON, ARIZONA 85721

Contemporary Mathematics
Volume 93 (1989)

On the Isomorphism of the Semi-Simple Group Algebras of the Primary Abelian Groups

TODOR ZH. MOLLOV

Let KG be a group algebra of the abelian p-group G over a field K of characteristic distinct from p, G^1 the subgroup of the elements of infinite height of the group G and E_n the primitive root of degree p^n of the unit in some extension of the field K. The field K is called a first kind field with respect to p if $K(E_n) \neq K(E_2)$ for some $n > 2$. Otherwise K is called a second kind field with respect to p (cf. **[1]** or **[6**, p. 684**]**). Berman **[1]** has found necessary and sufficient conditions for the isomorphism of the algebras KG and KH when the group G is countable and H is an abelian p-group, i.e. he has found a complete system of invariants of the algebra KG (defining the latter up to an isomorphism) when G is a countable abelian p-group. In **[4]** it is found a complete system of invariants of the algebra KG when a) K is a second kind field with respect to p and b) K is a first kind field with respect to p and the factor group G/G^1 is a direct product of cyclic groups. In the present paper it is proved that if the group G is uncountable its subgroup G^1 is countable, i.e. $|G^1| = \aleph_0$ and K is a first kind field with respect to p then the finding of the complete system of invariants of the algebra KG is reduced to a finding of a complete system of invariants of the algebra $K(G/G^1)$, i.e., to a study of the algebra KA where A is a separable abelian p-group.

Throughout in the work we shall assume that G is an abelian p-group, K is a field of characteristic distinct form p, $\Pi(G/F)$ is a complete system of representatives of G with respect to the subgroup F of the group G and $\langle f-1 | f \in F\rangle$ is the ideal of the algebra KG generated by the elements $f-1$ when f varies in F. If the set M is contained in the group H then by $\langle M\rangle$ we shall denote the subgroup of H which is generated by M. If A is a subgroup of the multiplicative group $U(KG)$ of the algebra KG and A is a basis of KG then A is called its group basis. In order to prove some preliminary results we use the notation $KM = \{\sum \alpha_i m_i | \alpha_i \in K, m_i \in M\}$ and the following well known fact:

LEMMA 1. *If H is a subgroup of the abelian group G and $I = \langle h-1 | h \in H \rangle$ is the ideal of the algebra KG then the set*

$$\{g + I | g \in \Pi(G/H) = \Pi\}$$

forms a group basis of the factor algebra KG/I and $K\Pi \cap I = \{0\}$.

Further we'll need some auxiliary assertions.

PROPOSITION 2. *Let K be a first kind field with respect to p and $S(KG)$ the Sylow p-subgroup of the normed multiplicative group of the algebra KG. Then the mapping $\varphi: S(KG) \to S(K(G/G^1))$ defined by*

$$\varphi\left(\sum_{g\in G} x_g g\right) = \sum_{g\in G} x_g(gG^1), \tag{1}$$

is a homomorphism onto the group $S(K(G/G^1))$ and its kernel Ker φ *satisfies the equalities*

$$\operatorname{Ker}\varphi = dS(KG) = S^1(KG),$$

where $dS(KG)$ is the maximal divisible subgroup of the group $S(KG)$ (and $S^1(KG)$ is the subgroup of the elements of infinite height of $S(KG)$).

The first part of this statement immediately follows from Lemma 18 of the paper **[5]** and the second one is established in the proof of the Theorem 19 in the same paper. Since $G^1 \subseteq S^1(KG)$ then this proposition allows us further to conclude that $g \in G^1$ implies $g \in dS(KG)$.

In the following statement we shall use the notation

$$n\left(\sum_{g\in G} x_g g\right) = \sum_{g\in G} x_g, \; x_g \in K.$$

PROPOSITION 3. *Let K be a first kind field with respect to p and $x \in S(KG)$. Then $x \in dS(KG)$ if and only if $x \equiv 1 \pmod{I}$, where $I = \langle g-1 | g \in G^1\rangle$.*

PROOF. Any $x \in KG$ can be uniquely written in the form

$$x = y_1 + y_2 g_2 + \cdots + y_k g_k, y_i \in KG^1, g_i \in \Pi(G/G^1), \; i = 1, \ldots, k. \tag{2}$$

Obviously

$$x + I = n(y_1)(1 + I) + n(y_2)(g_2 + I) + \cdots + n(y_k)(g_k + I). \tag{3}$$

Since, in view of the Proposition 2, $dS(KG) = \operatorname{Ker}\varphi$ then $x \in dS(KG)$ if and only if

$$n(y_1)G^1 + n(y_2)g_2G^1 + \cdots + n(y_k)g_kG^1 = \varphi(x) = G^1,$$

i.e., if and only if $n(y_1) = 1$, $n(y_2) = \cdots = n(y_k) = 0$. Therefore, in view of (3), the condition $x \in dS(KG)$ is equivalent to the equality $x + I = 1 + I$, i.e., equivalent to the congruence $x \equiv 1 \pmod{I}$. The proposition is proved.

Let $\mathbb{N}_0$ be the set of the non-negative integers. We shall say that an element a of the ring L has a p-height n, $n \in \mathbb{N}_0$ [**2**] if the equation $x^{p^m} = a$ has a solution for $m = n$ but has no solution for $m = n + 1$. If this equation has a solution for each positive integer m then a is called an element of an infinite p-height in L.

PROPOSITION 4. *Let K be a first kind field with respect to p. If the element $g \in G$ has an infinite p-height in KG then g has an infinite p-height in G.*

PROOF. Assume that g has a finite p-height in G, i.e., $g \in G \backslash G^1$. Let $I = \langle g - 1 | g \in G^1 \rangle$. Then $KG/I \cong K(G/G^1)$ (cf. [**1**, Lemma 1.1]). In view of [**3**, Corollary 1], there exists such an homomorphism θ of the group G/G^1 onto the finite group H that the restriction of θ onto the cyclic subgroup $\langle gG^1 \rangle$ is a monomorphism. Hence there exists such an homomorphism $\bar{\theta}\colon KG/I \to KH$ onto KH that $\bar{\theta}(g + I) \neq 1$. Since $g + I$ has an infinite p-height in KG/I then the non-unit p-element $\bar{\theta}(g + I)$ has an infinite p-height in the algebra KH which is impossible [**5**]. The proposition is proved.

Further we shall consider K being a first kind field with respect to p. If A is a subgroup of the group G then we shall write $A \leq G$. If $A \leq U(KG)$ then $KG = KA$ will mean that A is a group basis of the algebra KG and $KG = AK$ will mean that A is a K-generating set of the algebra KG. Direct product of the groups A and B we shall denote by $A \times B$ and the symbol $\cong$ will denote isomorphism of groups or K-algebras.

LEMMA 5. *Let F_1 be an abelian p-group, $F \leq F_1$ and $F_1^1 = F^1$. If $KF = K(P \times U)$, $P, U \leq S(KF)$, P is a divisible group, $U^1 = 1$ and $\Pi = \Pi(F/F^1)$ then there exist such a subgroup $V \leq S(KF)$ that $KF = K(P \times V)$, $V^1 = 1$, any element $\vartheta \in V$ has a univalent representation of the form*

$$\vartheta = fd, \ f \in \Pi, \ d \in dS(KF) \tag{4}$$

and the mapping $\varphi\colon V \to F/F^1$ defined by $\varphi(\vartheta) = fF^1$ is an isomorphism where f is defined by (4).

Let $KF_1 = K(P \times U_1)$, $V \subseteq U_1 \leq S(KF_1)$, $U_1^1 = 1$ and $\Pi_1 = \Pi(F_1/F_1^1) \supseteq \Pi$. Then can be found such a subgroup V_1 of the group $S(KF_1)$ that $V_1 \supseteq V$, $KF_1 = K(P \times V_1)$, and $\vartheta_1 \in V_1$ has a univalent representation of the form

$$\vartheta_1 = f_1 d_1, f_1 \in \Pi_1, d_1 \in dS(KF_1) \tag{5}$$

and the representation (4) of each $\vartheta \in V$ coincides with its representation of the kind (5) as an element of the group V_1. Besides, the mapping $\varphi_1\colon V_1 \to F_1/F^1$ defined by $\varphi_1(\vartheta_1) = f_1 F^1$ where f_1 is defined by (5) is an isomorphism continuing φ and $V \cap S^1(KF_1) = 1$.

PROOF. The ideal $I = \langle f - 1 | f \in F^1 \rangle$ is determined in an invariant manner in the algebra KF, i.e. independently of the choice of its basis, namely

$$I = \bigcap_{\Gamma \in M} \operatorname{Ker} \Gamma,$$

where M is the set of all irreducible finite-dimensional representations of the algebra KF over K [**3**, Lemma 4]. Therefore $I = \langle t-1 | t \in P\rangle$ in view of $(P \times U)^1 = P$. Thus, in view of Lemma 1, the sets

$$\{f + I | f \in \Pi\} \quad \text{and} \quad \{u + I | u \in U\}$$

are basis of the factor algebra KF/I. Hence for each $f+I$, $f \in \Pi$ there exist a unique element $\vartheta + I$ such that

$$f + I = \vartheta + I, \ \vartheta \in KU. \tag{6}$$

If $\vartheta_+ I = \vartheta_1 + I$, $\vartheta, \vartheta_1 \in KU$, then $\vartheta = \vartheta_1$ in view of Lemma 1. Consequently, if we put $\theta(f+I) = \vartheta$ for any $f \in \Pi$ where ϑ is defined by (6) then one obtains a mapping of the algebra KF/I into KU and this mapping is an injection keeping in mind again Lemma 1. Let

$$V = \{\theta(f + I) | f \in \Pi\}.$$

Then $V \subseteq S(KF)$. Indeed, formula (6) holds for any $\vartheta \in V$ and the order $O(f+I)$ of the element $f+I$ is equal to p^n, i.e. $\vartheta^{p^n} + I = 1 + I$. Therefore, in view of Lemma 1, $\vartheta^{p^n} = 1$, i.e. $V \subseteq S(KF)$. Let $\vartheta, \vartheta_1 \in V$ and

$$\vartheta + I = f + I, \vartheta_1 + I = f_1 + I, \ f f_1 = f_2 f'; \ f, f_1, f_2 \in \Pi, \ f' \in F^1. \tag{7}$$

Therefore

$$\vartheta\vartheta_1 + I = f_2 + I \ \ (f_2 \in \Pi). \tag{8}$$

It follows from (8) that $\theta(f_2 + I) = \vartheta\vartheta_1$ where $f_2 \in \Pi$, i.e. $\vartheta\vartheta_1 \in V$. Therefore $V \leq S(KF)$. From (6) we obtain $f^{-1}\vartheta \equiv 1 \pmod{I}$ whence, in view of Proposition 3, we conclude $f^{-1}\vartheta = d \in dS(KF)$, i.e. the equality (4) holds. For given $\vartheta \in V$ the representative $f \in \Pi$ from formula (6) is uniquely defined. Indeed, if

$$\vartheta + I = f + I, \ \vartheta_1 + I = f_1 + I, \ f, f_1 \in \Pi_1$$

then it follows from Lemma 1 that $f = f_1$. By this is proved the uniqueness of the element f in (4). Therefore, φ, defined by the way indicated in the lemma, is a mapping which, in addition, is a surjective one. If $\vartheta, \vartheta_1 \in V$ $(\vartheta \neq \vartheta_1)$ satisfy (7) then $f \neq f_1$, in view of Lemma 1. Therefore

$$\varphi(\vartheta) = fF^1 \neq f_1 F^1 = \varphi(\vartheta_1).$$

Thus φ is one-to-one mapping. Let the requirement (7) hold for $\vartheta, \vartheta_1 \in V$. Then it follows from (7) and (8) that

$$\varphi(\vartheta\vartheta_1) = f_2 F^1 = fF^1 \cdot f_1 F^1 = \varphi(\vartheta)\varphi(\vartheta_1),$$

i.e. φ is an isomorphism.

Let the representation (4) hold for $\vartheta \in V$. From $\vartheta + I = fd + I = f + I$ where the second equality follows from Proposition 3 it follows in view of Lemma 1 that the set $\{\vartheta + I | \vartheta \in V\}$ is a group basis of the algebra KF/I. Assume that KV is strictly contained in KU. Then there exists $u \in U \backslash KV$.

Therefore $u+I=\sum_i \alpha_i\vartheta_i+I$, $\alpha_i\in K$, $\vartheta_i\in V$. Hence $u-\sum_i\alpha_i\vartheta_i\in KU\cap I$, i.e., in view of Lemma 1 $u=\sum_i\alpha_i\vartheta_i\in KV$ which is a contradiction. Thus $KU=KV$. Since $P\times U$ is a group basis of the algebra KF then $KF=K(P\times V)$. The first part of the lemma is proved.

Now we shall prove the second part of the lemma. Similarly to that proved above, it is seen that if $I_1=\langle f-1|f\in F^1\rangle$ is an ideal of the algebra KF_1 then the sets

$$\{f_1+I|f_1\in\Pi_1\}\text{ and }\{u_1+I_1|u_1\in U_1\}$$

are basis of the algebra KF_1/I_1 and for each element f_1+I_1, $f_1\in\Pi_1$ there exists a unique element ϑ_1+I, such that

$$f_1+I_1=\vartheta_1+I_1,\ \vartheta_1\in KU_1\ \ (f\in\Pi_1). \tag{9}$$

Besides $\theta\colon KF_1/I_1\to KU_1$, defined for each f_1+I_1 ($f_1\in\prod_1$) by $\theta(f_1+I_1)=\vartheta_1$ where ϑ_1 is defined by (9), is an injective mapping, the set

$$V_1=\{\theta(f_1+I_1)|f\in\Pi\}$$

is a subgroup of the group $S(KF_1)$, the representation (5) holds for the element $\vartheta_1\in V_1$ and the mapping φ_1 is an isomorphism. Now we shall prove that $V_1\supseteq V$. Let $\vartheta\in V$. Then there exists such an element $f\in\Pi$ that $f+I=\vartheta+I$. From the definition of the group V_1, it follows that for the class $f+I$ there exists $\vartheta_1\in V_1\subseteq KU_1$, for which $f+I_1=\vartheta_1+I_1$. From the last two equalities we obtain $\vartheta+I_1=\vartheta_1+I_1$. Since $\vartheta\in V\subseteq U_1$ and $\vartheta_1\in V_1\subseteq KU_1$, then $\vartheta-\vartheta_1\in KU_1\cap I_1$. Therefore in view of Lemma 1, $\vartheta=\vartheta_1\in V_1$. Thus $V\subseteq V_1$ and the representation (4) of the element ϑ coincides with its representation of the form (5) in the group V_1.

Now we shall prove that the isomorphism φ_1 continues the isomorphism φ. Let $\vartheta\in V$,

$$\varphi(\theta)=fF^1,\ \varphi_1(\vartheta)=f_1F^1,\ f\in\Pi,\ f_1\in\Pi_1.$$

Then $f+I=\vartheta+I$ and $f_1+I_1=\vartheta+I_1$. These equalities imply $f_1+I_1=\vartheta+I_1$, whence it follows $f_1=f$ in view of Lemma 1. Thus $\varphi_1(\vartheta)=\varphi(\vartheta)$ for any $v\in V$, i.e. φ_1 is a continuation of φ. The lemma is proved.

Further we shall say that the triples (F,P,U) and (F_1,P,U_1) satisfy the condition of Lemma 5, the group V, the representation (4) of its elements and the isomorphism φ are obtained by the triple (F,P,U), the group V_1, the representation (5) of its elements and the isomorphism φ_1 are obtained by (F_1,P,U_1).

LEMMA 6. *Suppose G is an abelian p-group, F is a subgroup of the group G, such that $F^1=G^1$, P and U are subgroups of the group $S(KF)$, the triple (F,P,U) satisfies the conditions of the first part of Lemma 5 and the group V, the representation* (4) *of its element and the isomorphism φ are obtained by (F,P,U), $\Pi=\Pi(F/F^1)$ and $\overline{\Pi}=\Pi(G/F')\supseteq\Pi$. Then there exists a subgroup*

L of the group $S(KG)$, such that $KG = K(P \times L)$, $L \supseteq V$, for each $l \in L$ there exists a uniquely defined element $a \in \overline{\Pi}$ such that

$$(10) \qquad l = ad, \; d \in dS(KG)$$

and the mapping $\bar{\varphi}\colon L \to G/F^1$ defined by $\varphi(l) = aF^1$ where a is defined by (10) *is an isomorphism continuing the isomorphism φ.*

PROOF. Let us completely order the set $G \backslash F$ denoting by g_1 its least element. Put $G_0 = G$, $\Pi_0 = \Pi$, $G_\alpha = \langle G_{\alpha-1}, g_\alpha \rangle$ if α is a non-limit ordinal number and $G_\alpha = \bigcup_{\beta<\alpha} G_\beta$ otherwise. Let τ be the first ordinal for which $G_\tau = G$, $\Pi_\alpha = \Pi(G_\alpha/F^1)$ and $\Pi_\alpha \supseteq \Pi_\beta$ for each β, $\beta < \alpha \leq \tau$. We shall inductively prove that for any ordinal $\alpha \in [0, \tau]$ there exists sequences of subgroups U_α and V_α which satisfy the following six conditions:

1) $U_\alpha \leq S(KG_\alpha)$, $V_\alpha \leq S(KG_\alpha)$ and $KG_\alpha = K(P \times U_\alpha) = K(P \times V_\alpha)$.

2) If β is an ordinal and $\beta < \alpha$ then $V_\beta \subseteq V_\alpha$, $V_\beta \subseteq U_\alpha$ and $U_\alpha^1 = 1$.

3) If α is a non-limit ordinal and $\alpha \leq \tau$ then there exists such an element d_α that

$$U_\alpha = \langle V_{\alpha-1}, \tilde{g}_\alpha \rangle, \; \tilde{g}_\alpha = g_\alpha d_\alpha, \; d_\alpha \in dS(KG_{\alpha-1}).$$

4) For any ordinal $\beta, \beta \leq \alpha$ and $\vartheta \in V_\beta$ there exist uniquely determined $a \in \Pi_\beta$ and $d \in dS(KG_\beta)$ for which $\vartheta = ad$.

5) Let $\varphi_\alpha\colon V_\alpha \to G_\alpha/F^1$ be the following mapping: if $\vartheta \in V_\alpha$ then $\varphi_\alpha(\vartheta) = aF^1$ where a is the corresponding element to ϑ determined in the condition 4). Then φ_α is an isomorphism continuing the isomorphism φ_β for each $\beta < \alpha$.

6) If there exists $\alpha - 1$ then the triples $(G_{\alpha-1}, P, U_{\alpha-1})$ and (G_α, P, U_α) satisfy the condition of Lemma 5, the group $V_{\alpha-1}$, the representation (4) of its element and the isomorphism $\varphi_{\alpha-1}$, are obtained by the first triple and the group V_α, the representation of the kind (10) of its elements and the isomorphism φ_α are obtained by the second one.

For $\alpha = 0$ the group $U_0 = V_0 = V$ and the isomorphism $\varphi_0 = \varphi$ satisfy the conditions 1)–6), the triple (G_0, P, U_0) satisfies the first part of the conditions of Lemma 5 and the group V_0, the representation (4) of its elements and the isomorphism $\varphi_0 = \varphi$ are obtained by this triple.

Let α be any ordinal, $0 < \alpha \leq \tau$. Assume that for each ordinal $\alpha', \alpha' < \alpha$ the groups $U_{\alpha'}$ and $V_{\alpha'}$ (and the isomorphism $\varphi_{\alpha'}$) satisfying the conditions 1)–6) are already constructed. We shall construct the groups U_α and V_α satisfying conditions 1)–6). Consider both cases: I. $\alpha - 1$ does not exist and II. $\alpha - 1$ does exist.

I. Let $\alpha - 1$ does not exist. Put

$$U_\alpha = \bigcup_{\beta<\alpha} U_\beta \text{ and } V_\alpha = \bigcup_{\beta<\alpha} V_\beta.$$

Then

$$KG_\alpha = \left(\bigcup_{\beta<\alpha} G_\beta\right) K = \left(\bigcup_{\beta<\alpha} (P \times U_\beta)\right) K = \left(P \times \bigcup_{\beta<\alpha} U_\beta\right) K = (P \times U_\alpha)K.$$

One trivially proves the linear independence of the elements of the group $P \times U_\alpha$. Therefore $KG_\alpha = K(P \times U_\alpha)$. Similarly $KG_\alpha = K(P \times V_\alpha)$. Thus for the groups U_α and V_α the condition 1) is valid. The remaining conditions 2)–6) are obviously fulfilled for these groups.

II. Let $\alpha - 1$ exist. Let p^{n_α} be the order $O(g_\alpha G_{\alpha-1})$ of the coset $g_\alpha G_{\alpha-1}$ in the factor group $G_\alpha / G_{\alpha-1}$. Then there exist a uniquely determined element $a \in \prod_{\alpha-1}$ and $f_1 \in F^1 \subseteq dS(KF)$ for which:

$$g_\alpha^{p^{n_\alpha}} = af_1 \; (a \in \Pi_{\alpha-1}, f_1 \in dS(KF)).$$

Since, in view of the inductive condition 5), the mapping $\varphi_{\alpha-1}\colon V_{\alpha-1} \to G_{\alpha-1}/F^1$ is an isomorphism then follows from condition 4) the existence of such a uniquely determined element $\vartheta \in V_{\alpha-1}$ that $a = \vartheta d$, $d \in dS(KG_{\alpha-1})$. Hence there exists such an element d_α that

$$(g_\alpha d_\alpha)^{p^{n_\alpha}} = \vartheta, \; d_\alpha \in dS(KG_{\alpha-1}),$$

i.e. if we denote $\tilde{g}_\alpha = g_\alpha d_\alpha$ then we obtain

$$\tilde{g}_\alpha^{p^{n}\alpha} = \vartheta \; (\vartheta \in V_{\alpha-1}). \tag{11}$$

From here, in particular, it follows that

$$p^{n_\alpha} = O(g_\alpha G_{\alpha-1}) \geq O(\tilde{g}_\alpha V_{\alpha-1}) = p^{m_\alpha}.$$

Put

$$U_\alpha = \langle V_{\alpha-1}, \tilde{g}_\alpha \rangle. \tag{12}$$

By this the condition 3) is checked. Obviously $U_\alpha \leq S(KG_\alpha)$ and $U_\alpha \supseteq V_{\alpha-1}$. Now we show that any $\vartheta \in U_\alpha \backslash V_{\alpha-1}$ has a finite height in $S(KG_\alpha)$. From $\vartheta \in U_\alpha \backslash V_{\alpha-1}$ and (12) it follows that there exist such a uniquely determined number i and an element $\vartheta_{\alpha-1}$ that

$$\vartheta = \tilde{g}_\alpha^i \vartheta_{\alpha-1} = g_\alpha^i d_\alpha^i \vartheta_{\alpha-1}, \; 0 < i < p^{n_\alpha}, \; \vartheta_{\alpha-1} \in V_{\alpha-1}$$
$$(d_\alpha \in dS(KG_{\alpha-1})).$$

Besides

$$g_\alpha^i = f_\alpha d', \; f_\alpha \in \Pi_\alpha \backslash \Pi_{\alpha-1}, \; d' \in dS(KF).$$

From the inductive condition 4) follows the existence of such a uniquely determined element $g_{\alpha-1}$ that

$$\vartheta_{\alpha-1} = g_{\alpha-1} d'', \; g_{\alpha-1} \in \Pi_{\alpha-1}, \; d'' \in dS\,(KG_{\alpha-1}).$$

Then it is easy to see, in view of the above obtained representation of f, g_α^i and $\vartheta_{\alpha-1}$, that there exist a uniquely determined element a_α for which

$$\vartheta = a_\alpha d, a_\alpha \in \Pi_\alpha \backslash \Pi_{\alpha-1}, \; d \in dS(KG_{\alpha-1}).$$

As far as a_α has a finite height in G_α then, in view of Proposition 4, a_α has a finite p-height in KG_α too. Since the p-height of the element d in KG_α is infinite then the last representation of the element ϑ implies that ϑ has a finite height in $S(KG_\alpha)$. Now suppose that $\vartheta \in V_{\alpha-1}$ and $\vartheta \neq 1$. Then in the univalent representation (4) of ϑ we have $f \notin F^1$ since, in view of the inductive condition 5) the mapping $\varphi_{\alpha-1}$ for which $\varphi_{\alpha-1}(\vartheta) = fF^1$ is an isomorphism. As far as $f \notin F^1 = G^1_\alpha$ then the height of f in G_α is finite. Therefore the height of f in KG_α is finite. Then the height of ϑ in KG_α is finite too. Thus $U_\alpha \cap S^1(KG_\alpha) = 1$ which implies $U^1_\alpha = 1$. By this is checked a part of the condition 2).

From (12) and (11), in view of the inductive condition $P \cap V_{\alpha-1} = 1$ it follows from $P \cap U_\alpha = 1$. Using (12) it is easy to prove that $KG_\alpha = K(P \times U_\alpha)$.

Now we can apply Lemma 5 for the triples $(G_{\alpha-1}, P, U_{\alpha-1})$ and (G_α, P, U_α). From this lemma it follows that there exists a subgroup V_α of the group $S(KG_\alpha)$ and the remaining parts of the inductive properties 1), 2), 4), 5) and 6) hold. This completes the Induction. Putting

$$L = \bigcup_{\alpha \leq \tau} V_\alpha$$

and $\bar{\varphi} = \varphi_\tau$ we verify that the conclusion of the lemma are fulfilled.

LEMMA 7. *Let G be a reduced abelian p-group and $|G^1| = \aleph_0$. Then there exists such a countable subgroup F of the group G that $F^1 = G^1$, $KF = K(P \times U)$, where $P, U \leq S(KF)$, P is a divisible group which contains at least two quasicyclic groups $Z(p^\infty)$ in its direct decomposition into irreducible factors and $U^1 = 1$.*

PROOF. Equations

$$x_{n,g}^{p^n} = g,$$

where g and n vary in G^1 and $\mathbb{N}$ correspondingly have solutions $f_{n,g} \in G$. Let

$$F = \langle f_{n,g} | n \in \mathbb{N}, g \in G^1 \rangle.$$

Then F is a countable p-subgroup of G, $F^1 = G^1$ and the order of the elements of F/F^1 are unbounded as a whole. Therefore $|F^1| = \aleph_0$. Then from [**1**, Th. 2.2, point 9] follows the existence of the subgroups P and U of the group $S(KF)$ which satisfy the conclusion of the lemma.

In the following assertions $\prod_\alpha Z(p^\infty)$ denotes a (bounded) direct product of α groups $Z(p^\infty)$.

THEOREM 8. *Let G and H be reduced abelian p-groups, the subgroup G^1 countable, and K a first kind field with respect to p. Then*

(i) *the K-isomorphism $KH \cong KG$ holds if and only if*

(ii) *the subgroup H^1 is countable and $K(H/H^1) \cong K(G/G^1)$.*

PROOF. (i) implies (ii). Indeed we may regard H as a second group basis of the algebra KG, i.e. $KG = KH$. Let $I = \langle g - 1 | g \in G^1 \rangle$ be an ideal of the algebra KG. Since I is invariantly determined in $KG = KH$ then $I = \langle h - 1 | h \in H^1 \rangle$. Therefore

$$K(G/G^1) \cong KG/I = KH/I \cong K(H/H^1),$$

i.e. $K(G/G^1) \cong K(H/H^1)$. Since, in view of (4), the cardinality of the minimal system of generators of the ideal I is equal to $|G^1|$ on one hand and to $|H^1|$ on the other hand, and I is determined in an invariant manner in KG, then $|G^1| = |H^1|$ i.e. $|H^1| = \aleph_0$.

(ii) implies (i). Indeed since G is a reduced abelian p-group and $|G^1| = \aleph_0$, then in view of Lemma 7 there exists such a countable subgroup F of the group G that $F^1 = G^1$, $KF = K(P \times U)$, $P, U \leq S(KF)$, P is a divisible group containing at least two groups $Z(p^\infty)$ in its direct decomposition of irreducible factors and $U^1 = 1$. Then, in view of Lemma 5, there exist such a subgroup V of the group $S(KF)$ that $KF = K(P \times V)$, $V^1 = 1$, each $\vartheta \in V$ has a univalent representation of the form (4) and the mapping $\varphi\colon V \to F/F^1$ defined by $\varphi(\vartheta) = fF^1$ where f is determined by (4) is an isomorphism. Therefore applying Lemma 6 we obtain such a subgroup L of the group $S(KG)$ that

$$KG = K(P \times L), P \cong \prod_\alpha Z(p^\infty),\ \alpha \geq 2,\ L^1 = 1, \tag{13}$$

where the last equality follows from $L \cong G/F^1$ and $G^1 = F^1$. For the algebra KH we similarly obtain the existence of such subgroups P_1 and L_1 of the group $S(KH)$ that

$$KH = K(P_1 \times L_1), P_1 \cong \prod_{\alpha_1} Z(p^\infty), \alpha_1 \geq 2, L_1^1 = 1. \tag{14}$$

Now we shall prove that $KP \cong KP_1$. Indeed, since the cardinality $|G^1|$ is determined in an invariant manner in the algebra $KG = K(P \times L)$ **[4]** then

$$\aleph_0 = |G^1| = |(P \times L)^1| = |P|.$$

We similarly obtain that $\aleph_0 = |P_1|$. Since the groups P and P_1 are countable then in view of (13) and (14) it follows from **[1]** that $KP \cong KP_1$.

Now we shall prove that $KL \cong KL_1$. Let

$$I = \langle g - 1 | g \in G^1 \rangle \text{ and } I_1 = \langle h - 1 | h \in H^1 \rangle$$

be ideals of the algebras KG and KH correspondingly. Using (13) we obtain

$$K(G/G^1) \cong KG/I = K(P \times L)/I \cong K((P \times L)/P) \cong KL,$$

i.e. $K(G/G^1) \cong KL$. Analogously it follows from (14) that $K(H/H^1) \cong KL_1$. Since $K(G/G^1) \cong K(H/H^1)$ then $KL \cong KL_1$.

From $KP \cong KP_1$ and $KL \cong KL_1$ it follows

$$KG \cong KP \otimes_K KL \cong KP_1 \otimes_K KL_1 \cong KH,$$

i.e. $KG \cong KH$. The theorem is proved.

It is seen from the proof of the second part of the Theorem 8 that the following lemma holds.

LEMMA 9. *Let G be an uncountable abelian p-group which satisfies the following condition:*

(a) *$|G^1| = \aleph_0$ and there exists such a countable subgroup F of G that $F^1 = G^1$ and besides*

$$KF = K(P \times U), P \cong \prod_{\alpha} Z(p^\infty), \alpha \geq 2, U^1 = 1, U \leq S(KF). \tag{15}$$

If H is an abelian p-group satisfying a condition of the kind (a) *and $K(H/H^1) \cong K(G/G^1)$ then $KG \cong KH$.*

LEMMA 10. *Let G be an uncountable abelian p-group, $|G^1| = \aleph_0$ and there exist such a subgroup F of G that $F^1 = G^1$ and $|F/F^1| = \aleph_0$. Let H be an abelian p-group, $|H^1| = \aleph_0$ and there exist such a subgroup F_1 of H that $F_1^1 = H^1$ and $|F_1/F_1^1| = \aleph_0$. If $K(H/H^1) \cong K(G/G^1)$ then $KH \cong KG$.*

PROOF. Obviously $|F^1| = \aleph_0$. We shall prove that formulas (15) hold for F. Indeed, if the orders of the elements of the group F/F^1 are unbounded as a whole then it follows from [**1**, Th. 2.2., point 9] that (15) hold. Otherwise F is not a reduced group and from [**1**, Th. 2.2., point 7] it follows that (15) are fulfilled again. Since, in view of the symmetry, for the subgroup F_1 of the group H the formulas of the kind (15) are also fulfilled then Lemma 9 immediately implies Lemma 10.

THEOREM 11. *Let G be such an uncountable abelian p-group that its subgroup G^1 is countable, H an abelian p-group and K a first kind field with respect to p. Then*

(i) *the K-isomorphism $KH \cong KG$ holds if and only if*

(ii) *the subgroup H^1 is countable and $K(H/H^1) \cong K(G/G^1)$.*

PROOF. (i) implies (ii) is proved as in Theorem 8.

(ii) implies (i). Indeed we shall prove that it is possible to chosen such subgroups F and F_1 of the groups G and H correspondingly which satisfy the conditions of Lemma 10. Consider at first the group G. Let $G = P \times R$ where P is a divisible group and R is a reduced subgroup of G. Then $G^1 = P \times R^1$. Consider both cases 1) $R^1 \neq 1$ and 2) $R^1 = 1$.

1) Let $R^1 \neq 1$. Then exists such elements $g_{n,a} \in R$ that $g_{n,a}^{p^n} = a$, $a \in R^1$, $n \in \mathbb{N}$. Let

$$T = \langle g_{n,a} | n \in \mathbb{N}, a \in R^1 \rangle.$$

Then $T \leq R$, $|T| = \aleph_0$, $T^1 = R^1$ and the order of the elements of the group T/T^1 are unbounded as a whole. Therefore $|T/T^1| = \aleph_0$ and $F = P \times T$ is the desired subgroup of the group G.

2) Let $R^1 = 1$. Then $G^1 = P$, $|P| = \aleph_0$ and R is an uncountable abelian p-group. Let T be an arbitrary countable subgroup of the group R. Then $F = P \times T$ is the desired subgroup of G since $|F| = \aleph_0$, $F^1 = P = G^1$ and $|F/F^1| = |T/T^1| = \aleph_0$. In view of $|H| = |G|$ and the symmetry of the conditions for the groups G and H we can choose such a subgroup F_1 of the

group H that for G and H and for their subgroups F and F_1 correspondingly the conditions of Lemma 10 are fulfilled. Therefore $KG \cong KH$. The theorem is proved.

Note that in view of [**1**, Th. 2.2] the last result is not valid if G is a countable abelian p-group which is not a reduced group.

References

[**1**] С. Д. Берман, Групповые алгебры счетных абелевых p-групп, Publ. Math. **14**, 1967, 365–405.

[**2**] С. Д. Берман, Т. Ж. Моллов, О групповых кольцах абелевых p-групп любой мощности, Матем. заметки, **6** No. 4, 1969, 381–392.

[**3**] С. Д. Берман, Т. Ж. Моллов, Расщепляемость базисов и изоморфизм групповых колец абелевых групп, Плиска, **2** (1981), 6–22.

[**4**] С. Д. Берман, Т. Ж. Моллов, Изоморфизм полупростых групповых алгебр абелевых групп, Плиска, **8** (1986), 107–121.

[**5**] Т. Ж. Моллов, Силовские p-подгруппы групп нормированных единиц полупростых групповых алгебр несчетных абелевых p-групп, Плиска, **8** (1986), 34–46.

[**6**] D. S. Passman, *Algebraic structure of group rings*, Interscience, New York, 1977.

Prof. Todor Zhelyazkov Mollov
Dept. Algebra
Plovdiv University "P. Hilendarski"
4000 Plovdiv, Bulgaria

Contemporary Mathematics
Volume **93** (1989)

On the Canonical Irreducible Representations of Countable Locally Finite Groups and Infinite Frobenius Group

S. I. OSTROVSKAYA

The problems considered in this paper were posed by Professor S. D. Berman. He introduced the author to this area of research and helped her greatly in her work.

1°. Let G be a countable locally finite group. Then G is a union of increasing sequence of finite groups $G = \bigcup_{i=1}^{\infty} G_i$, $G_i \subseteq G_{i+1}$, $|G_i| < \infty$, $i = 1, 2, \dots$.

Let K be the field of complex numbers, KG and KG_i be group rings of G and G_i respectively over K.

Let M be an irreducible KG-module. Then $M \cong KG/V$ where V is some maximal ideal of KG. The irreducible representation of G corresponding to KG-module M will be called canonical if $V_i = V \cap KG_i$ is a maximal ideal of KG_i (i.e. V is a union of increasing sequence of maximal ideals of KG_i, $i = 1, 2, \dots$).

A method to construct the canonical representations of locally finite groups using the minimal idempotents of group rings KG_i is described below.

Consider the two-sided KG-module consisting of all formal linear combinations of elements of G: $\Gamma(K, G) = \{\sum_{g \in G} \alpha_g g | \alpha_g \in K\}$. (Operations of summing and multiplication by elements of KG can be defined naturally.)

If $x \in \Gamma(K, G)$, $x = \sum_{g \in G} \alpha_g g$ then for any subgroup $H \leq G$ let $x(H) = \sum_{g \in H} \alpha_g g$.

Call the sequence of idempotents $\{e_i\}_{i=1}^{\infty}$, $e_i \in KG_i$ special if the following conditions are satisfied for all $i = 1, 2, \dots$.

1) e_i is a minimal idempotent of KG_i;

2) $e_i e_{i+1} = e_{i+1} e_{i+1}$.

To obtain some properties of special sequences we need the following lemma:

Lemma 1. *Let H be a finite group, $F \leq H$, $e \neq 0$ and e_1 be minimal idempotents of KF and KH respectively, and let $ee_1 = e_1 e = e_1$. Then $e_1(F) = \lambda e$ where $0 \neq \lambda \in K$.*

Proof. Let $e_1 = \sum_{h \in H} \lambda_h h$. Then ([**1**]) $\lambda_1 = \frac{n}{|H|}$, where $n = \dim_K(KHe_1)$. Write e_1 as $e_1 = u + t$ where $u = e_1(F)$, $t = \sum_{h \in H \setminus F} \lambda_h h$. From assumption of lemma we have $e_1 = ee_1 e = eue + ete$. Since e is a minimal idempotent of KF then $eue = \mu e$, $\mu \in K$. Element ete is a linear combination of only such elements of H which do not belong to F. Then $e_1(F) = \mu e$. Comparing in the last equation the coefficients at unit at right and left sides one obtains that $\mu \neq 0$.

Lemma is proved.

The following statement points out the relation between the special idempotent sequences and irreducible representations.

Lemma 2. *Every special idempotent sequence $\{e_i\}_{i=1}^{\infty}$ of KG defines some irreducible K-representation of G.*

Proof. According to Lemma 1 $e_{i+1}(G_i) = \lambda_i e_i$, $0 \neq \lambda_i \in K$ for all $i = 1, 2, \ldots$. So the nonzero numbers $\{\mu_i\}_{i=1}^{\infty}$, $\mu_i \in K$ such that for elements $u_i = \mu_i e_i$ the conditions $u_{1 \neq i}(G_i) = u_i$ are satisfied exist.

Thus there exists the element $u \in \Gamma(K, G)$ such that $u(G_i) = u_i$ for all $i = 1, 2, \ldots$. We shall call u the "long" element of special sequence $\{e_i\}_{i=1}^{\infty}$.

Show that KGu is irreducible submodule of $\Gamma(K, G)$.

Really $e_i u_i = e_i \mu_i e_i = \mu_i e_i = \mu_i$ $(i = 1, 2, \ldots)$. Hence $e_i u = u$ because $e_i u(G_j) = u(G_j)$ for all $j \geq i$.

Let now x KG and $xu \neq 0$. Then there exists such index i that $x \in KG_i$ and $xu = xe_i u$. Since e_i is a minimal idempotent of KG_i and $xe_i \neq 0$ then $yxe_i = e_i$ for some $y \in KG_i$. But then $yxu = yxe_i u = e_i u = u$ what yields the irreducibility of KGu.

Lemma is proved.

Theorem 1. *Let $M \neq 0$ be an irreducible KG-module, $\dim_K M = n < \infty$. Then such special idempotent sequence $\{e_i\}_{i=1}^{\infty}$ with "long" element u that $M \cong KGu$ exists.*

Proof. Since M is irreducible KG-module then $M = (x)$ where $0 \neq x \in M$.

Let $1 = u_1 + \cdots + u_r$ be a decomposition of unit into the sum of minimal pairwise orthogonal idempotents of KG_1. Since $x \neq 0$ then $u_j x \neq 0$ for some $1 \leq j \leq r$. Let $e_1 = u_j$. Decompose e_1 into the sum of minimal pairwise orthogonal idempotents of KG_2: $e_1 = v_1 + \cdots + v_s$. Since $e_1 x \neq 0$ then $v_k x \neq 0$ for some $1 \leq k \leq s$. Let $v_2 = v_k$. Continuing this process we shall obtain an idempotent sequence $\{e_i\}_{i=1}^{\infty}$ which is special by the construction. Let u be its "long" element. Show that $M \cong KGu$.

Consider the sequence of nonzero elements $\{e_i x\}_{i=1}^{\infty}$ of module M. Since $\dim_K M < \infty$ then there exists number $m < \infty$ such that all $e_i x$ are linear combinations of the elements $e_1x, \dots, e_mx$: $e_ix = z_1^{(i)}e_1x + \dots + z_m^{(i)}e_mx$, $i = 1, 2, \dots$ (z_i are complex). For $i \geq m$ we obtain: $e_ix = e_ie_mx = e_m(z_1^{(i)}e_1x + \dots + z_m^{(i)}e_mx) = (z_1^{(i)} + \dots + z_m^{(i)})e_mx = a_ie_mx$, $a_i \neq 0$.

Since $e_mx \neq 0$ then $M = (e_mx)$.

To prove the isomorphism of M and KGu it is enough to show that $\mathrm{Ann}_{KG}(u) = \mathrm{Ann}_{KG}(e_mx)$. Let $y \in \mathrm{Ann}_{KG}(u)$, $y \in KG_n$. Then $ye_n = 0$. Show that $ye_mx = 0$. If $n \leq m$ then $ye_mx = ye_ne_mx = 0$. If $n > m$ then $e_nx = a_ne_mx$ ($a_n \neq 0$) or $e_mx = \frac{1}{a_m}e_nx$. But then $ye_mx = 0$.

Therefore we obtain $\mathrm{Ann}_{KG}(u) \subseteq \mathrm{Ann}_{KG}(e_mx)$. Since $\mathrm{Ann}_{KG}(u)$ and $\mathrm{Ann}_{KG}(e_mx)$ are maximal ideals of KG then the last formula means their coincidence. This implies the required isomorphism of the modules.

Theorem is proved.

THEOREM 2. *In order the irreducible representation of G with module M to be canonical it is necessary and sufficient that $M \cong KGu$, where u is a "long" element of some special idempotent sequence of KG.*

PROOF. 1) Sufficiency. Let $\{e_i\}_{i=1}^{\infty}$ be a special idempotent sequence with "long" element u. Then KGu is an irreducible KG-module and $KGu \cong KG/V$ where $V = \mathrm{Ann}_{KG}(u)$ is a maximal ideal of KG.

Let $V_i = KG_i(1 - e_i)$, $i = 1, 2, \dots$. Then V_i is a maximal ideal of KG_i and $V_i \subseteq V_{i+1}$ ($i = 1, 2, \dots$). Hence $\bigcup_{i=1}^{\infty} V_i$ is a maximal ideal of KG. Show that $\bigcup_{i=1}^{\infty} V_i = v$. Let $x \in V_i$ for some i. Then $xe_i = 0$ from which follows $xu = xe_iu = 0$ i.e. $x \in \mathrm{Ann}_{KG}(u)$.

Therefore $\bigcup_{i=1}^{\infty} V_i \subseteq V$ and the maximality of both ideals implies their coincidence.

2) Necessity. Let $V \subset KG$ be a maximal ideal, $V = \bigcup_{i=1}^{\infty} V_i$, where V_i is a maximal ideal of KG_i. Show that a special idempotent sequence $\{e_i\}_{i=1}^{\infty}$ show that $\mathrm{Ann}_{KG_i}(e_i) = V_i$, $1 = 1, 2, \dots$ exists.

It is enough to show that if $V_i \subset V_{i \neq 1}$ are maximal ideals, e_i is a minimal idempotent of KG_i and $\mathrm{Ann}_{KG_i}(e_i) = V_i$ then minimal idempotent e_{i+1} of KG_{i+1} such that $e_ie_{i+1} = e_{i+1}e_i = e_{i+1}$ and $\mathrm{Ann}_{KG_{i+1}}(e_{i+1}) = V_{i+1}$ exists.

Decompose the ideal $KG_{i+1}e_{i+1}$ into direct sum of minimal left ideals of KG_{i+1}: $KG_{i+1}e_i = L_1 \oplus \dots \oplus L_r$. For every L_j ($1 \leq j \leq r$) one have either $L_j \subseteq V_{i+1}$ or $L_j \cap V_{i+1} = \{0\}$. If all $L_j \subseteq V_{i+1}$ then $KG_{i+1}e_{i+1} \subseteq V_{i+1}$ whence a contradiction $KG_{i+1} \subseteq V_{i+1}$ follows. Hence $L_j \cap V_{i+1} = \{0\}$ for some j ($1 \leq j \leq r$). Then $KG_{i+1} = L_j \oplus V_{i+1}$. Let $1 = e_{i+1} + f_{i+1}$ ($e_{i+1} \in L_j$, $f_{i+1} \in V_{i+1}$). As $L_j \subseteq KG_{i+1}e_i$, then $e_{i+1}e_i = e_{i+1}$. On the other hand since $V_i \subseteq V_{i+1}$ then $(1 - e_i)(1 - e_{i+1}) = (1 - e_i)$, i.e. $e_ie_{i+1} = e_{i+1}$. From the construction it is clear that $V_{i+1} = \mathrm{Ann}_{KG_{i+1}}(e_{i+1})$.

Theorem is proved.

Therefore every canonical representation of countable locally finite group may be realised by some special idempotent sequence. Thus we have two equivalent methods to construct the canonical irreducible representations of countable locally finite groups.

2°. In this section the description of all canonical representations of infinite Frobenius group is given.

By an infinite Frobenius group we shall mean the group of the following kind: $G = \bigcup_{i=1}^{\infty} G_i$, $G_i \subseteq G_{i+1}$, $G_i = \Phi_{p^{a_i}} = \left\{\left(\begin{smallmatrix} a & \beta \\ 0 & 1 \end{smallmatrix}\right) \mid \alpha, \beta \in GF(p^a), \alpha \neq 0\right\}$ (to satisfy the condition $G_i \subset G_{i+1}$ it is necessary and sufficient that $a_i | a_{i+1}$, $i = 1, 2, \dots$). As known (see e.g. [2]) the group G_i is a semidirect product of normal subgroup

$$N_i = \left\{\begin{pmatrix} 1 & \beta \\ 0 & 1 \end{pmatrix} \mid \beta \in GF(p^{a_i})\right\}$$

by subgroup

$$H_i = \left\{\begin{pmatrix} \alpha & 0 \\ 0 & 1 \end{pmatrix} \mid \alpha \in GF(p^{\alpha_i}),\ \alpha \neq 0\right\}$$

where N_i is an elementary abelian group and H_i is a cyclic one.

Let F be a field which is a union of fields $GF(p^{a_i})$: $F = \bigcup_{i=1}^{\infty} GF(p^{a_i})$. Then

$$G = \left\{\begin{pmatrix} \alpha & \beta \\ 0 & 1 \end{pmatrix} \mid \alpha, \beta \in F,\ \alpha \neq 0\right\},$$

and $G = NH$, where $N = \bigcup_{i=1}^{\infty} N_i = \left\{\left(\begin{smallmatrix} 1 & \beta \\ 0 & 1 \end{smallmatrix}\right) \mid \beta \in F\right\}$; $H = \bigcup_{i=1}^{\infty} \left\{\left(\begin{smallmatrix} \alpha & 0 \\ 0 & 1 \end{smallmatrix}\right) \mid \alpha \in F,\ \alpha \neq 0\right\}$.

LEMMA 3. *Let G be a finite Frobenius group, $|G| = p^a(p^a - 1)$ and let A be a prime subalgebra of KG ($\dim_K A \neq 1$), $e = \frac{1}{|N|}\sum_{n\in N} \zeta(n)n$ (ζ is a linear nonprincipal character of N).*

Then any minimal idempotent $f \in A$ has a form $f = \rho e \tau$, where $\rho = \sum_{h\in H} \lambda_h h$, $\tau = \sum_{h\in H} \mu_h h$ and $\sum_{h\in H} \lambda_h \mu_{h^{-1}} = 1$.

PROOF. Let $f \in A$ be a minimal idempotent. As $KGe \cong KGf$ then $KGe(xf) = KGf$ for some $x \in KG$. Hence $f = yexf$ ($y \in KG$). But $ye = \rho e$ ($\rho \in KH$), $ext = e\tau$ ($\tau \in KH$), whence $f = \rho e \tau$.

Let $\rho = \sum_{h\in H} \lambda_h h$, $\tau = \sum_{h\in H} \mu_h h$. Consider $f^2 = \rho e \tau \cdot \rho e \tau = f$. The product $\rho\tau = \sum_{h,f\in H} \lambda_h \mu_f hf = \sum_{f\in H}(\sum_{h\in H} \lambda_h \mu_{fh^{-1}}) f = \sum_{f\in H} \nu_f f$; $e\tau\rho e = e\sum_{f\in H} \nu_f fe = \nu_1 e$.

Consequently $f^2 = \nu_1 f$ i.e. f is an idempotent if and only if $\nu_1 = \sum_{h\in H} \lambda_h \mu_{h^{-1}} = 1$.

Lemma is proved.

REMARK. Lemma 3 can be extended on any finite supersolvable group.

Let G be infinite Frobenius group covered by finite Frobenius groups G_i, $G_i \subset G_{i+1}$, $i = 1, 2, \dots$. Let χ be a linear character of N, $\chi_{N_i} = \zeta_i \neq 1_{N_i}$ for all $i = 1, 2, \dots$. Let $e_i = \frac{1}{|N_i|}\sum_{n\in N_i} \zeta_i(n)n$. Note $N_{H_i}(\zeta_{i-1}) = \{h \in H_i | \zeta_i(h_n h^{-1}) = \zeta_i(n), \forall n \in N_{i-1}\}$. Name $N_{H_i}(\zeta_{i-1})$ the "generalized normalizer" of linear character ζ_{i-1} in H_i. It is easy to show that "generalized normalizer" has the following properties:

1) $|N_{H_i}(\zeta_{i-1})| = (H_i : H_{i-1})$;

2) $h \in N_{H_i}(\zeta_{i-1}) \Leftrightarrow h^{-1} \in N_{H_i}(\zeta_{i-1})$;

3) $N_{H_i}(\zeta_{i-1}) \cap H_{i-1} = \{1\}$.

LEMMA 4. *The idempotent sequence* $\{f_i\}_{i=1}^{\infty}$ *of* KG *(G is infinite Frobenius group) is special if and only if* f_i *has the form:* $f_i = \rho_1 \dots \rho_i e_i \tau_1 \dots \tau_i$, *where* ρ_k, τ_k *are the linear combinations of elements of* $N_{H_k}(\zeta_{k-1})$, *besides if* $\rho_k = \sum_{h \in H_k} \lambda_h h$, $\tau_k = \sum_{h \in H_k} \mu_h h$ *then* $\sum_{h \in H_k} \lambda_h \mu_{h^{-1}} = 1$.

PROOF. Let us prove the statement by the induction with respect to i. Let $f_k = \rho_1 \dots \rho_k e_k \tau_1 \dots \tau_k$ for $k \leq i$ where ρ_j, τ_j are of the form stated above. By Lemma 3 $f_{i+1} = \rho e_{i+1} \tau$ $(\rho, \tau \in KH_{i+1})$, and $f_{i+1} = f_i f_{i+1} = \rho_1 \dots \rho_i e_i \tau_1 \dots \tau_i \rho e_{i+1} \tau$. Let $\tau_1 \dots \tau_i \rho = \sum_{h \in H_{1+i}} \alpha_h h$. Then $e_i \tau_1 \dots \tau_i \rho e_{i+1} = e_i (\sum_{h \in H_{i+1}} \alpha_h h) e_{i+1} = (\sum_{h \in N_{H_{i+1}}(\zeta i)} \alpha_h h) e_i e_{i+1} = (\sum_{h \in N_{H_{i+1}}(\zeta_i)} \alpha_h h) e_{i+1}$. Denoting $\sum_{h \in N_{H_{i+1}}(\zeta_i)} \alpha_h h = \rho_{i+1}$ one obtains $f_{i+1} = \rho_1 \dots \rho_{i+1} e_{i+1} \tau$. From $f_{i+1} f_i = f_{i+1}$ one obtains $f_{i+1} = \rho_1 \dots \rho_{i+1} e_{i+1} \tau_1 \dots \tau_{i+1}$, where ρ_k, τ_k are linear combinations of elements of $N_{H_k}(\zeta_{k-1})$, $k = 1, 2, \dots i+1$.

The condition for the coefficients can be easily verified directly.

Lemma is proved.

Thus for fixed linear character χ any special idempotent sequence $\{f_i\}_{i=1}^{\infty}$ can be uniquely represented in the form: $f_i = \rho_1 \dots \rho_i e_i \tau_1 \dots \tau_i$.

If $\{e_i\}_{i=1}^{\infty}$ is a special idempotent sequence of the form $e_i = \frac{1}{|N_i|} \sum_{n \in N_i} \chi(n) n$ (χ is a linear character of N) then call the sequence $\{\rho_i\}_{i=1}^{\infty}$, $\rho_i \in KH_i$ the adjoint sequence for $\{e_i\}_{i=1}^{\infty}$ if $\rho_i = 1 + \lambda_1 h_1 + \dots + \lambda_r h_r$, where $h_1, \dots h_r \in N_{H_i}(\zeta_{i-1})$, $\lambda_1, \dots \lambda_r \in K$, $i = 1, 2, \dots$.

LEMMA 5. *Let* $e_i = \frac{1}{|N_i|} \sum_{n \in N_i} \chi(n) n$, *i.e.* $\{e_i\}_{i=1}^{\infty}$ *is a special idempotent sequence. The sequence of the form* $\{t_i e_i\}_{i=1}^{\infty}$ *is special if and only if* $t_i e_i = \rho_1 \dots \rho_i e_i$, *where* $\{\rho_i\}_{i=1}^{\infty}$ *is an adjoint sequence for* $\{e_i\}_{i=1}^{\infty}$.

PROOF. The statement follows directly from Lemma 4.

LEMMA 6. *Let* $\{f_i\}_{i=1}$ *be a special idempotent sequence of infinite Frobenius group algebra* KG. *Then there exists linear character* $\tilde{\chi}$ *of* N *such that* $f_i = \rho_1 \dots \rho_i \varphi_i \tau_1 \dots \tau_i$, *where* $\varphi_i = \frac{1}{|N_i|} \sum_{n \in N_i} \tilde{\chi}(n) n$ *and* $\{\rho_i\}_{i=1}^{\infty}$ *is an adjoint sequence for the sequence* $\{\varphi_i\}_{i=1}^{\infty}$.

PROOF. Let χ be a linear character of N, $\chi_{N_i} \neq 1_{N_i}$ for all $i = 1, 2, \dots$ and $e_i = \frac{1}{|N_i|} \sum_{n \in N_i} \chi(n) n$. By Lemma 4 $f_i = \rho'_1 \dots \rho'_i e_i \tau'_i \dots \tau'_i$, where ρ'_k, τ'_k are linear combinations of the elements of $N_{H_k}(\zeta_{k-1})$. Let $\rho'_k = \sum_{h \in N_{H_k}(\xi_{k-1})} \alpha_h h$.

Because $\rho_k \neq 0$ there is $h = \xi_k$ such that $\alpha_h = \alpha_{\xi_k} \neq 0$. Then $\rho'_k = \alpha_{\xi_k} \xi_k \rho_k$, where $\rho_k = \frac{1}{\alpha_{\xi_k}} \xi_k^{-1} \rho'_k$. Let $\varphi_k = \xi_1 \dots \xi_k e_k \xi_k^{-1} \dots \xi_1^{-1} = e_k^{(\xi_1 \dots \xi_k)^{-1}}$. Show that $\{\varphi_k\}_{k=1}^{\infty}$ is special idempotent sequence and $\{\rho_k\}_{k=1}^{\infty}$ is its adjoint sequence. Obviously all φ_k are minimal idempotents of KG_k algebras. We have $\varphi_k \varphi_{k+1} = \xi_1 \dots \xi_k e_k \xi_k^{-1} \dots \xi_1^{-1} \cdot \xi_1 \dots \xi_{k+1} e_{k+1} \xi_{k+1}^{-1} \dots \xi_1^{-1} = \xi_1 \dots \xi_k e_k \dots \xi_{k+1} e_{k+1} \xi_1^{-1} \dots \xi_{k+1}^{-1} = \xi_1 \dots \xi_{k+1} e_{k+1} \xi_{-1}^{-1} \dots \xi_{k+1}^{-1} = \varphi_{k+1}$ (we have used the fact that $e_k \xi_{k+1} e_{k+1} = \xi_{k+1} e_{k+1}$). By the same way we have $\varphi_{k+1} \varphi_k = \varphi_{k+1}$, i.e. the idempotent sequence $\{\varphi_k\}_{k=1}^{\infty}$ is special.

Let us show that $\{\rho_k\}_{k=1}^{\infty}$ is adjoint sequence for $\{\varphi_k\}_{k=1}^{\infty} \cdot \rho_k$ is a linear combination of the elements of the form $\xi_k^{-1}h$, where $h \in N_{H_k}(\zeta_{k-1})$. It is obvious that the coefficient at unit of ρ_k is 1. Let $\varphi_k = \frac{1}{|N_k|}\sum_{n\in N_k}\omega_k(n)n$, and let $\tilde{\chi}$ be a linear character of N, $\tilde{\chi}_{N_k} = \omega_k$. Since $\varphi_k = e_k^{(\xi_1\dots\xi_k)^{-1}}$ then $\omega_k(n) = \zeta_k(n^{\xi_1\dots\xi_k})$, $n \in N_k$. Thus

$$\begin{aligned}\omega_k(\xi_k h^{-1} n h \xi_k^{-1}) &= \zeta_k(\xi_1^{-1}\dots\xi_k^{-1}\xi_k h^{-1} n h \xi_1 \dots \xi_k \xi_k^{-1})\\ &= \zeta_k(h^{-1}\xi_1^{-1}\dots\xi_{k-1}^{-1} n \xi_1 \dots \xi_{k-1} h)\\ &= \zeta_k(\xi_1^{-1}\dots\xi_{k-1}^{-1} n \xi_1 \dots \xi_{k-1}) = \omega_k(n), \quad \forall n \in N_{k-1}.\end{aligned}$$

Thus $\xi_k^{-1}h \in N_{H_k}(\omega_{k-1})$ which implies that $\{\rho_k\}_{k=1}^{\infty}$ is an adjoint sequence for $\{\varphi_k\}_{k=1}^{\infty}$. Therefore

$$\begin{aligned}f_i &= \alpha_{\xi_1}\dots\alpha_{\xi_i}\xi_1\rho_1\dots\xi_i\rho i e_i \xi_1^{-1}(\xi_1\tau_1)\dots\xi_i^{-1}(\xi_i\tau_i)\\ &= \alpha_{\xi_1}\dots\alpha_{\xi_i}\rho_1\dots\rho_i(\xi_1\dots\xi_i e_i \xi_1^{-1}\dots\xi_i^{-1})\xi_1\tau_1\dots\xi_i\tau_i\\ &= \rho_1\dots\rho_i\varphi_i(\alpha_{\xi_1}\xi_1\tau_1)\dots(\alpha_{\xi_i}\xi_i\tau_i)\\ &= \rho_1\dots\rho_i\varphi_i\tau_1\dots\tau_i.\end{aligned}$$

Lemma is proved.

LEMMA 7. *Let $\{e_i\}_{i=1}^{\infty}$ and $\{e_i t_i\}_{i=1}^{\infty} b \cdot (e_i, t_i \in KG_i)$ be special sequences with "long" elements u and v resp[ectively. Then $KGu \cong KGv$.*

PROOF. Since $\{e_i\}_{i=1}^{\infty}$ and $\{e_i t_i\}_{i=1}^{\infty}$ are special sequences then modules KGu and KGv are irreducible and $KGu \cong KG/V_1$, $KGv \cong KG/V_2$ where $V_1 = \mathrm{Ann}_{KG}(u)$, $V_2 = \mathrm{Ann}_{KG}(v)$ are maximal ideals of KG. Let us show that $V_1 = V_2$.

Let $x \in V_1$ and $x \in KG_n$. Then $xe_i = 0$ for all $i \geq n$. Hence $xe_i t_i = 0$ for all $i \geq n$. It means that $xv = 0$, i.e. $x \in V_2$. So $V_1 \subseteq V_2$ what implies the equality of these ideals.

Lemma is proved.

LEMMA 8. *Let $\{f_i\}_{i=1}^{\infty}$ be a special idempotent sequence with "long" element u. Then there exists linear character of subgroup $N(e_i = \frac{1}{|N_i|}\sum_{n\in N_i}\chi(n)n)$ and adjoint sequence $\{\rho_i\}_{i=1}^{\infty}$ for the idempotent sequence $\{e_i\}_{i=1}^{\infty}$ such that $KGu \cong KGv$, where v is a "long" element of the sequence $\{\rho_1\dots\rho_i e_i\}_{i=1}^{\infty}$.*

PROOF. The statement of lemma is the direct consequence of Lemmas 6 and 7.

So we have obtained the following result:

THEOREM 3. *Every canonical representation of the infinite Frobenius group is equivalent to some representation obtained by a special idempotent sequence of the form $\{t_i e_i\}_{i=1}^{\infty}$ where χ is a linear character N, $e_i = \frac{1}{|N_i|}\sum_{n\in N_1}\chi(n)n$, $t_i = \rho_1\dots\rho_i$, where $\{\rho_i\}_{i=1}^{\infty}$ is an adjoint sequence for $\{e_i\}_{i=1}^{\infty}$.*

Let χ be a linear character of N, $\chi_{N_i} \neq 1_{N_i}$ for all $i = 1, 2, \dots$. Let $e_i = \frac{1}{|N_i|}\sum_{n\in N_i}\chi(n)n$. Consider the special sequence $\{t_i e_i\}_{i=1}^{\infty}(t_i \in KG_i)$. By Lemma 4 $t_i e_i = \tau_1\dots\tau_i e_i$, where $\{\tau_i\}_{i=1}^{\infty}$ is an adjoint sequence for $\{e_i\}_{i=1}$.

THEOREM 4. *Let* $\{t_ie_i\}_{i=1}^{\infty}$ *and* $\{s_ie_i\}_{i=1}^{\infty}$ *be the special idempotent sequences* ($t_i = \prod_{j=1}^{i} \tau_j$, $s_i = \prod_{j=1}^{i} \delta_j$, *and* $\{\tau_j\}_{j=1}^{\infty}$, $\{\delta_j\}_{j=1}^{\infty}$ *are adjoint sequences for* $\{e_i\}_{i=1}^{\infty}$) *with "long" elements tu and su respectively. Modules KGtu and KGsu are isomorphic if and only if* $\exists N_0 < \infty$ *such that* $\tau_i = \delta_i$ *for all* $i \geq N_0$.

PROOF. 1) Necessity. Let $KGtu \cong KGsu$ and let θ be an isomorphism of the modules. Let $\theta(tu) = xsu$ ($x \in KG_n$). Since $(t_ie_i)tu = tu$ ($i = 1, 2, \dots$) then $t_ie_ixsu = sxu$ for all $i = 1, 2, \dots$.

Let $i > n$. Then $e_ixsu(G_i) = e_ixs_ie_i = \lambda_ie_i$ ($\lambda_i \in K$), hence $xsu = \lambda tu$ ($\lambda \in K$), i.e. $tu = x_1su$, $x_1 \in KG_n$. Show that the element x_1su has the form $x_1su = (\sum_{h\in H} \alpha_h h)u$, $\alpha_h \in K$. Let $x_1 = \sum_{m\in N_n, \varphi\in H_n} a_{m,\varphi} m\varphi$ ($a_{m,\varphi} \in K$), $su = (\sum_{h\in H} \lambda_h h)u$, $x_1su = \sum_{\rho\in H, \nu\in N} \beta_{\rho,\nu} \rho\nu$. Calculate the coefficients $\beta_{\rho,\nu}$: $\beta_{\rho,\nu} = \sum_{m,\varphi} a_{m,\varphi}\lambda_{\rho\varphi^{-1}}\chi(\nu(m^{-1})^{\rho}) = \chi(\nu)\sum_{m,\varphi} a_{m,\varphi}\lambda_{\rho\varphi^{-1}}\chi^{\rho}(m^{-1}) = \chi(\nu)\beta_{1,\rho} = \chi(\nu)\alpha_{\rho}$. Hence $x_1su = (\sum_{h\in H} \alpha_h h)u$. Show that the equality $tu = x_1su$ ($x_1 \in KG_n$) implies $\tau_i = \delta_i$ for all $i \geq n+1$ (i.e. take $N_0 = n+1$). Let $tu = (\sum_{h\in H} \mu_h h)u$.

It is enough to prove that $\lambda_h = \mu_h$ for all $h \in N_{H_{k+1}}(\zeta_k)$, $k \geq n$. After calculations we obtain coefficients $\mu_h = \sum_{m,\varphi} a_{m,\varphi}\lambda_{h\varphi^{-1}}\chi^h(m^{-1})$. When $h = 1$ we have $\mu_1 = \sum_{m,\varphi} a_{m,\varphi}\lambda_{\varphi^{-1}}\chi(m^{-1}) = 1$. When $h \in N_{H_{k+1}}(\zeta_k)$ $k \geq n$ we have $\lambda_{h\varphi^{-1}} = \lambda_h\lambda_{\varphi-1}, \chi^h(m^{-1}) = \chi(m^{-1})$. Hence $\mu_h = \lambda_h \sum_{m,\varphi} a_{m,\varphi}\lambda_{\varphi^{-1}}\chi(m^{-1}) = \lambda_h$.

Theorem is proved.

3°. Let us study the problem of the equivalence of the canonical representations of the infinite Frobenius group.

Let χ_1, χ_2 be the linear characters of N, $\chi_1|_{N_i} \neq 1_{N_i}$, $\chi_2|_{N_i} \neq 1_{N_i}$, $\chi_1|_{N_i} = \zeta_i$, $\chi_2|_{N_i} = \rho_i$; $e_i = \frac{1}{|N_i|}\sum_{n\in N_i} \zeta_i(n)n$, $f_i = \frac{1}{|N_i|}\sum_{n\in N_i} \rho_i(n)n$. Let $\{\tau_i\}_{i=1}^{\infty}$, $\{\delta_i\}_{i=1}^{\infty}$ be adjoint sequences for the idempotent sequences $\{e_i\}_{i=1}^{\infty}$ and $\{f_i\}_{i=1}^{\infty}$ respectively.

If $t_i = \prod_{k=1}^{i} \tau_k$, $r_i = \prod_{k=1}^{i} \delta_k$ then $\{t_ie_i\}_{i=1}^{\infty}$ and $\{r_if_i\}_{i=1}^{\infty}$ are the special sequences with "long" elements $tu_1 = (\sum_{h\in H} \alpha_h h)(\sum_{n\in N} \chi_1(n)n) = (\prod_{i=1}^{\infty} \tau_i)(\sum_{n\in N} \chi_1(n)n)$ and $ru_2 = (\prod_{i=1}^{\infty} \delta_i)(\sum_{n\in N} \chi_2(n)n)$. One has $f_i = e_i^{h_i}$ ($h_i \in H_i$) for every $i = 1, 2, \dots$. Using the speciality of the sequences $\{e_i\}_{i=1}^{\infty}$ and $\{f_i\}_{i=1}^{\infty}$ it is easy to show that $h_i = \xi_1 \dots \xi_i$, where $\xi_j \in N_{H_j}(\zeta_{j-1})$.

THEOREM 5. *The modules* $KGtu_1$ *and* $KGru_2$ *are isomorphic if and only if there exists* $N_0 < \infty$ *such that for all* $i \geq N_0$ *the condition*

$$\frac{1}{\alpha_{\xi_i^{-1}}}\tau_i\xi_i = \delta_i \tag{1}$$

is satisfied.

REMARK. If $\alpha_{\xi_i^{-1}} = 0$ for some i then we assume that condition (1) is not satisfied for the same i.

PROOF. 1) Necessity. Let $KGtu_1 \cong KGru_2$ and let θ be an isomorphism of the modules. Let $\theta(tu_1) = xru_2 = su_2 = (\sum_{h\in H} \beta_h h)u_2$.

Since $(t_ie_i)tu_1 = tu_1$ for all $i = 1, 2, \ldots$ then $su_2 = \theta(tu_1) = t_ie_i\theta(tu_1) = t_ie_isu_2$ for all $i = 1, 2, \ldots$. Hence for all $i = 1, 2, \ldots$ we obtain $(t_ie_i)su_2(G_i) = (su_2)(G_i)$, whence $t_ie_is_if_i = s_if_i$. Since $s \neq 0$ then $\exists i_0 < \infty$ such that $s_i \neq 0$ for $i \geq i_0$. Let $i \geq i_0$. Then $e_is_if_i = e_i(\sum_{h\in H_i}\beta_hh)f_i = \sum_{h\in H_i}\beta_h(e_ihf_i) = \sum_{h\in H_i}\beta_h(e_ihf_ih^{-1})h$. Product

$$e_ihf_ih^{-1} = \begin{cases} \theta, & h \neq h_i \\ e_i, & h = h_i \end{cases}.$$

That is why $e_is_if_i = \beta_{h_i}e_ih_i = \beta_{h_i}h_if_i$. Consequently $s_if_i = t_ie_is_if_i = \beta_{h_i}t_ih_if_i$ for $i \geq i_0$. Therefore $s_i = \lambda_ih_it_i$, $\lambda_i \in K$ for $i \geq i_0$.

Let us clarify the relation between coefficients λ_i for $i \geq i_0$. Recall that $s_{i+1}(H_i) = s_i$ and $s_{i+1} = \lambda_{i+1}h_{i+1}t_{i+1}$. Then

$$\begin{aligned} s_{i+1}(H_i) &= \lambda_{i+1}(t_i + t_i(\tau_{i+1} - 1))h_i\xi_{i+1}(H_i) \\ &= \lambda_{i+1}t_ih_i\xi_{i+1}(H_i) \neq \lambda_{i+1}t_ih_i(\tau_{i+1} - 1)\xi_{i+L}(H_i) \\ &= \begin{cases} \lambda_{i+1}t_ih_i\xi_{i+1} = \lambda_{i+1}t_ih_i, & \text{if } \xi_{i+1} = 1, \\ \lambda_{i+1}t_ih_i\alpha_{\xi_{i+1}^{-1}}, & \text{if } \xi_{i+1} \neq 1 \end{cases} \\ &= \lambda_{i+1}\alpha_{\xi_{i+1}^{-1}}t_ih_i = \lambda_it_ih_i. \end{aligned}$$

Thus we have $\lambda_i = \lambda_{i+1}\alpha_{\xi_{i+1}^{-1}}$ for $i \geq i_0$. Since all $\lambda_i \neq 0$ for $i \geq i_0$ then all $\alpha_{\xi_{i+1}^{-1}} \neq 0$ for $i \geq i_0$ and therefore $\lambda_i = \lambda_{i_0}\cdot\frac{1}{\alpha_{\xi_{i_0+1}^{-1}}}\cdots\frac{1}{\alpha_{\xi_i^{-1}}}$, for $i \geq i_0$. Thus we obtain $s_i = \lambda_it_ih_i = \lambda_{i_0}\cdot\frac{1}{\alpha_{\xi_{i+1}^{-1}}}\cdots\frac{1}{\alpha_{\xi_i^{-1}}}(\prod_{j=1}^{i}\tau_j)\xi_1\ldots\xi_i = \lambda_{i_0}t_{i_0}h_{i_0}(\prod_{j=i_0+1}^{i}\frac{1}{\alpha_{\xi_j^{-1}}}\tau_j\xi_j) = S_{i_0}(\prod_{j=i_0}^{i}\tau'_j)$. The elements τ'_j form the adjoint sequence for the idempotent sequence $\{f_i\}_{i=1}^{\infty}$. Hence $s_i = s_{i_0}\prod_{j=i_0+1}^{i}\tau'_j$ for $i \geq i_0$. That is why $KGru_2 = KGsu_2 = KGs_{i_0}t'_{i_0}u_2 = KGt'_{i_0}u_2$, where $t'_{i_0} = \prod_{j=i_0+1}^{\infty}\tau'_j$; thus we have $ru_2 = yt_{i_0}u_2$. Consequently by Theorem 4 there exists number $N_0 < \infty$ such that $\tau'_j = \delta_j$ for all $j \geq N_0$.

2) Sufficiency. Let there exists $N_0 < \infty$such that for all $i \geq N_0$ the condition (1) is satisfied:

$$\delta_i = \frac{1}{\alpha_{\xi_i^{-1}}}\tau_i\xi_i = \tau'_i.$$

Let $t' = (1 + \xi_1^{-1})(1 + \xi_2^{-1})\ldots(1 + \xi_{N_0-1})\prod_{i=N_0}^{\infty}\tau_i$, $s = (1 + \xi_1)\ldots(1 + \xi_{N_0-1})\prod_{i=N_0}^{\infty}\delta_i$. It is evident that $KGtu_1 = KGt'u_1$ and $KGsu_2 = KGru_2$. To prove the statement of the theorem it is enough to show that $\operatorname{Ann}_{KG}(t'u_1) = \operatorname{Ann}_{KG}(su_2)$. Note that $s_i = s(G_i) = \lambda_it'_ih_i$ where $\lambda_i \neq 0$ for all $i \geq 1$. Let $x = \sum_{m\in N_k,\varphi\in H_k}a_{m,\varphi}m\varphi \in \operatorname{Ann}_{KG}(t'u_1)$. It is the same that for all $\rho \in H$ $\sum_{m,\varphi}a_{m,\varphi}a'_{\varphi^{-1}\rho}\chi_1^{\rho}(m^{-1}) = 0$. Let ρ be an arbitrary element of H. Let $\rho \in H_n$, $n \geq k$. If $s = \sum_{h\in H}\beta_hh$ then $\beta_{\varphi^{-1}\rho} = \lambda_n\alpha'_{\varphi^{-1}\rho h_h^{-1}}$. Besides if $m \in N_n$ then $\chi_2^{\rho}(m^{-1}) = \chi_1^{h_n^{-1}\rho}(m^{-1})$ and consequently $\sum_{m,\varphi}a_{m,\varphi}\beta_{\varphi^{-1}\rho}\chi_2^{\rho}(m^{-1}) = \lambda_n\sum_{m,\varphi}a_{m,\varphi}\alpha'_{\varphi^{-1}(\rho h_h^{-1})}\chi_1^{(\rho h_n^{-1})}(m^{-1}) = 0$. Since element $\rho \in H$ is an arbitrary one then we have $\sum_{m,\varphi}a_{m,\varphi}\beta_{\varphi^{-1}\rho}\chi_2^{\rho}(m^{-1}) = 0$ for all $\rho \in H$, i.e.

$x \in \mathrm{Ann}_{KG}(su_2)$. Thus $\mathrm{Ann}_{KG}(t'u_1) \subseteq \mathrm{Ann}_{KG}(su_2)$ and the maximality of both ideals implies their coincidence. It means that $KGtu_1 \cong KGru_2$.

Theorem is proved.

REFERENCES

1. S. D. Berman, *On the equation* $x^m = 1$ *in an integral group ring*, Ukrainsk. Matem. Zh. **3** (1955), 253–261.

2. B. Huppert, *Endliche Gruppen I*, Springer-Verlag No. 134, Berlin, 1967.

Contemporary Mathematics
Volume 93 (1989)

Generators of Subgroups of $U(ZG)^*$

JÜRGEN RITTER AND SUDARSHAN K. SEHGAL

1. Introduction. Let $U = U(\mathbb{Z}G)$ be the group of units, having augmentation one, of the integral group ring $\mathbb{Z}G$ of a finite group G, namely,

$$U = \left\{u = \sum a_g g \in \mathbb{Z}G \middle| \sum a_g = 1 \text{ and there exists } v \in \mathbb{Z}G \text{ with } uv = 1\right\}.$$

Obviously, the group elements $g \in G$ belong to U; they are called trivial units. We also have for $a, b \in G$

$$u_{ab} = 1 + (a-1)b\hat{a} \in U, \text{ where } \hat{a} = 1 + a + \cdots + a^{o(a)-1};$$

of course,

$$u_{ab}^{-1} = 1 - (a-1)b\hat{a}.$$

It is a difficult problem to find generators of U. It might be easier to do the same for a subgroup of finite index. This paper is a contribution towards the latter problem. Let $\zeta(U)$ be the centre of U and H the subgroup of U generated by $\zeta(U)$ and the u_{ab}, $a, b \in G$. We denote by $\tilde{H}$ the normal closure of H in U. Throughout the paper, we fix the above meaning for U, H and $\tilde{H}$. The first result is a more explicit version of Theorem 1 of an earlier paper **[11]**.

THEOREM 1. *Suppose that the rational group algebra $\mathbb{Q}G$ does not have any simple components of the following type:*

a) *2×2 matrix ring over a division algebra which is a totally definite quaternion algebra or the centre of which is the rational field or an imaginary quadratic field.*

b) *a division ring which is a totally definite quaternion algebra.*

$(U : \tilde{H})$ is finite if and only if G has no nonabelian homomorphic image which is fixed point free.

A group F is said to be fixed point free if it has a complex irreducible representation ρ such that for every nonidentity element f of F, $\rho(f)$ has all eigenvalues different from one. These groups were characterized in a

*Research supported by the Humboldt Stiftung and NSERC.

fundamental paper [**15**] by Zassenhaus and are well known. Also see [**6**, p. 73] and [**14**, p. 96 and 204]. If a group G is odd, then $\mathbb{Q}G$ does not have simple components a) or b) of Theorem 1 and we have

COROLLARY. *If G is odd, then $(U: \tilde{H})$ is finite if and only if G has no nonabelian homomorphic image which is fixed point free.*

We expect to be able to prove that for odd p-groups G, actually, one has a much stronger result, namely, $(U: H) < \infty$. For now, we have

THEOREM 2. *If p is odd and G is a p-group all the complex irreducible representations of which have degree dividing p, then $(U: H)$ is finite.*

THEOREM 3. *If G is the dihedral group of order $2n$, then $(U: H)$ is finite.*

2. Notations. We begin by recalling some notations which were already used in [**11**]. $\mathbb{Q}G$ is the rational group algebra of the finite group G and $\mathbb{Z}G$ is the integral group ring. Λ is a fixed maximal order of $\mathbb{Q}G$, $\Lambda^\times$ its group of units and C its group of central units. Corresponding to the decomposition

$$\mathbb{Q}G = S_1 \oplus \cdots \oplus S_t$$

of $\mathbb{Q}G$ into its simple Wedderburn components S_i we have the decompositions

$$\Lambda = \Lambda_1 \oplus \cdots \oplus \Lambda_t$$
$$\Lambda^\times = \Lambda_1^\times \times \cdots \times \Lambda_t^\times$$
$$C = C_1 \times \cdots \times C_t$$

where Λ_i is a maximal order in S_i, $\Lambda_i^\times$ its group of units and C_i its group of central units. We identify S_i with $(D_i)_{n_i \times n_i}$, the full $n_i \times n_i$ matrix ring over the division algebra D_i the centre of which is denote by K_i. Observe that K_i is the character field $\mathbb{Q}(\chi_i)$ of an absolutely irreducible character χ_i of G that does not vanish on S_i. Let O_i be the ring of integers in K_i and let $\mathcal{O}_i$ be some maximal order in D_i. Then $(\mathcal{O}_i)_{n_i \times n_i}$ is a second maximal order in S_i. We shall write GL_i or $GL(n_i, \mathcal{O}_i)$ for its group of units and SL_i or $SL(n_i, \mathcal{O}_i)$ for the subgroup in GL_i that consists of all elements α having reduced norm one.

For an ideal Q of $\mathcal{O}_i$ we denote by $E(Q)$ the subgroup of SL_i generated by all Q-elementary matrices

$$I + ge_{lm}, g \in Q, l \neq m, e_{lm} \text{ a matrix unit}$$

and by $\tilde{E}(Q)$ its normal closure in SL_i. It is well known that if $n_i \geq 3$ then the normal subgroup of $E(\mathcal{O}_i)$ generated by $E(Q)$ is normal in SL_i and hence coincides with $\tilde{E}(Q)$. With this notation the congruence theorem of Bass-Milnor-Serre and Bak-Rehmann-Vaserstein ([**1**], [**12**] and [**9**, page 200]) reads as

LEMMA 2.1. *Assume that either $n_i \geq 3$ or $n_i = 2$ and S_i is not a 2×2 matrix ring over a division algebra which is a totally definite quaternion algebra or the centre of which is the rational field or an imaginary quadratic field.*

Then

(1) $(SL_i\colon \tilde{E}(Q))$ *is finite for any non zero ideal* Q *of* $\mathcal{O}_i$.

(2) *Every noncentral subgroup of* SL_i *normalized by a subgroup of finite index contains* $\tilde{E}(Q)$ *for some non zero ideal* Q *of* $\mathcal{O}_i$.

We shall also need the following result due to Vaserstein **[12]** and **[13]**.

LEMMA 2.2.

(i) *If* $n_i \geq 3$, *then* $\tilde{E}(Q^2) \subseteq E(Q)$, *in particular* $(SL(n_i, \mathcal{O}_i)\colon E(Q)) < \infty$.

(ii) *If* $n_i = 2$ *and if* $\mathcal{O}_i$ *is an order in the ring of integers of an algebraic numberfield which is not rational or imaginary quadratic, then* $(SL(2,\mathcal{O}_i)\colon E(Q)) < \infty$.

PROOF. The first part is easy **[12]** and we shall include its proof. For the second part we refer the reader to **[13]** and **[8]**. It is enough to prove

$$A \in E(\mathcal{O}_i), B \in E(Q^2) \Rightarrow A^{-1}BA \in E(Q).$$

Fix $x, y \in Q$ and $r \in \mathcal{O}_i$. We have to prove that

$$(1 - re_{kl})(1 + xye_{ij})(1 + re_{kl}) \in E(Q).$$

We have

$$\begin{aligned}(1 - re_{kl})(1 + xye_{ij})(1 + re_{kl}) &= 1 + xye_{ij} - rxye_{kl}e_{ij} + rxye_{ij}e_{kl} \\ &\quad - r^2xye_{kl}e_{ij}e_{kl} \\ &= \begin{cases} 1 + xye_{ij} + rxye_{ij}e_{kl} & \text{if } l \neq i \\ 1 + xye_{ij} - rxye_{kj} & \text{if } l = i, k \neq j \end{cases}\end{aligned}$$

which belongs to $E(Q)$ unless $l = i$, $k = j$ in which case we need to compute

$$(*) \qquad (1 - re_{ji})(1 + xye_{ij})(1 + re_{ji}).$$

Since $n \geq 3$, we can choose $m \neq i, j$. Then $1 + xye_{ij} = [1 + xe_{im}, 1 + ye_{mj}] = (1 + xe_{im})(1 + ye_{mj})(1 - xe_{im}) = (1 - ye_{mj})$. Therefore, $(*)$ equals

$$[(1 + xe_{im})^{1+re_{ji}}, (1 + ye_{mj})^{1+re_{ji}}].$$

But

$$(1 + xe_{im})^{1+re_{ji}} = (1 - re_{ji})(1 + xe_{im})(1 + re_{ji}) = 1 + xe_{im} - rxe_{jm}$$

$$\text{and } (1 + ye_{mj})^{1+re_{ji}} = (1 - re_{ji})(1 + ye_{mj})(1 + re_{ji}) = 1 + ye_{mj} - yre_{mi}$$

both belong to $E(Q)$. Thus $(*)$ belongs to $E(Q)$ and the lemma is proved.

We conclude this section by recalling the definition of $H = \langle \zeta(U), u_{ab} | a, b \in G\rangle$ with $\zeta(U) =$ centre of U and $u_{ab} = 1 + (a-1)b\hat{a}$, $\hat{a} = 1 + \cdots + a^{o(a)-1}$. $\tilde{H}$ is the normal closure of H in U. p will always be an odd prime.

3. Proof of Theorem 1.

(a) First suppose that G has no nonabelian homomorphic image which is fixed point free. We wish to prove that $(U\colon \tilde{H}) < \infty$. We claim that $\mathbb{Q}G$ *has no noncommutative division ring as a simple component.* On the contrary, suppose that D is such a component. Then a homomorphic image $G/K = F$ is embedded in $D^{\times}$ and is noncommutative. This F is fixed point free as seen below. Suppose F has a complex representation ψ with $\psi(f)$, $1 \neq f \in F$, having an eigenvalue one. Then there exists a column vector v such that $\psi(f)v = v$ and $\psi(f-1)v = 0$, $f - 1 \in D^{\times}$. Multiplying by $\psi(f-1)^{-1}$ we see that $v = 0$.

Thus the claim is established and all division ring components of $\mathbb{Q}G$ are commutative. Since the centre of $\Lambda_i^{\times}$ coincides with the centre of GL_i, due to Lemma 4 of **[11]** we have only to prove

$$(\Lambda_i^{\times} : \tilde{H}_i) < \infty. \tag{3.1}$$

Let T be a non linear complex irreducible representation of G. We claim

(3.2) *There exist elements a and b of G such that $T(u_{ab})$ is noncentral.* Write $\overline{G} = T(G) \simeq G/\operatorname{Ker} T$. Then by hypothesis, there exists an $a \in G$ such that $T(a) \neq I$ and has an eigenvalue one. Diagonalizing $T(a)$ suitably we can write

$$T((1-a)b\hat{a}) = \begin{bmatrix} O & & O & \\ & \alpha_1 & & \\ O & & \ddots & \\ & & & \alpha_r \end{bmatrix} \begin{bmatrix} B_1 & B_2 \\ B_3 & B_4 \end{bmatrix} \begin{bmatrix} nI & O \\ O & O \end{bmatrix}$$
$$= \begin{bmatrix} O & O \\ \alpha_i n b_{ij} & O \end{bmatrix}, \quad n = o(a),\ \alpha_i \neq 0,\ B_3 = [b_{ij}].$$

If this matrix is central, being nilpotent, it is zero. It follows that $B_3 = 0$. Since the representation is irreducible, there exists a b such that $B_3 \neq 0$, and $T(u_{ab})$ is noncentral as desired.

Let S_i be the simple component of $\mathbb{Q}G$ such that $T(S_i) \neq 0$. Since the T^{σ} with σ running through the Galois group of K_i over $\mathbb{Q}$ are precisely the irreducible complex representations of G that do not vanish on S_i, we have

(3.3) Projection on S_i of u_{ab} of (3.2) is noncentral.

(3.4) If $T(u_{ab}) = 1$ for some $a, b \in G$, then the projection of u_{ab} on S_i is also one.

We next assert

(3.5) $\tilde{H}_i \cap SL_i$ *is noncentral*

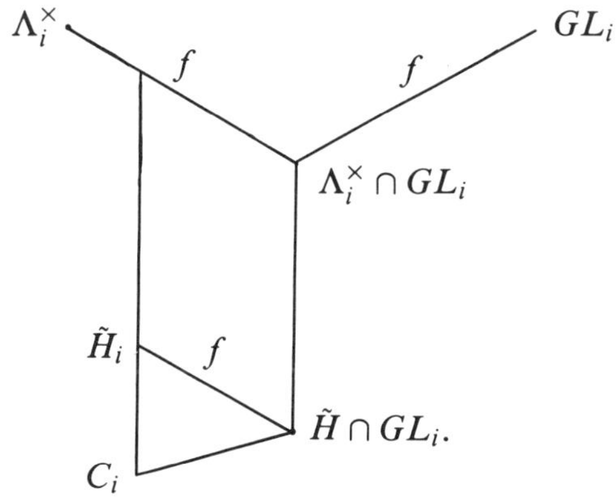

Suppose on the contrary that $\tilde{H}_i \cap SL_i$ is central. With respect to the diagram above compare the proof of Theorem 2 in **[11]**. If $y \in GL_i \cap \tilde{H}_i$ we have $y^m \in SL_iC_i \cap \tilde{H}_i$ where $m = (GL_i\colon SL_iC_i)$. Write $y^m = sc$, $s \in SL_i$, $c \in C_i$. Then $y^m c^{-1} = s \in SL_i \cap \tilde{H}_i$ is central. Since $(\tilde{H}_i\colon \tilde{H}_i \cap GL_i) = k$ is finite, we conclude that $\tilde{H}_i$ is of finite exponent t over C_i. Therefore, for every

$$h = 1 + (x-1)y\hat{x}$$

we have $T(h^t) = T(1 + t(x-1)y\hat{x})$ central which implies that $T(t(x-1)y\hat{x})$ is central. Thus $T(h)$ is central for all $x, y \in G$, which contradicts (3.2).

It follows by (2.1) that $(\tilde{H}_i \cap SL_i)$ is of finite index in SL_i. It is easy to see that $(\tilde{H}_i \cap GL_i)$ is of finite index in GL_i and consequently $(\Lambda_i^\times : \tilde{H}_i) < \infty$ as desired.

(b) Now, suppose that $(U\colon \tilde{H}) < \infty$, we shall prove that no nonabelian homomorhic image of G is fixed point free. Let T be a nonlinear complex irreducible representation of G. Suppose that $T(G)$ is a fixed point free group. We may suppose that for all $a \in G$, $T(a) \neq I$ implies that $T(a)$ has no eigenvalue 1. Thus on diagonalizing we find that $T(\hat{a}) = 0$ and thus $T(u_{ab}) = I$. It follows by (3.4) that if S is the rational component of $\mathbb{Q}G$ with $T(S) \neq 0$ then the projection of $\tilde{H}$ on S is central. It follows that if S is a matrix ring then $(\Lambda^\times : \tilde{H}) = \infty$, $(U\colon \tilde{H}) = \infty$. This is a contradiction.

We have to consider the projections of $\tilde{H}$ onto division ring components D_i of $\mathbb{Q}G$. Since $((a-1)b\hat{a})^2 = 0$, these projections are all central. Then $(U\colon \tilde{H}) < \infty$ implies that $(\mathcal{O}_i^\times : O_i^\times) < \infty$. Therefore by Lemma 2 of **[11]**, D_i is a totally definite quaternion algebra. This is not allowed by assumption. Thus the theorem is proved.

REMARK 3.6. For theproof of the implication $(\Rightarrow)$ only the assumption (a) has been used whereas for the converse the assumption (b) alone is used.

REMARK 3.7. If $\mathbb{Q}G$ has a division algebra D as a simple component which is a totally definite quaternion algebra then G has a nonabelian homomorphic image in the real quaternions H. This is because the completion of D at any prime must be the quaternion algebra H over the real field $\mathbb{R}$ and we have

$$\mathbb{Q}G \to D \to H$$

inducing a homomorphic image $\overline{G}$ in H with $\mathbb{R}\overline{G} = H$. Such $\overline{G}$ are well known [**10**, p. 344].

COROLLARY 3.8. *If G is odd, then $(U\colon \tilde{H})$ is finite if and only if G has no nonabelian homomorphic image which is fixed point free.*

4. Some Lemmas.

We shall denote by ζ_n a primitive nth root of unity.

LEMMA 4.1. *Let G be a nonabelian p-group containing a cyclic subgroup of index p. Then*

$$G = \langle a^{p^n} = 1 = b^p, a^b = a^{1+p^{n-1}}, n \geq 2\rangle.$$

PROOF. See [**4**, page 187].

LEMMA 4.2. *Let G be the group of* (4.1). *The complex irreducible representations of G are given as follows:*

(a) *p^n linear representations of G/G', $G' = \langle a^{p^n-1}\rangle$.*

(b) *$p^{n-2}(p-1)$ p-dimensional representations which are rationally conjugate, the character field being $K = \mathbb{Q}(\zeta_{p^{n-1}})$.*

PROOF. Let τ be a character of $\langle a\rangle$ given by $\tau(a) = \zeta_{p^n}$. Then $T = \operatorname{ind}_{\langle a\rangle}^G(\tau)$ is an irreducible representation of G as seen by [**3**, page 329]. Write $j = 1 + p^{n-1}$. Then j induces an automorphism of $\mathbb{Q}(\zeta_{p^n})$, the fixed field of which is $K = \mathbb{Q}(\zeta_{p^{n-1}})$. Let χ be the character of T. We observe that

$$\chi(a^i b^j) = \begin{cases} 0 \text{ if } j \neq 0 \\ \operatorname{tr}_j(\tau(a^i)) = \operatorname{tr}_j(\zeta_{p^n}^i) \in K \end{cases}.$$

Moreover, $\chi(a^p) = p\zeta_{p^{n-1}}$. Thus the characterfield $Q(\chi) = K$, giving rise to $(K\colon \mathbb{Q}) = p^{n-2}(p-1)$ conjugates of T. Adding degrees of these representations together with the linear ones we have

$$p^n + p^2(p-1)p^{n-2} = p^{n+1}.$$

Thus we have accounted for all the representations.

COROLLARY 4.3. *Any absolutely irreducible representation T of the above group G can be realized over $K = \mathbb{Q}(\zeta_{p^{n-1}})$, and $T(b)$ has an eigenvalue one exactly once.*

PROOF.

$$T(a) = \begin{bmatrix} 0 & 1 & & & \\ & \ddots & 1 & & \\ & & \ddots & & \\ & & & \ddots & 1 \\ \zeta_{p^n}^p & \cdots & & \cdots & 0 \end{bmatrix}, \quad T(b) = \begin{bmatrix} 1 & & & & \\ & w & & & \\ & & w^2 & & \\ & & & \ddots & \\ & & & & w^{p^{n-1}} \end{bmatrix},$$

$w = \zeta_{p^n}^{p^{n-1}}$ satisfy the same relations as a and b. This representation T is of degree p and faithful. Hence, T is irreducible and is therefore one of the rationally conjugate representations above. Clearly $T(b)$ has eigenvalue one exactly once.

LEMMA 4.4. *If a p-group G has a faithful complex irreducible representation T of degree p, then G has an abelian subgroup A of index p such that* $T = \operatorname{ind}_A^G(\tau)$ *for some* $\tau \in \operatorname{Hom}(A, \mathbb{C}^\times)$. *Moreover, G is the group of* (4.1) *if and only if* $\ker\tau = 1$. *Further,* $\ker\tau \neq 1$ *implies that* $\mathbb{Q}(\chi) = \mathbb{Q}(\tau)$.

PROOF. Since G is a p-group we can say that $T = \operatorname{ind}_A^G(\tau)$ for some subgroup A of G and some abelian character τ of A. As T is p-dimensional, $(G\colon A) = p$ and thus A is normal in G. Therefore, the commutator group A' is normal in G. Since A' is contained in the kernel of τ, it follows that $A' \subseteq (\ker\tau)^x$ for all $x \in G$. Thus $A' \subseteq \ker T = 1$. We have proved that A is abelian.

Suppose that $\ker\tau = 1$. Then A is cyclic and has index p in G. Therefore, it is the group of (4.1). Conversely, for the group of (4.1) τ is faithful is seen in (4.2).

We now assume that $\ker\tau \neq 1$, i.e. G is not the group of (4.1). We shall show that $\mathbb{Q}(\chi) = \mathbb{Q}(\tau)$. We have

$$G \supset A \supset (\ker\tau \neq 1) \supset \ker T = 1 \text{ and } \mathbb{Q}(\tau) \supset \mathbb{Q}(\chi) \supset \mathbb{Q}.$$

Pick an element $b \in G$, $b \notin A$, $b^p \in A$. Let σ be an automorphism of $Q(\tau)$ over $Q(\chi)$. Then

$$\operatorname{ind}_A(\tau^\sigma) = (\operatorname{ind}_A \tau)^\sigma = \chi^\sigma = \chi.$$

Since $\operatorname{res}_A \chi = \tau + \tau^b + \tau^{b^2} + \cdots + \tau^{b^{p-1}}$, it follows that $\tau^\sigma = \tau^{b^k}$ for some k. Suppose $0 < k \leq (p-1)$. Then

$$\ker\tau = \ker\tau^\sigma = \ker\tau^{b^k} = (\ker\tau)^{b^k}$$

which implies that $\ker\tau$ is normal in G. Thus $\ker\tau = 1$ which is a contradiction. Thus $Q(\tau) = Q(\chi)$ as desired.

LEMMA 4.5. *Let K be an elementary abelian p-subgroup of a group G. Suppose that an element y of G does not normalize K but* y^p *normalizes K. Then*

$$K \not\subset \bigcup_{i=1}^{p-1} K^{y^i}.$$

PROOF. Suppose that $K \subset \bigcup_{i=1}^{p-1} K^{y^i}$. Then $K = \bigcup_1^{p-1}(K^{y^i} \cap K)$. Write $|K| = p^n$. Then if $K^{y^i} \cap K \neq K$ for all i we must have $|K^{y^i} \cap K| \leq p^{n-1}$ for all i. This is a contradiction by adding orders. Therefore, there exists an i such that $K^{y^i} = K$ and consequently y normalizes K. This proves the lemma.

LEMMA 4.6. *If a p-group G has a faithful irreducible complex representation T of degree p then there exists an element a of G such that T(a) has eigenvalue 1 exactly once.*

PROOF. We know by (4.4.) that $T = \mathrm{ind}_A(\tau)$, $\tau \in \mathrm{Hom}(A, \mathbf{C}^\times)$, A is an abelian subgroup of G of index p. Because of (4.3) we may assume that $\ker\tau \neq 1$. Let

$$K = \{a \in \ker\tau / a^p = 1\}.$$

Pick an element y of G, $y \notin A$. Since $\ker T = 1$ and A is abelian y does not normalize any nontrivial subgroup of $\ker\tau$. In particular, y does not normalize K. Thus by (4.5), there exists an element $a \in K$ such that $a \notin \bigcup_{i=1}^{p-1} K^{y^i}$. For this a we have

$$T(a) = \begin{bmatrix} \tau(a) & & \\ & \tau(a^b) & \\ & & \ddots \end{bmatrix}.$$

Since $\tau(a) = 1$, $\tau(a^{b^i}) \neq 1$ for $i > 1$ we are done.

5. Proof of Theorem 2.

Let G be a p-group all whose nonlinear complex irreducible representations are of degree p. Let Π_i be a projection of $\mathbb{Q}G$ on a noncommutative simple component S_i. Then there is a complex irreducible representation T such that $T(S_i) \neq 0$. This T can be realized over the characterfield $\mathbb{Q}(\chi)$ where $\mathbb{Q}$ is the character of T. Thus we may regard T as a projection of $\mathbb{Q}G$ on S_i. Write $\Pi_i(H) = H_i$ and $\Pi_i(U) = U_i$. We need the following two results the proof of which we give later.

LEMMA 5.1. *$(U_i\colon H_i)$ is finite.*

LEMMA 5.2. *Let T_1 and T_2 be two nonabelian complex irreducible representations of G which are not conjugate over $\mathbb{Q}$. Then there is an element h of H such that $T_1(h) = 1 + \alpha e_{ij}$ for some i, j where $0 \neq \alpha \in \mathbb{Q}(\chi_1)$, α integral and $T_2(h) = 1$.*

We shall prove that H contains a subgroup $(1, 1, \dots, N_i, 1, \dots)$ for every i with $(U_i\colon N_i) < \infty$. We know by (5.2) that given i, j, H contains an element $(\dots, 1, \dots, \varepsilon, \dots)$ with 1 in ith place and ε noncentral in jth place, $\varepsilon \in SL_j = SL(O_j)$. Then H contains

$$\begin{aligned}(\dots, 1, \dots, \varepsilon, \dots)^H &= (\dots, 1, \dots, \varepsilon^H, \dots)\\ &= (\dots, 1, \dots, \varepsilon^{H_j}, \dots).\end{aligned}$$

The normal subgroup of H_j that is generated by ε intersects SL_j noncentrally and is normal in the subgroup H_j which is of finite index in U_j and thus in $\Lambda_j^\times$. It follows by (2.1) that it contains a group of finite index in SL_j. Since H contains the centre of U it follows that H contains $(\dots, 1, \dots, N_j, \dots)$ where

N_j is of finite index in $\Lambda_j^\times$. Suppose we have proved by induction that H contains

$$\Big(\ldots,\overbrace{1,1,\ldots,1}^{i},\ldots,N_j,\ldots\Big)$$

where there are i 1's and N_j is of finite index in jth place. Of course the 1's are allowed to appear in any place. In particular, H contains

$$\Big(\ldots,\overbrace{1\ldots,1,1}^{(i-1)},\ldots,N_j',1,\ldots\Big)$$

with N_j' of finite index. Taking commutators with $M_j = N_j \cap N_j'$, H contains

$$\Big(\ldots,\overbrace{1,1,\ldots 1}^{i},[M_j,M_j],1,\ldots\Big).$$

Now M_j is of finite index and $[M_j,M_j] \lhd M_j < \Lambda_j^\times$. It follows as above that H contains a group with $(i+1)$ 1's and jth component of finite index. By induction H contains groups with a subgroup of finite index in one component and ones elsewhere. Multiplying these groups together we have proved that $(U\colon H)$ is finite.

PROOF OF (5.1.)Let T be the complex representation corresponding to the projection Π_i. We may assume without loss of generality that T is faithful. By (4.4), $T = \operatorname{ind}_A(\tau)$ where A is abelian of index p in G and $\tau \in \operatorname{Hom}(A, C^\times)$. We consider two cases separately. Case (1) $\ker\tau \neq 1$. By (4.6) there exists $a \in \ker\tau$ such that

$$T(a) = \begin{bmatrix} 1=\tau(a) & & & \\ & w_2 & & \\ & & \ddots & \\ & & & w_p \end{bmatrix} \quad 1 \neq w_i,\ w_i^p = 1.$$

Pick $b \in G$, $b \notin A$ and c a generator of $A/\ker\tau$. Then

$$T(b) = \begin{bmatrix} 0 & 1 & & \\ & \ddots & \ddots & \\ & & \ddots & 1 \\ \mu & & & 0 \end{bmatrix}, \quad \mu = \tau(b^p)$$

$$T(c) = \begin{bmatrix} \tau(c) = \zeta & & \\ & \tau(c^b) & \\ & & \ddots \end{bmatrix},$$

$$T(1 + (a-1)b\hat{a}) = I + \begin{bmatrix} 0 & & \\ & w_2 - 1 & \\ & & \ddots \end{bmatrix} \begin{bmatrix} 0 & 1 & 0 & \dots & 0 \\ 0 & 0 & 1 & \dots & 0 \\ 0 & \dots & \dots & 0 & 1 \\ \mu & 0 & \dots & \dots & 0 \end{bmatrix}$$

$$\begin{bmatrix} p & & & \\ & 0 & & \\ & & \ddots & \\ & & & 0 \end{bmatrix}$$

$$\begin{aligned} &= I + p\mu(w_p - 1)e_{p1} = I + \alpha_1 e_{p1} \\ T(1 + (a-1)bc^v\hat{a}) &= I + \zeta^v \alpha_1 e_{p1} \\ T(1 + (a-1)b^i c^v \hat{a}) &= I + \zeta^v (w_{p-i+1} - 1)e_{p-i+1,1} \text{ for } 1 \le i \le (p-1) \\ &= I + \zeta^v \alpha_i e_{p-i+1,1}. \end{aligned}$$

Observe that $(u_{ab})^x = u_{a^x b^x} \in H$ for $x \in G$. Thus H_i contains

$$T(1 + (a-1)b^i c^v \hat{a})^{b^j} = 1 + \zeta^v \alpha(i,j) e_{i',j'}, \quad \begin{matrix} 1 \le i \le p-1 \\ 0 \le j \le p-1 \end{matrix}$$

where $\alpha(i,j)$ is an integer of $\mathbb{Q}(\chi)$ and is independent of v. Moreover, (i', j') is a permutation of (i,j). Thus H_i contains all matrices $I + O\alpha(i,j)e_{ij}$, $O = \mathbb{Z}[\zeta]$. We conclude that $E(\beta O)$ is contained in H_i with $\beta = \prod_{i,j} \alpha(i,j)$. It follows by (2.2) that $\tilde{E}(\beta^2 O)$ is contained in H_i and the proof is complete due to (2.1) in this case.

Case (2) $\ker\tau = 1$. Then by (4.4) we have

$$G = \langle a, b | a^{p^n} = 1 = b^p, a^b = a^{1+p^{n-1}} \rangle.$$

Moreover, by (4.3), T is given up to rational equivalence by

$$T(A) = \begin{bmatrix} 0 & 1 & 0 & & 0 \\ & 0 & 1 & & \\ & & 0 & \ddots & 0 \\ & & & \ddots & \\ & & & & 1 \\ \zeta^p & 0 & 0 & & \end{bmatrix}, \quad T(b) = \begin{bmatrix} 1 & & & \\ & w_2 & & \\ & & \ddots & \\ & & & w_p \end{bmatrix}$$

$$\zeta = \zeta_{p^n}, w_i^p = 1, w_i \ne 1.$$

This is a representation over $K = \mathbb{Q}(\zeta^p) = \mathbb{Q}(\zeta_{p^{n-1}})$. Moreover, for $c = a^p$, $T(c) = \zeta^p I$ is central. It can now be checked that the elements $T(1 + (b-1)ac^v\hat{b})^{a^j}$ generate a subgroup of finite index in H_i as required. This completes the proof of (5.1).

PROOF OF (5.2.):Let T_1 and T_2 be two nonlinear complex irreducible representations of G which are not equivalent over $\mathbb{Q}$. By using induction we may assume that $\ker T_1 \cap \ker T_2 = 1$. We know from the proof of (5.1) that there exists $a, b \in G$ such that $T_1(u_{ab}) = 1 + \alpha e_{lm}$, $\alpha \neq 0$, $l \neq m$, $\alpha \in \mathbb{Q}(\chi_1)$, α integral. If for this element a either $T_2(a) = I$ or $T_2(a)$ has no eigenvalue equal to one then $T_2(u_{ab}) = 1$ as desired. So assume

$$T_2(a) \neq I \text{ and } T_2(a) \text{ has an eigenvalue equal to one.}$$

By diagonalizing $T_2(a)$ we may write

$$T_2(a) = \begin{bmatrix} 1 & & & & & \\ & \ddots & & & & \\ & & 1 & & & \\ & & & w_{r+1} & & \\ & & & & \ddots & \\ & & & & & w_p \end{bmatrix}, \quad 1 \neq w_i, \ 1 \leq r+1 \leq p.$$

Then

$$T_2(a-1)b\hat{a}$$

$$= \begin{bmatrix} 0 & & & & & \\ & \ddots & & & & \\ & & 0 & & & \\ & & & \pi_{r+1} & & \\ & & & & \ddots & \\ & & & & & \pi_p \end{bmatrix} \begin{bmatrix} \; & \; \\ \; & \; \end{bmatrix} \begin{bmatrix} p & & & & & \\ & \ddots & & & & \\ & & p & & & \\ & & & 0 & & \\ & & & & \ddots & \\ & & & & & 0 \end{bmatrix}$$

$$= \begin{bmatrix} O_{r\times r} & O \\ * & O \end{bmatrix}, \quad \pi_i = w_i - 1.$$

$T_2(u_{ab}) = I + \sum \alpha_{ij}e_{ij}$, $(r+1) \leq i \leq p$, $1 \leq j \leq r$. We consider separately various cases.

Case (i) $\ker T_1 \neq 1$: We pick $z \in \ker T_1$, central in G of order p. then by assumption $z \notin \ker T_2$. For $0 \leq v \leq p-1$, H contains the elements $h_v = 1 + (a-1)bz^v a$. Then

$$T_1(h_v) = 1 + \alpha e_{lm}$$

$$T_2(h_v) = 1 + \sum_{\substack{(r+1)\leq i \leq p \\ 1 \leq j \leq r}} \zeta^v \alpha_{ij} e_{ij}, \ \zeta \neq 1, \text{ a } p\text{th root of unity.}$$

Taking $h = \prod_{v=0}^{p-1} h_v$ we get $T_1(h) = 1 + p\alpha e_{lm}$, $T_2(h) = 1$. It remains to consider

Case (ii) $\ker T_1 = 1$: Then by (4.4) we have $T_1 = \operatorname{ind}_A^G(\tau_1)$, $\tau_1 \in \operatorname{Hom}(A, \mathbf{C}^\times)$ where A is an abelian normal subgroup of index p in G and $G = \langle A, b\rangle$. Then $\ker \tau_1 \neq 1$ as otherwise G is the group of (4.1) which has only one representation up to rational equivalence. Now by (4.6) we pick $a \in \ker \tau_1$ of order p such that

$$T_1(a) = \begin{bmatrix} 1 & & & \\ & w_2 & & \\ & & \ddots & \\ & & & w_p \end{bmatrix}, \quad T_1(b) = \begin{bmatrix} 0 & 1 & & \\ & \ddots & \ddots & \\ & & \ddots & 1 \\ \gamma & & & 0 \end{bmatrix}, \quad \gamma = \tau_1(b^p).$$

Then $T_1(u_{ab}) = I + p\gamma(w_p - 1)e_{p1}$. Recall from (4.4) that $\mathbb{Q}(\chi_1) = \mathbb{Q}(\tau_1)$. We consider separately the possibilities that $\ker \tau_1$ is contained in $\ker \tau_2$ or not.

(a) Assume $\ker \tau_1 \subseteq \ker \tau_2$. In this case $\tau_2 = \tau_1^s$ for an integer s. If s is a multiple of p

$$T_2(a) = \begin{bmatrix} \tau_2(a) & & \\ & \tau_2(a^b) & \\ & & \ddots \end{bmatrix} = \begin{bmatrix} \tau_1^s(a) & & \\ & \tau_1^s(a^b) & \\ & & \ddots \end{bmatrix} = I$$

as a is of order p and $T_2(u_{ab}) = 1$. So let $(p, s) = 1$. Then

$$T_2 = \operatorname*{ind}_A^G(\tau_1^s) = \left(\operatorname*{ind}_A^G \tau_1\right)^{(s)} = T_1^{(s)}$$

where (s) is the obvious automorphism of $\mathbb{Q}(\chi_1) = \mathbb{Q}(\tau_1)$. This is a contradiction. Therefore, we have the last case

(b) $\ker \tau_1 \not\subseteq \ker \tau_2$: Suppose we can find

(*) an element $z \in \ker \tau_1$, $z \notin \ker \tau_2$, $T_2(z)$ has no eigenvalue equal to one.

Then let $h_v = 1 + (a-1)bz^v\hat{a}$. We find that

$$T_1(h_v) = I + \begin{bmatrix} 1 & & & \\ & w_2 - 1 & & \\ & & \ddots & \\ & & & w_p - 1 \end{bmatrix} \begin{bmatrix} 0 & 1 & & \\ & \ddots & \ddots & \\ & & \ddots & 1 \\ \gamma & & & 0 \end{bmatrix} \times \begin{bmatrix} \zeta_1 & & \\ & \ddots & \\ & & \zeta_p \end{bmatrix} \begin{bmatrix} p & & & \\ & 0 & & \\ & & \ddots & \\ & & & 0 \end{bmatrix}$$

$$= I + p\gamma(w_p - 1)\zeta_1 e_{p1}.$$

$$T_2(h_v) = I + \begin{bmatrix} & \\ & \end{bmatrix} \begin{bmatrix} 0 & 1 & & \\ & \ddots & \ddots & \\ & & \ddots & 1 \\ \beta & & & 0 \end{bmatrix} \begin{bmatrix} \zeta_1^v & & \\ & \ddots & \\ & & \zeta_p^v \end{bmatrix} \begin{bmatrix} & \\ & \end{bmatrix}$$

$$= I + \sum p * \zeta_i^v e_{i,i+1},$$

where the sum is over such terms that if e_{lm} is present then e_{mk} is not present. Therefore,

$$T_2(\Pi h_v) = I + \sum_v \sum p * \zeta_i^v e_{i,i+1} = I.$$

We have only to find an element z as in $(*)$. We have

$$T_2 = \operatorname{ind}(\tau_2) = \operatorname{ind}(\tau_2^{b^i}).$$

Therefore, we may assume that there exist elements $z_i \in \ker\tau_1 \backslash \ker\tau_2^{b^i}$. First, say, we have $z \in \ker\tau_1\backslash\ker\tau_2$ such that $T_2(z)$ has exactly r eigenvalues equal to one in $i_1, i_2, \ldots, i_r$ places of the diagonal. We shall construct $\tilde{z} \in \ker\tau_1$, $T_2(\tilde{z})$ has at most $(r-1)$ eigenvalues equal to one. Choose one of the z_i's, say z_1 such that $T_2(z_1)$ has no one entry in i_1 place of the diagonal. Let $j_1, \ldots, j_s$ be integers 1 to p different form the i_v's. Let $T_2(z)$ have eigenvalues θ_i at j_i spots and $T_2(z_1)$ have α_i there. These are $\leq (p-1)$ places. Pick an integer k less than p such that $\theta_i^{-1} \neq \alpha_i^k$ for $i = j_1, \ldots, j_s$. Then $T_2(zz_1^k)$ has at most $(r-1)$ eigenvalues equal to one. We have found $\tilde{z} = zz_1^k$. Repeating this procedure we find $z^* \in \ker\tau_1\backslash\ker\tau_2$ such that $T_2(z^*)$ has no eigenvalue one. This completes the proof of (5.2) and the theorem.

REMARK. We should like to mention that from our proof also an estimate for $(U: H)$ may be derived; we leave this as an exercise.

6. Proof of Theorem 3. We have $G = D_{2n} = \langle a, b / a^n = 1 = b^2, a^b = a^{-1}\rangle$, the dihedral group of order $2n$. Then

$$\mathbb{Q}G = \mathbb{Q}(G/G') \bigoplus_{m \neq 1,2} (K_m)_{2\times 2}, \; m|n$$

where K_m is the real subfield of the mth cyclotomic field $\mathbb{Q}(\zeta_m)$. Writing $x_1 = \zeta_m + \zeta_m^{-1}$, the projection T_m on $(K_m)_{2\times 2}$ is given by

$$T_m(a) = \begin{bmatrix} 0 & 1 \\ -1 & x_1 \end{bmatrix}, \quad T_m(b) = \begin{bmatrix} 1 & 0 \\ x_1 & -1 \end{bmatrix}.$$

The integers of K_m are $\mathbb{Z}[x_1]$ (see [7]). As in the proof of the last theorem, it suffices to prove

(a) $T_m(H)$ is of finite index in $SL(2, \mathbb{Z}[x_1])$.
(b) For two divisors $d \neq m$ of n, there exists $h \in H$ such that $T_d(h) = 1$, $T_m(h)$ noncentral in $(\mathbb{Z}[x_1])_{2\times 2}$.

(a) By change of basis, the complex irreducible representation corresponding to T_m is given by

$$T(a) = \begin{bmatrix} \zeta_m & \\ & \zeta_m^{-1} \end{bmatrix}, \quad T(b) = \begin{bmatrix} 0 & 1 \\ 1 & 0 \end{bmatrix}.$$

In fact,

$$MT(a)M^{-1} = T_m(a), MT(b)M^{-1} = T_m(b)$$

where

$$M = \begin{bmatrix} 1 & 1 \\ \zeta_m & \zeta_m^{-1} \end{bmatrix}.$$

Then

$$T((1-b)a^i(1+b)) = \begin{bmatrix} \zeta_m^i - \zeta_m^i & \zeta_m^i - \zeta_m^{-i} \\ -\zeta_m^i + \zeta_m^{-i} & -\zeta_m^i + \zeta_m^{-i} \end{bmatrix},$$

$$T_m((1-b)a^i(1+b)) = \begin{bmatrix} 0 & 0 \\ x_{i+1} - x_{i-1} & 0 \end{bmatrix} \text{ where } x_i = \zeta_m^i + \zeta_m^{-i}.$$

Therefore,

$$T_m(u_{ba^i}) = \begin{bmatrix} 1 & 0 \\ x_{i+1} - x_{i-1} & 1 \end{bmatrix}.$$

We observe that

$$x_1(x_{i+1} - x_{i-1}) = x_{i+2} - x_{i-2} = (x_{i+2} - x_i) + (x_i - x_{i-2}) \text{ for } i > 1 \text{ and}$$
$$x_1(x_2 - x_0) = x_3 - x_1.$$

This means that the $\mathbb{Z}$-linear combinations of $\{(x_{i+1} - x_{i-1}), i \geq 1\}$, form an ideal Q of $\mathbb{Z}[x_1]$ and $T_m(H)$ contains $\begin{bmatrix} 1 & 0 \\ Q & 1 \end{bmatrix}$. Since H is normalized by G, conjugating by

$$T_m(ba) = \begin{bmatrix} 1 & 0 \\ x_1 & -1 \end{bmatrix} \begin{bmatrix} 0 & 1 \\ -1 & x_1 \end{bmatrix} = \begin{bmatrix} 0 & 1 \\ 1 & 0 \end{bmatrix} \text{ we also get } \begin{bmatrix} 1 & Q \\ 0 & 1 \end{bmatrix} \text{ in } T_m(H).$$

This proves (a) If $K_m \neq \mathbb{Q}$ or imaginary quadratic. We are left with the cases $m = 3, 6$ and $m = 4$, which we now treat separately.

(i) $m = 3$: then $x_1 = \zeta_3 + \zeta_3^{-1} = -1$. Write

$$T(a) = T_m(a) = \begin{bmatrix} 0 & 1 \\ -1 & -1 \end{bmatrix}, \quad T(b) = T_m(b) = \begin{bmatrix} 1 & 0 \\ -1 & -1 \end{bmatrix}.$$

Then $T(ab) = \begin{bmatrix} -1 & -1 \\ 0 & 1 \end{bmatrix}$, $T(ba) = \begin{bmatrix} 0 & 1 \\ 1 & 0 \end{bmatrix}$. We observe that

$$T(1-(1-b)a(1+b)) = \begin{bmatrix} 1 & 0 \\ 3 & 1 \end{bmatrix}$$

$$T(1-(1-b)a(1+b))^{ba} = \begin{bmatrix} 1 & 3 \\ 0 & 1 \end{bmatrix}$$

$$T(1+(1-ba)a(1+ba)) = \begin{bmatrix} 4 & 3 \\ -3 & -2 \end{bmatrix}.$$

Thus $T(H)$ contains the subgroup

$$K = \left\langle \alpha = \begin{bmatrix} 1 & 0 \\ 3 & 1 \end{bmatrix}, \beta = \begin{bmatrix} 1 & 3 \\ 0 & 1 \end{bmatrix}, \gamma = \begin{bmatrix} 4 & 3 \\ -3 & -2 \end{bmatrix} \right\rangle,$$

which is normal in

$$SL(2,\mathbb{Z}) = \left\langle S = \begin{bmatrix} 0 & -1 \\ 1 & 0 \end{bmatrix}, U = \begin{bmatrix} 0 & 1 \\ -1 & -1 \end{bmatrix} \right\rangle.$$

As a matter of fact,

$$\alpha^U = \beta^{-1}, \beta^U = \gamma, \gamma^U = \alpha^{-1}, \alpha^s = \beta^{-1}, \beta^s = \alpha^{-1}, \gamma^s = \alpha\gamma^{-1}\beta^{-1}.$$

We have proved that $T(H)$ contains $\tilde{E}(3\mathbb{Z})$ which coincides with the congruence subgroup mod 3 by **[2]**; and (a) follows for $m = 3$. Now let $m = 6$. Then we have only to observe that

$$T_6(a^2) = \begin{bmatrix} -1 & 1 \\ -1 & 0 \end{bmatrix}, \quad T_6(b) = \begin{bmatrix} 1 & 0 \\ 1 & -1 \end{bmatrix};$$

and that these matrices are conjugate by $\begin{bmatrix} 1 & 0 \\ -1 & 1 \end{bmatrix}$ to $T_3(a)$ and $T_3(b)$ respectively.

(ii) $m = 4$: then $x_1 = \zeta_4 + \zeta_4^{-1} = 0$. Write

$$T(a) = T_m(a) = \begin{bmatrix} 0 & 1 \\ -1 & 0 \end{bmatrix}; \; T(b) = T_m(b) = \begin{bmatrix} 1 & 0 \\ 0 & -1 \end{bmatrix}.$$

Then

$$T(ba) = \begin{bmatrix} 0 & 1 \\ 1 & 0 \end{bmatrix}, \quad T(ab) = \begin{bmatrix} 0 & -1 \\ -1 & 0 \end{bmatrix}.$$

We observe that

$$T((1+(1-a^2b)a(1+a^2b)) = \begin{bmatrix} 1 & 4 \\ 0 & 1 \end{bmatrix} = X$$

$$T(1-(1-b)a(1+b)) = \begin{bmatrix} 1 & 0 \\ 4 & 1 \end{bmatrix} = Y$$

$$T(1+(1-ab)b(1+ab)) = \begin{bmatrix} 3 & -2 \\ 2 & -1 \end{bmatrix} = A$$

$$T(1-(1-a^3b)a(1+a^3b)) = \begin{bmatrix} -1 & -2 \\ 2 & 3 \end{bmatrix} = B.$$

Thus $T(H)$ contains the group

$$\left\langle X = \begin{bmatrix} 1 & 4 \\ 0 & 1 \end{bmatrix}, Y = \begin{bmatrix} 1 & 0 \\ 4 & 1 \end{bmatrix}, A = \begin{bmatrix} 3 & -2 \\ 2 & -1 \end{bmatrix}, B = \begin{bmatrix} -1 & -2 \\ 2 & 3 \end{bmatrix} \right\rangle.$$

We claim that the subgroup

$$K = \langle X, Y, A^2, B^2, AB^{-1} \rangle$$

of $T(H)$ is normal in $SL(2, \mathbb{Z})$. This is seen by observing, with U and S as in the case $m = 3$ that

$$\begin{aligned} X^s &= Y^{-1}, Y^s = X^{-1}, (A^2)^s = B^2, (B^2)^s = A^2, (AB^{-1})^s = (AB^{-1})^{-1}; \\ Y^U &= X^{-1}, X^U = Y^{-2}, (A^2)^U = (AB^{-1} \cdot B^2)^{-1} X, \\ (B^2)^U &= Y^{-1}, (AB^{-1})^U = B^{-1}A = (A^{-2} \cdot AB^{-1} \cdot B^2)^{-1}. \end{aligned}$$

We have proved that K contains $\tilde{E}(4\mathbb{Z})$ which equals the congruence subgroup mod $4\mathbb{Z}$ by **[2]**. Thus K and hence $T(H)$ is of finite index in $SL(2, \mathbb{Z})$, as required. This completes the proof of (a).

(b) We fix $\zeta = \zeta_n$ and change notation to write

$$T_d(a) = \begin{bmatrix} \zeta^d & \\ & \zeta^{-d} \end{bmatrix} T_m(a) = \begin{bmatrix} \zeta^m & \\ & \zeta^{-m} \end{bmatrix}, \zeta^m \neq -1$$

$$T_d(b) = T_m(a) = \begin{bmatrix} 0 & 1 \\ 1 & 0 \end{bmatrix}.$$

Observe that

$$(1 + (1 - b)x(1 + b))(1 + (1 - b)y(1 + b)) = 1 + (1 - b)(x + y)(1 + b).$$

Therefore, for any polynomial f, $1 + (1 - b)f(a)(1 + b) \in H$. Let ϕ be the irreducible $\mathbb{Z}$-polynomial of ζ^d. Then take $h = 1 + (1 - b)\phi(a)(1 + b)$ and get

$$\begin{aligned} T_d(h) &= I \\ T_m(h) &= I + \begin{bmatrix} 1 & -1 \\ -1 & 1 \end{bmatrix} \begin{bmatrix} \alpha & 0 \\ 0 & \beta \end{bmatrix} \begin{bmatrix} 1 & 1 \\ 1 & 1 \end{bmatrix}, \ \alpha = \phi(\zeta^m), \ \beta = \phi(\zeta^{-m}) \\ &= I + \begin{bmatrix} \alpha & -\beta \\ -\alpha & \beta \end{bmatrix} \begin{bmatrix} 1 & 1 \\ 1 & 1 \end{bmatrix} = I + \begin{bmatrix} \alpha - \beta & \alpha - \beta \\ -\alpha + \beta & -\alpha + \beta \end{bmatrix} \\ &= I + (\alpha - \beta) \begin{bmatrix} 1 & 1 \\ -1 & -1 \end{bmatrix}. \end{aligned}$$

Then $T_m(h)$ is noncentral unless $\alpha = \beta$. If so, we change h to $h_1 = 1 + (1 - b)\phi(a)a(1 + b)$. In this case, $T_d(h_1) = 1$, $T_m(h_1) = I + \alpha(\zeta^m - \zeta^{-m}) \left[\begin{smallmatrix} 1 & 1 \\ -1 & -1 \end{smallmatrix}\right]$. We are done since $\zeta^m \neq \zeta^{-m}$. The theorem is proved.

Added in Proof: Theorem 2 has now been proved for all odd primes p and, in fact, for all nilpotent groups whose representations have Schur index one.

References

[1] A. Bak and U. Rehmann, *The congruence subgroup and metaplectic problems for $SL_{n \geq 2}$ of division algebras*, J. Algebra **78** (1982), 475–547.

[2] J. L. Brenner, *The linear homogeneous group. III*, Ann. of Math. (2) **78** (1982), 475–547.

[3] C. Curtis and I. Reiner, *Representation theory of finite groups and associative algebras*, Interscience, N. Y. (1962).

[4] M. Hall, *The theory of groups*, Macmillan, N. Y. (1959).

[5] B. Huppert, *EndLiche Gruppen I*, Springer, Berlin-Heidelberg-New York (1967).

[6] N. Ito, *Frobenius and Zassenhaus groups*, Lecture Notes, University of Illinois at Chicago Circle, (1969).

[7] J. Liang, *On the integral basis of the maximal real subfield of a cyclotomic field*, J. Reine Angew. Math. **286/287** (1976).

[8.] B. Liehl, *On the group SL_2 over orders of arithmetic type*, J. Reine Angew. Math. **323** (1981), 153–171.

[9] U. Rehmann, *A survey of the congruence subgroup problem, in "Algebraic K-Theory,"* Springer Lecture Notes in Math. **966**, 197–207.

[10] I. Reiner, *Maximal orders*, Academic Press, N. Y. (1975).

[11] J. Ritter and S. K. Sehgal, *Certain normal subgroups of $U(\mathbb{Z}G)$*, J. Reine Angew. Math. **381** (1987), 214–220.

[12] L. N. Vaserstein, *The structure of classic arithmetic groups of rank greater than one*, Math. USSR Sbornik **20** (1973), 465–492.

[13] ——, *On the group SL_2 over Dedekind rings of arithmetic type*, Math. USSR Sbornik **18** (1973), 321–332.

[14] D. S. Passman, *Permutation groups*, Benjamin, N. Y. (1968).

[15] H. Zassenhaus, *Über endliche Fastkörper*, Hamburg Abh. **11** (1935), 187–220.

INSTITUT FÜR MATHEMATIK DER UNIVERSITÄT AUGSBURG
D-8900 AUGSBURG, W.-GERMANY
DEPT. OF MATHEMATICS, UNIVERSITY OF ALBERTA
EDMONTON, T6G 2G1, CANADA

Contemporary Mathematics
Volume **93** (1989)

Connectivity of the Finite Group and Generalized Characters

E. M. ŽMUD′

Dedicated to the memory of S. D. Berman

1. In the paper the finite groups are only considered. The following notations will be used: G is a group of order h; $H \leq G$ $(H < G)$ is an H-subgroup (proper subgroup) of G; $o(g)$ is the order of the element $g \in G$; $\pi(g)$ is the set of all prime divisors of $o(g)$; if $X \subseteq G$, then $X^{\#} = X \backslash \{1\}$, $\hat{X} = X \cup \{1\}$, $X^G = \bigcup_{g \in G} X^g$–here $X^g = g^{-1}Xg$; $\pi(X) = \bigcup_{g \in X} \pi(g)$; if $\pi \subseteq \pi(G)$, then $\pi' = \pi(G) \backslash \pi$; G_π is the set of all π-elements of a group G; if $G^{\#} = A \cup B$, $A = G^{\#}_{\pi}$, $B = G^{\#}_{\pi'}$, then $\{A, B\}$ is a (π, π')-partition of G; $Z(G)$, $F(G)$, $S_c(G)$ are the centre, Fitting subgroup and the socle of a group G, respectively; $\mathrm{Char}(G)$ is the set of all complex characters of G; $X(G) = \{x' - x'' | x', x'' \in \mathrm{Char}(G)\}$ is the ring of generalized characters of a group G; $\mathrm{Irr}(G) = \{x \in \mathrm{Char}(G) | \chi \text{ is irreducible}\}$. The natural numbers set is denoted by $\mathbb{N}$, the ring of rational integer number–by $\mathbb{Z}$, the ring of algebraic integer numbers–by $\mathbb{R}$.

2. The following result from paper **[2]** will be refined and strengthened in the present work: a group G possesses a nontrivial (π, π')-partition (π is not fixed) if and only if it possesses a generalized character, which takes on $G^{\#}$ values 0 and 1.

The terminology similar to one accepted in papers **[1]** and **[2]** will be used. Elements $x, y \in G^{\#}$ will be called *weak permutable*, if there exists such a sequence $x_0 = x, x_1, \ldots, x_n = y$ of elements $x_i \in G^{\#}$, that $x_{i-1}x_i = x_i x_{i-1}$ $(i = 1, \ldots, n)$. The relation of weak permutability is the equivalence relation on $G^{\#}$. Classes of weak permutable elements are called *connectivity components* of a group G. If a number of connectivity components is > 1, then the group G is called *non-connected*; otherwise, the group G is *connected*. For instance, the non-connected group are: Frobenius groups, double Frobenius groups, Zassenhaus groups, $SL(3, 4)$, M_{11}.

0271-4132/89 $1.00 + $.25 per page

A subset $A \subseteq G^{\#}$ will be called a C-subset if $g \in A$ implies $C_G(g) \subseteq \hat{A}$. A subset $A \subseteq G^{\#}$ is a C-subset if and only if it could be presented as a union of a certain number of connectivity components of a group G. If A and B are C-subsets, then $A \cup B$, $A \cap B$ and $G^{\#} \setminus A$ are C-subsets as well. Therefore the partially ordered by inclusion set $\mathscr{L}_C(G)$ of all the C-subsets of the group G is a Boolean lattice. It is obvious that the connectivity components of the group G could be permuted by its inner automorphisms. If $A \in \mathscr{L}_C(G)$, then $A^g \in \mathscr{L}_C(G)$ by any $g \in G$. Normal (i.e. invariant to all the inner automorphisms) C-subsets will be called NC-subsets. A group G is non-connected if and onldy if it possesses a nontrivial (i.e. different from $\varnothing$ and $G^{\#}$) NC-subset. The set $\mathscr{L}_{NC}(G)$ of all the NC-subsets of the group G is a sublattice of the lattice $\mathscr{L}_C(G)$. Atoms of the lattice $\mathscr{L}_{NC}(G)$ have form D^G; by D is a connectivity component of G denoted.

Let $\theta \in X(G)$. Function $\theta^{\#} = \theta|_{G^{\#}}$ will be called a truncated generalized character of the group G. The truncated generalized characters of the group G form a ring $X^{\#}(G)$, isomorphous to $X(G)/(\rho)$, where $(\rho) = X(G)\rho = \mathbb{Z}\rho$ is an ideal, generated by the regular character ρ of the group G. An ideal $I \subseteq X^{\#}(G)$ will be called complementable if there exists such an ideal $I' \subseteq X^{\#}(G)$ that $X^{\#}(G) = I \oplus I'$. The ideal I is complementable if and only if it is generated by some idempotent element of the ring $X^{\#}(G)$. The set $\mathscr{J}(G)$ of all complementable ideals of the ring $X^{\#}(G)$ is a Boolean lattice.

Following statements are the main results of the present work:

THEOREM 1. $\mathscr{L}_{NC}(G) \cong \mathscr{J}(G)$.

COROLLARY. *The group G is connected if and only if the ring $X^{\#}(G)$ is indecomposable into a direct sum of nontrivial ideals.*

THEOREM 2. *The lattice $\mathscr{L}_{NC}(G)$ is uniquely defined, up to isomorphism, by the ring $X(G)$.*

3. First we determine a number of properties of non-connected groups. Part of them was announced in **[1]** (see also **[2]**).

PROPOSITION 1. *A group G is non-connected if and only if it possesses a nontrivial NC-subset.*

PROOF. Suppose, the group G is non-connected and D is its connectivity component. If $\{D_0 = D, D_1, \dots, D_{n-1}\}$ is a class of connectivity components, which are conjugate with D, then $A = D^G = UD_i$ is a nonempty NC-subset. Since $|D_i| = |D|$ $(i = 0, 1, \dots, n-1)$, then $|A| = n|D|$. As $n = |G\colon N_G(D)|$, then we have $n|(|A|, h)$, hence $A \neq G^{\#}$. Thus a non-connected group contains a nontrivial NC-subset. The converse statement is obviously true.

A partition $\{A_i, \dots, A_m\}$ of the group G is said to be given, if $G^{\#} = UA_i$ and $A_i \cap A_j = \varnothing$ for $i \neq j$. A partition $\{A_1, \dots, A_m\}$ is normal, if all the subsets A_i are normal. A partition $\{A_1, \dots, A_m\}$ will be called a C-partition, if $A_i \in \mathscr{L}_C(G)$ $(i = 1, \dots, m)$, and NC-partition, if $A_i \in \mathscr{L}_{NC}(G)$ $(i = 1, \dots, m)$.

Proposition 1 implies the equivalence of the following assertions: (I) The group G is non-connected; (II) There exists a nontrivial (i.e. such that $A \neq \varnothing \neq B$) C-partition $\{A, B\}$; (III) There exists a nontrivial partition.

PROPOSITION 2. *Assume $A, B \in \mathscr{L}_C(G)$, $H < G$ is nilpotent, $H \subseteq \hat{A} \cup \hat{B}$. Then $H \subseteq \hat{A}$ or $H \subseteq \hat{B}$.*

PROOF. If $H \neq \{1\}$, then $Z(H) \neq \{1\}$, hence, the latter statement follows.

COROLLARY. *If $\{A_1, \ldots, A_m\}$ is a C-partition of the group G and $H < G$ is nilpotent, then H is contained in one of subsets A_i.*

PROPOSITION 3. *Assume $A \in \mathscr{L}_C(G)$, $A \neq G^\#$, $N \triangleleft G$, $N \subseteq \hat{A}$. Then N is nilpotent.*

PROOF. Let $B = C^\# \backslash A$, $g \in B$, $p \in \pi(g)$. Then $s = g^{o(g)/p} \in B$, $o(s) = p$. The element s generates in N a regular automorphism $x \mapsto x^s$ of the prime order p. Hence the proposition follows.

PROPOSITION 4. *The partition $\{A, B\}$ of the group G is a NC-partition if and only if it is a (π, π')-partition by some $\pi \subseteq \pi(G)$.*

PROOF. 1°. Let $\{A, B\}$ be a nontrivial NC-partition, $x \in A$, $y \in B$. If $(o(x), o(y)) \neq 1$, then $p | (o(x), o(y))$ for some $p \in \pi(G)$. Since $x' = x^{o(x)/p} \in A$, $o(x') = p$, then, by the Corollary of Proposition 2, S_p-subgroup of the group G, which contains x', is contained in $\hat{A}$. By analogy, all the S_p-subgroups of the group G are contained in B. The contradiction shows that $\{A, B\}$ is a (π, π')-partition of the group G, where $\pi = \pi(A)$, $\pi' = \pi(B)$.

2°. Let $\{A, B\}$ be a nontrivial (π, π')-partition of the group G, $x \in A$, $y \in B$, $xy = yx$. Then $o(xy) = o(x)o(y)$, hence $xy \notin G_\pi \cup G_{\pi'} = G$. The contradiction shows that $C_G(x) \subseteq \hat{A}$. So $A \in \mathscr{L}_{NC}(G)$ and $\{A, B\}$ is a NC-partition.

By the *capacity* $c(A)$ of a subset $A \subseteq G$ will be called the least common multiple of the orders of subgroups, which are contained in $\hat{A}$, by $c_S(A)$ we denote the least common multiple of the orders of Sylow subgroups of the group G, which are contained in $\hat{A}$.

PROPOSITION 5. *If $A \in \mathscr{L}_C(G)$, then $c(A) = c_S(A)$.*

PROOF. It is obvious that $c_S(A) | c(A)$. Let $H < G$; $H \subseteq \hat{A}$, $P \in \mathrm{Syl}(H)$. Then $P \subseteq \hat{A}$, this implies $|P| | c_S(A)$. Therefore $|H| | c_S(A)$. Thus $c(A) = c_S(A)$.

COROLLARY. *Let $A = D^G$, where D is a connectivity component of the group G. Then $c(A) = c(D)$.*

PROOF. It follows from Proposition 2, from the relation $c(D^g) = c(D)$ $(g \in G)$ and from Proposition 5.

Proposition 6. *If* $A, B \in \mathscr{L}_{NG}(G)$, $A \cap B = \varnothing$, *then* $(c(A), c(B)) = 1$ *and* $c(A \cup B) = c(A) \cdot c(B)$.

Proof. Since $\{A, G^{\#}\backslash A\}$ is a NC-partition and $B \subseteq G^{\#}\backslash A$, then $(c(A), c(B)) = 1$ by Proposition 4. The definition implies $c(A)|c(A \cup B)$, $c(B)|c(A \cup B)$. Therefore $c(A)c(B)|c(A \cup B)$. Suppose $P \in \mathrm{Syl}(G)$, $P \subseteq \widehat{A \cup B}$. Then $P \subseteq \hat{A}$ or $P \subseteq \hat{B}$. Therefore $|P||\, c(A)$ or $|P||\, c(B)$ and so $|P||\, c(A)c(B)$. Proposition 5 implies $c(A \cup B)|c(A)c(B)$. Thus $c(A \cup B) = c(A)c(B)$.

Corollary 1. *If* $A, B \in \mathscr{L}_{NC}(G)$, *then* $c(A \cup B)c(A \cap B) = c(A)c(B)$.

Corollary 2. *Let* $\{A_1, \ldots, A_m\}$ *be a* NC*-partition of the group* G; $c_i = c(A_i)$ $(i = 1, \ldots, m)$. *Then* $(c_i, c_j) = 1$ *by* $i \neq j$ *and* $h = c_1 \ldots c_m$.

Proposition 7. *If* $A \in \mathscr{L}_{NC}(G)$, *then* $\hat{A} = \{x \in G | x^{c(A)} = 1\}$.

Proof. If $x \in \hat{A}$, then $\langle x \rangle \subseteq \hat{A}$. Therefore $o(x)|c(A)$, hence $x^{c(A)} = 1$. And conversely, suppose $x^{c(A)} = 1$. If $x \in G\backslash A = \hat{B}$, it follows from the proven statement that $x^{c(B)} = 1$. Since $(c(A), c(B)) = 1$, then $x = 1$, i.e. $x \in \hat{A}$.

Lemma 1. *(Frobenius* [**3**]*). Let* $n \in \mathbb{N}$, $n|h$. *Then* n *divides the number of solutions of the equation* $x^n = 1$ $(x \in G)$.

Proposition 8. *Let* $\{A_1, \ldots, A_m\}$ *be a* NC*-partition of the group* G; $c_i = c(A_i)$, $a_i = |A_i|$ $(i = 1, \ldots, m)$. *Then* $a_i \equiv -\delta_{ij} \pmod{c_i}$ $(i, j = 1, \ldots, m)$.

Proof. By Proposition 7 and Lemma 1 $a_i = |\hat{A}_i| - 1 \equiv -1 \pmod{c_i}$. Let $j \neq i$, $p \in \pi(G)$, $p|c_j$, $P \in \mathrm{Syl}_p(G)$. Then $|P| = p^\alpha$ and $P \subseteq \hat{A}_j$. Since P acts by conjugations on A_i semiregularly, then $p^\alpha | a_i$. Hence $a_i \equiv) \pmod{c_j}$.

Note. As $\sum a_i = |UA_i| = h - 1$, then $0 \leq a_i < h$ $(i = 1, \ldots, m)$. Therefore a_i is the solution, reduced by module h, of the system of linear congruences

$$(1) \qquad x \equiv -\delta_{ij} \pmod{c_j} \ (j = 1, \ldots, m).$$

That shows that the orders of subsets A_i can define the whole complex of $\{c_i\}$-their capacities. As is shown by examples, not every system $\{c_i\}$ of natural numbers, which are relatively prime in pairs, is a system of capacities, which corresponds to a NC-partition of some group (the necessary condition for it is the equality $\sum a_i = h - 1$, where a_i is a solution, reduced by module $h = \prod c_i$, of the system (1)).

Lemma 2. *(Frobenius* [**3**]*). Let* $n \in \mathbb{N}$, $n|h$ *and function* $\vartheta\colon G \to \mathbb{Z}$ *is defined in the way that*

$$\vartheta(x) = \begin{cases} \frac{h}{n}, & \text{if } x^n = 1 \\ 0, & \text{if } x^n \neq 1. \end{cases}$$

Then $\vartheta \in X(G)$.

Lemma 3. *Let* $\theta \in X(G)$; $x, y \in G$, $xy = yx$, $y^{p^\alpha} = 1$, *where* $p \in \pi(G)$; η *is a primitive* p^α*th root of* 1; $\lambda = 1 - \eta$. *Then the following congruence takes place in the ring* $\mathbb{R}$: $\theta(xy) \equiv \theta(x) \pmod{\lambda}$.

PROOF. Since $\theta = \chi_1 - \chi_2$, where $\chi_1, \chi_2 \in \mathrm{Char}(G)$, it is sufficient to prove the assertion for the case, when $\theta = \chi \in \mathrm{Char}(G)$. Let Γ be a matrix linear representation (of degree $f = \chi(1)$) of the group G, generating a character χ. By certain numeration of the spectra $\{\xi_1, \dots, \xi_f\}$, $\{\eta_1, \dots, \eta_f\}$ of matrices $\Gamma(x)$ and $\Gamma(y)$, the matrix $\Gamma(xy)$ has the spectrum $\{\xi_1\eta_1, \dots, \xi_f\eta_f\}$. Since $\eta_j = \eta^{m_j}$ ($m_j \in \mathbb{Z}$), then $\eta_j \equiv 1 \pmod{\lambda}$ ($j = 1, \dots, f$). That is why $\chi(xy) = \sum \xi_j\eta_j \equiv \sum \xi_j \equiv \chi(x) \pmod{\lambda}$, what actually proves the lemma.

Let $\varepsilon_\Delta \colon G^\# \to \{0, 1\}$ be a characteristic function of a subset $A \subseteq G^\#$.

PROPOSITION 9. *A subset $A \subseteq G^\#$ is a NC-subset if and only if $\varepsilon_A \in X^\#(G)$.*

PROOF. 1°. Suppose $A \in \mathscr{L}_{NC}(G)$, $B = G\backslash\hat{A}$. Then, by Proposition 6, $(c(A), c(B)) = 1$, $c(A)c(B) = h$. The functions $\vartheta_A \colon G \to \mathbb{Z}$ and $\vartheta_B \colon G \to \mathbb{Z}$ will be defined by putting

$$\vartheta_A(x) = \begin{cases} c(B), & \text{if } x \in \hat{A} \\ 0, & \text{if } x \in B \end{cases} \qquad \vartheta_B(x) = \begin{cases} 0, & \text{if } x \in A \\ c(A), & \text{if } x \in \hat{B} \end{cases}.$$

PROPOSITION 7 AND LEMMA 2. *(by $n = c(A)$ and $n = c(B)$) imply $\vartheta_A, \vartheta_B \in X(G)$. Let $\vartheta = c(A)1_G - \vartheta_B$, where 1_G is main character of the group G. It is easily seen that $\vartheta \in X(G)$, $\vartheta(x) = c(A)$, if $x \in A$; $\vartheta(x) = 0$, if $x \in B$. Using the fact that $(c(A), c(B)) = 1$, we hope to find $u, v \in \mathbb{Z}$, such that $c(B)u + c(A)v = 1$. Setting $\theta = u\vartheta_A + v\vartheta_B$, we obtain: $\theta \in X(G)$, $\varepsilon_A = \theta^\# \in X^\#(G)$.*

2°. Suppose $\varepsilon_A \in X^\#(G)$. Then $\varepsilon_A = \theta^\#$, where $\theta \in X(G)$. By setting $B = G\backslash\hat{A}$, we shall prove $\{A, B\}$ is a NC-partition. If it is not true, then $A \neq \varnothing \neq B$ and there exists such $x \in A$, $y \in B$, that $xy = yx$. Using Lemma 3, it is easy to prove that the primary components of element x are contained in A, and primary components of element y- in B. One may therefore consider that $o(x) = p^\alpha$, $o(y) = q^\beta$, where $p, q \in \pi(G)$. Let ξ and η be p^αth and q^βth roots of unity respectively, $\lambda = 1 - \eta$, $\mu = 1 - \xi$. By Lemma 3, the following congruence take place in the ring $\mathbb{R}$: $\theta(xy) \equiv \theta(x) \pmod{\lambda}$, $\theta(xy) \equiv \theta(y) \pmod{\mu}$. As $x \in A$, $y \in B$ and $\theta^\# = \varepsilon_A$, then $\theta(x) = 1$, $\theta(y) = 0$. Therefore $\theta(xy) \equiv 1 \pmod{\lambda}$, $\theta(xy) \equiv 0 \pmod{\mu}$. If $xy \in A$, then $1 = \theta(xy) \equiv 0 \pmod{\mu}$; if $xy \in B$, then $0 = \theta(xy) \equiv 1 \pmod{\lambda}$. As these congruences are impossible (in view of the non-invertibility of λ and μ in the ring $\mathbb{R}$), then a contradiction is obtained. So, $\{A, B\}$ is a NC-partition and, hence, $A \in \mathscr{L}_{NC}(G)$.

PROOF OF THEOREM 1. As seen from the Proposition 9, mapping $A \mapsto \varepsilon_A$ ($A \in \mathscr{L}_{NC}(G)$) is a bijection of the lattice $\mathscr{L}_{NC}(G)$ onto the set of all idempotent elements of the ring $X^\#(G)$. The mapping $A \mapsto \varepsilon_A$ is, evidently, an isomorphism of the lattice $\mathscr{L}_{NC}(G)$ onto the lattice $\mathscr{J}(G)$.

PROOF OF THEOREM 2. Let G and G^* be groups, for which $X(G) \cong X(G^*)$. We prove that $X^\#(G) \cong X^\#(G^*)$. To do this, it is sufficient to prove that there exists an isomorphism $X(G) \to X(G^*)$, converting the regular character ρ of the group G to the regular character ρ^* of the group G^*. Assume that $\mathrm{Irr}(G) = \{\chi_1, \dots, \chi_k\}$. If $0 \neq \theta \in X(G)$, $\theta = \sum_i m_i\chi_i$ ($m_i \in \mathbb{Z}, i = 1, \dots, k$), then $d(\theta) = (m_1, \ldots, m_k)$ is the maximum of the numbers $n \in \mathbb{N}$, such that $n^{-1}\theta \in X(G)$. Therefore if $f\colon X(G) \to X(G^*)$ is the isomorphism of rings, then $d(f(\theta)) = d(\theta)$. Denote $\sigma = f^{-1}(\rho^*)$. As $d(\rho^*) = 1$, by the preceding note, $d(\sigma) = 1$. Further, since $\rho^{*2} = h^*\rho^*$, where $h^* = |G^*|$, then $\sigma^2 = h^*\sigma$. If $\theta \in X(G)$, then $f(\sigma\theta) = \rho^*\theta^* = \theta^*(1)\rho^* = f(\theta^*(1)\sigma)$, where $\theta^* = f(\theta) \in X(G^*)$. Therefore $\sigma\theta = m(\theta)\sigma$, where $m(\theta) = \theta^*(1) \in \mathbb{Z}$. Denote $S = \{g \in G | \sigma(g) \neq 0\}$. Since $\sigma \neq 0$ and $\sigma^2 = h^*\sigma$, then

$$\sigma(g) = \begin{cases} h^*, & \text{if } g \in S \\ 0, & \text{if } g \in G\backslash S. \end{cases} \tag{2}$$

Since $\sigma\chi_i = m(\chi_i)\sigma$, then $\chi_i(g) = m(\chi_i)$ for all $g \in S$ and for every $i = 1, \ldots, k$. Therefore S is the conjugacy class of the group G. Let s be a fixed representative of class S. Compute $(\sigma, \chi_i) = h^{-1}\sum_{g\in G}\sigma(g)\overline{\chi_i(g)}$. By (2) $(\sigma, \chi_i) = h^{-1}h^*|S|\overline{\chi_i(s)} = h^*|C_G(s)|^{-1}\overline{\chi_i(s)}$. In view of the fact that $(\sigma, \chi_i) \in \mathbb{Z}$, hence it follows that $\chi_i(s) \in \mathbb{Z}(= 1, \ldots,)$ and that $d = h^*|C_G(s)|^{-1} = (\sigma, 1_G) \in \mathbb{N}$. Since $\sigma = \sum_i(\sigma, \chi_i)\chi_i = d\sum_i \chi_i(s)\chi_i$ then $d^{-1}\sigma \in X(G)$. Therefore $d|d(\sigma)$, it implies $d = 1$ and so, $\sigma = \sum_i^i \chi_i(s)\chi_i$. Besides this, $h^* = |C_G(s)|$, what yields $h^*|h$. Having changed the groups G and G^* by their places, we have $h|h^*$. Thus $h = h^*$ and $|G_C(s)| = h$, hence $s \in Z(G)$. Since $\chi_i(s) \in \mathbb{Z}$, then $\chi_i(s) = \varepsilon_i\chi_i(1)$, where $\varepsilon_i = \pm 1$. Hence $s^2 = 1$. Suppose $\theta \in X(G)$. The function θ' will be defined by setting $\theta'(g) = \theta(sg)$. The mapping $\varphi\colon \theta \mapsto \theta'$ is an automorphism of the ring $X(G)$. As $\rho = \sum_i \chi_i(1)\chi_i$, then $\varphi(\rho) = \sum_i \chi_i(1)\varphi(\chi_i) = \sum_i \chi_i(1)\varepsilon_i\chi_i = \sum_i \chi_i(s)\chi_i = \sigma$. The mapping $f \circ \varphi\colon X(G) \to X(G^*)$ is an isomorphism of the rings, the relation $(f \circ \varphi)(\rho) = \rho^*$ follows. So, the relation $X^\#(G) \cong X^\#(G^*)$ is proven in this way. Hence $\mathcal{J}(G) \cong \mathcal{J}(G^*)$. And $\mathcal{L}_{NC}(G) \cong \mathcal{L}_{NC}(G^*)$ by Theorem 1.

4. Assume $A \in \mathcal{L}_{NC}(G)$, $\theta \in X(G)$, $\theta^\# = \varepsilon_A$ (the existence of θ is assured by Proposition 9). Function θ is uniquely determined by the additional condition $(\theta, 1_G) = 1$. Indeed, if $\theta_0 \in X(G)$, $\theta_0^\# = \varepsilon_A$, then θ has the form $\theta_0 + k\rho$, where $k \in \mathbb{Z}$. Condition $(\theta, 1_G) = 1$ implies $k = 1 - (\theta_0, 1_G)$. Denote by θ_A a generalized character θ, satisfying the conditions $\theta^\# = \varepsilon_A$, $(\theta, 1_G) = 1$; besides this, set $f_A = \theta_A(1)$. Let $B = G^\#\backslash A$. Then $f_A = |\hat{B}| = h - |A|$, hence, in particular, $f_A \geq 1$. Indeed, since $(\theta, 1_G) = 1$, then $h = \sum_{g\in G}\theta_A(g) = f_A + \sum_{g\in G^\#}\varepsilon_A(g) = f_A + |A|$, hence $f_A = h - |A| = |\hat{B}| \geq 1$.

PROPOSITION 10. *If $H \leq G$ then $|A \cap H|$ is equal to the remainder of the division of $|A|$ by $|H|$.*

PROOF. We have $|H|(\theta_A|_H, 1_H) = \sum_{g\in H}\theta_A(g) = f_A + \sum_{g\in G^*}\varepsilon_A(g) = f_A + |A \cap H| = h - |A| + |A \cap H|$, hence $|A| \equiv |A \cap H| \pmod{|H|}$. It is evident that $0 \leq |A \cap H| < |H|$.

COROLLARY 1. *If $H \leq G$, then $|A\backslash H| \equiv 0 \pmod{|H|}$.*

COROLLARY 2. *If $H \leq G$, then the conditions $A \subset H$ and $|A| < |H|$ are equivalent. In particular, if $\langle A\rangle = G$ and $H < G$, then $|H| \leq |A|$. Hence* *it follows, if G is a Frobenius group with the kernel N and $H \leq G$ and if $|H| \geq |N|$, then $H \geq N$.*

Let $\{A_1, \dots, A_m\}$ be a NC-partition of the group G; $a_i = |A_i|$ $(i = 1, \dots, m)$. Let a system of the integer numbers $R(H) = \{r_1(H), \dots, r_m(H)\} \in \mathbb{Z}^m$ correspond to a subgroup $H \leq G$, here $r_i(H)$ is a remainder from the division of a_i by $|H|$. It results from Proposition 10 that $\sum_i r_i(H) = |H| - 1$.

COROLLARY 3. *If $H_1, H_2 \leq G$, then the conditions $R(H_1) = R(H_2)$ and $|H_1| = |H_2|$ are equivalent.*

The number $a_0 = 1$ being aggregated to the system $\{a_1, \dots, a_m\}$, we shall have $\sum_{i=0}^m a_i = h$. If $d|h$, $d > 0$, then we set $S(d) = \sum\{\frac{a_i}{d}\}$, where $\{\alpha\}$ is fractional part of the number α. The divisor $d > 0$ of the number h will be called *Lagrangean*, if the group G contains a subgroup of the order d. The following assertion, resulting from Corollary 3, presents the necessary (but non-sufficient, as it is shown by examples) condition, that the divisor d of the order of the non-connected group G is of Lagrangean type.

COROLLARY 4. *Assume $d|h$, $d > 1$. Then $S(d) \in \mathbb{N}$. If d is the Lagrangean divisor of the number h, then $S(d) = 1$.*

Using Corollary 1 of Proposition 10 it is easy to receive

COROLLARY 5. *Suppose, G is a non-connected simple group, m is the number of atom in the lattice $\mathscr{L}_{NC}(G)$. Then $m + 1 \leq |G\colon H|$ for every subgroup $H < G$.*

5. Assume, as above, that $A \in \mathscr{L}_{NC}(G)$, $H \leq G$. As it follows from Corollary 2 of Proposition 10, $H \not\supset A$ implies $|H| \leq |A|$. The case, when $|H| = |A|$ is of interest. The certain is contained in the following statements.

PROPOSITION 11. *Suppose $A \in \mathscr{L}_{NC}(G)$, $H < G$, $|H| = |A|$. Then*
(a) *A is the conjugacy class of the group G.*
(b) *there exists such $p \in \pi(G)$ that $\hat{A} = P^G$, where $P \in \mathrm{Syl}_p(G)$.*
(c) *Then the factorization $G = P \cdot H$ takes place.*
(d) *The S_p-subgroups of the group G are elementary abelian CC-subgroups (The subgroup $K < G$ is called the CC-subgroup, if $g \in K^\#$ implies $C_G(g) \leq K$).*
(e) *H is the maximal subgroup of the group G.*
(f) *If $H \ntriangleleft G$, then $N_G(P) \cong$ 2-transitive Frobenius group of the degree $q = |P|$; if $H \triangleleft G$, then $q = 2$ and G is a Frobenius group with kernel H.*
(g) *$h = (q-1)q(qk+1)$, where $k \in \mathbb{Z}$, $k \geq 0$.*
(h) *If $F(G) = \{1\}$, then $\langle A \rangle = Sc(G)$ and is a non-abelian simple group.*

PROOF. (a) Since $|H| = |A|$, then $|A \cap H| = 0$ by Proposition 10. Therefore $H \subseteq \hat{B} = G \backslash A$. Since A is normal, then we have the action of the group H on A by conjugations. As $A \in \mathscr{L}_{NC}(G)$ and $|A| = |H|$, the action of H on

A is regular. Therefore $A = g^H$, where g is any element from A; since A is normal, then A is the conjugacy class of G.

(b) Let $g \in A$, $p \in \pi(g)$. Then $g' = g^{o(g)/p} \in A$, $o(g') = p$. Hence, via (1), it results that all the elements from A have order p. By Proposition 4, the elements belonging to B are p'-elements. That is why A is the set of all non-trivial p-elements of the group G. Hence $A = P^G$, where $P \in \mathrm{Syl}_p(G)$.

(c) Since $A \in \mathscr{L}_{NC}(G)$, then $C_G(g) \subseteq \hat{A}$ for any $g \in A$. We obtain from (b) that $C_G(g)$ is a p-subgroup. Since $H \subseteq \hat{B}$, H is a p'-subgroup. $A = g^G$ implies $|H| = |A| = |G: C_G(g)|$. Therefore $C_G(g) \in \mathrm{Syl}_p(G)$ and, hence, $|H| = |G: P|$. It yields the factorization $G = PH$. Remark that $c(A) = |P|$ and (according to equality $c(A)c(B) = h$) $c(B) = |G: P| = |H|$.

(d) Since the centralizers of non-trivial p-elements of the group G are S_p-subgroups, then [**4**, Theorem C] the latter ones are abelian CC-subgroups. By (1), S_p-subgroups are elementary.

(e) If $H < K \leq G$, then $|K| > |H| = |A|$, hence, by Corollary 2 of Proposition 10 $K \supset A \supseteq P^{\#}$. Thus $K \geq P$, $K \geq H$, hence by (c), $K = G$. And so H is the maximal subgroup of the group G.

(f) Let $\mathrm{Syl}_p(G) = \{P_1 = P, \dots, P_r\}$. By (d), $P_i^{\#} \cap P_j^{\#} = \varnothing$, if $i \neq j$. As $A = UP_i^{\#}$, hence it results that $|A| = r|P^{\#}| = r(q-1)$. Since $|A| = |H|$, $r = |G: N_G(P)|$ and (by (c)) $|G| = q|H|$, then $|N_G(P): P| = q - 1$. If $H \ntriangleleft G$, then $q = |G: H| > 2$ and hence $N_G(P) \neq P$. By (b) it follows that $N_G(P)$ is a Frobenius group with the kernel P. As $|N_G(P): P| = |P^{\#}|$, then $N_G(P) \cong$ 2-transitive Frobenius group of the degree q. If $H \triangleleft G$, then, by (e) and (c) $q = p$ and $G = H\lambda P$. Hence it follows that H is the set $\hat{B}$ of all the p'-elements of the group G. Therefore $A = G \backslash H$; it implies $|H| = |G \backslash H|$ and hence $p = |G: H| = 2$. It is obvious now, G is a Frobenius group with the kernel H.

(g) Since $c(A) = q$, then from Proposition 8 $r(q-1) = |A| \equiv -1 \pmod q$, hence $r \equiv 1 \pmod q$. Putting $r = qk + 1$ $(k \in \mathbb{Z}, k \geq 0)$ we have $h = |G: N_G(P)| \cdot |N_G(P): P| \cdot |P| = (kq+1)(q-1)q$.

(h) Suppose, $F(G) = \{1\}$. As the socle of the group G is completely reducible, its simplicity means its indecomposability into the direct product of the non-trivial subgroups. Assume $Sc(G) = T \times U$, where $T, U < Sc(G)$. If $T \cap A \neq \varnothing$, then $U \leq C_G(T) \subseteq \hat{A}$ and hence, $T \leq C_G(U) \subseteq \hat{A}$. If $s = t \cdot u \in Sc(G)^{\#}$, $t \in T$, $u \in U$, then $s \in C_G(t) \cap C_G(u) \leq \hat{A}$. Therefore $Sc(G) \subseteq \hat{A}$. Since $Sc(G) \triangleleft G$, then by Proposition 3 $Sc(G)$ is nilpotent; this is at variance with triviality of $F(G)$. That is why $T \cap A = \varnothing$. It can be similarly proven that $T \cap B = \varnothing$. Hence $T = \{1\}$-contradiction. Thus $Sc(G)$ is a simple group. Since $Sc(G) \nleq H$, then by (e), $Sc(G) \cdot H = G$, hence it follows that $g^{|H|} \in Sc(G)$ by any $g \in G$. In particular, if $g \in A$, then $g^p = 1 \in Sc(G)$; as $(p, |H|) = 1$, then $g \in Sc(G)$. Thus $A \subset Sc(G)$, hence, in view of simplicity of $Sc(G)$ and normality of A, $\langle A \rangle = Sc(G)$.

Bibliography

1. A. I. Weizblit and È. M. Žmud′, "*Connectivity*" *of finite group and Zassenhaus group*, Soob. Acad. Nauk GSSR 2(90) (1978), 277–279. (Russian)

2. È. M. Žmud′, *On notion of* "*connectivity*" *of finite group*, Publ. Math., **29** fasc. 1-2, (1981), 177–189. (Russian)

3. G. Frobenius, *Über einen Fundamentalsatz der Gruppentheorie*, Sitzungsber. Preus. Akad. Wiss., (1907), 428–437.

4. Z. Arad and D. Chillag, *On finite groups with conditions on centralizers of elements*, J. Algebra **51** (1978), 164–172.

ABCDEFGHIJ — 89